Springer Series in Molecular Biology

Series Editor: Alexander Rich

Springer Series in Molecular Biology

Series Editor: Alexander Rich

Yeast Genetics
Fundamental and Applied Aspects
J.F.T. Spencer, Dorothy M. Spencer, A.R.W. Smith, eds.

Myxobacteria
Development and Cell Interactions
Eugene R. Rosenberg, ed.

DNA Methylation
Biochemistry and Biological Significance
Aharon Razin, Howard Cedar, Arthur D. Riggs, eds.

DNA Methylation

Biochemistry and Biological Significance

Edited by
Aharon Razin, Howard Cedar,
and Arthur D. Riggs

With Contributions by

K. F. Conklin, W. Doerfler, R. H. Grafstron, M. Groudine, D. L. Hamilton, R. Jaenisch, D. Jahner, P. A. Jones, S. V. Kelly, I. Kruczek, K-D. Langner, M. G. Marinus, D. Renz, A. Rich, C-K. J. Shen, H. O. Smith, M. Szyf, L. Vardimon, J. Yisraeli, R. Yuan

With 73 Figures

Springer-Verlag
New York Berlin Heidelberg Tokyo

Aharon Razin
Chairman, Department of Cellular Biology
The Hebrew University
Hadassah Medical School
Jerusalem, Israel 91-010

Howard Cedar
Department of Cellular Biochemistry
The Hebrew University
Hadassah Medical School
Jerusalem, Israel 91-010

Arthur D. Riggs
Beckman Research Institute of the City of Hope
Duarte, California 91010
U.S.A.

Series Editor:

Alexander Rich
Department of Biology
Massachusetts Institute of Technology
Cambridge, Massachusetts 02139 U.S.A.

Library of Congress Cataloging in Publication Data
Main entry under title:
DNA methylation and its biological significance.
(Springer series in molecular biology)
Includes bibliographies and index.
1. Deoxyribonucleic acid—Metabolism. 2. Methylation.
3. Gene expression. I. Razin, Aharon, 1935– .
II. Cedar, Howard, 1943– . III. Riggs, Arthur D.
IV. Series. [DNLM: 1. DNA. 2. Methylation.
3. Molecular biology. QU 58 D629]
QP624.D15 1984 574.87′3282 84-5475

Softcover reprint of the hardcover 1st edition 1984

Typeset by University Graphics, Inc., Atlantic Highlands, New Jersey.

9 8 7 6 5 4 3 2 1

ISBN-13: 978-1-4613-8521-9 e-ISBN-13: 978-1-4613-8519-6
DOI: 10.1007/978-1-4613-8519-6

Series Preface

During the past few decades we have witnessed an era of remarkable growth in the field of molecular biology. In 1950 very little was known of the chemical constitution of biological systems, the manner in which information was transmitted from one organism to another, or the extent to which the chemical basis of life is unified. The picture today is dramatically different. We have an almost bewildering variety of information detailing many different aspects of life at the molecular level. These great advances have brought with them some breath-taking insights into the molecular mechanisms used by nature for replicating, distributing and modifying biological information. We have learned a great deal about the chemical and physical nature of the macromolecular nucleic acids and proteins, and the manner in which carbohydrates, lipids and smaller molecules work together to provide the molecular setting of living systems. It might be said that these few decades have replaced a near vacuum of information with a very large surplus.

It is in the context of this flood of information that this series of monographs on molecular biology has been organized. The idea is to bring together in one place, between the covers of one book, a concise assessment of the state of the subject in a well-defined field. This will enable the reader to get a sense of historical perspective—what is known about the field today—and a description of the frontiers of research where our knowledge is increasing steadily. These monographs are designed to educate, perhaps to entertain, certainly to provide perspective on the growth and development of a field of science which has now come to occupy a central place in all biological phenomena.

The information in this series has value in several perspectives. It provides for a growth in our fundamental understanding of nature and the manner in which living processes utilize chemical materials to carry out a variety of activities. This information is also used in more applied areas. It promises to have a significant impact in the biomedical field where an understanding of disease processes at the molecular level may be the capstone which ultimately holds together the arch of clinical research and medical therapy. More recently in the field of biotechnology, there is another type of growth in which this science can be used with immense practical consequences and benefit in a variety of fields ranging from agriculture and chemical manufacture to the production of scarce biological compounds for a variety of application.

This field of science is young in years, but it has already become a mature science. These monographs are meant to clarify segments of this field for the readers.

Cambridge, Massachusetts

Alexander Rich
Series Editor

Preface

A growing interest in enzymatic DNA methylation, by biologists of many disciplines, convinced us that publication of a volume covering all aspects of this field is due. Only four years ago, this interest could be satisfied by a brief review in Science, but the rapid growth of information since then demands a much more thorough and wider discussion of the subject. Therefore, we asked several authors to write chapters dealing with aspects of DNA methylation in which they are experts. In addition to the introductions included in each chapter, we also have written an introductory chapter that may serve as a general overview of the field and a directory to the rest of the book. Thus, we hope that this book will help students who are being introduced to the subject of DNA methylation, as well as established investigators in methylation or related fields.

We are most grateful to our colleagues who spent much time and effort to contribute their specialty to this book. We also want to express our gratitude to Caroline Gopin for her help in preparing this book.

Aharon Razin
Howard Cedar
Arthur D. Riggs

Contents

1. **Introduction and General Overview**
Aharon Razin, Howard Cedar, and Arthur D. Riggs . . 1

2. **Type I and Type III Restriction-Modification Enzymes**
Robert Yuan and Daniel L. Hamilton 11

3. **Methylases of the Type II Restriction-Modification Systems**
Hamilton O. Smith and Samuel V. Kelly 39

4. **The Restriction and Modification DNA Methylases: An Overview**
Robert Yuan and Hamilton O. Smith 73

5. **Methylation of Prokaryotic DNA**
M. G. Marinus 81

6. **DNA Methylation: DNA Replication and Repair**
Robert H. Grafstrom, Daniel L. Hamilton, and Robert Yuan 111

7. **DNA Methylation Patterns: Formation and Biological Functions**
Aharon Razin 127

8. DNA Methylation and Gene Expression
Howard Cedar 147

9. Gene Activation by 5-Azacytidine
Peter A. Jones 165

10. DNA Methylation in Early Mammalian Development
Detlev Jähner and Rudolf Jaenisch 189

11. Specific Promoter Methylations Cause Gene Inactivation
Walter Doerfler, Klaus-Dieter Langner, Inge Kruczek, Lily Vardimon, and Doris Renz 221

12. DNA Methylation and Developmental Regulation of Eukaryotic Globin Gene Transcription
Che-Kun James Shen 249

13. X Inactivation, DNA Methylation, and Differentiation Revisited
Arthur D. Riggs 269

14. Left-handed Z-DNA and Methylation of d(CpG) Sequences
Alexander Rich 279

15. Chromatin Structure and Gene Expression
Kathleen F. Conklin and Mark Groudine 293

16. Gene Methylation Patterns and Expression
Joel Yisraeli and Moshe Szyf 353

Index 379

Contributors

HOWARD CEDAR Department of Cellular Biochemistry, The Hebrew University, Hadassah Medical School, Jerusalem, Israel 91-010

KATHLEEN F. CONKLIN Hutchinson Cancer Research Center, Seattle, Washington 98104 U.S.A.

WALTER DOERFLER Institute of Genetics, University of Cologne, Weyertal 121, 5000 Cologne 41, Federal Republic of Germany

ROBERT H. GRAFSTROM NCI-Frederick Cancer Research Facility, Frederick, Maryland 21701 U.S.A.

MARK GROUDINE Department of Radiation Oncology, University of Washington Hospital, Seattle, Washington 98105 U.S.A.

DANIEL L. HAMILTON NCI-Frederick Cancer Research Facility, Frederick, Maryland 21701 U.S.A.

RUDOLF JAENISCH Heinrich-Pette-Institut für Experimentelle Virologie und Immunologie an der Universität Hamburg, 2000 Hamburg 20, Federal Republic of Germany

DETLEV JÄHNER Heinrich-Pette-Institut für Experimentelle Virologie und Immunologie an der Universität Hamburg, 2000 Hamburg 20, Federal Republic of Germany

PETER A. JONES Department of Pediatrics and Biochemistry, University of Southern California Comprehensive Cancer Center, Los Angeles, California 90033 U.S.A.

SAMUEL V. KELLY Department of Molecular Biology and Genetics, The Johns Hopkins University School of Medicine, Baltimore, Maryland 21205 U.S.A.

INGE KRUCZEK Institute of Biochemistry, Munich University, Munich, Federal Republic of Germany

KLAUS-DIETER LANGNER Institute of Genetics, University of Cologne, Cologne, Federal Republic of Germany

M. G. MARINUS Department of Pharmacology, University of Massachusetts Medical School, Worcester, Massachusetts 01605 U.S.A.

AHARON RAZIN Department of Cellular Biochemistry, The Hebrew University, Hadassah Medical School, Jerusalem, Israel 91-010

DORIS RENZ Institute of Genetics, University of Cologne, Cologne, Federal Republic of Germany

ALEXANDER RICH Department of Biology, Massachusetts Institute of Technology, Cambridge, Massachusetts 02139 U.S.A.

ARTHUR D. RIGGS Beckman Research Institute of the City of Hope, Duarte, California 91010 U.S.A.

CHE-KUN JAMES SHEN Department of Genetics, University of California at Davis, Davis, California 95616 U.S.A

HAMILTON O. SMITH Department of Molecular Biology and Genetics, The Johns Hopkins University School of Medicine, Baltimore, Maryland 21205 U.S.A.

MOSHE SZYF Department of Cellular Biochemistry, The Hebrew University, Hadassah Medical School, Jerusalem, Israel 91-010

LILY VARDIMON Developmental Biology Lab, Cummings Life Science Center, University of Chicago, Chicago, Illinois 60637 U.S.A.

JOEL YISRAELI Department of Cellular Biochemistry, The Hebrew University, Hadassah Medical School, Jerusalem, Israel 91-010

ROBERT YUAN NCI-Frederick Cancer Research Facility, Frederick, Maryland 21701 U.S.A.

1

Introduction and General Overview

Aharon Razin*
Howard Cedar*
Arthur D. Riggs†

It has become increasingly clear that postreplication modification of DNA is relevant to many fields that at first glance may not seem related. Some knowledge of DNA modification is necessary to understand, to teach, or to experiment in the fields of bacterial genetics, prokaryotic gene regulation, recombinant DNA and molecular cloning, eukaryotic gene regulation, developmental biology, and (probably) cancer. During the last 10 years, a large body of information has been accumulated on enzymatic DNA modification, both in prokaryotes and eukaryotes. We thought it might be useful to bring all of this information together in one book, because the field of DNA methylation has become so large that searching the original literature has become a formidable task. In the past, cross-fertilization between studies on prokaryotes and eukaryotes has occurred in the methylation field. We hope this book will stimulate some additional cross-fertilization. No attempt will be made to discuss the evolution of the DNA methylation field from an historical perspective. However, to a certain extent, this is done in some of the chapters; also, the following list of recent reviews and some earlier key papers will serve to chronicle the development of the field.

Modified bases in DNA were first described by Hotchkiss (1948), Wyatt (1951), Dunn and Smith (1955), and Sinsheimer (1955). Methyl transferase activity was demonstrated in bacteria (Gold *et al.* 1963) and later in animals and plants (Sheid *et al.* 1968; Kalousek and Morris, 1968). The restriction-

*Department of Cellular Biochemistry, The Hebrew University, Hadassah Medical School, Jerusalem, Israel 91–010

†Beckman Research Institute of the City of Hope, Duarte, California 91010

modification systems and methylation of specific sites in DNA were reviewed by Arber and Linn (1969) and later by Arber (1974). A possible role of DNA methylation in gene regulation was first suggested in 1964 (Srinivasan and Borek, 1964), and then by Scarano (1971). This hypothesis was further developed by Riggs (1975) and Holliday and Pugh (1975). The introduction of recombinant DNA technology resulted in considerable progress in this field of research. Several key studies were performed by Bird (1978), Taylor and Jones (1979), Naveh-Many and Cedar (1981), and Gruenbaum *et al.* (1982). The rapidly expanding literature of the last five years has been reviewed in a number of articles: Razin and Riggs (1980), Burdon and Adams (1980), Drahovsky and Boehm (1980), Ehrlich and Wang (1981), Doerfler (1981), Razin and Friedman (1981), Hattman (1981), Doerfler (1983), Riggs and Jones (1983), Razin and Cedar (1984), Riggs and Smith (1984), and Riggs *et al.* (1984b).

Basic Facts and Structures

Since this book may be used by many people who are newly learning about enzymatic DNA methylation, we will now provide a very basic introduction to the field.

It has been known for some time that the DNA of most organisms contains minor bases (Hall, 1971.) Most bacteria and vertebrates have either or both of two so-called "minor" bases, N6-methyladenine (6-mAde or m^6Ade) and 5-methylcytosine (5-mCyt or m^5Cyt) (Figure 1.1). For example, *Escherichin coli* K-12 has both of these bases at levels of 0.5 mol% and 0.25 mol%, respectively. In mammalian DNA, only one modified base, 5-mCyt, has been detected so far; it is present at about 1 mol% of total bases.

For *E. coli,* mammalian cells, and most viruses, it is clear that during the process of DNA replication only the four major nucleotides (dATP, dGTP, dTTP, and dCTP) are incorporated into DNA. Modified bases are formed after replication by enzymes that act on the DNA duplex. All known DNA methyl transferases (methylases) use, as a source of the methyl group, S-adenosylmethionine (SAM or AdoMet) (Figure 1.2). The enzymes are specific in that only certain cytosine moieties or adenine residues are modified by meth-

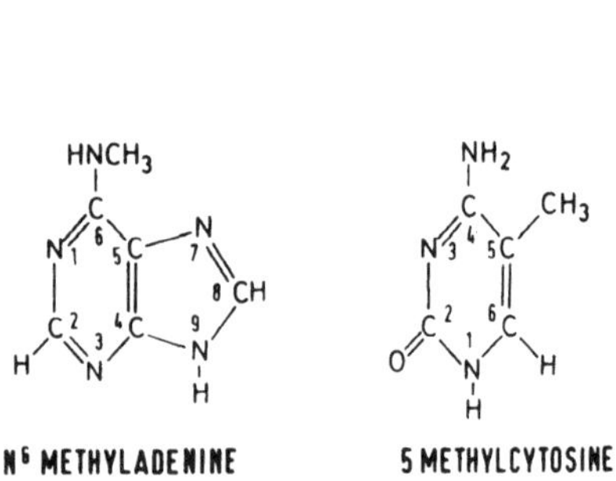

Figure 1.1 The methylated bases in DNA.

ylation. In mammals, most (but not all) cytosines followed at the 3'-side by a guanosine are methylated. The bacterial and bacteriophage methylases all have more complex sequence requirements (Razin, Chapter 7).

DNA Methyl Transferases

The detailed mechanism of action of methylases can be very complicated, and it is discussed in more detail in several chapters of this book. The bacterial restriction and modification methylases have been classified into three different categories according to their properties. Enzymes of type I are coded by genes in the bacterial chromosome, and their complex properties are discussed in detail by Yuan and Hamilton in Chapter 2. Type II methylases, whose accompanying restriction enzymes are commonly used in recombinant DNA work, usually are encoded in plasmid genomes; they are discussed in detail by Smith and Kelley in Chapter 3. Methylases of type III are similar to those of type I, but usually are encoded by bacteriophages. These enzymes are discussed and reviewed by Yuan and Hamilton in Chapter 2. In addition to the above enzymes, which are all part of restriction-modification systems, *E. coli* has two other methylase systems that are seemingly unassociated with restriction-modification. The *dam* gene encodes a methylase that forms 6mAde and the *dcm* gene encodes a methylase that forms 5mCyt. The properties and possible functions of these enzymes are discussed by Marinus in Chapter 5.

Studies of mammalian and vertebrate methylases are in a primitive state in relation to bacterial and phage methylases. Only very recently has a methylase been purified to homogeneity (Bestor and Ingram, 1983). The mammalian methylases and some of their properties are reviewed by Razin in Chapter 7. and by Grafstrom *et al.* in Chapter 6.

The maintenance methylase concept (Riggs, 1975; Holliday and Pugh, 1975) is central to most ideas with regard to the biological function of mammalian methylases, and is shown in Figure 1.3. A maintenance methylase is defined as a methylase whose preferred substrate is a hemimethylated, sym-

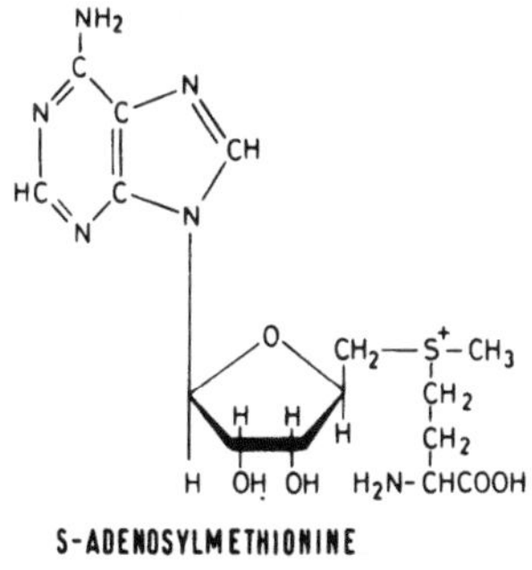

Figure 1.2 The methyl donor for transmethylations by DNA transmethylases.

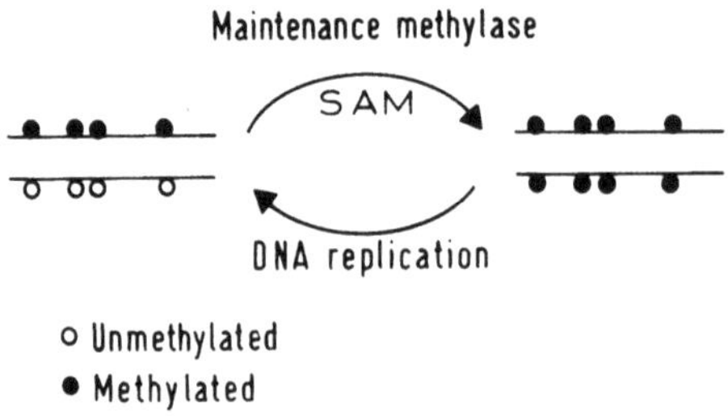

Figure 1.3 Maintenance methylation of DNA postreplication.

metrical site. If the activity of this methylase is minimal on an unmethylated site, then a mechanism for the somatic inheritance of methyl groups through DNA replication is readily apparent. As discussed by Razin in Chapter 7, and Cedar in Chapter 8, there now is strong experimental support for this type of enzyme being the predominant one in differentiated somatic cells. By definition, a *de novo* methylase is one that acts on unmethylated sites. There is experimental evidence for this type of methylase activity, especially in the early embryo (Jahner and Jaenisch, Chapter 10).

The Biological Significance of Methylation

An Information Encoding System

There is no doubt that modification of DNA adds an additional level of information to the DNA helix. What molecular purpose is served by DNA methylation? The notion that the primary biological function of DNA methylation is to strongly affect sequence-specific interactions of proteins with DNA (Riggs, 1975; Razin and Riggs, 1980) is now generally accepted, and it is strongly supported by a number of studies with bacteria and bacterial DNA-binding proteins (e.g., type I and type II modification enzymes, Chapters 2 and 3; see also Marinus, Chapter 5). Direct evidence for such an effect on mammalian proteins is lacking, since no *in vitro* systems are as yet responsive to methylation; however, it is assumed that chromatin structure will be affected by DNA methylation. Much further work will be needed to obtain experimental data on the effect of methylation on alternate DNA conformations (e.g., Z DNA; Rich, Chapter 14), DNase hypersensitive sites and other higher order chromatin structures (Conklin and Groudine, Chapter 15, and Cedar, Chapter 8).

Bacteria

It is now clear that *E. coli* can function without any detectable methylation; the triple mutant hsd$^-$ dcm$^-$ dam$^-$ strains are viable. Marinus discusses this in Chapter 5, as does Razin in Chapter 7. Therefore, methylation serves no essential function in DNA replication, which is a role often considered before these recent results (Chapter 6). Two functions seem to be rather well substantiated; the first is the classic one of DNA modification in restriction-modification (Chapter 4). Foreign DNA from a species or strain with a different methylation pattern is destroyed by the restriction enzyme of the host restriction-modification system. Thus, DNA methylation, along with restriction enzymes, functions to greatly reduce the efficiency of gene transfer between unrelated species. A second type of function that recently is receiving experimental support is mismatch repair. As discussed by Marinus in Chapter 5, the newly replicated (or repaired) strand is transiently unmethylated; thus, it can be distinguished from the old strand by the mismatch repair system, resulting in preferential correction to the sequence of the methylated (old) strand.

Eukaryotes

What is the function of methylation in eukaryotes? There is no evidence for restriction enzymes in higher eukaryotes. Also, in marked contrast to bacteria, many potentially methylatable sites are completely unmethylated. Unmethylated sites would be sensitive to restriction if restriction-modification systems similar to those in bacteria were in operation. Therefore, it is likely that methylation serves other functions in eukaryotes. Present evidence suggests that postreplication DNA modification by the formation of 5-methylcytosine plays a role in both vertebrate gene regulation and (possibly) embryonic development. There is a large body of evidence supporting such an involvement, some of which is outlined in Table 1.1. The table also indicates which chapters review and/or discuss the relevant experimental data.

Drosophila melanogaster, a well-studied and evolutionarily very successful eukaryote does not have any detectable methylated bases (Urieli-Shoval *et al.* 1982). This important fact clearly indicates that the majority of methylated sites in vertebrates are not likely to be involved in such critical functions as DNA replication. *Drosophila* also manages to undergo very complex development; therefore, methylation may not be essential at the most fundamental level in the control of gene expression during development. However, it is of interest to note that biologically significant methylation similar to that of the *E. coli* type I (*Eco*K) system would be undetectable with present methods. The *Eco*K methylase recognizes a seven-base sequence that occurs only about once in every 16,000 base pairs. For mammals, the evidence has become extremely strong that methylation is involved in gene control—for example, X chromosome inactivation, which is a differentiation event occurring in the early embryo (Chapter 13). We think that all available data can be best accommodated by assuming that mammalian differentiation is controlled by several independent mechanisms that probably operate autonomously. Some of these mechanisms are discussed by Conklin and Groudine in Chapter 15 and Cedar in Chapter 8. Since, in some cases, the level of transcription changes at least eight orders of magnitude during development (Ivarie *et al.* 1983), it is difficult to imagine that one single control mechanism would be adequate. Therefore,

Table 1.1 Evidence Supporting Gene Control by 5-Methylcytosine

Type of Experiment	Reference
1. mCyt affects protein-DNA interactions	Razin and Riggs (1980) Yuan and Hamilton (Chapter 2) Smith and Kelly (Chapter 3) Marinus (Chapter 5)
2. mCyt stabilizes Z-form DNA	Rich (Chapter 14)
3. Maintenance methylases exist	Razin (Chapter 7)
4. Methylation patterns exist and are tissue-specific	Razin (Chapter 7) Cedar (Chapter 8) Conklin and Groudine (Chapter 15) Shen (Chapter 12) Yisrael and Szyf (Chapter 16)
5. Methylation patterns are somatically heritable	Cedar (Chapter 8)
6. Gene activity correlates with hypomethylation	Cedar (Chapter 8) Yisraeli and Szyf (Chapter 16)
7. *In vitro* methylation reduces *in vivo* gene activity	Doerfler (Chapter 11) Jahner and Jaenisch (Chapter 10) Cedar (Chapter 8)
8. DNA methylation maintains X chromosome inactivation	Riggs (Chapter 13)
9. Inhibition of methylation activates many genes	Jones (Chapter 9)

in mammals and other higher organism, it may be advantageous to add another level of control beyond those used by *Drosophila.* The evidence for mammals is most supportive of the idea that methylation of DNA at critical sites (promotor and 5′-flanking sites) functions as a somatically heritable locking mechanism (Razin and Riggs, 1980). Methyl groups must be absent before efficient transcription can begin.

For many genes, hypomethylation of critical sites may be a necessary, but not-sufficient, condition for high-level expression. Work with inhibitors of DNA methylation (5-azacytidine in particular) have indicated that for a surprisingly high number of cellular and viral genes, hypomethylation may be a sufficient condition for transcriptional activity (reviewed by Jones, Chapter 9; see also Riggs, Chapter 13).

Methylation Patterns

As reviewed and discussed by Razin in Chapter 7, Cedar in Chapter 8, Shen in Chapter 12, and as summarized in Chapter 16, tissue-specific methylation patterns certainly exist in vertebrates. It is also firmly established that methylation patterns are faithfully maintained through repeated mitoses in somatic cells. How are these methylation patterns established and specifically changed? The DNA of most genes examined so far in sperm or the early embryo are highly methylated; therefore, the notion emerged (Singer *et al.* 1979; Razin and Riggs, 1980) that active *de novo* methylation (methylation of an unmethylated site) occurs in the early embryo, which leads (in mammals) to a fully methylated ground state in the embryo proper by late blastocyst. Chapter 10, by Jahner and Jaenisch, reviews recent experimental data on gene silencing by methylation in the early embryo.

If the ground state is in full methylation, then information relevant to development could be written by the removal of methyl groups. In bacteria, the enzymatic removal of methyl groups has not been observed and, as discussed by Razin in Chapter 7, there is no convincing evidence for a true demethylation mechanism in mammals—that is, the actual removal of a methyl group from the DNA duplex. It is most likely that the loss of methyl groups during development occurs by a passive process. The inhibition of the maintenance methylase with accompanying DNA replication would lead to the loss of methyl groups in the new strand. The proteins that interact specifically with DNA sequences and cause the formation of a methylation pattern have been called *determinator proteins* (Riggs and Jones, 1983); it is these proteins that are postulated as controlling methylation. The use of the word demethylation should be applied with caution when discussing changing methylation patterns because of the implied active process. We suggest that instead of stating that sites are demethylated, one should state they are hypomethylated; the latter term is more compatible with the probably passive nature of the event. Inhibitors of DNA methylation cause the switching of gene activity (Chapter 9), and in every case DNA replication is required. In at least one system, only one round of replication is required (Compere and Palmiter, 1981), suggesting that the generation of hemimethylated (one strand methylated) sites is sufficient to activate a gene silenced by methylation.

Because it is logically attractive to influence a developmental program by establishing a methylation pattern before the final step of overt transcription, several present models assume that methylation changes take place before transcription. It should be noted that this has not yet been convincingly verified. This point is discussed by Cedar in Chapter 8 and is worthy of increased experimental effort for verification or disapproval.

Combinatorial interactions of determinator (delayed-action) and differentiator (immediate-action) proteins very likely control the development of higher organisms. The key question for this book is: *How does DNA modification by methylation affect the interplay between these proteins and DNA?* We do not, as yet, have a detailed answer to the question, but present evidence does strongly suggest that methylation is a role player on the development stage.

Evidence suggesting a role in abnormal development (cancer) recently has been reviewed (Riggs and Jones, 1983). There is considerable experimental activity in this area, and since it is an area not covered in other chapters, a short summary will be given here. Several tissue culture variants recently have been shown not to be true mutants, but (rather) to have genes silenced by methylation (Harris, 1982). These variants can be efficiently reactivated by 5-azacytidine. Methylation changes, thus, can masquerade as mutations. 5-azacytidine does not function to significantly cause true mutations in mammalian cells (Landolph and Jones, 1982), but it does cause the transformed phenotype in tissue culture cells (see Jones, Chapter 9) as well as a wide variety of tumors in rats (Carr *et al.* 1984). It also causes changes in the immunogenic and metastatic properties of established mouse tumor lines (Kerbel *et al.* 1984; Olsson and Forschhammer, 1984). The majority of tumors are relatively hypomethylated, and carcinogen-damaged DNA is a poor substrate for DNA methylases (see Riggs and Jones, 1983). Thus, it is quite likely that DNA methylation may be significant for one or more of the steps of neoplastic progression.

References

Arber W, Linn S: DNA modification and restriction. *Ann Rev Biochem* 1969;38:467–500.

Arber W: DNA modification and restriction. *Prog Nucleic Acids Res Mol Biol* 1974;14:1–37.

Bestor TH, Ingram VM: Two DNA methyltransferases from murine erythroleukemia cells: Purification sequence specificity, and mode of interaction with DNA. *Proc. Natl. Acad. Sci. USA* 1983, 80:5559–5563.

Bird AP: DNA methylation. II. The symmetry of methylated sites supports semi-conservative copying of the methylation pattern. *J Mol Biol* 1978;118:49–60.

Burdon RH, Adams RLP: Eukaryotic DNA methylation. *Trends Biochem Sci*, 1980; 5:294–297.

Carr BI, Reilly JG, Smith SS, Riggs AD: The tumorigenicity of 5 azacytidine in the male Fischer rat (Unpublished data, 1984).

Compere SJ, Palmiter RD: DNA methylation controls the inducibility of the mouse metallothionein-I gene in lymphoid cells. *Cell* 1981;25:233–240.

Doerfler W: A regulatory signal in eukaryotic gene expression. *J Gen Virol*, 1981; 57:1–20.

Doerfler W: DNA methylation and gene activity. *Ann Rev Biochem* 1983;52:93–124.

Drahovsky D, Boehm TLJ: Enzymatic DNA methylation in higher eukaryotes. *Intern J Biochem* , 1980;12:523–528.

Dunn DB, Smith JD: The occurrence of 6 methylaminopurine in deoxyribonucleic acids. *Biochem J*, 1958;68:627–636.

Ehrlich M, Wang RYH: 5 methylcytosine in eukaryotic DNA. *Science*, 1981; 212:1350–1357.

Gold M, Hurwitz J, Andres M: The enzymatic methylation of RNA and DNA II. On the species specificity of the methylation enzymes. *Proc. Natl Acad Sci USA,* 1963; 50:164–169.

Gruenbaum Y, Cedar H, Razin A: Substrate and sequence specificity of a eukaryotic DNA methylase. *Nature*, 1982;295:620–622.

Hall RH: *The Modified Nucleosides in Nucleic Acids*. New York, Columbia University Press, 1971.

Harris, M: Induction of thymidine kinase in Enzyme-deficient Chinese hamster cells. *Cell* 1982;19:483–492.

Hattman S, *DNA Methylation in The Enzymes* 14 in Boyer PD (ed): New York and London, Academic Press, 1981;14:517–547.

Holliday R, Pugh JE: DNA modification mechanisms and gene activity during development. *Science*, 1975;187:226–232.

Hotchkiss RD: The quantitative separation of purines, pyrimidines and nucleosides by paper chromatography. *J Biol Chem* 1948;175:315–332.

Ivarie RD, Schachter BS, O'Farrell PH: The level of expression of the rat growth hormone gene in liver tumor cells is at least eight orders of magnitude less than that in anterior pituitary cells. *Mol Cell Biol* 1983;3:1460–1467.

Kalousek F, Morris NR: Deoxyribonucleic acid methylase activity in rat spleen. *J Biol Chem* 1968;243:2440–2443.

Kerbel RS, Frost P, Liteplo R, *et al.*: Induction of high frequency heritable changes in the tumorigenic and metastatic properties of tumor cell populations by 5-azacytidine treatment. *J Cell Physiol* (in press, 1984).

Landlph, JR, Jones PA: Mutagenieity of 5-azacytidine and related nucleosides in C3H/10½ C18 and V79 cells. *Cancer Res* 1982,42:817–823.

Naveh-Many T, Cedar H: Active gene sequences are undermethylated. *Proc Natl Acad Sci (USA)* 1981; 78:4246–4250.

Olsson L, Forschhammer J: Induction of the metastatic phenotype in a mouse tumor model by 5 azacytidine. *Proc Natl Acad Sci (USA), (in press, 1984)*

Razin A, Riggs AD: DNA methylation and gene function. *Science* 1980;210:604–610.

Razin A, Friedman J: DNA methylation and its possible biological roles. *Prog Nucleic Acids Res and Mol Biol* 1981;25:33–52.

Razin A, Cedar H: DNA methylation in eukaryotic cells. *Intern Rev Cytobiol* (in press, 1984).

Riggs, AD: X inactivation, differentiation and DNA methylation. *Cytogenet Cell Genet* 1975;14:9–11.

Riggs AD, Jones P: 5-methylcytosine, gene regulation and cancer. *Adv Cancer Res*, 1983;40:1–40.

Riggs AD, Smith SS: DNA cytosine methylation: a new mechanism in somatic heredity, in: *DNA Recombinant Technology*, vol. 2. Boca Raton, Florida, CRC Press, Inc., (in press, 1983).

Riggs AD, Singer-Sam J, Keith D, Carr BI:DNA methylation, *X-inactivation and Cancer*. 1984 Miami Winter Symposium: Human Genetic Disorders. ICSV Press (in press, 1984b).

Scarano E: The control of gene function in cell differentiation and in embryogenesis. *Adv Cytopharmacol* 1971;1:13–23.

Sheid B, Srinivasan PR, Borek E: Deoxyribonucleic acid methylase of mammalian tissues. *Biochemistry*, 1968;7:280–285.

Singer J, Robert-Ems J, Riggs AD: Methylation of mouse liver DNA studied by means of the restriction enzymes MspI and HpaII. *Science* 1979;203:1019–1023.

Sinsheimer RL. The action of pancreatic deoxyribonuclease. II. Isometric dinucleotides. *J Biol Chem* 1955;215:579–583.

Srinivasan PR, Borek E: Enzymatic alteration of Nucleic acids structure. Enzymes put finishing touches, characteristics of each species on RNA and DNA by insertion of methyl groups. *Science* 1964; 145:548–553.

Taylor SM, Jones PA: Multiple new phenotypes introduced in 10T1/2 and 3T3 cells treated with 5-azacytidine. *Cell* 1979; 17:771–779.

Urieli-Shoval S, Gruenbaum Y, Sedat J, Razin A: The absence of detectable methylated bases in Drosophila melanogaster DNA, FEBS Lett 1982;146:148–152.

Wyat GR: Recognition and estimation of 5-methylcytosine in nucleic acids. *Biochem J*, 1951;48:581–584.

2

Type I and Type III Restriction-Modification Enzymes

Robert Yuan*
Daniel L. Hamilton*

DNA methylation is the most ubiquitous form of DNA modification. It has been studied extensively in a wide variety of organisms. Although it generally is believed that DNA methylation is involved in mismatch repair in prokaryotic cells, and in the control of gene expression in eukaryotic cells, the precise nature of its biological functions remains largely unknown. One case in which the role of DNA methylation has been clearly defined is that of host-controlled restriction and modification in bacteria. This biological system consists of two highly specific enzymatic activities: an endonuclease and a DNA methylase. The endonuclease enables a given strain to both recognize and destroy foreign DNA by cutting both DNA strands at a limited number of sites. This function is defined as *restriction.* Kuhnlein and Arber (1972) showed that methylation at specific sequences protected the DNA from its homologous restriction endonuclease. This function is defined as *modification.*

The restriction-modification systems have now been classified into three types: I, II, and III (Yuan, 1981). Type II restriction enzymes, as exemplified by *Eco*RI, appear to have simple protein structures (*Eco*RI is a dimer) and only require Mg^{2+} for activity. The type II DNA methylases are separate proteins and also appear to have simple structures. The type I enzymes (e.g., those from *Escherichia coli* strains B and K) and the type III enzymes (e.g., those coded by the phage P1, the plasmid P15, and the bacterial strain *Haemophilus influenzae* Rf) are multifunctional proteins that are able to both cleave and methylate unmodified DNA. Type I and type III enzymes differ from each other in the number of subunits (three for type I, two for type III) and in the requirement for S-adenosylmethionine (AdoMet) in the endonuclease reaction (absolute for type 1, stimulatory for type III). Although both types of enzyme

*NCI-Frederick Cancer Research Facility, P.O. Box B, Frederick, Maryland 21701

require ATP for DNA cleavage, extensive ATP hydrolysis has been observed only for the type I enzymes. The DNA sequences recognize by these proteins are asymmetric; however, the type III sequences consist of five to six bases, whereas the type I sequences are hyphenated with two constant domains of three and four bases separated by a nonspecified spacer of six or eight bases. DNA methylation occurs at these sites, but DNA cleavage occurs 25–27 (type III) or several thousand (type 1) bases away. The properties of these various systems are summarized in Table 2.1.

These complex type I and type III multifunctional enzymes represent an excellent subject for the study of protein-DNA interactions. Although our emphasis will be on the methylase reactions, these systems cannot be discussed without referring in some detail to the related restriction activities. In this chapter, we review the genetic structure of the restriction modification loci and their transcription patterns, the nucleotide sequences recognized by the enzymes, and the multiple steps of their reaction mechanism. We also discuss the effect of certain phage antirestriction systems on DNA methylation.

Type I Restriction-Modification Systems

Structure: Genes and Enzymes

The classic experiments that defined host-controlled restriction-modification were carried out by Arber *et al.* on the type I systems of *E. coli* K and B (Arber, 1965), which are allelic and also are related to both the A system of *E. coli* 15 (Arber and Wauters-Willems, 1970) and the restriction-modification systems of some strains of *Salmonella* (Colson and Colson, 1971, 1972). Complementation experiments (Boyer and Roulland-Dussoix, 1969; Glover, 1970) were consistent with the presence of three genes in the restriction-modification locus (called *hsd*). The *hsdR* gene is responsible for restriction, the *hsdM* gene for modification, and the *hsdS* gene for recognition of the host specificity site on the DNA. A mutation in the *hsdR* gene resulted in the loss of restriction, but it did not affect modification (r^-m^+ phenotype). An *hsdS* mutation led to the loss of site recognition and resulted in a restriction-modification-deficient phenotype (r^-m^-). The *hsdM* mutation gave rather unexpected results, since it also had a r^-m^- phenotype—a fact that was only understood later in terms of the enzymatic mechanism (Hubacek and Glover, 1970).

Since the *hsdS* gene determines host specificity, there should be mutants and recombinants with new host specificities. None have been found in *E. coli*, but recombination experiments between *Salmonella typhimurium* (SB system) and *S. postdam* (SP system) resulted in a recombinant with a new host specificity called *SQ* (Bullas *et al.* 1976). All three *Salmonella* systems are related to those of *E. coli*.

The *hsd* locus originally was mapped in the vicinity of *serB* (Boyer, 1964).

Table 2.1 Characteristics of Restriction and Modification Systems

	Type I	Type II	Type III
Restriction and modification activities	Single multi-functional enzyme	Separate endonuclease and methylase	Single multifunctional enzyme
Protein structure	3 different subunits	Simple	2 different subunits
Requirements for restriction	ATP, AdoMet, and Mg^{2+}	Mg^{2+}	ATP (AdoMet, Mg^{2+}
Requirements for methylation	AdoMet (ATP, Mg^{2+})*	AdoMet	AdoMet, (ATP, Mg^{2+})[a]
Sequence for host specificity sites	sB: T-G-A-N_8-T-G-C-T; sK: A-A-C-N_6-G-T-G-C	Twofold symmetry	sP1:A-G-A-C-C; sP15:C-A-G-C-A-G; s*Hin*fIII:C-G-A-A-A-T;
Sites of DNA methylation	Host specificity sites	Host specificity sites	Host specificity sites
Sites of DNA cleavage	Possible random, at least 1000 bp from host specificity site	Host specificity sites	25–27 bp to 3′ of host specificity sites
Expression of restriction and modification activities	Mutually exclusive	Separate activities	Simultaneous

*Compounds in parentheses stimulate activity, but are not required.

The order of genes in the cloned *hsd* locus of *E. coli* K was shown to be *hsdR, hsdM, hsdS,* and transcription was shown to occur in the same direction for all three genes (Sain and Murray, 1980). The cloned *E. coli* K *hsd* region contains sequences of extensive homology with the *hsd* region of *E. coli* B but it did not hybridize with *E. coli* C DNA (a natural r^-m^- strain). The screening of a large number of *E. coli* strains resulted in the characterization of only one (*strain*) with a new specificity, called *E. coli* D. The *hsd* region of *E. coli* D was nearly homologous to those of *E. coli* K and B. The *hsdR* and *hsdM* genes of these three strains were highly conserved and functionally interchangeable (Murray *et al.* 1982). The DNA sequence of the *hsdS* gene showed small variations in length and little homology, except for a region of 100 base pairs (bp) in the midde of the gene and another region of 250 bp at the terminus. Gough and Murray (1983) proposed that the region at the carboxy end is involved in interacting with *hsdM* subunit and that the central region defines the active site.

There were two promoters in *hsd;* one regulates *hsdM* and *hsdS* and the other regulates *hsdR*. The *hsdM* termination codon overlaps with the initiation codon of *hsdS,* which allows for translational coupling. The existence of two separate promoters permits synthesis of a DNA methylase that is composed of the *hsdM* and *hsdS* peptides independent of synthesis of the endonuclease that is composed of all three peptides. These results are summarized in Figure 2.1.

These experiments provided the genetic basis for the biochemical character-

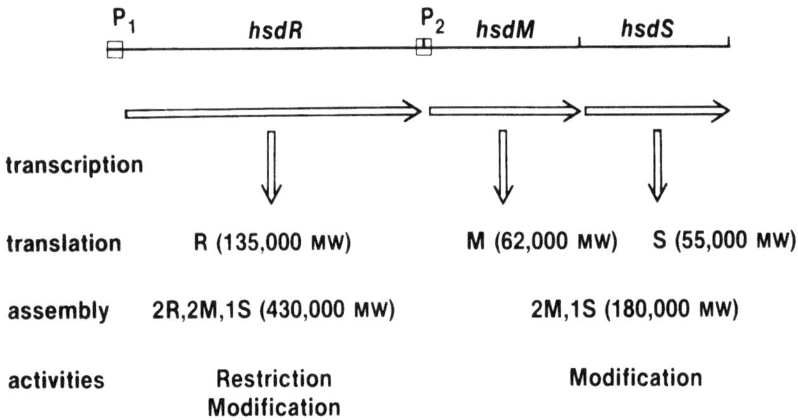

Figure 2.1. The *hsd* locus of *E. coli* K is composed of three genes: *hsdR* is responsible for restriction, *hsdM* for modification, and *hsdS* for recognition of the host specificity site. There are two promoters, P1 and P2. P1 controls transcription of *hsdR* and P2 that of both *hsdM* and *hsdS*. Translation of the transcripts yields polypeptides with a mol wt of 135,000 (*hsdR*), 62,000 (*hsdM*), and 55,000 (*hsdS*). These polypeptides can be associated into an endonuclease composed of all three subunits or into a methylase composed of the *hsdM* and *hsdS* subunits.

ization of the *Eco*K and *Eco*B enzymes* and led to two possible enzyme models. In one, there are separate methylase and endonuclease enzymes; the former is composed of the *hsdM* and *hsdS* gene products and the latter of all three gene products. The alternative is a single enzyme composed of all three polypeptides that catalyzes both restriction and modification

The purified *Eco*K was shown to catalyze three different reactions: 1) specific DNA cleavage in the presence of ATP, AdoMet, and Mg^{2+} (Meselson and Yuan, 1968); 2) DNA methylation in the presence of AdoMet (Haberman *et al.* 1972); and 3) ATP hydrolysis that is coupled to DNA cleavage (Yuan *et al.* 1972). The native enzyme had an approximate mol wt of 400,000 and had three subunits with mol wts of 135,000, 62,000, and 55,000 (in molar ratios of 2:2:1) (Messelson *et al.* 1972). Infection of ultraviolet (UV)-irradiated cells by λ containing the *hsd* locus resulted in the identification of the 135,000 mol wt polypeptide as the *hsdR* gene product, the 62,000 mol wt polypeptide as that of the *hsdM* gene, and the 55,000 mol wt polypeptide as that of the *hsdS* gene (Sain and Murray, 1980). The purified *Eco*B had similar enzymatic activities (Roulland-Dussoix and Boyer, 1969; Eskin and Linn, 1972a; Vovis *et al.* 1974) and subunit structure, although the protein preparation was more heterogeneous in size (Eskin and Linn, 1972b).

*The nomenclature in this chapter follows the rules proposed by Smith and Nathans (1973). *Eco*K refers to the restriction enzyme from *E. coli* K; *Eco*P15 is the enzyme coded by the plasmid P15. The host specificity sites are named s, followed by the name of the restriction-modification system.

Modification enzymes consisting of only the *hsdM* and *hsdS* polypeptides have been purified from both *E. coli* B (Lautenberger and Linn, 1972) and *E. coli* K (Bickle, 1983). The absence of the *hsdR* subunit renders them incapable of catalyzing the restriction reaction. The *E. coli* K DNA methylase catalyzed the reaction on unmodified DNA much more efficiently than the *Eco*K restriction enzyme, and it was neither stimulated nor inhibited by ATP. When hemimethylated DNA was used as a substrate, the methylase rates were approximately the same for both the methylase and the restriction enzyme (in the presence of ATP and Mg^{2+}). The methylase was not stimulated by either ATP or Mg^{2+}, as was the restriction enzyme (Suri *et al.* 1983). On the basis of the enzyme characteristics, it is impossible to decide whether the restriction enzyme or the methylase is responsible for modification *in vivo.* Given the existence of two different promoters, both alternatives are possible. The restriction enzyme could be responsible for the maintenance of the methylation pattern, but under some physiologic conditions, the presence of a modification activity devoid of the associated endonuclease might be crucial for the survival of the chromosomal DNA.

The identification of the three different subunits did not resolve the question of which reaction was catalyzed by each of the subunits. This difficult problem was approached by the characterization of enzymes from various mutant strains. It had been reported that crude extracts from r^-m^+ and r^-m^- strains of *E. coli* B complemented each other (Linn and Arber, 1968). This type of *in vitro* complementation was used to purify three mutant enzymes: 1) an *hsdR*$^-$ enzyme from a r^-m^+ strain; 2) an *hsdS*$^-$ enzyme from a r^-m^- strain; and 3) an *hsdM*$^-$ enzyme from a different r^-m^- strain (Hadi and Yuan, 1974; Buhler and Yuan, 1978). The *hsdR*$^-$ enzyme lacked endonuclease activity, did not bind unmodified DNA to filters, and had a low level of DNA-independent ATPase activity that required AdoMet; however, it did methylate DNA. The *hsdS* enzyme lacked the endonuclease, DNA-binding, and methylase activities, but it had an AdoMet-dependent ATPase activity. All four of these activities were absent in the *hsdM*$^-$ enzyme. However, the purified *hsdR*$^-$ enzyme complemented the hsdS$^-$ or *hsdM*$^-$ proteins, as proven by DNA cleavage, DNA binding to filters, restriction-dependent ATPase activity, and DNA methylation. The interaction of these mutant proteins with ATP, AdoMet, and DNA have been studied in detail. A mutation in the *hsdM* subunit resulted in the loss of AdoMet binding and, consequently, of DNA cleavage and methylation. A lesion in the *hsdS* subunit did not affect AdoMet binding, but it prevented enzyme activation and subsequent DNA sequence recognition. Both restriction and modification were lost. The *hsdR* mutation allowed all of the early events in the restriction mechanism, including DNA sequence recognition, but it blocked the ATP-induced conformational change required for restriction. DNA methylation was not affected. The effect of these changes can be best understood in the context of our knowledge of the reaction mechanism, which is presented in a later section of this chapter.

Antibodies against the *Eco*K subunits *hsdM* and *hsdR* have been used to determine the relationship between *Eco*K and the enzymes coded by *E. coli* strains B, C, and A and *S. typhimurium, S. postdam,* and *Salmonella* recombinant strain SQ (Murray *et al.* 1982). The antibodies reacted strongly with extracts from *E. coli* B, to a lesser degree with those from the three *Salmonella* strains, and not at all with those from *E. coli* strains C (which lacks a restriction-modification system) and A. This indicated the existence of a type 1 family consisting of *E. coli* strains B and K and the *Salmonella* strains SB, SP, and SQ; *E. coli* A belongs to a different family, even though it shares certain functional characteristics with the others. It also provides further proof that *E. coli* C lacks a restriction-modification system. Antiserum to the *hsdR* subunit cross-reacted with the *hsdM* subunit, indicating that these may have certain structural similarities.

The studies of the purified enzymes complemented the findings from genetic studies to a remarkable degree. Both provided evidence for a complex protein composed of three separate gene products responsible for restriction, modification, and site recognition, respectively. The *hsdR* and *hsdM* genes, which are responsible for the catalytic activities, are largely conserved; however, the same is not true for the *hsdS* gene, which defines the specificity of the system.

DNA Recognition Sequences

The susceptibility of a given genome to a restriction-modification system is determined by the presence of host specificity sites on the DNA. Such sites can be lost (Arber and Kuhnlein, 1967) or acquired (Sclair *et al.* 1973) by mutation. Their presence is required for methylation *in vivo* (Smith *et al.* 1972) and *in vitro* (Kuhnlein and Arber, 1972).

Although the type I enzymes recognize and methylate at these host specificity sites, they cleave the DNA elsewhere—presumably at random sites (Horiuchi and Zinder, 1972). Because the enzymes interact with both recognition sites and cleavage sites, their sequencing was a far from routine task. The recognition sites were mapped by standard genetic procedures (Lyons and Zinder, 1972; Murray *et al.* 1973), by a combination of *in vitro* DNA methylation with ^{3}H-AdoMet and restriction analysis (Kan *et al.* 1979), or by electron microscopy of enzyme-DNA complexes (Brack *et al.* 1976). The strategy for sequencing the specificity sites required small, fully sequenced genomes with a corresponding collection of host specificity site mutants. The sequences surrounding the sites were compared. The two wild-type sB1 and sB2 sequences in f1 were determined along with those of nine independent sB1 and four sB2 mutants (Ravetch *et al.* 1978). The same approach was used for the sB site on ϕ X-174 (Lautenberger *et al.* 1978). Both studies identified the sB sequence as 5′ T-G-A-N_8-T-G-C-T 3′. Loss of the host specificity site was associated with changes at the T and G of the trinucleotide and the G, C, and T of the tetranucleotide. The sequence of a ΦX-174 variant that was not restricted by *E. coli* B was subjected to computer analysis and shown to contain sequences that differed from that of sB at each of the seven constant positions, or by variations in the spacer ranging from 5–15 bases. Therefore, the two constant domains of T-G-A and T-G-C-T separated by an eight-base spacer define the sB site.

The product of *Eco*B methylation was N^6-methyladenine; two methyl groups were transferred to each sB site (Smith *et al.* 1972). Hemimethylated DNA (one strand modified and one unmodified) was the preferred substrate. The unmodified strand in either of the two hemimethylated DNAs was methylated *in vitro* (Vovis and Zinder, 1975). Partial sequences of *in vitro*-methylated fd DNA (vanOrmondt *et al.* 1973) and SV40 DNA (Dugaiczyk *et al.* 1974) were both interpreted as methylation of the adenine in the T-G-A sequence and the first adenine in the A-G-C-A sequence complementary to the second domain.

Sequence comparisons of the three sK sites on ΦXsK1, ΦXsK2, and G4 and the two sK sites on pBR322 (Kan *et al.* 1979) led to the identification of the sK site as 5′ A-A-C-N_6-G-T-G-C 3′, with two constant domains of three and four bases separated by a six-base spacer of variable sequence. Loss of the sK site was associated with a change in the C of the trinucleotide. The acquisition of a K host specificity site by mutation was due to a change from G-X-G-C to G-T-G-C. Computer analysis of the four genomes showed the presence of other sequences that differed from the sK site at each of the seven constant positions—none of which were functional sK sites. Methylation of the sK sites also occurs at an adenine, but no data is available on the methylated sequence.

The sequence of the sK and sB sites are very similar in that they are hyphenated with two constant domains of three and four bases, with a spacer of undefined sequence of eight bases for sB sites and six bases for sK sites. No symmetry is present, but the spacer places both constant domains on the same side of the helix. Kan and her collaborators (1979) aligned the two sequences in the following manner:

```
sB  5′  T-G-A*N-N-N-N-N-N-N-N-T - G- C- T  3′
        A-C -T - N-N-N-N-N-N-N-N-A*- C- G- A

sK  5′    A-A- C -N-N-N-N-N-N-G-T-G-C      3′
          T-T -G-N-N-N-N-N-N-C -A-C-G
```

Four of the seven conserved bases were the same, as if the same positions were always methylated, the second A in the A-A-C and the A in the C-A-C-G sequences would be the sites of *Eco*K methylation

The Reaction Mechanism

The ability of *Eco*K and *Eco*B to catalyze *in vitro* the opposing reactions of restriction and modification has posed certain basic questions: What factors determine the activity that will be expressed by a multifunctional protein? How can DNA methylation occur at a host specificity site and DNA cleavage at another unspecified site? What roles do ATP and AdoMet play in these events?

DNA methylation can be best understood in the context of the overall *Eco*K reaction mechanism. The scheme shown in Figure 2.2, originally was proposed for *Eco*K, but is equally applicable to *Eco*B. This consisted of several distinct events: 1) AdoMet is rapidly bound by *Eco*K; 2) *Eco*K slowly becomes activated (*Eco*K*); 3) *Eco*K* nonspecifically binds to DNA to form an initial complex; 4) *Eco*K* specifically binds to sK sites to form recognition complexes; 5) ATP interacts with the enzyme, resulting in release if the sK site is modified; 6) ATP induces a conformational change in *Eco*K* that is bound to unmodified sK sites, converting it into a form that leads to DNA retention on filters—no ATP hydrolysis is required for this step; 7) *Eco*K* remains bound to the sK site and translocates DNA past it in a reaction coupled to ATP hydrolysis; 8) recognition of certain undefined cleavage regions results first in the cleavage of one strand followed rapidly by a second cut in the complementary strand at a

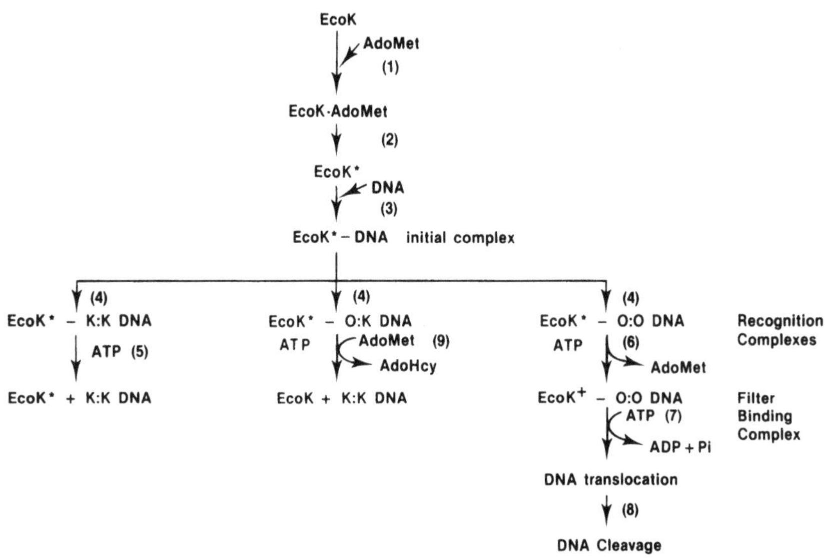

Figure 2.2 The reaction of *Eco*K is summarized in this flow chart. Binding of AdoMet, activation of *Eco*K*, and binding to DNA are common to all three types of DNA. Addition of ATP enables the enzyme to discriminate between the three forms of the recognition sequence. The methylated strand is indicated by K and the unmethylated one is indicated by O; thus, O:K represents a hemimethylated site. *Eco*K* will interact with each site to form unique enzyme complexes, which are identified by the following superscripts: k for methylated, h for hemimethylated, and o for unmethylated. In modified complexes, *Eco*K^{k} is released as *Eco*K* from the sK sites. In hemimethylated complexes, *Eco*K^{h} methylates the unmethylated strand; once the sK site is methylated, *Eco*K is released. In unmodified complexes, *Eco*K^{o} binds to the sK site and takes on a new form, *Eco*K^{+} which uses ATP hydrolysis to translocate the DNA until it reaches a cleavage region where DNA hydrolysis occurs.

site opposite the first; and 9) the enzyme rapidly methylates hemimethylated DNA, but does not nick or cleave it.

The essential features of this mechanism are that the early steps are common to both restriction and modification, and the *Eco*K will recognize and bind to all three forms of the sK site (methylated, unmethylated, and hemimethylated). It is this specific interaction that determines whether the enzyme will dissociate from the modified sK site, methylate the unmodified strand of the hemimethylated sK site, or translocate DNA and cleave it at a site distal to the unmethylated sK site.

The *Eco*B methylase activity required hemimethylated DNA and AdoMet, and it was stimulated by ATP and Mg^{2+} (Vovis *et al.* 1973). The same requirements were observed for *Eco*K. In the absence of ATP, *Eco*K also methylated unmodified DNA, but at an extremely slow rate (Haberman *et al.* 1972). The evidence for the mechanism of DNA methylation is reviewed below.

AdoMet Binding

AdoMet is required for two separate steps in DNA methylation: initially as an allosteric effector and subsequently as a methyl donor. The apparent Michaelis-Menten constant (Km) values for AdoMet in the methylation reaction were found to be 2×10^{-7}M (with ATP) and 2.2×10^{-7}M (without ATP), which compare favorably with the value of 3×10^{-7}M found for the restriction reaction. The minimum number of AdoMet binding sites was estimated as five (Burckhardt *et al.* 1981a). The *Eco*K-AdoMet complex was tested for its ability to methylate hemimethylated DNA in the absence of free AdoMet. Negligible methylation was observed, indicating that the enzyme-bound AdoMet was not acting as a methyl donor. The mutation in the *hsdM* subunit prevented this AdoMet binding and effectively blocked all subsequent steps (Buhler and Yuan, 1978).

Enzyme Activation

Having once bound AdoMet, there is a slow transition to an activated enzyme form, *Eco*K*. Attempts to measure this transition directly were unsuccessful, because AdoMet was required for both early and late steps in the reaction. Sinefungin, an AdoMet analog, inhibited the methylation by *Eco*K or *Eco*K*. However, *Eco*K* DNA complexes were relatively resistant to it (Burckhardt *et al.* 1981a). These results indicated that *Eco*K had two types of AdoMet binding sites: effector sites and methyl donor sites. The effector sites were able to bind both AdoMet and Sinefungin, whereas the methyl donor sites had a much higher affinity for AdoMet than for its analog. The effector sites were accessible to AdoMet (or Sinefungin) before the enzyme was bound to DNA, but not afterwards. The reverse was true of the methyl donor sites; these were able to interact with free AdoMet only after the formation of recognition complexes. By the use of AdoMet, enzyme activation was shown to be a first-order reaction with a rate constant of 1.2×10^{-2} sec^{-1} and a half-time of 57 seconds after a lag of 19 seconds. This lag was presumed to be due to binding of AdoMet. These values are similar to those obtained for enzyme activation in the restriction reaction ($k = 1.3 \times 10^{-2}$ sec^{-1} and $t_{1/2} = 54$ sec) with a lag of 15–20 seconds. This was consistent with the proposal that both restriction and modification share the early steps in the reaction mechanism. The mutation in the *hsdS* subunit allows normal AdoMet binding, but effectively blocks enzyme activation (Hadi and Yuan, 1974).

Formation of Initial Complexes

*Eco*K* interacts with any DNA regardless of the presence or absence of sK sites. Indirect evidence for this was provided by the stabilization of *Eco*K* with DNA lacking sK sites (the half-life of *Eco*K* was changed from 130 seconds to 6 minutes by the presence of λ DNA without the sites (Yuan *et al.* 1975). Direct evidence for the existence of initial complexes was provided by electron microscopy studies showing *Eco*K* bound to restriction fragments that lacked sK sites. The enzyme molecules were randomly located along the DNA (Brack *et al.* 1976).

Formation of Recognition Complexes

After the formation of initial complexes, the enzyme seeks and binds to sK sites. If this interaction with methylated, unmethylated, or hemimethylated sequences afffects the activity of the enzyme, the three types of recognition complexes would have different biochemical properties. These are summarized in Table 2.2. Most of the experiments have focused on unmethylated and hemimethylated DNA, because methylated DNA is not a substrate for either restriction or modification. Unmethylated and hemimethylated recognition complexes differ in: 1) their efficiency of reaction at different sK sites; 2) their sensitivity to various inhibitors; and 3) their interaction with ATP.

The efficiency of methylation at different sites was tested. λ DNA has five sK sites located at 18.0, 30.9, 33.7, 71.7, and 96.7 map units from the left end. Unmethylated and hemimethylated DNAs were digested with *Pst*I or *Hpa*I and were then methylated with *Eco*K. Those fragments carrying sK sites were methylated. However, although all five hemimethylated sK sites were methylated with equal efficiency, the efficiency of methylation differed among the unmethylated sites (Burckhardt *et al.* 1981a). In fact, there was a correlation between efficiencies of cleavage and methylation for unmethylated sites (e.g., fragment *Hpa*I-7a was poorly cut and poorly methylated). Therefore, recognition and/or methylation of hemimethylated sequences appeared to be independent of the neighboring sequences, in contrast to the situation with unmethylated sites. Finally, in the absence of ATP, the hemimethylated sites were methylated in 30 minutes, but methylation of the unmethylated sites required 12–16 hours.

A number of compounds differ in their inhibitory effects on recognition complexes. As pointed out earlier, Sinefungin is a strong inhibitor of DNA methylation. When Sinefungin was used in place of AdoMet in the restriction reaction, partial DNA cleavage was observed (Burckhardt *et al.* 1981b). This showed that *Eco*K-Sinefungin was able to interact effectively with unmethylated sK sites, but not with the hemimethylated ones.

Actinomycin D and heparin inhibit both endonuclease and methylase activities. When hemimethylated recognition complexes were incubated with either inhibitor before ATP addition, no DNA methylation was detected. When unmodified complexes were used in the same experiment, no inhibition of DNA cleavage took place. Therefore, *Eco*K bound to hemimethylated sites was susceptible to inhibition by Actinomycin D and heparin, whereas *Eco*K bound to unmethylated sites was resistant.

The Effect of ATP on Recognition Complexes

The interaction of ATP with the recognition complexes commits each one to its final course of action. DNA methylation by *Eco*K or *Eco*B does not require ATP, but is strongly stimulated by it. ATP and its nonhydrolyzable β,γ-imido-analog acted as allosteric effectors with apparent Km values of 3.2×10^{-8}M for ATP and 2.8×10^{-5}M for the analog, indicating a minimum number of two binding sites on *Eco*K. The corresponding value for ATP in the restriction reaction was 1-3 $\times 10^{-5}$M, i.e., three orders of magnitude higher than for modification.

Incubation with ATP converted hemimethylated recognition complexes into methylase complexes. Neither recognition nor methylase complexes were trapped on nitrocellulose filters (Burckhardt *et al.* 1981b). ATP (or its imido analog) converted unmodified recognition complexes into a new form that had lost its enzyme-bound AdoMet and could now be retained on filters. Exami-

Table 2.2 Properties of Recognition Complexes

	Recognition Complexes with Unmethylated DNA	Recognition Complexes with Hemimethylated DNA
Binding at sK sites	Yes	Yes
Reaction with sinefungin	Partial DNA cleavage	No DNA methylation
Inhibition with heparin or actinomycin	DNA cleavage by complexes is resistant	DNA methylation by complexes is sensitive
Effect of ATP addition	1. Formation of filter-binding complexes	No filter-binding complexes
	2. Change in apparent size	No change in enzyme
	3. ATP hydrolysis coupled to DNA cleavage	No ATP hydrolysis required for DNA methylation
	4. DNA translocation	No translocation
	5. DNA cleavage	DNA methylation

nation of the filter-binding complexes by electron microscopy showed an apparent decrease in the diameter of the enzyme, although it was impossible to determine whether this was due to a change in shape or to the dissociation of one or more subunits (Bickle *et al.* 1978). This conformational change was not observed with the methylase complexes. These methylase and filter-binding complexes also showed a differential sensitivity to ethidium bromide, which inhibits both DNA methylation and cleavage. If hemimethylated complexes were pulsed with ATP before the addition of ethidium bromide, no DNA methylation took place. DNA cleavage occurred when the same experiment was repeated with unmethylated complexes.

The fate of the enzyme during the methylase reaction was investigated in an experiment in which a methylase complex formed with *Hpa*I fragments and ^{3}H-AdoMet was isolated. The methylase complex showed no detectable methylation. If the complex was further incubated with ^{3}H-AdoMet and ATP for

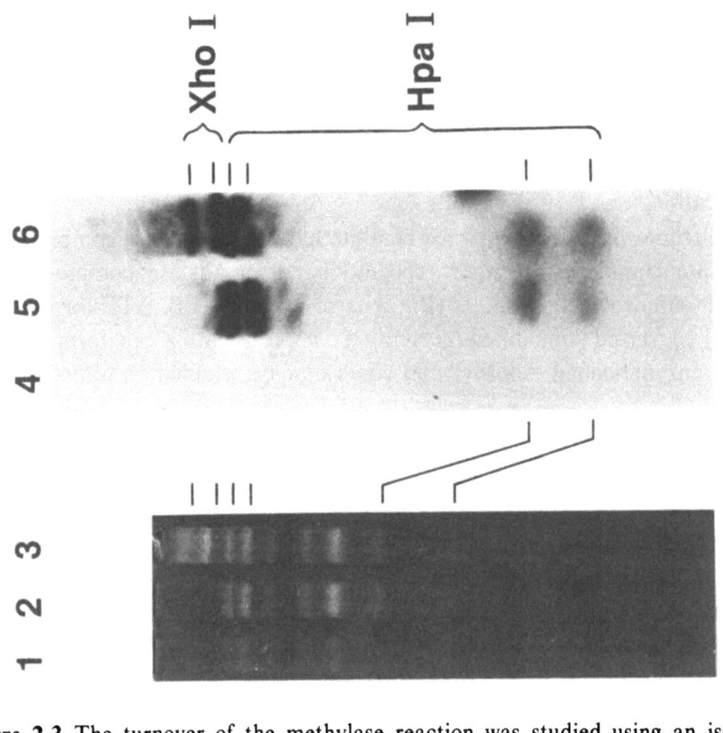

Figure 2.3 The turnover of the methylase reaction was studied using an isolated *Eco*K-DNA complex. Recognition complexes were prepared with hemimethylated *Hpa*I DNA fragments, 3H-AdoMet, and *Eco*K; they were isolated by gel filtration. The recognition complexes were then divided into three portions (1, 2 and 3). Sample 1 was not incubated further. Samples 2 and 3 were incubated in the presence of ^{3}H-AdoMet and ATP for 30 minutes 30°C. Sample 3 then had *Xho*I hemimethylated DNA fragments added, and the incubation was continued for a further 30 minutes. Lanes 1–3 were stained with ethidium bromide. Lanes 4–6 are autoradiographs of the same samples.

30 minutes, DNA methylation took place. Further addition of a new DNA substrate (*Xho*I fragments) resulted in its methylation. Figure 2.3 shows that no detectable methylation occurred in the methylase complex until it was incubated with both AdoMet and ATP. The second DNA was also methylated. In a second experiment, the enzyme molecules dissociated from the DNA as the methylation reaction progressed. It was concluded that the enzyme was bound to the hemimethylated sites until methylation was completed; then the enzyme was released from this modified site and was able to bind to a new hemimethylated site. This sequence of events is in sharp contrast to that with an unmethylated site; in the latter reaction, the enzyme remained bound to the sK site and translocated DNA past it in a reaction coupled to ATP hydrolysis. This process generated highly supertwisted loop structures (Yuan *et al.* 1980) and eventually resulted in DNA cleavage.

A Model

Having dissected all of the separate steps that make up the mechanism of DNA methylation by *Eco*K, we return once more to the question of how the methylated state of a nucleotide sequence determines the activity of an enzyme. The model shown in Figure 2.4 is consistent with all of the available data and also can be tested experimentally. The interaction of *Eco*K with the sK site determines the arrangement of the five subunits. The two constant domains of the sequence are on the same side of the helix, and the two methyl groups are present in the major groove. The *hsdS* subunit binds to the two constant domains regardless of their methylated stage (top three figures of Figure 2.4). Enzyme activation by AdoMet results in the positioning of the *hsdM* subunits into the major groove (middle three figures of Figure 2.4). At a methylated sK site, the two methylated adenines prevent them from entering the major groove, which results in an open configuration. At a hemimethylated site, one *hsdM* subunit enters the major groove at the unmethylated domain, while the other one is excluded by the methylated base at the other domain. The result is a partially open configuration. At unmethylated sK sites, both *hsdM* subunits enter the major groove, leading to a closed configuration. The bottom three figures show one hypothetical model for the grouping of thc five subunits around each of the sK sites. Most importantly, this hypothetical model can be tested by experiments that measure the protection of sK sites from nuclease digestion

These molecular details of DNA modification should be examined in the context of the larger biological picture. By using the restriction-modification enzyme complex, the cell can examine all host specificity sites. When it recognizes the methylated sequences as its own, it rapidly dissociates from them. Hemimethylated DNA is normally the product of semiconservative replication. As such, it should be resistant to the homologous restriction activity; however, it needs to be rapidly methylated or else the next round of replication will result in the loss of the methylation pattern. Unmethylated DNA typically is foreign DNA and, as such, is destined for degradation.

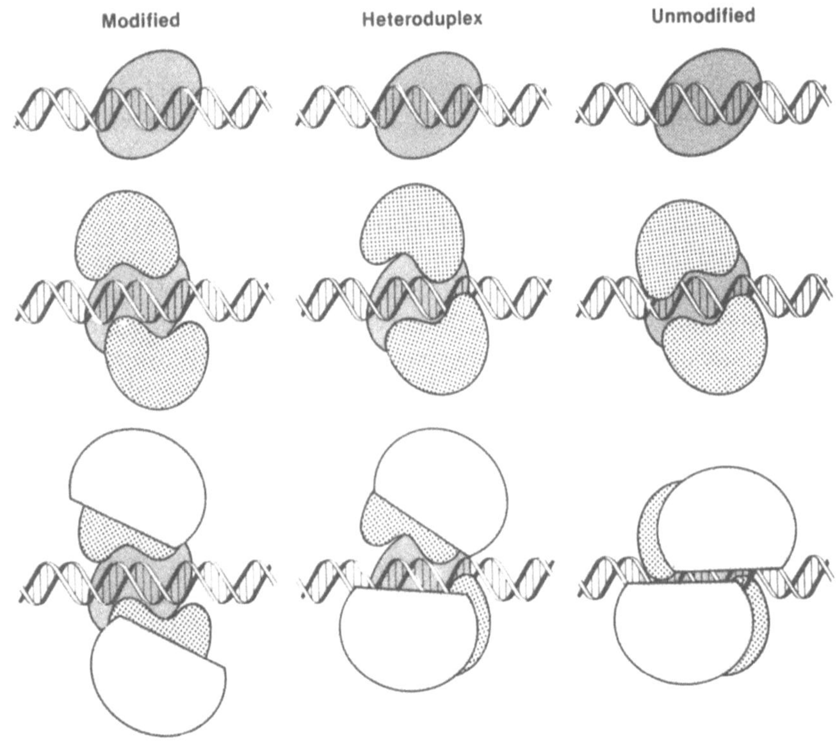

Figure 2.4 The conformation assumed by *Eco*K and the resulting enzymatic activity appears to be directed by the methylated state of the recognition sequence. The top row shows the binding of the S subunit (site recognition) to the three forms of the recognition sequence. The middle row shows the positioning of the M subunit (modification). In the modified complex, the M subunits cannot enter the major groove due to the presence of the methylated adenines leading to the open configuration *Eco*K^k, which does not act on the DNA. In the hemimethylated complex, an M subunit enters the major groove at the unmethylated domain and the other M subunit is blocked by the methyladenine. This leads to a half-open configuration, *Eco*K^h, which catalyzes the methylase reaction. In the unmodified complex, both M subunits enter the major groove. This results in a closed configuration, *Eco*K^o, which is able to cleave the DNA. The bottom row conceptualizes the binding of the restriction-modification form of *Eco*K to the various sites.

Type III Restriction and Modification Systems

Structure: Genes and Enzymes

The three type III systems that have been characterized to date are those from the phage P1, the plasmid P15, and the bacterial strain *H. influenzae* Rf. The P15 system is one of two restriction-modification systems carried by *E. coli* 15T$^-$ (Arber and Wauters-Willems, 1970). This plasmid is approximately 90%

homologous with P1, and its restriction-modification locus will recombine with that of P1. Transformation experiments with the *Hin*fIII gene showed that it was closely linked to the chromosomal marker for novobiocin resistance, suggesting that it is located on the chromosome (Kauc and Piekarowicz, 1978).

The possibility that P1 and P15 were not type I systems was suggested from experiments in which a methionine-requiring strain was grown in ethionine. After one round of replication, 50% of the DNA was degraded, but only in strains carrying P1 and P15. No DNA degradation was observed with strains carrying type I systems (Lark and Arber, 1970). The conclusion from these experiments was that type I systems require AdoMet for restriction, but that type 3 systems do not.

As in the case of type I systems, two classes of restriction-negative mutants were isolated: modification-proficient (r^-m^+) and modification-deficient (r^-m^-). It was believed that three genes also were required for restriction and modification. This matter has now been resolved by the experiments of Iida *et al.* (1983), who have characterized the two type III genes (called *res* and *mod*) on the *Bam*HI-4 fragments of P1 and P15. A combination of transposition mutagenesis, restriction mapping, and heteroduplex analysis showed the existence of a *res* gene of approximately 2.8 kilobase (kb) contiguous to a *mod* gene of approximately 2.2 kb. Mutations in the *res* gene resulted in the loss of restriction, but they did not affect modification. The *mod* gene is responsible for both site recognition and modification; mutations in *mod* resulted in the loss of both restriction and modification. Two promoters were identified by *in vitro* transcription—one in front of *mod* and the other preceding *res*. There also was some read-through transcription from the *mod* promoter into *res* (Figure 2.5). The *res* genes of P1 and P15 are homologous except for a region of partial homology towards the N terminus. The *mod* genes were much less homologous. The N and carboxyl termini appeared to be identical; however, towards the N terminus, there was a 1.2-kb region of nonhomology. Presumably, this region of nonhomology is responsible for sequence recognition.

The enzymes *Eco*PI (Haberman, 1974; Hadi *et al.* 1983), *Eco*P15 (Reiser and Yuan, 1977; Hadi *et al.* 1983), and *Hin*fIII (Kauc and Piekarowicz, 1978) all have been extensively purified and characterized. Although at first glance their enzymatic properties resembled those of type 1 enzymes, they differed from them in some important respects. The restriction reaction required ATP and Mg^{2+} but only was stimulated by AdoMet. The type III enzymes did not bind DNA to filters nor did they show any extensive ATP hydrolase activity. Their digests showed distinctive restriction fragments, but the digestion was incomplete in either the presence or absence of AdoMet. The enzymes also methylated unmodified DNA, both in the presence and absence of ATP. In the presence of ATP, the methylation was more efficient, but it acted in competition with the restriction reaction that generates larger fragments. This was in sharp contrast with the type I enzymes, which either restrict or modify the DNA, but do not do both simultaneously. These differences are important because they reflect the structure and reaction mechanism of this class of enzymes.

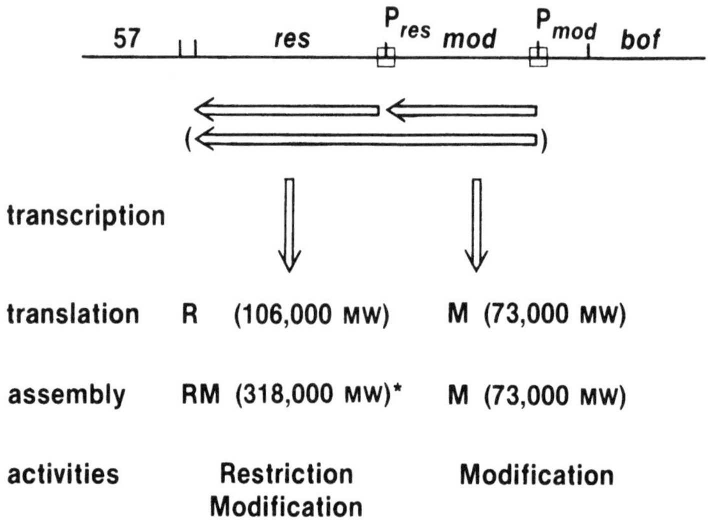

Figure 2.5 The restriction and modification locus of P1 is composed of two genes: *res* is responsible for restriction and *mod* is responsible for both sequence recogniton and modification. There are two promoters, P*res* and P*mod*. P*mod* controls transcription of *mod* and P*res* that of *res*. Translation of the transcripts yields polypeptides with a mol wt of 106,000 *(res)* and 75,000 *(mod)*. These polypeptides can be assembled either into an endonuclease composed of the two subunits or into a methylase composed solely of the *mod* subunit. It is not known whether the endonuclease is composed of one *res* and three *mod* or two of each.

*Hin*fIII was shown to have a mol wt of 200,000 and to be composed of subunits of 110,000 mol wt and 80,000 mol wt (Kauc and Piekarowicz, 1978). The purified *Eco*P15 had a mol wt of 318,000 and was composed of subunits of 106,000 mol wt and 73,000 mol wt (possibly in ratios of 1:3) (Hadi *et al.* 1983). *Eco*P1 was similar, but the smaller subunit had a mol wt of approximately 75,000. The size of the subunits and of the genes is consistent, with the larger one being the *res* gene product and the smaller one being the product of the *mod* gene.

The subunit functions of *Eco*P1 and *Eco*P15 were determined by experiments using an antibody against *Eco*P15 (Hadi *et al.* 1983). Purified *Eco*P15 and a crude extract from a res^- mod^+ deletion mutant of *Eco*P15 were electrophoresed in the presence of sodium dodecyl sulfate, and the gel was then reacted with the *Eco*P15 antibody. Although the two subunits were detected in *Eco*P15, only the smaller one was found in the mutant extract. The same experiment was repeated with *Eco*P1 and a res^-mod^+ mutant of P1. Two bands were seen when *Eco*P15 was reacted with the *Eco*P15 antibody, but only the smaller subunit was detected in the mutant extract. Therefore, *res* codes for the larger subunit and *mod* codes for the smaller one. The isolated *res* gene was cloned, and the resulting strain had a Res^-Mod^- phenotype. However, the *res* subunit was being actively synthesized, indicating that both *res* and *mod* subunits had to be present for restriction activity to be expressed. The reactivity of the *Eco*P15 antiserum with the *Eco*P1 subunits is consistent with the high degree of homology between the two sets of *res* and *mod* genes.

The modification enzymes of P1 and P15 were isolated from $res^{-}mod^{+}$ mutants. The enzymes were specific for unmodified DNA and retained the same sequence specificity. However, their methylase was less efficient than that of the restriction enzyme, as measured by their Km for AdoMet and by the kinetics of the reaction. In contrast to the restriction enzyme, they required Mg^{2+} but were not affected by the presence of ATP. As in the case of the type I systems, it is not clear whether modification *in vivo* is catalyzed by the complete restriction enzyme and the *mod* subunit or by either of them alone under the appropriate physiologic conditions.

DNA Recognition Sequences

The characterization of the sP1 sequences on SV40 DNA (Bachi *et al.* 1979), the sP15 sequences on pBR322 DNA (Hadi *et al.* 1979) and the s*Hinf*III sequences on fd DNA (Piekarowicz *et al.* 1981) was done by a procedure similar to that used for the type I systems. The DNA was methylated *in vitro* and subjected to restriction analysis. The sequences of the methylated fragments were then examined for the absence of the presumed recognition sequences. The sequences were shown to be

sP1	5′ A-G-A-C-C-3′
sP15	5′ C-A-G-C-A-G 3′
sHinfIII	5′ C-G-A-A-T 3′

All three sequences are five or six bases in length and are asymmetric, but, they do not show the hyphenated structure of the type I sites. The methylated base has been shown to be adenine. Two of the methylated fragments from *Eco*P1-modified SV40 DNA were sequenced. The results were consistent with methylation of the central adenine of A-G-A-C-C. No direct sequencing data is available for sP15 or s*Hinf*III sites, but separation of the DNA strands at all three sites have shown that only one of them was methylated. In any event, in two of the three systems, adenine is present on only one strand in the recognition sequence. *Eco*P15-DNA complexes have been studied by electron microscopy, and the enzyme molecules have been shown to be bound to the sP15 sites (Yuan *et al.* 1981). Although the lack of site mutants and the scarcity of direct sequencing data makes the evidence weaker, it does suggest that the type III enzymes recognize and methylate at the host specificity sites. Unfortunately, it also raises the problem of how newly replicated DNA becomes methylated, since one of the new daughter molecules will be unmethylated and, thus, subject to the restriction reaction.

DNA cleavage took place at the recognition site or in its vicinity. Unlike the type I enzymes, *Eco*P1 and *Eco*P15 were shown to always cleave specifically

25–27 bases in the 3′ direction from the recognition sequence (i.e., for any given site, the distance was fixed, but the distance varied from site to site). The efficiency of cleavage also varied from site to site. In the case of *Hinf*III, the situation was obscured by the presence of a low level of contaminating exonuclease; however, it was estimated that the cleavage site was 20–30 bases in the 3′ direction from s*Hinf*III.

The Reaction Mechanism

At the outset, it appeared that the mechanism of the type III enzymes would be simpler than that of type I; in fact, the opposite has turned out to be the case. This has been due, in part, to certain features such as the simultaneous expression of DNA cleavage and methylation, the incomplete nature of the endonuclease reaction, and the differential cleavage at various sites. Also, while the mechanisms of *Eco*P1 and *Eco*P15 are very similar, they differ in some important respects from that of *Hinf*III. It generally is agreed that those differences are due to changes in the enzymatic properties during extended storage. The various steps described here apply to *Eco*P1, *Eco*P15, and *Hinf*III, unless otherwise indicated.

Enzyme Activation

The experimental data available are consistent with *Hinf*III existing in three possible forms that we will call *Hinf*III, *Hinf*III*, and *Hinf*III$^+$. *Hinf*III was isolated in a form that had AdoMet bound to it that was called *Hinf*III* (Piekarowicz and Brzezinski, 1980). This was demonstrated by incubating unmodified λ DNA with *Hinf*III* alone, isolating the DNA, and methylating it with *Hinf*III* with ^{3}H-AdoMet. The amount of methylation detected was considerably lower than when the first incubation was omitted. The preincubation with *Hinf*III* resulted in the transfer of the methyl group from the enzyme-bound AdoMet to the DNA. *Hinf*III* lost its bound AdoMet upon storage and was converted into a different form called *Hinf*III. *Hinf*III is believed to be an irregular form of the enzyme, since it required high ATP concentration for DNA cleavage and did not seem to form stable complexes with DNA, even in the presence of AdoMet. *Hinf*III* was able to bind free AdoMet and to undergo a transition to a different enzyme form, *Hinf*III$^+$. Therefore, *Hinf*III is the free enzyme, *Hinf*III* is the enzyme that has bound AdoMet *in vivo*, and *Hinf*III$^+$ is a form that has bound additional AdoMet during the *in vitro* reaction.

Similar forms of *Eco*P15 also could exist. The purified *Eco*P15 used in the following experiments could be *Eco*P15* (i.e., it would have AdoMet bound to it). Incubation of this enzyme species with AdoMet would result in the formation of *Eco*P15$^+$ with additional AdoMet bound to it. This binding of AdoMet by *Eco*P15* resulted in allosteric changes; it was estimated that at least two AdoMet binding sites were involved (Yuan and Reiser, 1978).

Formation of Recognition Complexes

Both *Eco*P15* and *Eco*P15$^+$ bound to unmodified DNA, presumably by first forming complexes at random sites and then at the s15 sites. Electron microscopy studies showed that both *Eco*P15* and *Eco*P15$^+$ were located at s15 sites, although the frequency of binding at each site varied (Yuan *et al.* 1981). The kinetics of complex formation were the same for both enzyme forms; however, the *Eco*P15$^+$ complexes were more stable. All of these enzymes were inhibited by heparin. The resistance of the enzyme-DNA complexes to heparin has been used as an indicator of their stability. One difference between *Hin*fIII and *Eco*P15 was that *Hin*fIII$^+$-DNA complexes were heparin-resistant, while the *Hin*fIII* complexes were heparin-sensitive. Both *Eco*P15* and *Eco*P15$^+$ complexes were heparin-resistant.

In the type I reactions, the interaction between the enzyme and the host recognition site determines the activity that will be expressed: dissociation from methylated sites, methylation of hemimethylated sites, or cleavage at sites distal from unmethylated sites. Since only one strand can be methylated in type III sites, these sites can exist in only two forms: methylated and unmethylated. The unmethylated sites serve as substrates for both DNA cleavage and methylation. How can these two activities be modulated? Several experiments give a hint of the possible answer. The apparent Km for AdoMet in the endonuclease reaction was 10^{-8}M while that for the methylase reaction was 4×10^{-7}M (Reiser and Yuan, 1977). This 40-fold difference in Kms could provide a control point. The structure of the DNA substrate also could play a critical role (see below).

The Interaction of ATP with Recognition Complexes

ATP is known to stimulate DNA methylation by *Eco*P15; however, few details are available as to its role in the mechanism of modification. Cleavage of supercoiled DNA by *Hin*fIII was much faster than cleavage of linear DNA, but no such difference was detected in the rates of the methylation reaction (Brzezinski and Piekarowicz, 1982). These results might be somewhat suspicious, since *Hin*fIII could be an altered enzyme form. However, these results agree with kinetic experiments using *Eco*P15* and *Eco*P15$^+$ with supercoiled SV40 DNA, which showed an accumulation of linear recognition complexes that were not cleaved any further. The interpretation is that the first cleavage event on supercoiled DNA would be extremely fast, but subsequent scission of linear molecules would be slow and would be overtaken by DNA methylation.

The properties of the P15 modification subunit are an apparent Km for AdoMet that is 15 times higher than for the intact EcoP15, the slower methylation kinetics, and no stimulation by ATP (Hadi *et al.* 1983). Activated *Eco*P15 is sufficiently large for being able to interact with both the recognition site and the cleavage site. ATP would interact with *Eco*P15* or *Eco*$P15^+$ bound to the unmethylated site and would trigger a conformational change leading either to the fast methylation at the recognition site or to cleavage 25–27 bases downstream. The conformational changes apparently take place faster with *Eco*$P15^+$.

Our understanding of the mechanism of action of type III enzymes is sketchy, particularly as it relates to the relationship between DNA cleavage and methylation. Furthermore, the attempts to determine the sequence elements that characterize good sites for both DNA cleavage and methylation have been unsuccessful.

Antirestriction systems

A discussion of the mechanisms of type I and type III restriction-modification systems would be incomplete without mention of antirestriction systems. Many phages have developed a way of coping with the powerful restriction systems that they encounter when they infect their bacterial hosts. They have evolved numerous, diverse antirestriction mechanisms that can be divided into the following categories:

1. The postreplicative modification of DNA generating unusual bases. The T-even phages catalyze DNA glycosylation. The *mom* gene of phage mu in conjunction with the *dam* gene of *E. coli* produces an unusual modification of the adenine residue that contains a carboxyl group (Hattman, 1980) and interferes with restriction by both type I and type III systems (Toussaint, 1976).
2. Inhibition of restriction and/or modification. T3 and T7 code for proteins that block the type 1 systems.
3. Alteration of the modification reaction. The λ*ral* gene codes for a protein that stimulates the methylation of unmodified DNA.

In this section, we will focus on those systems that directly affect the mechanism of modification by type I enzymes. These are controlled by the T3, T7, and *ral* proteins and are summarized in Table 2.3.

T3 AdoMet Hydrolase System

The first T3 gene to be transcribed codes for an AdoMet hydrolase, which also is responsible for overcoming restriction by *E. coli* K and B. However, the hydrolysis of AdoMet is not essential for its antirestriction activity, as indicated by the finding that mutants lacking the AdoMet hydrolase activity are still able to suppress restriction (Studier and Movva, 1976). The T3 AdoMet hydrolase was purified extensively, and two separate forms of the enzyme have been characterized (Spoerel and Herrlich, 1979). The mechanism of this protein was studied *in vitro* by using *Eco*K (Spoerel *et al.* 1979). The AdoMet hydrolase did not bind to DNA, but, when added to an *Eco*K restriction reaction, it effectively inhibited the ATPase activity (which was taken as a measure of the endonuclease reaction). At AdoMet concentrations 10 times higher than required, 50% of the ATPase reaction was still observed. If the AdoMet hydrolase activity was inhibited with S-adenosylhomocysteine, the recognition complexes were still blocked for the ATPase reaction. An AdoMet hydrolase from a mutant that had lost both its hydrolase and antirestriction activities showed the loss of the same activities *in vitro*. An AdoMet hydrolase from a mutant that lacked the hydrolase activity, but still retained its antirestriction activity, showed the same properties *in vitro*. The conclusion was that the antirestriction activity was not due to the loss of the AdoMet that was required for the endonuclease reaction. The inhibition of restriction was due to a protein-protein interaction between the AdoMet hydrolase and *Eco*K. Given the similarities between the T3 and T7 systems, it is likely that the same protein-protein interaction blocks DNA methylation by *Eco*K (see below). In any event, the hydrolysis of AdoMet would remove the essential cofactor for both restriction and modification reactions.

Table 2.3 Antirestriction Systems that Affect DNA Methylation

	T3		T7	λ
Phage gene	AdoMet hydrolase		0.3	*ral*
Systems affected	*E. coli* K and B		*E. coli* K and B	*E. coli* K and B
Gene expression	Infection		Infection	Infection
Protein structure*	A: 17,000	17,000 20,000	13,000(2)	
	B: 170,000	17,000(1) 20,000(2) 49,000(3)		
Effect on DNA methylation	Inhibition		Inhibition	Stimulation of reaction with unmodified DNA
Effect on DNA cleavage	Inhibition		Inhibition	Probable inhibition
Other effects	AdoMet hydrolase			

*Two different forms of T3 AdoMet hydrolase were isolated. Form A has a mol wt of 17,000, both in its native and denatured forms, with minor amounts of a polypeptide of 20,000 mol wt. Form B had three different subunits in the stoichiometries indicated.

T7 0.3 Protein System

The 0.3 gene of phage T7 was the first to be transcribed (Studier, 1975). A mutation in this gene resulted in normal restriction by *E. coli* K and B and in partial modification.The product of this gene is made in large amounts, has been extensively purified, and has been shown to have a mol wt of 13,000 (Mark and Studier, 1981). *In vitro* studies showed that it inhibited all of the reactions catalyzed by *Eco*K. More recently, we have conducted studies to determine which steps of the *Eco*K reaction mechanism are affected by the 0.3 protein. The 0.3 protein did not bind to DNA. In the case of unmodified recognition complexes, the 0.3 protein interacted with *Eco*K and prevented the ATP-induced conformational change that leads to DNA cleavage (Burckhardt *et al.* 1981b). With hemimethylated recognition complexes, the 0.3 protein effectively blocked methyl transfer from free AdoMet to the sK sites. Both of these inhibitory effects were due to protein-protein interactions with *Eco*K already bound to the host specificity site.

λ*ral* Protein System

Phage λ carries a gene called *ral,* which prevents restriction of unmodified λ by *E. coli* K. Previous work suggested that the *ral* gene product stimulated the modification reaction in *E. coli* K (Zabeau *et al.* 1980). The effect of *ral* has been examined in restriction-deficient, modification-competent *E. coli* K (Ann Abeles, personal communication). Infection of r^-m^+ cells by unmodified *ral*$^+$ phage resulted in nearly 100% modification of the progeny phages. Infection by *ral*$^-$ phage under the same conditions produced phages in which fewer than 1% of the progeny were modified. If the lysogens were induced for 10–15 minutes before infection by unmodified *ral*$^-$ phage, then nearly 100% of the progeny were modified. Therefore, expression of the *ral* gene affected the efficiency with which unmodified phages became modified. These results are consistent with the suggestion that an interaction between the *ral* protein and *Eco*K permits the altered *Eco*K to methylate unmodified DNA at high efficiency.

These three antirestriction systems have certain characteristics in common: the genes are expressed very early and the proteins are produced in large amounts (around 10,000 molecules of the AdoMet hydrolase/cell) and are small in size. Their mechanism of action involves both binding to the restriction enzyme and alteration of its modification activity. This can be done either by inhibition of methylation or by a change in its substrate specificity (e.g., unmethylated DNA becomes an efficient substrate). Inhibition of DNA methylation, in association with continued DNA replication, results in effective demethylation. Methylation of unmethylated sequences is the characteristic predicted for a *de novo* methylase (maintenance methylases are defined as those that work only on newly replicated hemimethylated sequences). In effect, demethylation and *de novo* methylation can take place by modulation of the activity of existing methylases by small effector proteins.

Summary

Much of our knowledge of DNA methylation, both at the mechanistic and biological levels, has been derived from the studies of restriction and modification enzymes. The complex type I and type III restriction-modification systems have been studied extensively. The genetics, the structure of the enzymes and their DNA substrate, and the reaction mechanism (which have been discussed in detail) provide the basis for a comparison between the type I and type III systems. Similar work on the type II systems is presented in the next chapter. All of these results have been used to propose models for DNA methylation in higher organisms. An overview of the restriction-modification system is presented in Chapter 4 to allow the reader to better evaluate these models.

Acknowledgments

The authors would like to express their gratitude to those colleagues who provided them with copies of their preprints and unpublished results. They also would like to thank Dr. Ann Abeles and Dr. Ken Abremski for their careful reading of the manuscript and Dr. Hamilton O.Smith for useful suggestions.

Research was sponsored by the National Cancer Institute, DHHS under Contract No. NO1-CO-23909 with Litton Bionetics, Inc. The contents of this publication do not necessarily reflect the views or policies of the Department of Health and Human Services, nor does the mention of trade names commercial products, or organizations imply endorsement by the United States Government.

References

Arber W: Host controlled modification of bacteriophage. *Ann Rev Microbiol* 1965; 19:365–378.

Arber W, Kuhnlein U: Mutationeller Verlust B-Spezifischer Restriktion des Bakteriophagen fd. *Pathol Microbiol* 1967;30:946–952.

Arber W, Wauters-Willems D: The two restriction and modification systems of strain 15T$^-$. *Mol Gen Genetics* 1970;108:203–217.

Bächi B, Reiser J, Pirrotta V: Methylation and cleavage sequences of the *Eco*P1 restriction-modification enzyme. *J Mol Biol* 1979;128:143–163.

Bickle TA, Brack C, Yuan R: ATP-induced conformational changes in the restriction endonuclease from *Escherichia coli* K12. *Proc Natl Acad Sci* USA 1978;75:3099–4103.

Bickle TA: The ATP-dependent restriction endonucleases, in Linn SM, Roberts RJ (eds.): *Nucleases*. Cold Spring Harbor, NY, Cold Spring Harbor Laboratory, 1983∝ 85–108.

Boyer HW: Genetic control of restriction and modification in *Escherichia coli*. *J Bacteriol* 1964;88:1652–1660.

Boyer HW, Roulland-Dussoix D: A complementation analysis of the restriction and modification of DNA in *Escherichia coli*. *J Mol Biol*. 1969;41:459–472.

Brack C, Eberle H, Bickle TA, Yuan R: Mapping of recognition sites for the restriction endonuclease from *Escherichia coli* K12 on bacteriophage PM2 DNA. *J Mol Biol* 1976;108:583–593.

Brzezinski R, Piekarowicz A: Steps in the reaction mechanism of the *Haemophilus influenzae* Rf restriction endonuclease. *J Mol Biol*. 1982;154:615–627.

Bühler R, Yuan R: Characterization of a restriction enzyme from *Escherichia coli* K carrying a mutation on the modification subunit. *J Biol Chem* 1978;253:6756–6760.

Bullas LR, Colson C, Van Pel A: DNA restriction and modification systems in Salmonella. SQ, a new system derived by recombination between the SB system of Salmonella typhimurium and the SP system of Salmonella postdam. *J Gen Microbiol* 1976;95:166–172.

Burchhardt J K, Weisemann J, Yuan R: Characterization of the DNA methylase activity of the restriction enzyme from *Escherichia coli* K. *J Biol Chem* 1981a; 256:4024–4032.

Burckhardt J, Weisemann J, Hamilton DL, Yuan R: Complexes formed between the restriction endonuclease EcoK and heteroduplex DNA. *J Mol Biol* 1981b;153:425–440.

Colson C, Colson AM: A new Salmonella typhimurium DNA host specificity. *J Gen Microbiol* 1971;69:345–351.

Colson AM, Colson C: Expression of the Escherichia coli K, B and phage P1 DNA host specificities in Salmonella typhimurium. *J Gen Microbiol*, 1972;70:123–128.

Dugaiczyk A, Kimball M, Linn S, Goodman HM: Location and nucleotide sequence of the site on SV40 DNA methylated by the EcoB modification methylase. *Biochem Biophys Res Commun* 1974;61:1133–1140.

Eskin B, Linn S: The deoxyribonucleic acid modification and restriction enzymes of *Escherichia coli* B. II. Purification, subunit structure, and catalytic properties of the restriction endonuclease. *J Biol Chem* 1972b;247:6183–6191.

Eskin B, Linn S: The deoxyribonucleic acid modification and restriction enzymes of *Escherichia coli* B. II. Purification, subunit structure, and catalytic properties of the restriction endonuclease. *J Biol Chem* 1972b;247:6183–6191.

Glover SW: Functional analysis of host-specificity mutants in *Escherichia coli*. *Genet Res* 1970;15:237–250.

Gough JA, Murray, NE: Sequence diversity among related genes for recognition of specific targets in DNA molecules. *J Mol Biol* 1983;166:1–19.

Haberman A: The bacteriophage P1 restriction endonuclease. *J Mol Biol* 1974; 89:545–563.

Haberman A, Heywood J, Meselson M: DNA modification methylase activity of *Escherichia coli* restriction endonucleases K and P. *Proc Natl Acad Sci USA* 1972; 69:3138–3141.

Hadi SM, Yuan R: Complementation in vitro by mutant restriction enzymes from Escherichia coli K. *J Biol Chem* 1974;249:4580–4586.

Hadi SM, Bächi B, Shepherd JCW, Yuan R, Ineichen K, Bickle TA: DNA recognition and cleavage by the *Eco*P15 restriction endonuclease. *J Mol Biol* 1979;134:655–666.

Hadi SM, Bächi B, Iida S, Bickle TA: DNA restriction-modification enzymes of phage P1 and plasmid p15B: subunit functions and structural homologies. *J Mol Biol* 1983; 165:19–34.

Hattman S: Specificity of the bacteriophage Mu *mom*$^{+-}$ controlled DNA modification. *J Virol* 1980;34:277–279.

Horiuchi K, Zinder ND: Cleavage of bacteriophage fl DNA by the restriction enzyme of *Escherichia coli* B. *Proc Nat Acad Sci USA* 1972;69:3220–3224.

Hubacek J, Glover SW: Complementation analysis of temperature-sensitive host specificity mutations in *Escherichia coli. J Mol Biol* 1970; 50:111–127.

Iida S, Meyer B, Bächi B, Stolhammar-Carlemalm M, Schrickel S, Bickle TA, Arber W: DNA restriction-modification genes of phage P1 and plasmid p15B: Structure and *in vitro* transcription. *J Mol Biol* 1983; 165:1–18.

Kan NC, Lautenberger JA, Edgell MH, Hutchinson III CA: The nucleotide sequence recognized by the *Escherichia coli* K12 restriction and modification enzymes. *J Mol Biol* 1979;130:191–209.

Kauc L, Piekarowicz A: Purification and properties of a new restriction endonuclease from *Haemophilus influenzae* Rf. *Eur J Biochem* 1978;92:417–426.

Kuhnlein U, Arber W: The role of nucleotide methylation in in vitro B-specific modification. *J Mol Biol* 1972;63:9–19.

Lark C, Arber W: Host specificity of DNA produced by *Escherichia coli.* Breakdown of cellular DNA upon growth in ethionine of strains with r^{+}_{15}, r^{+}_{p1} or r^{+}_{N3} restriction phenotypes. *J Mol Biol* 1970;52:337–348.

Lautenberger JA, Linn S: The deoxyribonucleic acid modification and restriction enzymes of *Escherichia coli* B. I. Purification, subunit structure, and catalytic properties of the modification methylase. *J Biol Chem* 1972;247:6176–6182.

Lautenberger JA, Kan NC, Lackey D, Linn S, Edgell MH, Hutchinson III CA: Recogniton site of *Escherichia coli* B restriction enzyme on ΦXsBI and Simian virus 40 DNAs: An interrupted sequence. *Proc Natl Acad Sci USA* 1978;75:2271–2275.

Linn S, Arber W: In vitro restriction of phage fd replicative form. *Proc Natl Acad Sci USA* 1968;59:1300–1306.

Lyons LB, Zinder ND: The genetic map of the filamentous bacteriophage fl. *Virology* 1972;49:45–60.

Mark KK, Studier FW: Purification of the gene 0.3 protein of bacteriophage T7, an inhibitor of the DNA restriction system of *Escherichia coli. J Biol Chem* 1981; 256:2573–2578.

Meselson M, Yuan R: DNA restriction enzyme from *E. coli. Nature* 1968;217:1111–1113.

Meselson, M, Yuan R, Heywood J: Restriction and modification of DNA. *Ann Rev Biochem* 1972;41:447–466.

Murray NE, Manduca de Ritis P, Foster L: DNA targets for the *Escherichia coli* K restriction system analyzed genetically in recombinants between phages ϕ80 and lambda. *Mol Gen Genet* 1973;120:261–281.

Murray NE, Gough JA, Suri B, Bickle TA: Structural homologies among Type I restriction-modification systems. *The EMBO J* 1982;1:535–539

Piekarowicz A, Brzezinski R: Cleavage and methylation of DNA by the restriction endonuclease HinfIII isolated from *Haemophilus influenzae* Rf. *J Mol Biol* 1980; 144:415–429.

Piekarowicz A, Bickle TA, Shepherd JCW, Ineichen K: The DNA sequence recognized by the *Hinf*III restriction endonuclease. *J Mol Biol* 1981;146:167–172.

Ravetch JV, Horiuchi K, Zinder ND: Nucleotide sequence of the recognition site for the restriction-modification enzyme of *Escherichia coli* B. *Proc Natl Acad Sci USA* 1978;75:2266–2270.

Reiser J, Yuan R: Purification and properties of the P15 Specific restriction endonuclease from *Escherichia coli. J Biol Chem* 1977;252:451–456.

Roulland-Dussoix D, Boyer H: The *Escherichia coli* B restriction endonuclease. *Biochem Biophys Acta* 1969;195:219–229.

Sain B, Murray NE: The hsd (host specificity) genes of *Escherichia coli* K12. *Mol Gen Genet* 1980; 180:35–46.

Sclair M, Edgell MH, Hutchinson, CA III: Mapping of new *Escherichia coli* K and 15 restriction sites on specific fragments of bacteriophage Φ X174 DNA. *J Virol* 1973; 11:278–285.

Smith HO, Nathans D: A suggested nomenclaturefor bacterial host-modification and restriction systems and their enzymes. *J Mol Biol* 1973;81:419–423.

Smith J, Arber W, and Kühnlein U: Host specificity of DNA produced by *Escherichia coli* 15. The implication of nucleotide methylation in *in vitro* B-specific modification. *J Mol Biol* 1972;63:9–19.

Spoerel N, Herrlich P: Colivirus T3-coded S-adenosylmethionine hydrolase. *Eur J Biochem* 1979; 95:227–233.

Spoerel N, Herrlich PA, Bickle TA: A novel bacteriophage defence mechanism: the anti-restriction protein. *Nature (London)* 1979;278:30–34.

Studier FW: Gene 0.3 of bacteriophage T7 acts to overcome the DNA restriction enzyme of the host. *J Mol Biol* 1975; 94:283–295.

Studier FW, Movva NR: SAMase gene of bacteriophage T3 is responsible for overcoming host restriction. *J Virol* 1976;19:136–145.

Suri B, Nagaraja V, Bickle TA: Bacterial DNA modification. *Curr Topics Microbiol Immunol* (vol 102, in press, 1983).

Toussaint A: The DNA modification function of temperate phage Mu-1. *Virology* 1976;70:17–27.

van Ormondt H, Lautenberger JA, Linn S, deWaard A: Methylated oligonucleotides derived from bacteriophage fd Rf DNA modified in vitro by *E. coli* B modification methylase. *FEBS Letters* 1973;33:177–180.

Vovis GF, Zinder ND: Methylation of f1 DNA by a restriction endonuclease from *Escherichia coli* B. *J Mol Biol* 1975:95:557–568.

Vovis GF, Horiuchi K, Hartman N, Zinder ND: Restriction endonuclease B and f1 heteroduplex DNA. *Nature New Biol* 1973;246:13–16.

Vovis GF, Horiuchi K Zinder ND: Kinetics of methylation by a restriciton endonuclease from *Escherichia coli* B. *Proc Natl Acad Sci USA* 1974;71:3810–3813.

Yuan R: Structure and mechanism aof multifunctional restriction endonucleases. *Ann Rev Biochem.* 1981;50:285–315.

Yuan R, Reiser J: Steps in the reaction mechanism of the *Escherichia coli* plasmid P15-specific restriction endonuclease. *J Mol Biol* 1978;122:433–445.

Yuan R, Heywood J Messelson M: ATP hydrolysis by restriction endonuclease from *E. coli* K. *Nature New Biol* 1972;240:42–43.

Yuan R, Bickle TA, Ebbers W, Brack C: Multiple steps in DNA recognition by restriction endonuclease from *E. coli* K. *Nature* 1975;256:556–560.

Yuan R, Hamilton DL, Hadi SM, Bickle TA: Role of ATP in the cleavage mechanism of the EcoP15 restriction endonuclease. *J Mol Biol* 1980:144:501–519.

Yuan R, Burckhardt J, Weisemann J, Hamilton DL: The mechanism of DNA methylation by the restriction endonuclease from *E. coli* K. in Usdin E, Borchardt RT, Creveling CR (eds.): *Biochemistry of S-adenosylmethionine and related compounds.* London, MacMillan Press, Ltd, 1981, pp. 239–247.

Zabeau M, Friedman SF, VanMontagu M, Schell J: The ral gene of phage λ. I: Identification of a non-essential gene that modulates restriction and modification in *E. coli. Mol Gen. Genet* 1980;179:63–73.

3

Methylases of the Type II Restriction-Modification Systems

Hamilton O. Smith*
Samuel V. Kelly*

Most bacteria contain a small fraction (0.5–2%) of methylated cytosine or adenine bases in their chromosomes. Some of this methylation apparently plays a role in directing mismatch repair systems to the correct strands in newly replicated DNA. However, in many bacterial strains, a substantial fraction can be identified with host-specific restriction-modification (RM) systems. These ubiquitous systems play an important biological role in protecting bacteria against viral infections. Each system has two functional components: 1) a restriction endonuclease capable of recognizing sequence-specific sites in DNA and producing double-stranded cleavage; and 2) a modification enzyme recognizing the same DNA sites as the restriction enzyme and protecting them by modification. So far, all modifications found are either 5-methylcytosine or N6-methyladenine. The modification enzymes are methyltransferases, and AdoMet appears to be the exclusive methyl group donor. (For reviews, see Arber, 1974; Modrich, 1979; Smith, 1979; Modrich and Roberts, 1982).

Restriction-modification systems have been classified into types I, II, and III, based on differences in enzyme structure and reaction mechanisms. The type I and III enzymes (described in Chapter 2) are complex, multifunctional proteins that require both ATP and AdoMet as cofactors and are able to both cleave and methylate DNA. The type I enzymes recognize hyphenated, asymmetric sites in DNA (e.g., *Eco*K recognizes 5′AACNNNNNNGTGC); if the site is unmethylated, they cleave randomly at distances up to several thousand base pairs away. Cleavage at a distance is achieved by an ATP-requiring trans-

*Department of Molecular Biology and Genetics, The Johns Hopkins Unniversity, School of Medicine, Baltimore, Maryland 21205

[1]Abbreviations: AdoMet, S-adenosylmethionine; Ap^r, ampicillin resistance; N-AcdG, N-acetyldeoxyguanosine; dx, deoxyxanthosine; kd, kilodalton; kb, kilobase; bp, basepair.

translocation mechanism. Hemimethylated sites, which arise naturally during DNA replication, are interpreted as substrates for methylation, rather than cleavage. The known type III enzymes recognize nonhyphenated, asymmetric sites (e.g., *Eco*PI recognizes 5′AGACC) and cleave at a relatively precise distance of 24–26 base pairs away. Methylation competes with restriction for these enzymes.

Type II enzymes are structurally and mechanistically simpler. Typically, the restriction endonuclease is a dimer of a single type of subunit (although a few appear to be monomers), requires Mg^{2+} for activity, and cleaves within the site or only a few base pairs away. The modification methylase is a separate protein, usually a monomer, and catalyzes transfer of a methyl group from AdoMet to one of the bases (A or C) on each straind of the site. As a possible exception, certain asymmetric sites may be methylated on only one strand. Recognition sites either are symmetric (e.g., 5′G-G-C-C or 5′G-A-A-T-T-C), degenerate and symmetric (e.g., 5′GTPyPuAC) hyphenated and symmetric (e.g., 5′G-A-N-T-C or 5′G-C-C-N-N-N-N-N-G-G-C) or asymmetric (e.g., 5′G-A-C-G-C).

The type II enzymes appear to be far more common than the type I and III variety. Less than 12 of the latter have been identified from among nearly 400 restriction enzymes listed in a recent compendium by Roberts (1983). However, this proportion could be misleading, because most bacterial strains have been examined by using assays based on specific cleavage that would not detect the type I enzymes. It generally is assumed that each of the many restriction enzymes in Roberts' list is associated with a cytosine or adenine methylase of identical specificity, but this has actually been demonstrated in only a few cases. One must consider other possibilities, such as broader (less) specificity of the methylase, unusual base methylations, modifications other than methylation, or even no modification (a possibility for periplasmic location of the restriction enzyme). To address these possibilities, Brooks and Roberts (1982) examined 26 bacterial species and found: 1) no instance of a restriction enzyme cleaving the DNA of its host strain, which implies that the appropriate sequences were always modified; 2) identical specificity of both the restricting and modifying activities (in no case did less specific modification involving, for example, recognition of the central tetramer of a hexamer sequence exist); and 3) no modification other than cytosine and adenine methylation. Thus, the simple picture of a modification enzyme as a cytosine or adenine methylase of specificity identical to that of its restriction enzyme partner seems to be universal.

The type II restriction endonucleases have been extremely useful in recombinant DNA work; hence, most reviews have focused on their properties, rather than on the methylases. However, recent studies in eukaryotes indicate an important role for DNA methylation in gene expression. As a result, there has been a renewed interest in the methylases. They also are potentially useful for study of sequence-specific, protein-nucleic acid interactions. Thus, it is timely and appropriate that this chapter focuses on the type II modification methylases. In some cases, both enzymes of an RM system will be discussed. Whenever possible confusion might arise, the restriction enzyme will be designated by its system name alone or with an R·prefix (e.g., *Eco*RI or R·*Eco*RI) and the methylase with an M·prefix (e.g., M·*Eco*RI).

Type II Methylases

Assay

Sequence specific methylases can be assayed in crude extracts. There are two commonly used methods: one measures protection of DNA against subsequent restriction and the other measures incorporation of ^{3}H-methyl groups into DNA.

A good description of the protection assay is given by Levy and Welker (1981). Methylation of a plasmid or viral DNA is first carried out in excess unlabeled AdoMet. The DNA is then restricted and analyzed by gel electrophoresis. Appropriate ethidium-stained bands are selected for photographic quantitation of the extent of their decrease in comparison to an unmethylated control. Units of enzyme activity are calculated from the initial velocity of the protective methylation reaction and may be expressed as the percent decrease of a band per minute. Units also may be expressed as pmoles of methyl groups that were transferred by using an appropriate factor derived from the amount of DNA in the reaction and the number of sites per microgram of DNA.

For assay by ^{3}H-methyl group incorporation, a typical reaction mixture (50 μl) contains 50 mM Tris-HCl (pH8), 10 mM Na_2EDTA, 10 mM 2-mercaptoethanol, 1–10 μg substrate DNA, 1–10 μM [methyl-^{3}H]AdoMet, and 1–10 μl of enzyme extract. After incubation at an optimal temperature, incorporation of ^{3}H-radioactivity into acid-precipitable material is determined (Roy and Smith, 1973d). However, there are several possible problems with this assay. Even though the reaction is routinely carried out in EDTA, restriction endonuclease present in a crude extract can bind to sites and competitively inhibit methylation. Other nonspecific DNA binding proteins also may cause inhibition. Protein or RNA methylases may cause a high background of acid-precipitable incorporation. Finally, it may be difficult to sort out one particular DNA methylation specificity from among several others that may be present in the extract.

These problems can be overcome in several ways. If DNA substrate is saturating, free sites will be available despite the presence of endonuclease. Extraction of DNA after completion of the reaction will remove spurious counts that are incorporated into protein (or RNA). Specificity can be determined directly by gel electrophoresis of defined substrate molecules (e.g., plasmid DNA) after cleavage with the appropriate restriction endonuclease. Loss of cleavage products indicates methylation of specific sites.

Determination of Specificity

A full description of methylation specificity requires identification of both the sequence modified and the particular bases within the sequence that are modified. For the type II methylases, the sequence specificity is most readily determined for the corresponding restriction endonuclease. Then, one can prove identity of the methylase and endonuclease sites by demonstrating: 1) that cleavage inhibits methylation; and conversely 2) that methylation inhibits cleavage.

A next step is determination of the modified base. The type II methylases can be classed as either adenine or cytosine methylases. Adenine methylation is readily differentiated from cytosine methylation by 2-dimensional chromatography of the radioactively labeled 5-methylcytosine and N6-methyladenine bases released by treatment of ^{3}H-methylated DNA with hot formic acid (Roy and Smith, 1973a). High-pressure liquid chromatography (HPLC) also is useful for analysis and accurate quantitation of the bases (Roy and Smith, 1973b).

The position of the methylated adenine(s) or cytosine(s) in a site can be determined by bilateral nearest neighbor analysis (Roy and Smith, 1973b); or in some cases. simply by knowing whether the methylated base is an A or C, since a number of sites contain only one such base on each strand in the site (e.g., M · *Taq*I is an adenine methylase and must methylate the unique A in the sequence 5′TCGA). More generally, the position of the methylated A or C in a site is best determined by analyzing partial digestion products of ^{3}H-methylated synthetic oligomer substrate sites. For example, the octomer pTGAATTCA can be methylated using M · *Eco*RI and [methyl-^{3}H] AdoMet. After partial digestion with snake venomphosphodiesterase and separation of the oligomers by one-dimensional electrophoresis on DEAE-paper at pH 1.9, ^{3}H-counts are found to be associated with pTGAA and longer oligomers—but not with pTGA or pTG. Thus, M · *Eco*RI methylates the 5′ distal A in the site (see Table 3.1; Greene *et al.* 1975). Another method based on the Maxam and Gilbert sequencing method also is quite useful for cytosine methylases. In that procedure, 5′-methylcytosine is resistant to the limited hydrazine cleavage reaction, and the methylated C residue is revealed by a gap on the sequencing ladder (Ohmori *et al.* 1978; Walder *et al* 1983).

In Table 3.1, a number of Type II methylases are listed for which methylation specificity has been rigorously determined. Since the sites are all symmetric, the position of methylation is indicated on only one strand; however, one can assume that the complementary strand in the duplex site will also be methylated at the corresponding position, although this has been documented in only a few cases. In general, the Type II methylases introduce only a single methyl group onto each strand in a site; this is sufficient to completely protect against cleavage by the corresponding restriction enzyme. However, it is an intriguing possibility that some methylases may be capable of introducing more than one methyl group per strand in a site. This was first thought to be the case with M · *Msp*I. Extracts of *Moraxella* species, from which the enzyme is derived, methylate both Cs of the sequence CCGG (Jentsch *et al.* 1981). How-

Table 3.1 Methylation Specificity of Some Type II Methylases

Microorganism	Methylase	Sequence*	References
Arthrobacter luteus	M·*Alu*I	AG|mCT	Kramarov and Smolyaninov (1982)
*Bacillus amyloliquifaciens*H	M·*Bam*HI	G|GATmCC	Hattman *et al.* (1978)
*Bacillus brevis*S	M·BbvISI	GmC($^{A}_{T}$)GC	Vanyushin and Dobritsa (1975)
B. centrosporus	M·*Bcn*I	CmCGGG	Petrusyte and Janulaitis (1981)
*B. sphaericus*R	M·*Bsp*RI	GG|mCC	Venetianer and Kiss (1981) Feher *et al.* (1983)
*B. stearothermophilus*1503	M·*Bst*1503I	G|GmATCC	Levy and Welker (1981)
*B. subtilis*R	M·*Bsu*RI†	GG|mCC	Günthert *et al.* (1978; 1981a,b)
*B. subtilis*168	M·*Bsu*M	CTmCGAG	S. Jentsch (personal communication)
*B. subtilis*1231	M·*Bsu*F	mCCGG	S. Jentsch (personal communication)
Brevibacterium albidum	M·*Bal*I	TGG|mCCA	T. A. Trautner (unpublished)
Caulobacter fusiformis	M·*Cfu*I	GmA|TC	Hughes and Murray (1980)
Caryophanon latum	M·*Cla*I	AT|CGmAT	McClelland (1981)
Diplococcus pneumoniae	M·*Dpn*II	GmA|TC‡	Lacks and Greenberg (1977)
Escherichia coli RY13	M·*Eco*RI	G|AmATTC	Greene *et al.* (1975)
*E. coli*R245	M·*Eco*RII	CmC($^{A}_{T}$)GG	Boyer *et al.* (1973)
Haemophilus aegyptius	M·*Hae*III	GG|mCC	Mann and Smith (1977)
H. haemolyticus	M·*Hha*I	GmCG|C	Mann and Smith (1979)
H. haemolyticus	M·*Hha*II	G|mANTC	Mann *et al.* (1978) S. Kelly (unpublished)
H. influenzae d	M·*Hin*dII	GTPy|PumAC	Roy and Smith (1973a,b)
H. influenzae d	M·*Hin*dIII	mA|AGCTT	Roy and Smith (1973a,b)
H. parainfluenzae	M·*Hpa*I	GTT|AmAC	Yoo *et al.* (1982)
H. parainfluenzae	M·*Hpa*II†	C|mCGG	Mann and Smith (1977)
Moraxella species	M·*Msp*I	mC|CGG	Walder *et al.* (1983)
*Providencia stuartii*164	M·*Pst*I	CTGCmA|G	Walder *et al.* (1981) J. A. Walder (personal communication)
Pseudomonas aeruginosa	M·*Pae*R7	C|TCGmAG	Gingeras and Brooks (1983)
*Thermus aquaticus*YTI	M·*Taq*I	T|CGmA	Sato *et al.* (1980)
T. aquaticus	M·*Taq*XI	CmC|AGG	Grachev *et al.* (1981)
*T. thermophilus*HB8	M·*Tth*HB8I	TCGmA	Sato *et al.* (1980) McClelland (1981)

*Restriction endonuclease cleavage positions indicated by vertical lines are referenced in Roberts (1983).

†Purified as two species (see Table 3.2).

‡Denotes probable specificity.

ever, the M·*Msp*I gene recently was cloned (Walder *et al.* 1983); and enzyme from this clone methylates only the outer C of the site. one must suppose then, that *Moraxella* extracts contain two separate enzymes, with each acting independently to produce the double methylation; but, the second methylase with *Hpa*II-type (i.e., CmCGG) specificity has not yet been reported.

A more convincing case for double methylation is found with the inducible methylase of the lysogenic phage SPβ found in *Bacillus subtilis* R (renamed

A more convincing case for double methylation is found with the inducible methylase of the lysogenic phage SPβ found in *Bacillus subtilis* R (renamed SPR by Noyer-Weidner *et al.* 1983) *Bacillus subtilis* R carries its own constitutive chromosomal genes for the *Bsu*R (GGĊC) system. (The asterisk indicates the methylation position.) Strains lysogenic for SPβ express inducible methylases that protect against *Hae*III (GGĊC), *Hpa*II (CĊGG), and *Msp*I (ĊCGG) cleavages. Two classes of phage mutants have been isolated: those acquiring sensitivity to *Hae*III and those acquiring simultaneous sensitivity to all three enzymes. Two separate genes responsible for *Hpa*II and *Msp*I protection could not be resolved by mutational analysis or phage crosses (Jentsch *et al.* 1981). Thus, a single gene and a single protein could be responsible for the double methylation (mCmCGG). Kiss and Baldauf (1983) have suggested that this gene also might determine the GGĊC methylation specificity; hence, a triple specificity for a single methylase. Their primary evidence came from cloning a short segment of SPβ DNA, which specified both activities. They were able to show by subcloning that a single *Sal*I cleavage simultaneously destroys the GGCC and CCGG activities (see section on "Methylase Gene Cloning"). Finally, P. Venetianer (personal communication) has found that a single protein is responsible for both GGCC- and CCGG-specific methylation. Further studies of this very interesting multiple specificity should be enlightening.

The type II methylases that recognize asymmetric sites are best dealt with as a separate class. Unfortunately, none belonging to this class has yet been studied, although nine restriction enzymes recognizing eight different four- to six-base pair asymmetric sites are listed by Roberts (1983). Methylation of asymmetric sites presents some interesting problems. If both strands are methylated, the modified base will be at a different relative position on each strand. For several systems (e.g. *Mbo*I [GAAGA] and *Mnl*I [CCTC]), the problem is even more severe, since an A would need to be modified on one strand and a C on the other. Since the methylase presumably can only bind to the asymmetric site with one polarity, it would appear that it must have two active sites. Another possible solution would be two separate enzymes or two different subunits in a single enzyme. Alternatively, only one strand might be methylated, as is seen with type III methylases; but, how then do newly replicated unmethylated sites in the host cell become protected before cleavage?

Structure and Properties

The type II methylases listed in Table 3.2 have been purified and partially characterized. In general, the most effective purification step is affinity chromatography on phosphocellulose, heparin-agarose, or DNA-agarose. When used early, purifications of 10- to 100-fold are usual. Hydrophobic chromatography on phenyl sepharose or related resins may be an excellent second step. Hydroxyapatite and various standard ion exchange resins (e.g., DEAE-cellulose, CM cellulose, and BioRex 70) give good results in some cases. Gel exclu-

Table 3.2 Properties of Some Purified Type II Methylases

Methylase	Structure	Km DNA Site	Km AdoMet	pI	pH Optimum	Temperature Optimum	Turnover Number	References
M · *Bam*HI	56Kd monomer	4nM						Nardone *et al.* (1983)
M · *Bsp*RI	48.3Kd* monomer							Venetianer and Kiss (1981) Venetianer (personal communication)
M · *Bst*1503I	425Kd tetramer			5.6	8.1–9.3	54–61°C		Levy and Welker (1981)
M · *Bsu*RIa	41Kd monomer	2.7nM	0.7μM		8.4	40–43°C		Günthert *et al.* (1981a,b)
M · *Bsu*RIb	42Kd monomer	2.7nM	0.7μM		8.3	40–43°C		
M · *Eco*RI	38Kd* monomer	1.3nM	0.3μM	8.7		37°C	3/min/monomer	Rubin and Modrich (1977)
M · *Hha*II	19.8Kd* monomer							Schoner *et al.* (1983); Kelly (unpublished)
M · *Hpa*I	37Kd monomer	27nM	0.1μ		6.8–7.0		0.9/min/monomer	Yoo et al. (1982)
M · *Hpa*II	38.5Kd monomer		1.4μM		7.5–8.2	40°C		Yoo and Agarwal (1980); Quint and Cedar (1981)
M · *Hpa*II′	41.5Kd monomer				7.5–8.2	40°C		Yoo and Agarwal (1980)
M · *Tth*HB8I	41Kd monomer	10μg/mlλ (∽100nM sites)	0.8μM	8.3	7.4	70°C		Sato *et al.* (1980)

*Size determined from nucleotide sequence of the gene.

sion chromatography also is useful and gives information for calculating the molecular weight (mol wt) of the native enzyme.

All except one of the methylases of Table 3.2 exist in solution as monomers under the usual concentrations and conditions of assay. The exception is M·*Bst*1503I, which appears to be a tetramer of a 105 Kd subunit. The monomeric enzymes range in size from 37–56 Kd, except for M·*Hha*II, which is only 19.8 Kd. In most cases, methylases are larger in size than their corresponding restriction enzyme polypeptide subunits. For example, the subunit polypeptide mol wt of *Bam*HI is 25 Kd compared to 56 Kd for the methylase, and *Eco*RI is 28.5 Kd compared to 38 Kd for the methylase. Two cases of larger restriction enzyme subunits are *Bsu*RI (68 Kd) and *Hha*II (25.9 Kd), compared to 42 Kd and 19.8 Kd for the corresponding methylases. The restriction enzymes most frequently exist as dimers or tetramers in solution, although there are at least two examples—*Bgl*I and *Bsu*RI— where only monomers have been found (see Table 3 in Modrich and Roberts, 1982).

Two forms of M·*Bsu*RI have been isolated from *B. subtilis* OG3R (Günthert *et al.* 1981a, b). M·*Bsu*RIa has a mol wt of 41 Kd by SDS-gel electrophoresis, elutes at 0.21 M KCl on phosphocellulose chromatography, and accounts for 10–20% of the activity in exponential cells. M·*Bsu* RIb elutes at 0.32 M KCl on phosphocellulose, and it represents 80–90% of the activity. (It also splits into a doublet band by SDS-polyacrylamide gel electrophoresis.) By gel filtration, the Ia and Ib species elute as 37 Kd and 40 Kd protein respectively; by sedimentation, they both have an $S_{20,w}$ of 3.55 corresponding to a 43 Kd protein. Thus, both are monomeric species.

It has not yet been determined whether a single gene codes for the two enzymes that simply represent different forms (one perhaps being a precursor of the other) or whether distinct genes code for each enzyme. However, the two enzymes differ in some respects. The Ia form is slightly more heat-stable and has a lower ionic optimum (0.1 M Tris-HCl, pH 8.4) than the Ib form (0.44 M Tris-HCl, pH 8.4). The two methylases are similar in pH optimum (pH 8.4 and 8.3, respectively), Km for DNA sites (2.7 nM 5′GGCC), and Km for AdoMet (0.7 μM).

Yoo and Agarwal (1980) also have observed two forms of *Hpa*II-specific methylase. These are separable by DEAE-Sephadex chromatography. M·*Hpa*II elutes at 0.18 M NaCl and is a 38.5 Kd monomer. M·*Hpa*II′ elutes in a second peak at 0.27 M NaCl and is a 41.5 Kd monomer. The two forms have the same site specificity and their pH optima, temperature optima, and ionic requirements are identical. However, M·*Hpa*II is slightly less stable for storage than M·*Hpa*II′. Preliminary one-dimensional peptide maps are similar, suggesting that one form is a precursor of the other.

M·*Bst*1503I is rather distinct in structure from the other methylase. *In vivo,* it is membrane-associated, and membrane fractions can be prepared with enzyme still bound (Levy and Welker, 1981). However, to assay the enzyme, it appears necessary to release it by sonication. The *Bst*1503I restriction enzyme, conversely, is cytoplasmic (Catteral and Welker, 1977). Homogeneous methylase, free of membrane components, has been prepared and is an acidic protein with a subunit mol wt of 105 Kd. It is present as a tetramer in solution. It is unclear why this particular methylase is located in the membrane; however, one possibility is that it might be closely associated with the replication complex to ensure protection of the chromosome. Another possibility is that it is stabilized against thermal denaturation by its membrane association. In regard to this latter possibility, *Bacillus stearothermophilus* 1503-4R, from which the methylase is prepared, grows optimally at 63°–67°C; while purified, R·*Bst*1503I is maximally active in a similar temperature range (60°–65°C), purified M·*Bst*1503I is more heat-labile under *in vitro* assay conditions and exhibits maximal activity at 54°–61°C. It is completely inactivated at 55°C for 6–7 hours or at 60°C for 1.5 hours. However, membrane-associated methylase is fully active after 3 hours at 60°–70°C. In addition, 0.1 M monovalent ions (Na^+, K^+, NH_4^+) dramatically improve the heat stability. Thus, the relative heat sensitivity of purified enzyme appears to be an artifact of its *in vitro* environment.

No information is currently available on the three-dimensional structure or functional domains of the methylases. However, Nardone *et al.* (1983) have probed the AdoMet and DNA binding sites of *Bam*HI methylase by using specific inhibitors. The arginine-specific probe butanedione inactivates the enzyme; but, preincubation with DNA efficiently protects against inactivation, while AdoMet does not. Dinucleotide subsets of the *Bam*HI recognition sequence (GGATCC) specifically protect the enzyme from the butanedione inactivation in the order: GG>GA>AT>TC>CC, with GG being the best protector by far. Thus, arginine residues probably participate in the DNA binding site. N-ethylmaleimide and p-chloromercuribenzoate inactivate the enzyme, indicating that cysteine residues are important for activity. DNA gives minimal (10%) protection, but AdoMet and S-adenosylhomocysteine efficiently protect against such inactivation. The protection by AdoMet is abolished in the presence of DNA that contains no *Bam*HI sites; protection by S-adenosylhomocysteine is abolished by DNA that contains sites (J. Chirikjian, personal comunication). In the case of M·*Eco*RI, thiol reagents are required during purification and assay to maintain acitvity. Treatment with 3 mM N-ethylmaleimide for 5 minutes at 15°C abolishes activity; however, this sensitivity has not been localized to a specific binding domain. Further studies of this type could be useful in defining active sites.

Only limited information is available for the type II methylases concerning optimal reaction conditions or thermodynamic and kinetic parameters. Most of the methylases of Table 3.2 have been shown to obey Michaelis-Menten kinetics with respect to DNA sites and AdoMet. Kms for DNA sites fall in the nanomolar range, while Kms for AdoMet are in the micromolar range. The only case of atypical behavior is M·*Bst*1503I, which obeys Michaelis-Menten kinetics for DNA, but not for AdoMet. This is not surprising in view of its tetrameric structure.

Optima for pH range from neutral to slightly basic, which is not a surprising finding in view of the lability of AdoMet at alkaline pH (Zappia *et al.* 1979). Ionic requirements are quite variable from one enzyme to another. M·*Bam*HI is inhibited by > 5 mM Mg^{2+} ion and > 300 mM KCl; however, some monovalent ion is required for optimal activity. Activity increases approximately three-fold in the presence of 100 mM KCl (Nardone *et al.* 1983). M·*Bsu*RIa and M·*Bsu*RIb activities differ markedly in their sensitivity to ionic concentrations; the Ia form has a distinct optimum in 80–100 mM Tris-HCl (pH8.4) or in 90–100 mM KCl, while the Ib form has a broad optimum at 400 mM Tris-HCl(pH 8.4) and two optima in KCl—one at 50 mM and a second at 200 mM KCl. M·*Hpa*I activity is maximal in the absence of NaCl and completely inhibited by 150 mM NaCl. The two forms of M·*Hpa*II are maximally active in the absence of NaCl and completely inhibited by 200 mM NaCl. M·*Tth*HB81 is most active at monovalent ion (NaCl, KCl, or NH_4Cl) concentrations under 20 mM, and it is almost completely inhibited in 200 mM NaCl or KCl. None of the methylases have been shown to require divalent cations, and several are inhibited by them. Temperature optima for the various methylases range from 37°–70°C and reflect the optimal growth temperatures for the host bacteria.

Cloned Type II Restriction-Modification Genes

Extensive biochemical and structural studies of many of the type II enzymes will depend on gene cloning. Clones can be engineered to overproduce the enzymes, nucleotide sequencing can reveal the primary amino acid sequences of the enzymes, and the cloned genes are accessible for both mutational analysis and studies of expression. Thus, a number of laboratories have committed considerable effort to cloning selected systems. One conclusion from these efforts is that it has been surprisingly difficult to clone certain restriction-modification systems in *E. coli.* Part of the difficulty may arise from the mode of expression of the genes. If they compose a constitutively expressed operon, they may be refractory to cloning because of degradation of the host cell chromosome on introducing the restriction gene. It seems unlikely that the modification enzyme could completely protect the chromosomal sites before extensive restriction had occurred. Conversely, some plasmid and virally encoded systems routinely become efficiently established on transfer to new host cells. In such cases, one can assume that two forms of expression exist: a start-up mode and a maintenance mode. Start-up would require sequential expression so that modification could be completed before turning on restriction. Once established, the genes could switch to the constitutive or maintenance mode. (In some cases, there may be additional regulation during maintenance. The specific activity of M·*Tth*HB81 peaks in early-to-middle exponential phase, decreases in early stationary phase, and almost vanishes in late stationary phase [Sato *et al.* 1980]).

Inability to clone also could result from inherent incompatibility of the gene products with *E. coli*. For example, certain modification genes might methylate and inactivate essential regulatory sites of the host genome. Another possibility is that the genes might be unlinked, making it virtually impossible to obtain both genes in a single clone. However, methylase genes sometimes can be independently cloned, thus making it possible to introduce a restriction gene on a second compatible plasmid.

Several plasmid-encoded systems (*Eco*RII, *Eco*RI, *Eco*RV, and *Pae*R7) have been subcloned in *E. coli* (see Figure 3.1). These systems are designed for cell-to-cell transfer, and selection of the proper clones from the relatively small plasmid genome library is relatively easy. For chromosomally encoded systems, conversely, strong selection is needed to detect restricting or modifying phe-

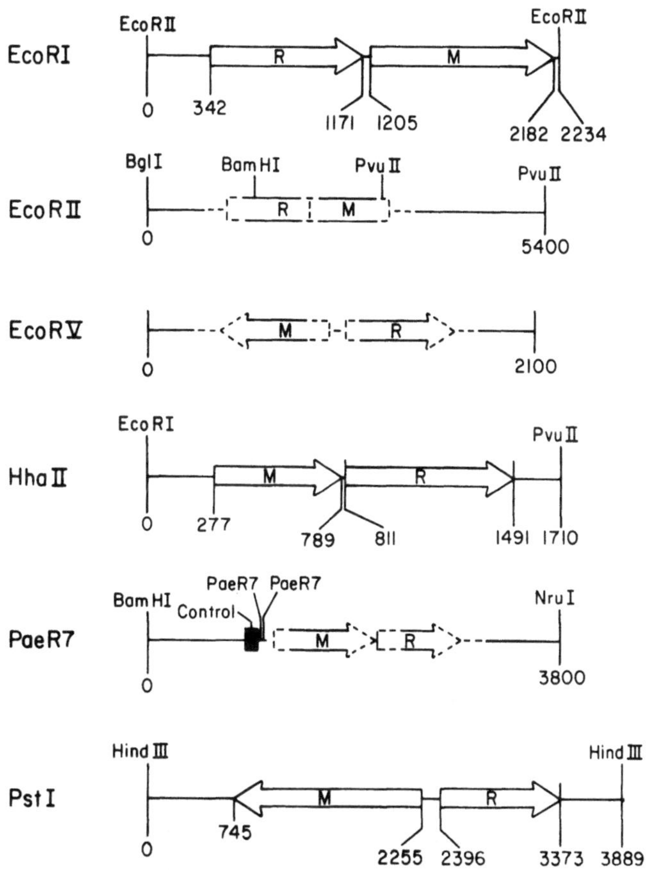

Figure 3.1 Maps of some cloned RM systems. Genes are indicated by arrows (showing direction of transcription and translation) or by rectangular boxes. Dotted lines indicate uncertainty as to lengths of coding regions. Drawings are to scale and numbers indicate distances in base pairs.

notypes from among the several thousand clones in a typical genome library. Two selection strategies have been successfully used: 1) biochemical selection for modification phenotype, and 2) biological selection for restriction phenotype.

Cloning of the *Msp*I modification gene (Walder *et al.* 1983) is a good illustration of the first approach. Genomic libraries of partial *Eco*RI and *Hin*dIII restriction fragments of *Moraxella* DNA ligated into pBR322 were constructed and propagated in *E. coli* strain HB101. Plasmid DNA from either the *Eco*RI or *Hin*dIII library then was restricted exhaustively with *Msp*I and used to retransform *E. coli* cells. There are 26 *Msp*I sites in pB322; thus, there should be little chance for survival of a plasmid molecule unless it carries the *Msp*I *m* gene and has become appropriately modified. In fact, over 50% of the clones surviving enzymatic selection tested positively for *Msp*I-specific modification.

It is readily appreciated that the strength of this kind of selection is directly related to the number of sites in the plasmid. The authors point out that efforts to clone the *Sma*I *m* gene (recognizing CCCGGG) using pBR313, which has only one *Sma*I site, have been unsuccessful due to a high background of plasmid survivors. Another possible cause of failure is incompatibility of the cloned *m* gene with cell survival. This could be due either to nonspecific effects of a DNA binding protein present in large amounts in the cell or to interference with gene expression by binding and/or modification of critical regulatory sequences in the host genome. A distinct advantage of the biochemical selection procedure is that *m* genes can be cloned independently of their proximity to *r* genes.

The second approach, based on selection for restriction phenotype, is illustrated by the cloning of the *Hha*II system from *Haemophilus haemolyticus* (Mann *et al.* 1978); 1,400 clones of a pBR322 library (in *E. coli* HB101) that contained partial *Pst*I fragments of *H. haemolyticus* DNA joined by GC-homopolymer extensions into the *Pst*I site of the Ap^r gene were screened. Clones were replica-plated onto nutrient agar dishes spread with 10^8 λvir plaque-forming units. Control HB101 colonies were sensitive to the phage and did not grow on the replicas, nor did the pBR322 vector confer resistance. Two types of survivors were expected: those acquiring resistance to absorption of λ , and those acquiring a new restriction-modification system capable of restricting λ. The former can be eliminated by selection against an unrelated phage (e.g., T5). From among the 1,400 clones, one was found that had simultaneously acquired resistance to five different phages; this proved to contain the *Hha*II genes on a 3-kb *H. haemolyticus* DNA insert in pBR322.

Success with this approach depends on several factors, including close linkage of the genes, expression of modification activity in advance of restriction activity, and absence of inactivation of essential cellular control sites by modification. The *r* and *m* genes also must function so they will confer classic restriction phenotype to the cells, since this is the basis for selection.

From Figure 3.1, it is evident that several gene arrangements and transcription patterns are possible for RM systems. The *Hha*II and *Pae*R7 genes are closely linked and contranscribed in the order *m* to *r*. In the case of *Hha*II, there seems to be no functional promoter on the cloned fragment, even though 800 bp of *Haemophilus* DNA is found 5′ to the genes in the original plasmid clone, pDI10. Expression is constitutive and originates from the ampicillin and antitet promoters of pBR322 (Schoner *et al.* 1983). Since the *Hha*II genes are readily introduced into new host cells, start-up expression may be translationally or posttranslationally controlled. One could imagine a scenario in which the ribosome binding site for the *r* gene is in a stem-loop structure, thus preventing ribosome loading. Methylase initially would be bound to DNA sites. Once methylation was completed, free methylates could bind to and destabilize the stem-loop structure, thus activating *r* translation. However, there currently is no evidence for this type of mechanism, and other possibilities exist.

The *Pae*R7 genes are found on a 42-kb plasmid (pMG7) carried in *Pseudomonas aeroginosa*; recently, they have been subcloned on a 3.8-kb *Bam*HI to *Nru*I fragment in pBR322. Clones confer a high level restriction (10^{-5}) and modification phenotype to *E. coli* host cells (Gingeras and Brooks, 1983). A set of BAL31 nuclease deletions from the *Bam*HI site (see Figure 3.1) have defined a 5′ control region, and sequence analysis locates a promoter in this region. Two *Pae*R7 sites occur about 24 base pairs downstream of the promoter, raising the possibility that methylation of these sites might modulate expression in a start-up situation (T. R. Gingeras, personal communication). More extensive deletions into the *m* gene give an unusual set of clones, which make endonuclease—but not methylase—as determined by *in vitro* assays. However, they neither restrict nor modify phage $\phi 80$ in *in vivo* tests. There are several possible explanations for this phenomenon. First, it may be that the restriction enzyme is membrane-bound or secreted into the periplasmic space. Second, the enzyme might be altered and no longer active under physiologic conditions. A third possibility is that the methylase or another gene product might be necessary for *in vivo* restriction activity.

The genes of the *Eco*RI system are cotranscribed (Rosenberg *et al.* 1981) in the order *r* to *m* (Newman *et al.* 1981; Greene *et al.* 1981) from a promoter 5′ to the *r* gene. However, O'Connor and Humphreys (1982) also have located an independent *m* gene promoter. Insertion of transcriptional terminators in the *r* gene do not abolish methylase activity, and a 406-bp *Sau*3A fragment containing the intergenic region activates Cm-resistance in the promoter probe plasmid pPV502 described by Close and Rodriguez (1982).

The *Eco*RII genes have been subcloned (Kosikh *et al.* 1979) and their order has been determined on a restriction map (Figure 3.1). The *Bam*HI site is in the endonuclease gene and the *Pvu*II site is in the methylase gene (T.R. Gingeras, personal communication). The mode of transcription is not yet known. The *Eco*RII gene sequences show weak homology with the *E. coli dcm* gene and strong homology to the *Ecl*II system from *Enterobacter cloaca* by Southern hybridization. The latter also appears to be plasmid encoded.

The *Pst*I (J. Donelson, personal communication) and *Eco*RV (M. Zabeau, personal communication) systems show a more novel arrangement. The genes in each system are closely linked, but they are divergently transcribed from promoters situated in an intergenic region, thus allowing possibilities for either independent or coordinately controlled expression.

The M·*Bsp*RI (5′GGCC) chromosomal gene from *Bacillus sphaericus* has been cloned into pBR322 on a 2.5-kb *Eco*RI fragment; it expresses in *E. coli* from its own promoter (Szomolanyi *et al.* 1980). The gene was recently sequenced (P. Venetianer, personal communication), and it specifies a 424-amino acid protein with calculated mol wt of 48.3 Kd in reasonable agreement with previous estimates of 50–52 Kd for the protein. The initiation point of transcription was localized by *in vitro* experiments with *E. coli* polymerase. The open reading frame starts 29-bp downstream from the transcriptional initiation site, and it is preceded by a sequence showing extensive Shine-Dalgarno complementarity. Removal of this ribosome binding site results in secondary starts 29 amino acids downstream; the shortened protein retains its methylase activity. It is not known whether both forms normally are made *in vivo.* However, it is interesting to speculate that a situation similar to this could possibly account for both the two forms of *Hpa*II methylase found by Yoo *et al.* (1982) and the two forms of *Bsu*RI methylase reported by Günthert *et al.* (1981,a,b). The biological rationale for the existence of two forms of methylase is not clear; however, it is conceivable that one form might be closely associated with the replication complex to assure chromosomal protection and the other might be a cytoplasmic form.

Two modification methylase genes of *B. subtilis* R have been isolated by cloning in *E. coli* (Kiss and Baldauf, 1983). One gene is part of the constitutive *Bsu*RI RM system with specificity for the sequence 5′-GGCC. The other is an inducible gene associated with the chromosomally integrated lysogenic phage SPβB (retermed phage SPR by Noyer-Weidner *et al.* 1983). Clones carrying this phage methylase gene produce the methylase M·*Bsu*PβBI with 5′GGCC specificity, but they also produce a second methylase activity, M·*Bsu*PβBII, with 5′CCGG specificity. Kiss and Baldauf (1983) found that both activities were simultaneously abolished by deleting either to the right or left of a single *Sal*I site in the cloned insert; it suggested the possibility that a single gene accounted for both CCGG and GGCC methylase specificities. Recent unpublished work (P. Venetianer, personal communication) appears to bear out their prediction that a single protein determines both specificities. This leaves the rather intriguing theoretical problem as to how two different sequences can be recognized. Perhaps the protein has two independent domains each with the requisite specificity. It is not readily apparent as to how a single protein site could accomplish the double recognition.

A number of *Bacillus* RM systems recognize GGCC and CCGG sites, and one wonders if, perhaps, they are related systems in some cases. Cross-hybridizations have been done between the genes for M·*Bsu*RI, M·*Bsu*PβBI (or II), and *Bsp*RI without finding any detectable homology. However, it is possible that they will show some amino acid sequence homology. It is interesting that the phage methylase gene is apparently not associated with a restriction gene. The phage genes might, in fact, function alone as antirestriction systems (Trautner *et al.* 1980).

Janulaitis *et al.* (1982) cloned the M·*Bcn*I gene of *B. centrosporus* in *E. coli*. It is contained on a 3.2-kb *Hin*dIII fragment, is apparently expressed from its own promoter, and is unassociated with restriction activity. Walder *et al.* (1983) cloned the *Msp*I methylase gene in *E. coli* and localized it on a 4.5-kb *Hin*dIII to *Eco*RI fragment, but they were unable to find the restriction gene. Recently, R. J. Roberts (personal communication) cloned a 15-kb *Eco*RI fragment (partial) containing both the *Msp*I modification and restriction genes; however, the clone does not restrict phage λ and the relative positions of the two genes are not known. The sequences of several of these genes are currently being determined.

Reaction Mechanisms of Type II Enzymes

There is considerable interest in the reaction mechanisms of the type II enzymes, because they constitute the purest examples of sequence-specific DNA enzymes. A good part of the work has focused on the mechanism of specific binding; furthermore, most of the experimental studies have been done with *Eco*RI, which is the most available enzyme. However, now that several systems have been cloned, it is expected that these will yield interesting results for comparison. Ultimately, it is hoped that some general picture will emerge of how these enzymes precisely recognize specific DNA sites and cleave or methylate them.

The reaction mechanism of type II methylases can be separated into two steps: binding and catalysis. In the binding step, the methylase can be thought to search along the DNA molecule until it locates a correct site to which it binds tightly and specifically. Catalysis involves step-wise transfer of methyl groups from AdoMet to specific base residues in the site on each strand. The enzyme then dissociates and is free to find and methylate other sites. The restriction endonuclease mechanism is similar in basic outline, except that catalysis results in phosphodiester bond cleavage rather than methylation. The important questions to be answered have to do with how a correct site is located, how an enzyme interacts with a site to achieve accurate base-specific recognition, and the nature of the catalytic mechanism. Very few studies along these lines have been done with the methylases. However, if we assume that sequence-specific DNA binding proteins find their sites and bind to them by basically similar mechanisms, it is possible to extrapolate from studies of the restriction enzymes and other sequence-specific proteins to gain a clear picture of how the methylases might work. *Eco*RI has been the most thoroughly studied and will serve as our model system.

Thermodynamic and Kinetic Mechanism of Specific Binding

EcoRI and *Hin*fI have been shown to form site-specific complexes with DNA in the absence of a divalent cation—conditions under which cleavage does not occur. The specific complex association constant for *Eco*RI on pBR322 DNA (one site) in buffer of ionic strength 0.07–0.15 M at 37°C is in the range of 10^{10}–10^{11} M^{-1} by using nitrocellulose filter-binding assays (Rosenberg et al. 1981); for *Hin*fI an apparent association constant of about 2×10^{10} M^{-1} is obtained under standard reaction buffer conditions (minus Mg^{2+}) by equilibrium analysis on gel filtration columns (Frankel, 1983). Several other restriction endonucleases also appear to bind specifically in the absence of divalent ions. Thus, it is possible to study binding independently of catalysis.

The methylases, conversely, may require AdoMet for specific complex formation. Attempts to measure specific binding constants of *Eco*RI methylase in the absence of AdoMet have not been successful (Modrich, 1979; Jack *et al.* 1981). Günthert *et al.* (1981b) have shown that AdoMet binds to *Bsu*RI methylase in the absence of DNA; they have proposed that *Bsu*RI-AdoMet complexes may be necessary for specific binding to DNA sites. On theoretical grounds, it seems reasonable that methylase-AdoMet complexes should be the active binding forms, since there would be no obvious purpose for binding in the absence of cofactor. However, more experimentation will be necessary to confirm this concept and to show its generality.

The restriction endonucleases and methylases also bind nonspecifically to DNA, although with an affinity several orders of magnitude lower than for specific binding. For example, *Hin*fI binds to *Hin*fI cleavage fragments with an association constant of 5×10^{4} M^{-1} (Frankel, 1983), which is about five to six orders of magnitude less than for the same DNA containing uncleaved sites. *Eco*RI shows similar decreases in affinity for nonspecific sequences or for DNA containing methylated sites (Modrich and Roberts, 1982). This nonspecific binding presumably involves interactions with the DNA backbone, and it may be important in the mechanism by which the enzymes find their sites on DNA.

With regard to this latter point, a number of sequence-specific DNA proteins appear to find their sites more rapidly than can be accounted for by their diffusion rates in solution. It seems necessary to propose some sort of facilitated diffusion mechanism to account for these rapid rates of binding. Two basic models that have been described in some detail (Berg *et al.* 1981) are: 1) a sliding mechanism in which the protein is envisioned as binding nonspecifically and then executing a random one-dimensional walk along the DNA until the site is encountered; and 2) a hopping mechanism, in which the protein translocates itself from one point to another along the DNA chain by rapid dissociation and reassociation within the domain of the randomly coiled molecule. Experiments with both *lac* repressor (Winger *et al.* 1981) and *Eco*RI (Jack *et al.* 1982) favor the sliding mechanism. The data with *Eco*RI seems to be particularly convincing. Kinetic rates of formation and dissociation of specific filter-binding endonuclease-DNA complexes were measured for nine linear DNA fragments ranging from 34–6,200 base pairs and containing the *Eco*RI site in a central location. The first-order dissociation rate constant increased from a value of 8×10^{-5} sec^{-1} ($t_{1/2}$ = 140 min) for the 34-bp fragment to 7×10^{-4} sec^{-1} ($t_{1/2}$ = 15 min) for molecules of 4,000 bp and longer. Thus, sequences external to the site enhance the rate of dissociation; but, there is a limit to the range of the effects. Measurements of formation rates by competition between small and large DNA fragments showed a similar enhancement with increasing size. Thus, DNA sequences outside of the recognition site proportionally enhance the rate at which *Eco*RI locates or leaves its site—with the equilibrium binding constant remaining independent of size. Put in intuitive terms, one can imagine that an enzyme will bind more rapidly to larger DNA molecules because the target for initial nonspecific binding is larger. However, when dissociating from the specific site, the enzyme can slide further away and lose itself along the molecule; hence, the more rapid dissociation. For short pieces of DNA, the forward rate is slow because of the smaller target, but enzyme dissociating from a site is trapped on a smaller molecule and is more likely to refind the site. By fitting mathematically to the sliding model, the average DNA length that was scanned was calculated to be about 1,300 base pairs for *Eco*RI. Similar experiments have not been performed with the methylase, but one could imagine that it might employ a similar mechanism.

Site Recognition

Since the *Eco*RI methylase and restriction endonuclease both recognize the same 6-bp site (5′-GAATTC), one might anticipate that they would make similar contacts with the site and that they might even share some amino acid sequence homology in their DNA binding site regions as a result of a common ancestry through gene duplication; however, just the opposite seems to be true. They share no amino acid sequence homology (Greene *et al.* 1981), and they appear to interact with their common DNA site in dramatically different fashions. Currently available data reveal no common contacts except for the 6-amino group of the adenine residue at the position of methylation.

Three types of studies have been useful in defining DNA recognition determinants in the site region: 1) effect of base analog substitutions on catalytic rates; 2) interference with enzyme binding by chemical modification of the DNA site; or 3) protection against chemical modification by bound enzyme.

Berkner and Folk (1977) found that DNA containing glucosylated hydroxymethylcytidine is methylated as well as normal DNA by the *Eco*RI methylase—but it is not restricted. DNA containing uridine in place of thymine is very poorly methylated, while the rate of restriction and the Km for DNA is normal. DNA containing hydroxymethyluridine also is poorly methylated; in this case, however, the rate of restriction is decreased 20-fold in addition. However, the Km for DNA is unaffected, suggesting that catalysis rather than binding is affected. Since both thymines are altered in a site, effects cannot be assigned to a given residue. Modrich and Rubin (1977) found that sites containing dI in place of dG (loss of the 2-amino group in the minor groove) were modified very poorly while restriction was normal. Finally, the methylase must interact with the 6-amino group of adenine at the position of methyl group transfer. Since this interferes with both *Eco*RI binding (Jack *et al.* 1981) and cleavage, the 6-amino group could be considered a common recognition determinant for the two enzymes.

Binding of R·*Eco*RI to its site has been studied in some detail by using the alkylation interference and protection methods developed by Gilbert *et al*, (Siebenlist *et al.* 1980). The main findings are summarized in Figure 3.2. Ethylation of four phosphates along each chain in the region of the site significantly interferes with the amount of specific enzyme-DNA complexes formed and retained on nitrocellulose membrane filters (Lu *et al.* 1981). Interestingly, two of the phosphate groups contacted by the enzyme are immediately 5′ to, and outside of, the site. Thus, *Eco*RI binds to a region of about 10 base pairs, and nonbase-specific electrostatic interactions appear to contribute significantly to the binding. Methylation of dA and dG residues by dimethylsulfate treatment

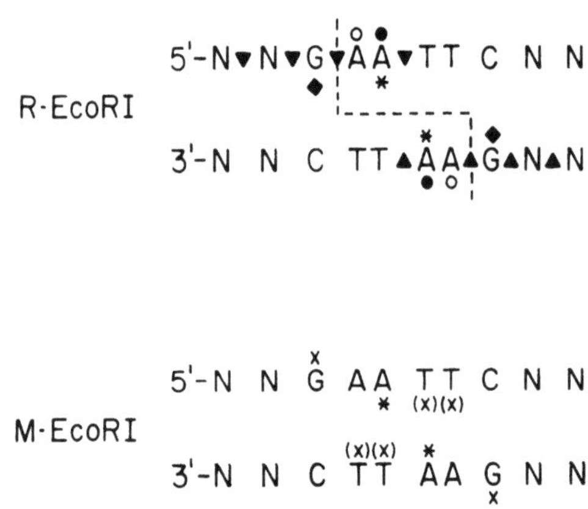

Figure 3.2 Comparison of potential contacts between the *Eco*RI restriction and modification enzymes and their DNA recognition site. The dashed line indicates the cleavage domains; ▼, phosphates identified on the basis of ethylation interference by ethylnitrosourea treatment; ●, N3 of dA residues identified on basis of methylation (by dimethylsulfate treatment) interference. These residues also were protected against methylation with DMS in the presence of bound *Eco*RI; ○, N3 of dA residues protected against methylation by bound *Eco*RI, but not interfering when premethylated; ♦, N7 of dG residues identified on basis of both methylation interference and protection;*, 6-NH_2 groups of dA residues, which are modified by M *Eco*RI and also interfere with binding of R *Eco*RI; X, base determinants (2-NH_2 of dG and 5-CH_3 groups of dT), which are potential points of contact of M *Eco*RI, but not of R *Eco*RI. Parentheses indicate uncertainty as to whether one or both dT residues act as contacts in methylase interaction. External symbols are minor groove contacts and internal symbols are major groove contacts.

was examined for effects on *Eco*RI binding. Methylation of the N7 of dG (major groove), and of the N3 of the inner dA residue (minor groove) on each strand interfered with complex formation. In protection experiments, bound endonuclease gave significant reduction of methylation at the N7 of dG and the N3 of both dA residues. These results suggest the presence of protein interactions in both grooves in the 5′GAA region. Specific methylation of the *Eco*RI site by M·*Eco*RI eliminated measurable complex formation; thus, the 6-amino position of the inner adenine residue on each chain is a probable contact point.

In summarizing these experiments, it is particularly striking that most of the determinants for both enzymes are accessible on one side of the helix. The few minor groove contacts are just around the lip of the sugar-phosphate backbone. Whether the enzymes use a twofold symmetric pair of α-helices for recognition, as suggested for the *cro* and λ-repressor proteins (Pabo and Lewis, 1982; Ohlendorf *et al.*, 1982; Sauer *et al.*, 1982), will have to await crystallographic structural studies. Again, the marked difference in the apparently recognized determinants by the endonuclease and methylase is surprising and should evoke thought about its possible functional significance. However, one should keep in mind that the methylase data is based on catalytic, rather than binding, assays; therefore, it may be misleading. Detailed studies of other RM system enzymes will be necessary to confirm the general picture of site recognition gained from the *Eco*RI studies. As yet, little has been published, but several systems are now cloned and under study by various laboratories.

Catalysis

Once the *Eco*RI has recognized and engaged the correct DNA site, catalysis can begin. At this stage, the enzyme either may nick the opposite strand to complete the duplex cleavage or it may dissociate from the nicked site. A second enzyme molecule then must bind to complete the double-stranded cleavage. The fraction of intermediate dissociations that take place depends on the reaction conditions and, to some extent, on the site. For example, with colEI DNA at 37 °C, duplex breaks are made in one binding event, while at 0 °C the nicked product predominates (Modrich and Zabel, 1976; Rubin and Modrich, 1978). With SV40 DNA at 37 °C, a mixture of 25% nicks due to intermediate dissociation and 75% duplex cleavage without dissociation is obtained (Rubin and Modrich, 1978). Studies at 30 °C with colEI DNA show a half-time of about 1 second for the first cleavage and 3 seconds for the second cleavage. Comparison of these with the catalytic constant (0.7/min) makes release of dimer from the end-product the rate-limiting step (Modrich and Zabel, 1976).

The *Eco*RI methylase not only recognizes different groups within the *Eco*RI site to achieve recognition, it differs in mechanism. It is easy to understand how the restriction endonuclease dimer is able to cleave each strand in one binding event. The methylase, however, exists as a monomer at concentrations of the typical reaction. Initial rate of methyl group transfer is first order with respect to enzyme and obeys Michaelis-Menten kinetics, indicating that the monomer is the functionally active species. Thus, it is not surprising that the methylase obligatorily transfers one methyl group at a time, with dissociation in between (Rubin and Modrich, 1977). This is not a generality, however, since the *Bam*HI methylase transfers both methyl groups in a single binding event (Nardone *et al.* 1983). Steady-state kinetic characteristics of *Eco*RI methylase for addition of the first and second methyl groups also are quite similar—unlike the case of the type I methylases, where addition of a second methyl group is as much as 100 times greater (Vovis *et al.* 1974). Günthert *et al.* (1981b) also found a somewhat more rapid addition of a second methyl group to hemimethylated *Bsu*RI sites, although binding affinity of the *Bsu*RI methylase to DNA is unchanged.

Relaxation of Specificity ("Star" Activity)

Under appropriate reaction conditions, the *Eco*RI restriction enzyme is capable of cleaving sites closely related to the canonical site, GAATTC (Polisky *et al.* 1975). This relaxed specificity is referred to as *Eco*RI* activity. Conditions favoring *Eco*RI* are low ionic strength and elevated pH (Polisky *et al.* 1975), substitution of Mn^{2+} for Mg^{2+} (Hsu and Berg, 1978), and inclusion of small, polar organic molecules (ethylene glycol, glycerol, dimethylsulfoxide, or formamide) in the buffer (Mayer, 1978). Trivial explanations for *Eco*RI* activity (e.g., a contaminating second enzyme) have been ruled out. Even more convincingly, such effects are not limited to *Eco*RI, but also have been observed with several other restriction enzymes; these include *Bam*HI (George *et al.* 1980; Malyguine, 1980), *Bst*I (Clark and Hartly, 1979), *Bsu*I (Heininger *et al.* 1977), *Hha*I (Malyguine *et al.* 1980), and *Hin*dIII (Hsu and Berg, 1978). In addition, the *Eco*RI methylase manifests a broadened specificity under standard buffer conditions when excess enzyme is allowed to react with DNA for extensive times (Berkner and Folk, 1978).

The altered recognition process in *Eco*RI* activity has been extensively studied by Woodbury and von Hippel (1981) and Rosenberg and Greene (1982) with the hope that it will reveal important clues to the sorts of interactions involved in sequence-specific recognition. The basic observations can be briefly summarized. The canonical site, GAATTC, is always cleaved much more rapidly than *Eco*RI* sites. The *Eco*RI* sites can be arranged in hierarchies based on their cleavage rates as follows: $G>>A>T>>C$ at the first position (Goodman *et al.* 1977), $A>>[G,C]>>T$ at the second and third positions (Gardner *et al.* 1982), and the corresponding complements at the last positions. Thus, cleavage rates are GAATTC > AAATTC > CAATTC, and so forth.

Woodbury and von Hippel (1981) proposed a plausible model to explain the effects based on consideration of various H-bond possibilities available with each base pair. Tacitly in their model is the assumption that effects are at the level of binding, rather than catalysis. They propose that recognition involves bridging acceptor and donor combinations of H-bond contacts between stacked pairs of base pairs in the upper half (GAA) of the site on each strand. For example, they suggest the following two pairs of bridging contacts for the canonical *Eco*RI site: 1) N7 of dG to N6 of dA; and 2) N7 of dA to N6 of dA. Under *Eco*RI* conditions, the enzyme is thought to undergo sufficient structural perturbation to now fit a different pair of bridging contacts that are available in the star sites. The amount of structural perturbation required to fit these sites determines the hierarchic order. With AAA, the enzyme could fit N7 of dA to N6 of dA at each level; with GGA, it could fit 06 of dG to N4 of dC and N7 of dG to N6 of dA, and so forth.

With this sort of a model for base-specific interactions in mind, it is not hard to imagine that many sequence-specific enzymes might show similar effects. Thus, it is not surprising to find an *Eco*RI* methylase activity that closely parallels that of the restriction enzyme. ΦX174RF DNA contains no canonical *Eco*RI site, but it becomes extensively methylated (90 methyl groups per molecule) when incubated for a long period with excess methylase. The DNA becomes almost completely protected against *Eco*RI* cleavage activity; however, a few sites remain cleavable (Woodbury *et al.* 1980). Thus, the endonuclease and methylase star activities overlap, but do not exactly coincide, which is not a surprising result considering the differences in their structures and manner of site recognition.

Aberrant Methylation and Hemimethylated Sites

Normally, a specific adenine or cytosine residue is methylated on each strand of a symmetric site by the modification methylase of an RM system that recognizes that site. This methylation confers a high degree of protection against the corresponding endonuclease and is referred to as the standard methylation for that site. However, methylation sometimes may be found at other positions on a site due to other methylases in the cell acting on the same or overlapping sites. This is referred to as aberrant methylation of the site. For example, the *dam*$^+$ (GmATC) and *dcm*$^+$ (CmCNGG) methylases of *E. coli* modify *Bam*HI (GG$\overset{*}{\text{A}}$TCC) and *Bst*NI (C$\overset{*}{\text{C}}$NGG) sites at the positions of the asterisks, respectively. In both of these cases, the methylations are at aberrant (nonstandard) positions, and they do not protect against cleavage by the respective restriction enzymes (Hattman *et al.* 1978; McClelland, 1981,1983). In other cases, methylation at aberrant positions gives protection, although few (if any) quantitative studies of the extent of protection have been done. Table 3.3 lists a number of restriction enzymes for which there is information concerning cleavage of various aberrant methylation patterns in their recognition sequences. McClelland (1983) has compiled a more extensive list that probably will be updated as new information arises. Such information is particularly useful in analyzing methylation patterns in eukaryotic DNA by restriction analysis.

Table 3.3 Cleavage of Methylated and Hemimethylated Sites

Enzyme	Methylated Sites Cut[a]	Methylated Sites Uncut	References
*Hpa*II	m C C G G G G C | C	m m m m C C G G C C G G[f] C C G G G G C C G G C C G G C C m m	Walder *et al.* (1983) Gruenbaum *et al.* (1981)
*Msp*I	m m m C | C G G C C G G G G C | C G G C | C m	m m m C C G G[f] C CG G[c] G GC CG G[d] G G C C G GC C C CG GC C m m	Walder *et al.* (1983) Gruenbaum *et al.* (1981) Kesket and Cedar (1983) Busslinger *et al.* (1983)
*Dpn*I	m G A | T C C T | A G m	cuts only methylated sites	Lacks and Greenberg (1977)
*Sau*3A	m m G | A T C G A T C C T A | G C | T A G m	m[e] G A T C C T A G m	Streeck (1980) Sussenbach *et al.* (1976)
*Hha*I		m m m m G C G C[f] G C G C G C G C C G C G C G C G C G C G m m	Mann and Smith (1979) Smith (1979) Gruenbaum *et al.* (1981)
*Hae*III	m m m G G | C C G G C C C C | G G C C | G G m	m G G C C[f] C C G G m	Mann and Smith (1977) Gruenbaum *et al.* (1981)
*Taq*I	m m T | C G A T | C G A A G C | T A G C | T m	m[f] T C G A A G C T m	Streeck (1980) Gruenbaum *et al.* (1981) Sato et al. (1980)

<table>
<tr><td>TthI</td><td>m
T C G A
A G C T
m</td><td>m[f]
T C G A
A G C T
m</td><td>Sato et al. (1980)
McClelland (1983)</td></tr>
<tr><td>HinfI</td><td>m
G|A N T C
C T N A|G</td><td>m
G A N T C[e]
C T N A G
m</td><td>Gruenbaum et al. (1981)
S. Kelly (unpublished)</td></tr>
<tr><td>BstNI</td><td>mm m m
CC|N GG CC | N|GG CC|N GG
GG N|CC GG N CC GG N|CC
m m</td><td>??</td><td>Gruenbaum et al. (1981)
McClelland (1983)</td></tr>
<tr><td>EcoRII</td><td>m
|C C N G G
G G N C C|</td><td>m m m
C C N G G[f] C C N G G
G G N C C G G N C C
m</td><td>Boyer et al. (1973)
Gruenbaum (1981)</td></tr>
<tr><td>TaqXI</td><td>m
C C|N G G
G G N|C C
m</td><td>m
C C N G G[f]
G G N C C
m</td><td>Grachev et al. (1981)</td></tr>
<tr><td>HindII
(HincII)</td><td>m |
G T Py|Pu A C
C A Pu|Py T G</td><td>m
G T Py Pu A C[f]
C A Pu Py T G
m</td><td>Roy and Smith (1973b)
Gruenbaum et al. (1981)</td></tr>
<tr><td>XhoII</td><td>m
Pu|G A T C Py
Py C T A G|Pu
m</td><td>m
Pu G A T C Py
Py C T A G Pu
m</td><td>Brooks and Roberts (1982)
McClelland (1983)</td></tr>
</table>

Table 3.3 *(continued)*

<table>
<tr><th>Enzyme</th><th>Methylated Sites Cut[a]</th><th>Methylated Sites Uncut</th><th>References</th></tr>
<tr><td>BglII
(PvuI)
(XorI)</td><td>m
A|G A T C T
T C T A G|A
m</td><td>m
A G A T C T
T C T A G A
m</td><td>Brooks and Roberts (1982)</td></tr>
<tr><td>BamHI</td><td>m m
G|G A T C C[b] G|G A T C C
C C T A G|G C C T A G|G
m m</td><td>m
G G A T C C[f]
C C T A G G
m</td><td>Brooks and Roberts (1982)
Hattman et al. (1978)
Mann and Smith (1977)</td></tr>
<tr><td>HpaI</td><td>m
G T T|A A C
C A A|T T G</td><td>m
G T T A A C[f]
C A A T T G
m</td><td>Yoo et al. (1982)
Gruenbaum et al. (1981)</td></tr>
<tr><td>BclI</td><td>m
T|G A T C A
A C T A G|T
m</td><td>m
T G A T C A
A C T A G T
m</td><td>Brooks and Roberts (1982)</td></tr>
</table>

[a]Cut positions are from Roberts (1983) and are assumed to be unchanged by aberrant methylations. This seems highly likely, but is unproven in most cases.

[b]Levy and Welker (1981) find that adenine methylation protects against *Bam*HI cleavage; it may be a question of rates.

[c]Gruenbaum *et al*. (1981) were able to show partial cleavage of the ummodified strand in hemimethylated sites; Walder *et al*. (1983) found none.

[d]The site is cut at very high enzyme concentrations.

[e]Probable standard modification specificity.

[f]Proven standard modification specificity.

Hemimethylated (half-modified) sites are transiently generated at the replication fork in actively growing cells—even in the presence of modification enzymes. The question arises as to whether these sites sometimes might be subject to nicking or duplex cleavage. Demonstration of such an event *in vivo* would be difficult, but action of the restriction enzymes on such sites has been looked at *in vitro* in a few cases. Duplex cleavage appears to be ruled out in all cases examined. However, for a few RM systems, the unmodified strand can be nicked—while for others, the site is completely resistant. Gruenbaum *et al.* (1981) prepared hemimethylated ϕX174 duplex DNA substrate labeled on either the methylated or the unmethylated strand. The methylated strand in each case was synthesized by using 5mdCTP and, thus, was totally modified for dC residues. Using this substrate, they showed that *Hae*III nicks the unmodified strand in the GGCC site and that *Msp*I partially nicks the unmodified strand in the CCGG site at very high enzyme-to-substrate ratios. The latter was in contradiction to the findings of Walder *et al.* (1983) that hemimethylated (mCCGG) sites were not nicked; however, they may not have used extreme levels of enzyme. *Sau*3A also nicks hemimethylated (GATmC) sites (Streeck, 1980). *Hpa*II nicks unmethylated strands in (mCCGG) hemimethylated sites (Walder *et al.* 1983), but not when both cytosines are methylated (mCmCGG) on one strand (Gruenbaum *et al.* 1981). *Hha*I, *Hae*III, and *Msp*I have a limited capacity to digest single-stranded DNA when in large enzyme excess. This might account for some of their ability to nick hemimethylated sites (i.e., it may reflect a recognition mechanism oriented mainly to groups along one strand of the DNA). As expected, none of these three enzymes can cleave methylated single-stranded DNA.

Sequence context may modulate the effects of aberrant methyl groups in sites, as has been shown recently for *Msp*I (Busslinger *et al.* 1983; Keshet and Cedar, 1983). This enzyme fails to cut a GGCmCGG sequence in the 5′ region of the γ globin gene at enzyme concentrations that give total digestion both of other CmCGG sites and of the unmethylated CCGG and GGCCGG sites. The resistance is not total, however, since greater than 100-fold excesses of *Msp*I do cleave the site. Secondary structural features of the GC-rich region, perhaps, account for the rather bizarre behavior of the site. It would be interesting to examine *Msp*I cleavage of some defined methylated or unmethylated synthetic oligonucleotides to determine how the combination of methylation and sequence context can cause inhibition of cleavage while neither alone inhibits cleavage. These effects of sequence context do not appear to be unique to *Msp*I. S. Jentsch (personal communication) has found that M·*Bsu*F methylates CCGG sites to *Msp*I resistance as expected from the position of methylation (mCCGG); interestingly, however, when these sites are examined with *Hpa*II, those which are in a GC-rich environment tend to be *Hpa*II-resistant while the majority are sensitive to cleavage. Thus, the observation of Mann and Smith (1978) that hemimethylated mCCGG sites are resistant, and of Walder *et al.* (1983) that fully methylated mCCGG sites are resistant to *Hpa*II cleavage, may depend strongly on the particular sites examined.

The enzymes in Table 3.3 all act at symmetric sites. Little is known concerning methylation of asymmetric sites. Two type III enzymes are known to methylate only one strand of their recognition site. A similar situation could hold for some of the type II enzymes that recognize asymmetric sites. In this regard, it is noteworthy that *Tth*1112 (CAAPuCA) has no dA or dC residues on the complementary strand of its site and that *Mbo*I (GAAGA) and *Mnl*I (CCTC) have only dA residues on one strand and dC residues on the opposite strand of their sites. It seems likely that these sites might be modified on only one strand. If so, then how does the unmodified, newly replicated daughter strand in the host cell escape restriction? One must postulate a close coupling of methylation to replication, or a low accessibility of the restriction enzyme to the replication region.

Restriction and Modification of Single-Stranded DNA

It is well documented that certain restriction enzymes (*Hha*I, *Hae*III, or *Msp*I) will cleave single-stranded DNA and that the specificity is identical to that for duplex DNA (Blakesley and Wells, 1975; Horiuchi and Zinder, 1975; Godson and Roberts, 1976). However, the rate of cleavage is much slower. It is, as yet, unresolved whether these enzymes actually bind and cleave the sites in the single-stranded form, or whether transient duplex formation occurs during catalysis. The latter mechanism implies a second-order reaction with respect to substrate DNA; this has not yet been examined. Blakesley *et al.* (1977) have obtained results with *Hha*I and *Hae*III supporting transient duplex formation as the mechanism. Cleavage of single-stranded ϕX174 DNA was sensitive to inhibition by low levels of netropsin and actinomycin D. These antibiotics bind preferentially to duplex DNA. More convincingly, cleavage of single strands by *Hae*III occurred at a lower optimum temperature (47°C) than for duplex DNA (72°C).

Yoo and Agarwal (1980) demonstrated that *Msp*I will slowly cleave the individual deoxyoligonucleotides TCTCCGGTT and GAACCGGAGA; neither of which is likely to form a stable duplex with itself. However, the rate of cleavage diminished markedly as temperature was increased from 4–37°C; this again suggests transient formation of CCGG duplexes at the moment of catalysis. It should be noted that sequence-specific cleavage of actual single-stranded sites implies that the enzyme interacts with sufficient groups on one strand of the helix to provide full discrimination.

In contrast to the restriction endonucleases that usually are dimeric and active only on duplex sites, some methylases appear to be quite active on single-stranded DNA. Both *Hae*III and *Hpa*II methylases (and also the *dam* methylase of *E. coli*) have been found to methylate denatured salmon sperm and T7 DNAs at only slightly reduced rates (Mann and Smith, 1979; Herman and Modrich, 1981). The *Hha*I methylase methylates the random copolymers d(G,C)n, d(N-AcG,C)n, and d(X,C)n at relative rates of 100%, 127%, and

66%, respectively (Mann and Smith, 1979). Also, the latter two substrates do not show any thermal melting transition, suggesting a lack of secondary Watson-Crick structure. Thus, methylases that are monomeric may, in some cases, interact asymmetrically with sufficient groups along one strand of the site to provide accurate discrimination. It should be noted, however, that not all methylases act this way, since M·*Bam*HI requires duplex DNA as a substrate (J. Chirikjian, personal communication) and M·*Eco*RI and M·*Hpa*I do not methylate single-stranded DNA (Modrich and Roberts, 1982).

Expression of Bacterial Methylases in Eukaryotes

The DNA of most eukaryotes contains a small fraction of methylated cytosine residues. The most common sites for methylation are CG dinucleotides; *Hpa*II and *Msp*I restriction enzymes have been widely used to probe the methylation patterns of these sites adjacent to particular genes. There is now fairly general agreement that certain sites in the regions of activated genes become undermethylated relative to their inactivated states (Razin and Riggs, 1980; Vardimon *et al.* 1982). A natural question arises as to whether introduction of bacterial methylases into eukaryotes would perturb gene expression, perhaps in some controllable or reproducible fashion.

Feher *et al.* (1983) inserted the 2.5-kb *Eco*RI fragment containing the M·*Bsp*RI gene into the yeast-*E. coli* shuttle vector, YEp6, and established this plasmid in *Saccharomyces cerevisiae* (which normally contains no detectable mC residues). Plasmid reisolated from the yeast cells is protected against restriction at its GGCC sites. Chromosomal *his*3 and *leu*2 gene regions were specifically tested and found to be protected, but the bulk of chromosomal sites remained unmethylated. It is possible that the plasmid methylation is perpetuated by yeast maintenance methylases. However, the plasmid-borne *Bsp*RI methylase clearly works in *trans*, because introduction of a second compatible plasmid resulted in its efficient methylation. The different behaviors of plasmid and chromosomal DNAs may reflect differences in their chromatin structure.

Brooks *et al.* (1983) have cloned and sequenced the *E. coli dam* gene, which codes for a DNA adenine methylase of specificity GmATC. This recently has been placed in a shuttle vector and introduced into yeast, where it replicates as a minichromosome in the nucleus (T. R. Gingeras, personal communication). All of the plasmid GATC sites become methylated, while only a fraction ($<$ 30%) of the chromosomal sites are methylated. Again, the dichotomy between plasmid and chromosomal accessibility is presumed to reflect some feature of the chromatin structure. Several metabolic pathways have been examined for effects of the adenine methylation, but no changes have been detected. Yeast normally carries no methylated adenine residues in its DNA, and it seems to be oblivious to the new *dam*-specific decoration of its DNA. As new bacterial methylases become available for introduction into yeast and other eukaryotic cells, it will be interesting to observe their effects.

One result of these studies in eukaryotic systems has been a new and revived interest in the DNA methylases as important enzymes to be studied. It is possible that their usefulness as probes and modulators of gene expression, and as objects for study of sequence-specific protein-nucleic acid interactions, may—in the future—bring them into a prominence equal to that of the restriction endonucleases.

Acknowledgments

The authors acknowledge support from American Cancer Society Grant No. NP-258 and National Institutes of Health Grant No. 5-Pol-CA16519. Hamilton O. Smith is an American Cancer Society Research Professor. We especially thank Mildred Kahler for her careful and patient preparation of the manuscript.

References

Arber W: DNA modification and restrictions in Cohn WE (ed): *Progress in Nucleic Acid Research and Molecular Biology*. New York, Academic Press, 1974, vol 14, pp 1–37.

Berg OG, Winter RB, von Hippel PH: Diffusion-driven mechanisms of protein translocation on nucleic acids. 1. Models and theory. *Biochemistry* 1981;20:6929–6948.

Berkner KL, Folk WR: *Eco*RI cleavage and methylation of DNAs containing modified pyrimidines in the recognition sequence. *J Bio Chem* 1977;252:3185–3193.

Berkner KL, Folk WR: Over methylation of DNAs by the *Eco*RI methylase. *Nucl Acids Res* 1978;5:435–450.

Blakesley RW, Wells RD: "Single-stranded" DNA from ϕX174 and M13 is cleaved by certain restriction endonucleases. *Nature* 1975;257:421–422.

Boyer HW, Chow LT, Dugaiczyk A, Hedgpeth J, Goodman HM: DNA substrate site for the *Eco*RII restriction endonuclease and modification methylase. *Nature New Biol* 1973;244:40–43.

Brooks JE, Roberts RJ: Modification profiles of bacterial genomes. *Nucl Acids Res* 1982;10:913–934.

Brooks JE, Blumenthal RM, Gingeras TR: The isolation and characterization of the *Escherichia coli* DNA adenine methylase *(dam)* gene. *Nucl Acids Res* 1983; 11:837–851.

Busslinger M, deBoer E, Wright S, Grosveld FF, Flavell RA: The sequence of GGCmCGG is resistant to *Msp*I cleavage. *Nucl Acids Res* 1983;11:3559–3569.

Catterall JF, Welker NE: Isolation and properties of a thermostable restriction endonuclease (Endo R·*Bst* 1503). *J Bacteriol* 1977;129:1110–1120.

Cheng, S-C, Modrich P: Positive-selection cloning vehicle useful for overproduction of hybrid proteins. *J Bacteriol* 1983;154:1005–1008.

Clark CM, Hartley BS: Purification, properties, and specificity of the restriction endonuclease from *Bacillus* stearothermophilus. *Biochem J* 1979;177:49–62.

Close TJ, Rodriguez RL: Construction and characterization of the chloramphenicol resistance gene cartridge: A new approach to the transcriptional mapping of extra-chromosomal elements. *Gene* 1982;20:305–316.

Cregg JM, Nguyen AH, Ito J: DNA modification induced during infection of *Bacillus subtilis* by phage ϕ3T. *Gene* 1980;12:17–24.

Dobritsa AP, Dobritsa SV: DNA protection with the DNA methylase M·*Bbv*I from *Bacillus brevis* var. GB against cleavage by the restriction endonucleases *Pst*I and *Pvu*II. *Gene* 1980;10:105–112.

Dreiseikelmann B, Eichenlaub R, Wackernagel W: The effect of differential methylation by *Escherichia coli* of plasmid DNA and phage T7 and λ DNA on the cleavage by restriction endonuclease *Mbo*I from *Moraxella bovis*. Biochem et Biophys. *Acta* 1979;562:418–428.

Dugaiczyk A, Hedgpeth J, Boyer HW, Goodman HM: Physical identity of the SV40 deoxyribonucleic acid sequence recognized by the *Eco*RI restriction endonuclease and modification methylase. *Biochem* 1974;13:503–512.

Ehrlich M, Wang RY-H: 5-Methylcytosine in eukaryotic DNA. *Science* 1981;212:1350–1357.

Feher Zs, Kiss A, Venetianer P: Expression of a bacterial modification methylase gene in yeast. *Nature* 1983;302:266–268.

Frankel A: Sequence-specific recognition of DNA by the *Hinf*I restriction endonuclease. Ph.D. Thesis, Baltimore, Johns Hopkins University, 1983.

Gardner RC, Howarth AJ, Messing J, Shepherd RJ: Cloning and sequencing of restriction fragments generated by *Eco*RI* *DNA* 1982;1:109–115.

George J, Blakesley RW, Chirikjian JG: Sequence-specific endonuclease *Bam*HI: Effect of hydrophobic reagents on sequence recognition and catalysis. *J Biol Chem* 1980;255:6521–6524.

Gingeras TR, Brooks JE: Cloned restriction/modification system from *Pseudomonas aeruginosa*. *Proc Nat Acad Sci USA* 1983;80:402–406.

Godson GN, Roberts RJ: A catalogue of cleavages of ϕX174, S13, G4, and ST-1 DNA by 26 different restriction endonucleases. *Virology* 1976;73:561–567.

Goodman HM, Greene PJ, Garfin DE, Boyer HW: DNA site recognition by the *Eco*RI restriction endonuclease and modification methylase, in Vogel HJ (ed): *Nucleic Acid-Protein Recognition*. New York, Academic Press, 1977, pp 239–259.

Grachev SA, Mamaev SV, Gurevich AL, Lgoshun AV, Kolosov MN, Slyusar AG: Restriction endonuclease *Taq*XI from *Thermus aquaticus*. *Bioorg Khim* 1981;7:628–631.

Greene PJ, Poonian MS, Nussbaum AL, et al: Restriction and modification of a self-complementary octanucleotide containing the *Eco*RI substrate. *J Mol Biol* 1975;99:237–261.

Greene PJ, Gupta M, Boyer HW, Tobias L, Garfin DE, Boyer HW, Goodman HM: Sequence analysis of the DNA encoding the *Eco*RI endonuclease and methylase. *J Biol Chem* 1981;256:2143–2153.

Gruenbaum Y, Cedar H, Aharon R: Restriction enzyme digestion of hemimethylated DNA. *Nucl Acids Res* 1981;9:2509–2515.

Günthert U, Storm K, Bald R: Restriction and modification in *Bacillus subtilis*. Localization of the methylated nucleotide in the *Bsu*RI recognition sequence. *Eur J Biochem* 1978;90:581–583.

Günthert U, Freund M, Trautner TA: Restriction and modification in *Bacillus subtilis:* Two DNA methyltransferases with *Bsu*RI specificity. *J Biol Chem* 1981a; 256:9340–9345.

Günthert U, Jentsch S, Freund M: Restriction and modification in *Bacillus subtilis:* Two DNA methyltransferases with *Bsu*RI specificity. *J Biol Chem* 1981b; 256:9346–9351.

Hattman S, Keister T, Gottehrer A: Sequence specificity of DNA methylases from *Bacillus amyloliquifaciens* and *Bacillus brevis. J Mol Biol* 1978;124:701–711.

Heininger K, Horz W, Zachau HG: Specificity of cleavage by a restriction nuclease from *Bacillus subtilis. Gene* 1977;1:291–303.

Herman GE, Modrich P: *Escherichia coli dam* methylase: Physical and catalytic properties of the homogeneous enzyme. *J Biol Chem* 1981;257:2605–2612.

Horiuchi K, Zinder ND: Site-specific cleavage of single-stranded DNA by a *Haemophilus* restriction endonuclease. *Proc Nat Acad Sci USA* 1975;72:2555–2559.

Hsu M, Berg P: Altering the specificity of restriction endonuclease: Effect of replacing Mg^{2+} with Mn^{2+}. *Biochemistry* 1978;17:131–138.

Hughes SG, Murray K: The nucleotide sequences recognized by endonucleases *Ava*I and *Ava*II from *Anabaena variabilis. Biochem J* 1980;185:65–75.

Jack WE, Rubin RA, Newman A, Modrich P: Structures and mechanisms of *Eco*RI DNA restriction and modification enzymes, in Chirikjian JG (ed): *Gene Amplification and Analysis: Restriction Endonucleases.* New York, Elsevier-North Holland, 1981, vol 1, pp 165–181.

Jack WE, Terry BJ, Modrich P: Involvement of outside DNA sequences in the major kinetic path by which *Eco*RI endonuclease locates and leaves its recognition sequence. *Proc Nat Acad Sci USA* 1982;79:4010–4014.

Janulaitis A, Povilionis P, Sasnauskas K: Cloning of the modification methylase gene of *Bacillus centrosporus* in *Escherichia coli.* 1982;*Gene* 20;197–204.

Jentsch S, Günthert U, Trautner TA: DNA methyltransferases affecting the sequence 5′CCGG. *Nucl Acids Res* 1981;9:2753–2759.

Keshet E, Cedar H: Effect of CpG methylation on *Msp*I. *Nucl Acid Res* 1983;11:3571–3580.

Kiss A, Boldauf F: Molecular cloning and expression in *Escherichia coli* of two modification methylase genes of *Bacillus subtilis. Gene* 1983;21:111–119.

Kosikh VG, Buryanov YI, Bayer AA: Cloning of the genes of *Eco*RII endonuclease and methylase. *Dokl Akad Nauk USSR* 1979;247:1269–1271.

Kramarov VM, Smolyaninov VV: DNA methylase from *Arthrobacter luteus* screens DNA from the action of site-specific endonuclease *Alu*I. *Biokhimiya* 1981;46:1526–1529.

Lacks S, Greenberg B: Complementary specificity of restriction endonucleases of *Diplococcus pneumoniae* with respect to DNA methylation. *J Mol Biol* 1977;114:153–168.

Levy WP, Welker NE: Deoxyribonucleic acid modification methylase from *Bacillus stearothermophilus. Biochemistry* 1981;20:1120–1127.

Lu A-L, Jack WE, Modrich P: DNA determinants important in sequence recognition by *Eco*RI endonuclease. *J Biol Chem* 1981;256:13200–13206.

Malyguine E, Vannier P, Yot P: Alteration of the specificity of restriction endonucleases in the presence of organic solvents. *Gene* 1980;8:163–177.

Mann MB, Smith HO: Specificity of *Hpa*II and *Hae*III DNA methylases. *Nucl Acids Res* 1977;4:4211–4222.

Mann MB, Rao RN, Smith HO: Cloning of restriction and modification genes in *E. coli:* The *Hha*II system from *Haemophilus haemolyticus.* 1978; *Gene* 3:97–112.

Mann MB, Smith HO: Specificity of DNA methylases from *Haemophilus sp,* in Usdin E, Borchardt R, Kreveling C (ed): *Transmethylation.* New York, Elsevier-North Holland, 1979, p 483.

Mayer H: Optimization of the *Eco*RI*-activity of *Eco*RI endonuclease. *FEBS Letters* 1978;90:341–344.

McClelland M: The effect of sequence specific DNA methylation on restriction endonuclease cleavage. *Nucl Acids Res* 1981;9:5859–5866.

McClelland M: Purification and characterization of two new modification methylases: M·*Cla*I from *Caryophanon latum* L and M·*Taq*I from *Thermus aquaticus* YT1. *Nucl Acids Res* 1981;9:6795–6804.

McClelland M: The effect of site specific methylation on restriction endonuclease cleavage (update). *Nucl Acids Res* 1983;11:r169–r173.

Modrich p. Zabel D:*Eco*RI endonuclease:Physical and catalytic properties of the homogenous enzyme. *J Biol Chem* 1976;2515866-5874.

Modrich P, Rubin RA: Role of the 2-amino group of deoxyguanosine in sequence recognition by *Eco*RI restriction and modification enzymes. *J Biol Chem* 1977;252:7273–7278.

Modrich P: Structures and mechanisms of DNA restriction and modification enzymes. *Q Rev Biophys* 1979;12:315–369.

Modrich P: Structures and mechanisms of DNA restriction and modification enzymes. *Q Rev Biophys* 1979;12:315–369.

Modrich P, Roberts RJ: Type II restriction and modification enzymes, in Linn SM, Roberts RJ (eds): *Nucleases.* Cold Spring Harbor Laboratory, New York, 1982; pp 109–154.

Nardone G, George J, Chirikjian J: Properties and substrate specificity of *Bam*HI methylase and endonuclease. *Fed Proc Abstr* 1983;42:21–51.

Newman AK, Rubin RA, Kim S-H, Modrich P: DNA sequence of structural genes for *Eco*RI DNA restriction and modification enzymes. *J Biol Chem* 1981;256:2131–2137.

Noyer-Weidner M, Jentsch S, Pawleh B, Günthert U, Trantner TA: Restriction and modification in *Bacillus subtilis:* DNa methylation potential of the related bacteriophages Z, SPR, SPβ, ϕ3T, and p11. *J Virol* 1983;46:446–453.

O'Connor CD, Humphreys GO: Expression of the *Eco*RI restriction-modification system and the construction of positive-selection cloning vectors. *Gene* 1982;20:219–229.

Ohlendorf DH, Anderson WF, Fisher RG, Takeda Y, Mathews BW: The molecular basis of DNA protein recognition inferred from the structure of cro repressor. *Nature* 1982;298:718–723.

Ohmuri H, Tomizawa J, Maxam AM: Detection of 5-methylcytosine in DNA sequences. *Nucl Acid Res* 1978;5:1479–1485.

Pabo CO, Lewis M: The operator-binding domain of λ-repressor, structure and DNA recognition. *Nature* 1982;298:443–447.

Petrusyte MP, Janulaitis AA: Specific methylase from *Bacillus centrosporus*. *Bioorg Khim USSR* 1981;7:1885–1887.

Polisky B, Greene P, Garfin DE, McCarthy BJ, Goodman HM, Boyer HW: Specificity of substrate recognition by the *Eco*RI restriction endonuclease. *Proc Nat Acad Sci USA* 1975;72:3310–3314.

Quint A, Cedar H. In vitro methylation of DNA with *Hpa*II methylase. *Nucl Acids Res* 1981;9:633–646.

Razin A, Riggs AD: DNA methylation and gene function. *Science* 1980;210:604–610.

Razin A, Urieli S, Pollack Y, Gruenbaum Y, Glaser G: Studies on the biological role of DNA methylation; IV. Mode of methylation of DNA in *E. coli* cells. *Nucl Acids Res* 1980;8:1783–1797.

Roberts RJ: Restriction and modification enzymes and their recognition sequences. *Nucl Acids Res* 1983;11:r135–r173.

Rosenberg JM, Boyer HW, Greene PJ: The structure and function of the *Eco*RI restriction endonuclease, in Chirikjian JG (ed): *Gene Amplification and Analysis*. New York, Elsevier-North Holland, 1981; vol 1. pp 131–164.

Rosenberg JM, Greene P: *Eco*RI* specificity and hydrogen bonding. *DNA* 1982;1:117–124.

Roy PH, Smith HO: The DNA methylases of *Haemophilus influenzae* Rd. I. Purification and properties. *J Mol Biol* 1973a;81:427–444.

Roy PH, Smith HO: The DNA methylases of *Haemophilus influenzae* Rd. II. Partial recognition site base sequences. *J Mol Biol* 1973b;81:445–459.

Rubin RA, Modrich P: *Eco*RI methylase. Physical and catalytic properties of the homogeneous enzyme. *J Biol Chem* 1977;252:7265–7272.

Rubin RA, Modrich P, Vanaman TC: Substrate dependence of the mechanism of *Eco*RI endonuclease. *Nucl Acids Res* 1978;5:2991–2997.

Sato S, Nakazawa K, Shinomija T: A DNA methylase from *Thermus thermophilus* HB8. *J Biochem* 1980;88:737–747.

Sauer RT, Yocum RR, Doolittle RF, Lewis M, Pabo CO: Homology among DNA-binding proteins suggests use of a conserved super-secondary structure. *Nature* 1982;298:447–451.

Schoner B, Kelly S, Smith HO: The nucleotide sequence of the *Hha*II restriction and modification genes from *Haemophilus haemolyticus*. *Gene* 1983;24:227–236.

Siebenlist U, Simpson RB, Gilbert W: *E. coli* RNA polymerase interacts homologously with two different promoters. *Cell* 1980;20:269–281.

Smith HO: Nucleotide sequence specificity of restriction endonucleases. *Science* 1979;205:455–466.

Streeck RE: Single-strand and double-strand cleavage at half-modified and fully modified recognition sites for the restriction nucleases *Sau*3A and *Taq*I. *Gene* 1980;12:267–275.

Sussenbach JS, Monfoort CH, Schiphof R, Stobberingh EE: A restriction endonuclease from *Staphylococcus aureus*. *Nucl Acids Res* 1976;3:3193–3202.

Szomolanyi E, Kiss A, Venetianer P: Cloning the modification methylase gene of *Bacillus sphaericus* R in *Escherichia coli*. *Gene* 1980;10:219–225.

Trautner TA, Pawlek B, Bron S, Anagnostopoulos C: Restriction and modification in *B. subtilis:* Biological aspects. *Mol Gen Genet* 1974;131:181–191.

Trautner TA, Pawlek B, Günthert U, Canosi U, Jentsch S, Freund M: Restriction and modification in *Bacillus subtilis:* Identification of a gene in the temperate phage SPβ coding for a *Bsu*R specific modification methyltransferase. *Mol Gen Genet* 1980;180:361–367.

vander Ploeg LHT, Flavell RA: DNA methylation in the human $\gamma\delta\beta$-globin locus in erythroid and nonerythroid tissues. *Cell* 1980;19:947–958.

Vanyushin BF, Dobritsa AP: On the nature of the cytosine-methylated sequence in DNA of *Bacillus brevis* var. G.-B. Biochim. et Biophys. *Acta* 1975;407:61–72.

Vardimon L, Kressman A, Cedar H, Maechler M, Doerfler W: Expression of a cloned adenovirus gene is inhibited by in vitro methylation. *Proc Nat Acad Sci USA* 1982;79:1073–1077.

Venetianer P, Kiss A: The restriction-modification enzymes of *Bacillus sphaericus* R. *Nucl Acids Res* 1981;9:209–215.

Vovis GF, Horiuchi K, Zinder ND: Kinetics of methylation by a restriction endonuclease from *Escherichia coli* B. *Proc Nat Acad Sci USA* 1974;71:3810–3813.

Walder RY, Hartley JL, Donelson JE, Walder JA: Cloning and expression of the *Pst*I restriction-modification system in *Escherichia coli. Proc Nat Acad Sci USA* 1981;78:1503–1507.

Walder RY, Langtimm CJ, Chatterjee R, et al: Cloning of the *Msp*I modification enzyme. *J Biol Chem* 1983;258:1235–1241.

Winter RB, Berg OG, vonHippel PH: Diffusion-driven mechanisms of protein translocation on nucleic acids. 3. The *Escherichia coli lac* repressor-operator interaction: Kinetic measurements and conclusions. *Biochemistry* 1981;20:6961–6977.

Woodbury CP Jr, Hägenbuckle O, vonHippel PH: DNA site recognition and reduced specificity of the *Eco*RI endonuclease. *J Biol Chem* 1980;255:11534–11546.

Woodbury CP Jr, vonHippel PH: Relaxed sequence specificities of *Eco*RI endonuclease and methylase: Mechanisms, possible practical applications, and uses in defining protein-nucleic acid recognition systems, in Chirikjian JG (ed): *Gene Amplifications and Analysis.* New York, Elsevier-North Holland, Inc, 1981, vol 1, pp 181–207.

Yoo OJ, Agarwal KL: Isolation and characterization of two proteins possessing *Hpa*II methylase activity. *J Biol Chem* 1980;255:6445–6449.

Yoo OJ, Dwyer-Hallquist P, Agarawal KL: Purification and properties of the *Hpa*I methylase. *Nucl Acids Res* 1982;10:6511–6519.

Zappia V, Carteni-Farina M, Porcelli M: Biochemical and chemical aspects of decarboxylated S-adenosylmethionine, in Usdin E, Borchardt RT, Creveling CR (eds): *Transmethylation.* New York, Elsevier-North Holland, Inc, 1979; pp 95–104.

4

The Restriction and Modification DNA Methylases: An Overview

Robert Yuan*
Hamilton O. Smith†

The preceding two chapters have presented in detail all that is known about the genetic and enzymatic mechanisms of the modification DNA methylases and their relationship to the homologous restriction endonucleases. Although the number of systems that have been characterized is limited, sufficient information is available to allow a comparison of the three types of restriction-modification systems from both biological and mechanistic viewpoints.

Biological Function

What is the biological function of these enzymatic systems? The type I and type III systems originally were characterized as classic restriction-modification systems; that is, they degraded infecting phage DNA, thereby greatly decreasing the efficiency of phage infection. The type I (*Escherichia coli* K and B) restriction-modification genes are located on the chromosome, whereas those of type III can be present on a phage (P1), a plasmid (P15), or the chromosome (*Haemophilus influenzae* Rf). The complexity of the type I enzymes and the reactions that they catalyze has resulted in suggestions that they are involved in other functions (e.g., transposition or recombination). No direct evidence for these alternate functions has emerged. Restriction-modification mutants showed no other detectable phenotype. Furthermore, *E. coli* C (a strain naturally lacking both restriction and modification) showed no unusual

*NCI-Frederick Cancer Research Facility, P. O. Box B, Frederick, Maryland 21701
†Department of Moleculer Biology and Genetics, The Johns Hopkins University School of Medicine, Baltimore, Maryland 21205

phenotype related to transposition, recombination, and other DNA processes. An analogous situation exists with the type III systems for which there is also no evidence of alternate functions. Conversely, the biological function of the type II enzymes has remained more obscure. These enzymes, in general, have been characterized biochemically as sequence-specific endonucleases and have been classified as restriction enzymes. The corresponding methylases have also been detected in a number of instances. In only a few cases (e.g., *Eco*RI and *Eco*RII) have the systems been shown to be involved in classic restriction-modification. Conversely, the *Hha*I and *Pst*I systems have been cloned in *E. coli,* and they restrict phage DNA very efficiently in the new host. Thus, while these systems seem capable of establishing a biological restriction-modification phenotype, in most cases this has not been examined in the strains from which they were isolated.

One curious fact is that the restriction-modification systems are relatively rare. Only one new type I system was found out of some 40 *E. coli* isolates (Murray *et al.* 1982), and only 250 type II restriction enzymes (and presumably the corresponding methylases) were found in the approximately 700 strains screened (R. Roberts, personal communication). In a larger context, the frequency of restriction-modification systems is of the same order as other genes that confer resistance to antibiotics and toxic metals; the surprising fact is their chromosomal location. Therefore, the methylases involved in restriction-modification do not appear to be essential for the survival of the cell. Their primary function is to regulate the flow of genetic information from one strain to another. The existence of such a system may depend solely on the presence of phages in the ecologic niche occupied by the particular strain.

Genetic and Protein Structure

What is the genetic organization of the restriction-modification loci and how is it reflected in the structure of the encoded enzymes? The type II systems are the simplest; they are composed of two genes—one specifying the restriction enzyme and the other the modification enzyme. The two genes can be arranged in different configurations: totally separate, contiguous with separate promoters oriented in the same (e.g., *Eco*RI) or in opposite (e.g., *Pst*I) directions, contiguous with a single promoter (e.g., *Pae* R7), or noncontiguous (e.g., the *Bacillus* systems). It originally had been assumed that the restriction-modification genes in a given strain had arisen from a common ancestor and could be expected to have large regions of homology (Boyer, 1971); however, in the type II systems that have been characterized to date, no homology between the restriction and methylase genes has been detected at either the DNA or protein levels. Furthermore, different enzyme systems (isoschizomers) that recognize the same DNA sequence do not seem to be related to each other. This strongly suggests that different organisms have evolved to produce different proteins that interact with the same nucleotide sequence. Therefore, a given DNA sequence can interact with protein conformations generated by different amino acid sequences.

The type I and type III systems consist of a single genetic locus but they differ in the number of genes. The type I *hsd* locus has three genes (*hsdR*, *hsdS*, and *hsdM*) and two separate promoters—one controlling *hsdR* and the other regulating both *hsdM* and *hsdS*. Transcripts of *hsdR* and *hsdM-hsdS*, as well as a polycistronic transcript of all three genes, have been detected (Sain and Murray, 1980). The most important feature of these transcripts is the presence of the specificity element in one gene, *hsdS*. The *hsdR* and *hsdM* genes of different systems in *E. coli* are functionally interchangeable and basically homologous. The *hsdS* genes have only two short conserved sequences; one is in the middle and the other one codes for the carboxyl end of the protein. These two short conserved sequences have been assumed to be involved in the binding of the *hsdS* subunit to the *hsdR* and *hsdM* subunits.

The type III restriction-modification locus has two genes, *res* and *mod*, with separate promoters. Transcripts of the two individual genes, as well as a polycistronic transcript, have been detected (Iida *et al.* 1983). In contrast to the type I system, the recognition function is located in the *mod* gene. The *res* genes from different systems are homologous, whereas the *mod* genes have a large region that is not homologous except for the amino and carboxyl termini. Presumably, this nonhomologous region is responsible for DNA sequence specificity. There is no apparent homology between the type I and type III systems.

At the level of gene structure and expression, all three systems are capable of synthesizing individual restriction and methylase activities. The type II systems, which are the simplest, produce independent gene products; each one is capable of recognizing the same nucleotide sequence. Increasing complexity occurs when the recognition function is encoded in the *mod* gene and is shared with the *res* gene (type III). The *res* gene by itself cannot restrict, whereas the *mod* gene by itself can only modify, but not restrict. In the type I system, the recognition function is on a gene by itself (*hsdS*). Modification requires *hsdS* and *hsdM*, but restriction requires all three genes.

At the level of protein structure, the type II methylases are the more simple ones. Available evidence indicates that they act as monomers, while their counterpart restriction enzymes act as dimers. The type III methylases, as well as the type I methylases, can act in two oligomeric forms—a simpler one that catalyzes only methylation and a more complex one that has both endonuclease and methylase activities (Table 4.1). The structure and function of these enzymes is consistent with their gene structure. The biological significance of these transcriptional patterns and the various protein oligomers is not clear; however, in all systems, they allow DNA methylation to take place independently of restriction. The presence of the genes on a phage or plasmid would allow modification of the host chromosome before it could be restricted. The two oligomeric forms of the type I and type III methylase activities differ in

Table 4.1 Relationship Between Enzyme Structure and Enzyme Activities

Type	Enzyme Structure	Enzyme Activity
I	(a) Modification methylase: recognition and modification subunits	DNA methylation
	(b) Restriction endonucleases: restriction, modification and recognition subunits	DNA methylation and DNA cleavage
II	Modification methylase	DNA methylation
III	(a) Modification methylase: modification subunit	DNA methylation
	(b) Restriction endonuclease: restriction and modification subunits	DNA methylation and DNA cleavage

their catalytic properties (see section in this chapter on "Reaction Mechanism"); this may allow the modification reaction to be modulated in terms of specificity and reaction rates.

Evolution of Restriction-Modification Systems

How have these systems evolved? It is difficult to imagine a way in which both the restriction and modification enzymes of a system could have evolved simultaneously. Therefore, speculation on the possible evolutionary pathways can follow one of two directions: 1) modification preceding restriction or 2) restriction preceding modification.

Let us assume that a cell possesses a specific DNA methylase. This type of enzyme could function by modulating binding of a regulatory protein to its DNA site. Duplication of the gene for such a protein, along with other evolutionary alterations, could result in the appearance of an endonuclease with the same sequence specificity as the methylase. The end-result would be a new restriction-modification system capable of restricting foreign DNA and simultaneously protecting its own DNA. This and other similar scenarios seem possible, but suffer from a lack of direct evidence.

Alternatively, the restriction-modification systems could have originated with a specific phage-coded DNAse that served to digest host DNA. A bacterial strain with a DNA methylase whose specificity overlapped or coincided with that of the phage DNAse would have had an increased probability of survival. This strain could have slowly evolved into one in which the chromosome contained both a DNAse gene and a methylase gene that recognized the same DNA sequence. Differential rates of transcription then might have provided selective pressure for a single genetic locus encompassing both genes. In this case, however, mutation of the methylase gene would lead to a r^+m^- phenotype and to cell death. Further changes would lead to strains in which the restriction gene no longer coded for an active enzyme by itself. Then, the protein coded by the restriction gene might lack a recognition function and, therefore, act only by binding to the modification polypeptide and by using its recognition domain. Such an oligomeric protein would have the disadvantage of expressing both DNA cleavage and methylation simultaneously. By evolving one step further, the DNA recognition function would reside in a single gene, and a mutation in that gene would result in the simultaneous loss of restriction and modification. The same result would occur with a mutation in the modification gene. Furthermore, the recognition polypeptide would be able to discriminate between a methylated, an unmethylated, and a hemimethylated DNA sequence. By this process, certain bacterial strains could have acquired the capability both to protect their own DNA and to degrade foreign DNA by using a mechanism provided with a series of fail-safe features that protect its own DNA during DNA replication or following mutations in the restriction-modification locus.

The first of these evolutionary pathways would have resulted in restriction-modification genes with regions of homology. No homology has been found, which suggests the absence of a common ancestral system and the possibility of a convergent pathway as described above.

DNA Sequences

What are the structural features of the DNA sequences that are recognized by these systems? Table 4.2 shows the general sequences that are recognized by the various systems. All systems that have been examined to date are methylated at the recognition sequences. The methylation of type I and type II sites can take place on either unmethylated or hemimethylated sites (i.e., by methyl transfer to either one or both strands). The type III sequences are asymmetric and can only be methylated on one strand. Six type II sequences of the XXXXX class and one of the XXXX class have three sequences in which A is present on only one strand and five sequences in which C is present on only one strand. Unfortunately, no methylation data is available; however, it is not

Table 4.2 Structure of DNA Recognition Sequences

Type	General Structure of DNA Sequence
I	X X X (N_6 or N_8) X X X X
II	(a) X Y Z Z′Y′X′
	(b) X Y Z (N_1–N_6) Z′Y′X′
	(c) X X X X or X X X X X
III	X X X X X or X X X X X X

X, Y, and Z are defined bases and X′, Y′, and Z′ are the complementary ones. They are specified for each restriction-modification system. N_1–N_6 represent spacers of defined length, but of random sequence.

unlikely that methylation could occur on only one strand as it does in the type III sites.

If one looks at DNA cleavage by the corresponding restriction activity, there is a correlation between the structure of the recognition site and the site of DNA cleavage. The asymmetric, hyphenated structure of type I sites leads to cleavage thousands of bases downstream. Symmetric (hyphenated or nonhyphenated) sequences of type II sites result in cleavage within the sequence, whereas asymmetric sequences of the type II sites result in cleavage five to eight bases downstream. Asymmetric sequences of type III result in cleavage 25–27 bases downstream. Cleavage that is more than one helical turn away coincides with the appearance of a requirement for ATP as a cofactor. Cleavage at a distance of more than 1,000 base pairs away not only requires ATP, but it is also dependent on its hydrolysis. If we believe that the type I system is a more highly evolved one, is cleavage at a distance more efficient? There is some indication that this may be the case. In the type I systems, the recognition sites are not destroyed by the restriction reaction; potentially, they can be used more than once. Also, they produce termini that are not amenable to enzymatic ligation. These two aspects of the type I mechanism may result in a more effective inactivation of foreign DNA *in vivo.*

Enzyme Mechanism

How many different mechanisms of DNA methylation are known and how do they relate to biological function? All three methylase systems use AdoMet as the methyl donor; in all three systems, there is good evidence that AdoMet binding by the methylase is the initial step in the reaction. In the case of type I and type III methylases, this AdoMet binding may be a requirement for recognition of the DNA sequence.

The actual methyl transfer is defined by the DNA substrate. In type I systems, hemimethylated DNA is a good substrate for both the complete restriction enzyme and the methylase; unmethylated DNA is not. In type II systems both hemimethylated and unmethylated DNAs are good substrates. In type III systems, the DNA substrate is by definition unmethylated, since only one strand can be methylated. The methylation of type II sequences is carried out by a methylase monomer, and one methyl group is transferred per binding event. Therefore, the conversion of an unmethylated sequence to a methylated sequence requires two independent events. The methylation of type I and type III sites requires only the transfer of one methyl group per binding event to result in a methylated DNA. This reaction is strongly stimulated by ATP. This ATP effect seems to be related to a fast release of the enzyme from the product.

The specificity of each enzyme system for hemimethylated versus unmethylated sequences probably reflects the temporal relationship between DNA synthesis and methylation. If a methylase system acts only on hemimethylated or unmethylated DNA of the type III class, then it must be fairly tightly coupled to DNA synthesis. Otherwise, two rounds of replication would lead to both a loss of the methylation pattern in 50% of the DNA molecules and the resulting cleavage by the homologous restriction enzyme. This kind of tight coupling would be particularly advantageous to chromosomal replication. There is evidence to show that type I methylation is incomplete in lambda phage, suggesting that phage replication has outpaced methylation. Conversely, if one assumes that type II methylases originated with phages, it would be clearly advantageous for them to act on totally unmethylated DNA, which would consist of unmodified host DNA and rapidly replicating phage DNA.

Antirestriction Systems

What mechanisms exist for the suppression of restriction systems? As discussed earlier, many phages have mechanisms for overcoming restriction. Although these mechanisms have been shown to occur mostly with types I and III, there is no reason to believe that they are not as common with type II systems. In fact, all three kinds of antirestriction mechanisms occur for the three restriction-modification systems: 1) the presence of unusual bases on the DNA; 2) the inhibition of restriction activity; and 3) expression of methylase activity. The most dramatic form of the last mechanism relates to the type II system. The *Bacillus subtilis* phages SPβ and ϕ3T carry a methylase gene that has the same specificity as the host methylase. The expression of the phage gene results in the protection of the phage DNA from the corresponding restriction activity. Interestingly enough, the chromosomal and phage methylases are different, even though they share the same specificity.

In a sense, this brings us full cycle from the presumed phage origins of the restriction-modification system and its adaptation by the host for its own protection. The phages now have further evolved to overcome the restriction system of the host. Two of the mechanisms used by phages to do this are alteration of the host methylase system and development of their own methylase with a specificity similar to that of the host.

The DNA methylases involved in restriction-modification have been the most extensively studied methylase enzymes, both at the genetic and biochemical levels. They provide precedents for alternative mechanisms of DNA modification that must be thoughtfully considered as researchers focus on the role of DNA methylation in eukaryotic cells. For example, in some systems, the methylase acts specifically on hemimethylated DNAs; whereas in other systems, the methylase can modify both unmethylated and hemimethylated DNAs equally well. Therefore, it is conceivable that some eukaryotic cells might also contain methylases that can act on both hemimethylated and unmethylated DNAs, and that others might contain separate *de novo* and maintenance methylases. The *ral* system of lambda also shows that a methylase system can be altered from one type of specificity to the other. Although there has been no evidence for demethylase activity in prokaryotes, the existence of antirestriction systems that inhibit or alter methylase activity could provide, in conjunction with DNA replication, an example of demethylation that might also occur in higher organisms. Lastly, but most importantly, these bacterial methylases provide impressive evidence for the large biological effects that result from the presence or absence of methyl groups on the DNA.

Acknowledgments

Research was sponsored by the National Cancer Institute, Department of Health and Human Services, under Contract No. NO1-CO-23909 with Litton Bionetics, Inc. The contents of this publication do not necessarily reflect the views or policies of the Department of Health and Human Services, nor does the mention of trade names, commercial products, or organizations imply endorsement by the United States Government.

References

Boyer HW: DNA restriction and modification mechanisms in bacteria. *Annu Rev Microbiol* 1971;25:153–176.

Iida S, Meyer B, Bächi B, Stolhammar-Carlemalm M, Schrickel S, Bickle TA, Arber W: DNA restriction-modification genes of phage P1 and plasmid p15B: Structure and *in vitro* transcription. *J Mol Biol* 1983;165:1–18.

Murray NE, Gough JA, Suri B, Bickle TA: Structural homologies among Type I restriction-modification systems. *EMBO J* 1982;1:535–539.

Sain B, Murray NE: The *hsd* (host specificity) genes of *Escherichia coli* K12. *Mol Gen Genet* 1980;180:35–46.

5

Methylation of Prokaryotic DNA

M. G. Marinus*

The biological function of methylated bases in DNA of prokaryotes appears to be quite different than that of eukaryotes. In this chapter, most of the information presented is derived from studies with *Escherichia coli* K-12 simply because more is known about DNA methylation in this organism than in any other one. Some data from certain *E. coli* bacteriophages also will be reviewed, in addition to selected aspects about DNA methylation in certain other prokaryotes. Other recent reviews that complement this one are by Razin and Friedman (1981) and Hattman (1981).

General Aspects of *E. coli* DNA Methylation

Methylated Bases in DNA

Two methylated bases have been detected in *E. coli* DNA (Dunn and Smith, 1955; Doskocil and Sormova, 1965): 5-methylcytosine (5-meCyt) and 6-methyladenine (6-meAde). These bases are the products of reactions catalyzed by three enzymes that are specified by the *hsd* (host specificity), *dam* (DNA adenine methylation), and *dcm* (DNA cytosine methylation) genes (Table 5.1). Since the *hsd* methylase has been described in Chapter 2 of this book, it will not be discussed here. All the 6-meAde and 5-meCyt in *E. coli* K-12 can be accounted for by the action of the above methylases. This conclusion is drawn from a mutant strain devoid of these enzymes, in which there is no detectable

*Department of Pharmacology, University of Massachusetts Medical School, 55 Lake Avenue North, Worcester, Massachusetts 01605

Abbreviations: Cyt = cytosine, C = cytidine, Ade = Adenine, and 5-aza-C = 5 azacytidine.

DNA methylation (Marinus *et al.* 1984). Table 5.1 also shows that the *dam* recognition sequence is the most frequent, while that for *hsd* is the least frequent sequence. This, in turn, is reflected by the amount of methylated base found *in vivo,* the 6-meAde produced by *dam* being the most abundant, and that produced by the *hsd* methylase being the least abundant.

Strain and Sequence Specificity of DNA Methylases

Early studies of the enzymology of DNA methylation (reviewed by Borek and Srinivasan, 1966) revealed that methylation was strain- and species-specific; that is, DNA methylases isolated from *E. coli* could not methylate DNA from *E. coli in vitro* (all sites were methylated), but they could act on DNA from unrelated organisms (Gold *et al.* 1963; Fujimoto *et al.* 1965). The discovery that *relA* mutants of *E. coli,* when starved for methionine, continued to replicate DNA—but not to methylate it—allowed the isolation of undermethylated DNA. This undermethylated DNA was a substrate for the homologous enzymes. The methylation reaction occurred on double-stranded DNA in the presence of S-adenosyl-L-methionine (SAM), and it resulted in the formation of 6-meAde and 5-meCyt. These key observations led to a model in which methylation occurs at or near the replication fork on double-stranded DNA subsequent to DNA synthesis. Later studies have validated this model.

In these early studies, the enzyme activities for the production of 5-meCyt and 6-meAde could not be purified away from each other. This may indicate that the methylases exist in a complex *in vivo,* possibly together with other modifying or replication enzymes.

As mentioned above, DNA methylases from *E. coli* are species-specific. The fact that DNA from unrelated organisms (e.g., calf thymus) could be methylated to saturation indicated that the enzymes acted in a *de novo* manner to methylate sites on both DNA strands; that is, totally unmethylated (with respect to adenine) DNA was as efficient a substrate as hemimethylated DNA (isolated from *relA* mutants). This indicated that the *dam* methylase (and the *dcm* methylase) did not require hemimethylated DNA for activity and, thus,

Table 5.1 DNA Methylases of *E. coli* K-12

Gene(s)	Recognition Sequence	Theoretical Frequency	Product
hsdS,R,M	–AA*[a]C(N_6)GTCG–	1/46,000	6-meAde
dam	–GA*TC–	1/256	6-meAde
dcm	–CC*(A/T)GG–	1/512	5-meCyt

[a]The asterisks denote the site of methylation.

were not "maintenance" methylases (Borek and Srinivasan, 1966). Since homologous (i.e., *E. coli*) DNA could not be methylated by the *E. coli* enzymes, it was concluded that at least 99% of the *dam* and *dcm* sequences were methylated *in vivo* (Borek and Srinivasan, 1966).

Methylation as a Postreplicative Event

The *in vitro* enzymologic studies referred to above were compatible with a model in which methylation occurred as a postreplicative process and resulted in the methylation of all available sites (Borek and Srinivasan, 1966). Studies by Billen (1968) and Lark (1968a) confirmed the model *in vivo* in *E. coli* by analyzing DNA on cesium chloride gradients after various treatments. It was found that the nascent daughter strand is methylated as DNA is replicated. Since less than 3% of the total DNA had been replicated in the minimum time allowed for resolution of the method, it was concluded that methylation normally occurs at/or near the replication fork and only on the daughter DNA strand. If *E. coli* was starved of methionine, DNA synthesis continued, but such DNA was not methylated. On restoration of methionine, DNA methylation occurred before subsequent replication. In normal cells (not starved), DNA methylation was concomitant with synthesis; hence, the two processes probably are coupled.

In summary; 1) there are three DNA methylases in *E. coli* K-12 that can account for all the 5-meCyt and 6-meAde in DNA; 2) these enzymes act at/ or near the replication fork and transfer the methyl group from SAM to specific sequences in the daughter DNA strand; 3) they are species- and strain-specific and can initiate methylation *de novo;* and 4) at least 99% of the substrate sequences are fully methylated *in vivo.*

Demethylation

There is no evidence for enzymatic demethylation of 5-meCyt or 6-meAde from *E. coli* DNA either *in vivo* or *in vitro.* However, agents interfering with DNA methylation may cause loss of methyl groups. Methionine analogs (e.g., ethionine) block DNA methylation by preventing formation of SAM and by inhibition of methylase activity by S-adenosyl ethionine. Use of such agents blocks both DNA methylation *in vivo* and all other methylation reactions that are dependent on SAM. Similarly, nicotinamide blocks DNA methylation *in vivo* and *in vitro* in *E. coli* C and in phage ϕX174-infected cells (Razin *et al.* 1975). This compound does not block DNA methylation in *E. coli* K-12, and it presumably blocks growth by another mechanism (Marinus, unpublished data). Use of methionine analogs or nicotinamide may lead to incorrect interpretations of data, because the agents are not specific inhibitors of DNA methylation in the K-12 strain of *E. coli.*

DNA Methylation Mutants of *E. coli* K-12

The information above, although extremely useful, did not shed any light on the biological functions of the *dam* and *dcm* methylases. In order to uncover some of these functions, Marinus and Morris (1973) and Hattman *et al.* (1973) isolated mutant strains of *E. coli* K-12 that were deficient in one or the other methylase. Hattman *et al.* (1973) isolated *dcm* (= *mec*) mutants by using a screening method based on the observation that phage lambda grown on *E. coli* K-12 is weakly restricted by cells carrying plasmid N3 (= *Eco*RII restriction system), whereas phage grown on *E. coli* B is severely restricted. This is due to the fact that *E. coli* B does not contain a *dcm* methylase (and, hence, has no 5-meCyt in DNA); consequently, phage propagated on it are restricted by the *Eco*RII enzyme of pN3. Phage propagated on *E. coli* K-12, however, do contain 5-meCyt in their DNA and, therefore, are partially restricted by pN3. *dcm* mutants of K-12 were recognized by the restriction of phage propagated on them by pN3.

In a different approach, Marinus and Morris (1973) screened DNA from mutagenized survivors because of their ability to accept methyl groups from a crude enzyme extract of wild-type *E. coli.* As mentioned above, no methyl transfer to DNA can occur under these conditions (because it already is fully methylated), but DNA from a mutant strain lacking *dam* or *dcm* methylase should allow methyl group transfer. This method allowed the isolation of both *dam* and *dcm* mutants and also of two in particular that carried the *dam-3* and *dcm-6* alleles. The strain described by Hattman *et al.* (1973) has been designated *dcm-12* (= *mec*). Strains with these three alleles have been widely used and have little, if any, residual enzyme activity (Marinus and Morris, 1973; Hattman *et al.* 1973; Lacks and Greenberg, 1977; Bale *et al.* 1979). There is one report of residual activity in the *dcm-6* strain (Szyf *et al.* 1982). Mutation *dcm-13* is a deletion that encompasses *supD* to *fla* (Kondoh and Ozeki, 1976), and it also was shown to be deleted for *dcm* (Bale *et al.* 1979). The viability of this strain indicates that the *dcm* gene product is not necessary for viability of *E. coli* K-12.

More *dam* mutants were isolated by screening for hyper-recombination (Marinus and Konrad, 1976; Konrad, 1977), by localized mutagenesis (Bale *et al.* 1979), and by increased precise excision of Tn*10* (Lundblad and Kleckner, 1982). The difficulties of fine structure mapping precluded a classical genetic analysis of these mutations to try to characterize the mutational defect. The *dam-3* strain originally was thought to contain some residual activity, but subsequent analysis revealed that there was no *dam* methylase activity above the limit of detection (Lacks and Greenberg, 1976; Bale *et al.* 1979). We recently have isolated and characterized two new *dam* mutations (Marinus *et al.* 1983). One of these has Mu*d* (*lac,* Ap) inserted into the *dam* gene (*dam-12*::Mu*d*) and the other Tn*9* (*dam-13*::Tn*9*). Strains containing these mutations have no detectable DNA adenine methylase *in vitro* or *in vivo.* Since these strains are viable, the data indicate that the *dam* gene is dispensable. Analysis by Southern blotting indicates that in the *dam-12*::Mu*d* strain, the *dam* gene has been disrupted. Using the *dam-12*::Mu*d* mutation, we have constructed a *dam*$^-$ *dcm*$^-$ *hsd*$^-$ strain that is viable and contains no detectable methylated bases in DNA. This indicates that DNA methylation in *E. coli* K-12 is not essential for viability.

The *dcm* mutations are mapped in a single gene and are recessive (Marinus, 1973). The gene order is *supD-dcm-fla,* which was determined by two factor crosses: F-prime and deletion mapping. All the *dam* mutations except one are mapped in a single complementation group and are recessive to wild type (Bale *et al.* 1979). The gene order is *trpS-dam-aroB* (Marinus, 1973).

Testing for Methylase Deficiency

The most simple and most reliable test for *dam* or *dcm* methylase deficiency is to examine the susceptibility of cellular, plasmid, or phage DNA to *MboI* or *Eco*RII enzymes, respectively. Methylated DNA molecules are resistant to digestion with these enzymes, but methyl deficient molecules are not. Another test for *dcm* methylase deficiency is restriction of phage grown on *dcm*$^-$ strains by cells harboring group N plasmids (see above).

Additional rapid tests for *dam* mutants include sensitivity to 2-aminopurine, N-methyl-N-nitro-N-nitrosoguanidine, or ultraviolet (UV) light. Alternatively, *mom* mutants of phage Mu grown on *dam*$^-$ strains are restricted by Pl lysogens to a greater extent than those grown in *dam*$^+$ bacteria (Touissaint, 1978). A papillation test on MacConkey-Lactose media that uses reversion of *lacZ118,* hyper-recombination in *lac,* or increased precise excision of transposons inserted in *lacZ* is also useful.

Use in Recombinant DNA Technology

Plasmid and phage DNA molecules do not contain methylated bases at *dam* and *dcm* sites when propagated on methylation-deficient hosts (Marinus and Morris, 1975). This allows DNA to be cleaved by restriction endonucleases whose activity may be inhibited by methylation. These restriction endonucleases include: *Atu*BI, *Atu*II, *Atu*CI, *Bcl*I, *Bss*GII, *Bst*GI, *Bst*GII, *Bsl*XI, *Bst*EIII, *Cpa*I, *Cpe*I, *Dpn*II, *Eca*II, *Ecl*II, *Eco*RII, *Fnu*AII, *Fnu*CI, *Mbo*I, *Mno*III, *Mos*I, *Mph*I, and *Nde*II (Roberts, 1982). Strains used for this purpose include GM31 (*dcm*$^-$), GM33 (*dam*$^-$), GM48 (*dam*$^-$ *dcm*$^-$), GM119 (*dam*$^-$ *dcm*$^-$), GM161 (*dam*$^-$ *hsdS*$^-$), GM215 (*dam*$^-$ *endA*$^-$), GM271 (*dcm*$^-$ *hsdR*$^-$), and GM272 (*dam*$^-$ *dcm*$^-$ *hsdS*$^-$). A *dam*$^-$ *recF*$^-$ strain can be used to prevent plasmid recombination (Marinus, unpublished data,).

Plasmid DNA, prepared by amplification using antibiotics, is fully methylated at *dam* and *dcm* sites in the case of pBR322 (Sutcliffe, 1978) and pBR325 (Marinus, unpublished data,). DNA sequencing by the Maxam-Gilbert procedure may be complicated by the presence of 5-meCyt, which is less reactive than cytosine to hydrazine. Bands corresponding to 5-meCyt do not appear in the cleavage patterns (Ohmori *et al.* 1978). In general, the use of methylation-deficient strains is useful in determining unknown DNA sequence or for restriction enzyme sites.

In site-specific mutagenesis, complications may arise due to mismatch repair (Kramer *et al.* 1982). This can be circumvented by: 1) using DNA molecules propagated in *dam* mutants; 2) transfecting DNA molecules into *dam*-over-producing strains; 3) use of *mutH, L,* or *S* recipients that are deficient in mismatch repair; or 4) enzymatic methylation of the product.

DNA Cytosine Methylation

dcm Methylase Substrate Sequence

The DNA sequence methylated by the *dcm* enzyme is –CC(A/T)GG–; it is the internal cytosine residue that is methylated (May and Hattman, 1975a, b). This DNA sequence is the same as that for the *Eco*RII methylase/endonuclease (Boyer *et al.* 1973). Unmethylated *dcm* sites are cleaved by *Eco*RII, while methylated ones are not, thus making this assay a convenient one for the presence or absence of methylation at these sites (Hattman, 1977). In *E. coli* K-12, the amount of 5-meCyt has been found to correlate with the theoretical frequency of –CC[A/T]GG-substrate sites (twice the amount of 6-meAde; Marinus and Morris, 1973; Hattman *et al,* 1973) although Szyf *et al* (1982) have found the level to be overrepresented. Therefore, it is unlikely that other substrate sites (e.g., –C[A/T]GG–) are methylated. Indeed there is no report of 5-meCyt sites in sequences other than –CC(A/T)GG– in *E. coli* K-12 or *E. coli* C.

The *dcm* and *Eco*RII methylases have been partially purified and shown to have similar properties (Hattman, 1977). It will be worthwhile to molecularly clone the *dcm* gene to obtain large amounts of enzyme for purification. It also would be interesting to compare the DNA sequence maps of the *dcm* and *Eco* RII methylase genes.

dcm Mutant Phenotype

One way to try to determine the function of 5-meCyt in DNA would be to compare cells that are devoid of this base to those with normal amounts. As mentioned above, *dcm* mutants devoid of 5-meCyt have been isolated. To date, no phenotypic abnormality has been associated specifically with the *dcm* mutation in *E. coli* K-12. *Escherichia coli* B cells also do not contain the *dcm* methylase. Since it has been demonstrated that the *dcm* methylase is not required for viability (see above), and that the *Eco*RII and *dcm* enzymes are very similar, it is possible that—in an *E. coli* K-12 ancestor—a group N plasmid (or part of it) integrated into the host chromosome. This integration left intact the *Eco*RII methylase that we now recognize as the *dcm* gene. This event occurred after divergence of the K-12 and B strains, or else the gene has been lost from B strains. DNA sequence analysis and the restriction map of the *dcm* chromosomal region may be useful in testing this prediction.

Mutagenic Deamination

Cytosine residues can spontaneously deaminate to form uracil. These mispaired uracil residues are removed by a specific repair system (Lindahl, 1982). Deamination of 5-meCyt, however, yields thymine that is not subject to repair by the uracil repair system; after the next round of replication, a G/C to A/T mutation should result. This will change the -CCAGG-sequence to -CTAGG-. If the UAG triplet is read in-frame, it will cause a nonsense mutation. Analysis of the *lacI* gene uncovered two major mutagenic hotspots for spontaneous amber mutations, both occurring in -CCAGG- sequences (Coulondre *et al.* 1978). Similar hotspots have been found in the *cI* gene of phage lambda (Lieb, 1981), but not in the *rpoB* gene of *E. coli* (Nene and Glass, 1982). Mutagenic hotspots are not found in *dcm*$^{-}$ strains, and creation of new *dcm* sites results in formation of mutagenic hotspots (Coulondre *et al.* 1978). Mutant strains lacking uracil-N-glycosylase activity (required for repair) do not show mutagenic hotspots (Duncan and Miller, 1980). In summary, there is considerable evidence for mutagenic deamination of 5-meCyt, although other factors also may be involved, since one *dcm* site in *lacI* is not a hotspot nor are any sites in *rpoB*.

Regulation of Gene Activity

The -35 region of the *araBAD* promoter contains the sequence -TCAGG-. DNA sequence analysis of nine promoter mutations (*araI*^{c}) indicated that, in each, the sequence was changed to -CCAGG- (Horwitz and Wilcox, unpublished data). These experiments were done in *E. coli* B (which is *dcm*$^{-}$), and no *araI*^{c} promoter mutations have been isolated from *E. coli* K-12. This suggests that methylation of the -CCAGG- sequence may be preventing gene expression of *araI*^{c} mutants in *E. coli* K-12. Preliminary experiments (Horwitz and Wilcox, unpublished data), indicate that a plasmid containing the *araI*^{c} mutation is expressed in *E. coli* B, but not in K-12. These results show that methylation of bases that are in critical sites may influence activity of nonessential genes. This type of regulation may be useful in inducible systems, where it is necessary to rapidly establish maximal gene expression.

The *lexA* repressor controls many regulatory circuits in *E. coli* (Little and Mount, 1982). The repressor can bind to the two SOS boxes in front of the *lexA* gene. The second SOS box contains a *dcm* site, and methylation at this site decreases the affinity of the repressor (R. Brent, personal communication). The biological significance of this observation is not yet clear, but it may be important in regulating the level of *lexA* repressor. Measurements of *lexA* repressor activity in *dcm*$^+$ and *dcm*$^-$ strains should answer this question.

Recombination and Repair

Among a collection of hyper-recombinogenic strains (Konrad, 1977), *arl* (accumulation of recombinogenic lesions) bacteria were identified (Hays and Korba, 1979). These strains and plasmids maintained in them are partially deficient in 5-meCyt (Korba and Hays, 1982a,b). The hyper-recombination phenotype is not present when phages are grown on *dcm*$^-$ or *dcm*$^-$*arl*$^-$ strains, thus implicating an interaction between *dcm* and *arl*. Heteroduplex lambda DNA—containing one strand that is methylated and the other that is not—are highly recombinogenic, suggesting that hemimethylated *dcm* sites could be the basis for enhanced recombination. The nature of the *dcm* and *arl* interaction may prove to be very interesting.

One way to generate mismatched sites *in vivo* is by genetic recombination. An amber mutation *(am6)* at a *dcm* site has been described in the *cI* gene of phage lambda, and it recombines much more frequently than expected (Lieb, 1983). Recombinants between *am6* and adjacent markers usually retain the parental flanking configuration. This suggests that *am6*/+heteroduplexes (i.e., -CTAGG-/-GGTCC-) are very susceptible to mismatch repair. The frequency of coconversion of *am6* with closely linked mutations suggests that the repair tracts extend no farther than 20 base pairs to either side of the mismatch (Lieb, 1983). The use of *in vitro*-constructed heteroduplexes containing T/G mismatches may yield more information as to the mechanism of repair. At present, it is not clear what the relationship is between the repair system and the hyper-recombination system studied by Korba and Hays. It may be that the repair enzyme recognizes hemimethylated sites as well as mismatched sites.

5-Azacytidine

5-Azacytidine (5-aza-C) is bacteriostatic for *E. coli*. This cannot be due solely to inhibition of the *dcm* enzyme, since it is already known that the enzyme is dispensable for growth. Furthermore, *dcm*$^-$ and *dcm*$^+$ strains show similar sensitivities to the agent (Friedman, 1979). Nevertheless, 5-aza-C is interesting, because it appears to be an irreversible inhibitor of DNA cytosine methylases (Friedman, 1979;1981;1982); thus, it affords an opportunity to study the enzyme's mechanism of action (Santi *et al.* 1983). DNA that contains 5-aza-C inactivates *dcm, Hpa*II, or *Eco*RII methylases. This loss can be prevented by cleavage of the DNA with micrococcal nuclease or the corresponding restriction enzyme, or by addition of normal substrate DNA. Digestion with *Hpa*II nuclease does not destroy inhibitory activity for *Eco*RII methylase. In recent *in vivo* experiments (Friedman, 1982), cells containing the *Eco*RII restriction-modification system on a plasmid were exposed briefly to 5-aza-C. These cells did not resume growth after removal of drug, whereas plasmidless cells did. Mutants defective in the restriction function were sensitive, but a mutant defective in modification function lost 5-aza-C sensitivity. This means the effect is due to the methylase enzyme and not to the restriction enzyme. It may be that irreversible inactivation of the enzyme causes the replication machinery to malfunction and cause subsequent cell death (Friedman, 1982).

DNA Replication

It had been suggested that *dcm* methylation is essential for replication of phage ϕX174 (Razin *et al.* 1975). A subsequent study, however, did not support this idea (Hattman *et al.* 1979).

DNA Adenine Methylation

dam Methylase Substrate Sequence

The *dam* enzyme methylates the sequence –GATC– and all these sites are methylated in *E. coli* K-12 DNA (Lacks and Greenberg, 1977; Hattman *et al.* 1978; Geier and Modrich, 1979) and in *E. coli* C (Szyf *et al.* 1982). This confirms the earlier enzymologic work on the strain and species specificity of the *E. coli* enzymes (Borek and Srinivasan, 1966).

As mentioned above, methylation of DNA daughter strands occurs at/or near the replication fork. An examination of Okazaki fragments in *lig* (ligase) or *polA* (polymerase) *lig* strains of *E. coli* (which accumulate such fragments) showed that the fragments were undermethylated compared to bulk DNA (Marinus, 1976). This experiment could not be done in wild-type *E. coli*, because not enough labeled material could be obtained for analysis. The data from the *lig* mutants, however, have been interpreted to favor a model in which methylation occurs subsequent to synthesis and mismatch repair (Glickman and Radman, 1980). In a recent study, however, no detectable hemimethylated *dam* sites in newly replicated DNA were found (Szyf *et al.* 1982). This conclusion was based on *Dpn*I—but not *Mbo*I—sensitivity of the DNA, which indicates that both strands of the pulse-labeled DNA were methylated. It also was concluded that methylation does not lag behind the replication fork for more than 3,000 base pairs. A potential problem with these experiments, however, are the long pulse times used that may lead to labeling of large stretches of DNA away from the replication fork.

Endonuclease *Mbo*I (an isoschizomer of *Dpn*II) cleaves only at unmethylated -GATC- sites, and we have found that this is the most reliable method to determine whether cells are Dam^+ or Dam^-. The *Dpn*I enzyme is also very useful, since it cleaves only methylated *dam* sites (Lacks and Greenberg, 1977).

The sequence -GATC- should occur once every 256 base pairs, if it is randomly distributed in bulk DNA; this is the case in *E. coli* (Lacks and Greenberg, 1977; Razin *et al.* 1980). However, the distribution of these sites is not random at the origin of DNA replication (see below).

The *dam* Gene

A fragment of chromosomal DNA containing the *dam* gene was identified in the Clarke and Carbon (1976) collection by its proximity to *trpS* (Geier and Modrich, 1981; Hall and Yanofsky, 1981). Subcloning from this fragment has allowed both a restriction map of the DNA in the *dam-trpS* region to be constructed (Arraj and Marinus, 1983; Hall and Yanofsky, 1981) and the nucleotide sequence of the 834-base pair *dam* gene to be determined (Brooks *et al.* 1983). The availability of a variety of plasmids that bear *dam* will allow investigation into how the *dam* gene is regulated, and it will allow researchers to vary the amount of methylase in the cell.

We have recently constructed plasmids in which the β-galactosidase gene is under control of the *dam* promoter (Arraj and Marinus, unpublished data, 1984). β-Galactosidase levels are similar in a variety of mutant strains (dam^+, dam^-, $recA^-$, $recB^-$, $recC^-$, $lexA^-$) that contain these plasmids, and they do not appear to vary with the age of the cells. The availability of this plasmid also should allow researchers to isolate mutations in the promoter region of *dam*. In another plasmid construction, we have placed the strong *tac* promoter in front of the *dam* gene (Marinus, Poteete, and Arraj, unpublished data), which allows for a 200–300-fold overproduction of the *dam* gene product (see below). The cloned *dam* gene should allow directed mutations to be constructed and to determine the effect of these mutations on enzyme activity and specificity. The following deletions all result in loss of activity: 1) the proximal 316 base pairs; 2) base pair 316 to the end of the gene (base pair 834); 3) 692–834; 4) 205–323, and 5) 323–834 base pairs (Brooks *et al.* 1983; Arraj and Marinus, unpublished data). Further studies along these lines should allow critical enzyme sites to be identified in addition to those that alter specificity.

Properties of the Purified Enzyme

The *dam* methylase from *E. coli* K-12 has been purified 3,000-fold to 95% purity by Herman and Modrich (1982). The enzyme is a single polypeptide chain of molecular weight (mol wt) 31,000, has an $S_{20,w}$ of 2.8 S, a Stokes radius of 24A, and exists in solution as a monomer. The mol wt agrees with that predicted from the DNA sequence. S-adenosyl-L-methionine has no effect on the state of aggregation of the enzyme. The enzyme has a turnover number of 19 methyl transfers per minute, and the apparent Km for –GATC– sites in ColE1 DNA is 3.6nmols. Double-stranded DNA is a better methyl acceptor than denatured DNA, and there is little difference in the rate of methylation between unmethylated or hemimethylated DNA. The enzyme transfers one methyl group per DNA-binding event. The properties of the *dam* methylase are similar to other type 2 modification methylases, such as *Eco*RI and *Hpa*II. A partially purified *dam* enzyme preparation from *E. coli* C has similar properties to the K-12 enzyme (Urieli-Shoval *et al.* 1983).

To detect the purified *dam* protein on SDS-polyacrylamide gels, the samples were heated at 100°C for 20 minutes in the presence of 2% dithiothreitol and 1% SDS to reduce highly resistant disulfide in the protein (Herman and Modrich, 1982). It is interesting that there is one report (Goze and Sedgewick, 1978) of a *dam*-specific band on an SDS-polyacrylamide gel from crude *E. coli* extracts and two negative reports (Arraj and Marinus, 1983; Brooks *et al.* 1983). Using the denaturing conditions above, plus a strain that overproduces the methylase several hundred fold, no *dam*-specific protein from cell extracts could be detected on gels (Arraj and Marinus, unpublished data). A hybrid protein composed of the amino terminal end of the *dam* methylase fused to β-galactosidase also cannot be detected (Arraj and Marinus, unpublished data). It is unclear as to why the *dam* protein is so elusive; perhaps, an immunologic method will help clarify the puzzle.

Biological Functions

Before reviewing the phenotypes of *dam* mutants, some proposed models for *dam* function will be presented.

Mismatch Repair

The most direct and convincing evidence for the involvement of methylation in mismatch repair comes from *in vitro*-constructed heteroduplexes of phage lambda DNA (Pukkila *et al.* 1983; Ryokowski and Meselson, 1984). Heteroduplexes containing a mismatch in *c*I were constructed with one strand methylated, both strands methylated, or neither strand methylated. If neither strand was methylated, repair occurred on either DNA strand. The unmethylated strand was preferentially repaired in heteroduplexes containing one methylated and one unmethylated strand. Surprisingly, no repair was observed when both strands of the heteroduplex were methylated (Pukkila *et al.* 1983; Ryokowski and Meselson, 1984). DNA adenine methylation, therefore, determines strand selectivity for mismatch repair.

Mismatch repair usually is considered in the context of genetic recombination. Although it can be argued that methylated/unmethylated heteroduplex regions in DNA can be generated via recombination in *E. coli,* the replication fork region of *E. coli* should contain such regions if methylation lags behind replication. If the newly synthesized daughter strand is transiently unmethylated, methyl-directed mismatch repair would be a mechanism for ensuring fidelity of genetic information (Wagner and Meselson, 1975; Glickman and Radman, 1980). The evidence for postreplicative mismatch repair in *E. coli* DNA is not as direct as that for phage lambda. Such evidence includes: 1) that *dam* mutants show a higher spontaneous mutation frequency (Marinus and Morris, 1974); 2) that *dam* mutants are hypermutable by base analog (Glickman *et al.* 1978, but see below), and that such base analogs are bactericidal for *dam*$^-$, but not *dam*$^+$ strains (Glickman and Radman, 1980); 3) that inactivation of mismatch repair suppresses the inviability of *dam*$^-$ *recA*$^-$ strains (McGraw and Marinus, 1980; Glickman and Radman, 1980); 4) that overproduction of the *dam* methylase would lead to methylation of daughter strand DNA before repair that would result in an increased spontaneous mutation rate (Herman and Modrich, 1981; Marinus, Poteete, and Arraj, unpublished data); 5) that *mutH, L,* or *S* mutant alleles suppress *dam* phenotypes (McGraw and Marinus, 1980); 6) and that the *mut* strains show high spontaneous mutation rates and are deficient in mismatch repair (Rydberg, 1978). These points will be discussed more fully below. It is also necessary to remember that phage lambda is unusual in that its DNA is only partially methylated and that the question of undermethylation of newly synthesized DNA in *E. coli* is in dispute (see above).

Glickman and Radman (1980) have postulated that the production of double-strand breaks in DNA occurs in *dam* mutants due to overlapping mismatch repair tracts on both parental and daughter strands. These double-strand breaks have been proposed to explain the phenotypes of *dam* mutants (Glickman, 1982). As experimental support for the model, advantage was taken of the observation that *dam*$^-$ cells are sensitive to 2-aminopurine (2-AP) and that 2-AP-resistant *dam*$^-$ derivatives contained mutations in the *mutH, L,* or *S* genes. Mutation in these *mut* genes reduces mismatch repair (Rydberg, 1978).

McGraw and Marinus (1980) took advantage of the lethality of *dam*$^-$*recA*$^-$ bacteria to isolate derivatives that were viable. These included *dam*$^+$, *dam*$^-$*mutL*$^-$, *dam*$^-$*mutS*$^-$, and *dam*$^-$*sinA*$^-$ strains. The isolation of the *mut*$^-$ derivatives can be accommodated by the mismatch repair model, but the *sinA*$^-$ cells repair mismatches normally.

If it is accepted that methyl-directed repair is important for phage lambda and *E. coli,* can this be a general mechanism for reducing spontaneous mutations? The answer is probably not. There is no evidence for such a mechanism in eukaryotic cells. Among prokaryotes, there are many gram-positive genera that do not have 6-meAde in DNA, and there are species with no detectable methylated bases in DNA (Hall, 1971). Phage T7 has an asymmetric mismatch repair system that is dependent on *mutH, L,* or *S* genes of *E. coli* (Bauer *et al.* 1981), but this phage has few methylated *dam* sites (Lacks and Greenberg, 1977). Phage T2 *dam* mutants do not show increased reversion frequencies for three markers (Hattman, 1981). It is possible that methyl-directed repair is restricted to the Enterobacteria.

A more general model for mismatch repair envisions single-strand breaks in the target strand for initiating repair (Lacks *et al.* 1982). In this attractive model, DNA strands containing single-strand breaks would be repaired, while the unbroken strand would not. Single-strand breaks could arise by: 1) discontinuous replication on both parental strands; or 2) repair of (for example) uracil residues in DNA. In *dam* strains, DNA breaks in the parental molecule would lead to repair and, hence, the production of mutations. Increased single-strand breaks in *dam*$^-$ *lig*$^-$ strains have consistently been found (Marinus and Morris, 1975; Bale *et al.* 1979). The model proposed by Lacks *et al.* (1982) has the usefulness of being applied to any organism, and it can explain directed mismatch repair in the absence of methylation. In *E. coli,* it is possible that both methyl-directed and methyl-independent mismatch repair occur, although experimental evidence thus far suggests that only a methyl-directed repair system is present (Pukkila *et al.* 1983).

An important step in the development of an *in vitro* mismatch repair system has been made by Lu *et al.* (1983). A mismatch repair activity has been found in crude extracts of *E. coli,* which is greatly diminished if fractions are prepared from, *mutH, L,* or *S* mutant derivatives. The activity is dependent on the state of *dam* methylation in that fully methylated, mismatched duplexes are subject to reduced repair. A striking finding is that at least some repair events appear to initiate at/or near *dam* sites.

Regulation of Gene Activity

The presence or absence of a methylated base in a DNA sequence might be expected to alter the binding of a particular protein. If such a DNA binding protein were involved in regulating gene activity, DNA methylation either might hinder or be required for gene expression. An example of the latter is that the expression of the *mom* gene of phage Mu requires an intact host *dam* gene (Touissaint, 1977). Hattman (1982) showed that the level of *mom* RNA in *dam*$^-$ strains was decreased to at least 5% of the level in *dam*$^+$ strains. The effect was specific for the *mom* transcript, since no difference in total Mu transcripts could be detected between *dam*$^+$ or *dam*$^-$ strains. Plasterk *et al.* (1983) showed that the *dam*-specific effect lies in a region preceding the *mom* gene, and also that when the *mom* gene is under control of another promoter (pL of phage lambda), *dam* dependence is lost. Although it appeared that the methylated –GATC– sequences overlapped the promoter region (Plasterk *et al.* 1983), it is now clear that the methylated sequences lie upstream of the *mom* promoter (Plasterk; Bukhari, personal communication).

An example of methylation at –GATC– sequences that inhibit gene expression is the inward promoter of the element Tn*10,* the product of which is involved in transposition. One class of host mutants that increases Tn*10* transposition are *dam*$^-$ cells. A -GATC- sequence overlaps the -10 region of the promoter, and its demethylation leads to increased gene expression (Roberts and Kleckner, personal communication).

No essential genes of *E. coli* can be regulated *via* methylation, since *dam* methylase-deficient strains are viable. It may be a useful mechanism, however, for augmenting control of gene activity. It should be kept in mind, however, that *dam* sites are fully methylated *in vivo;* this poses the problem of how *dam*-regulated genes can be turned on in the absence of methylation.

Given the examples above, several laboratories are attempting to identify genes whose activity might be influenced by methylation. Some of the phenotypes of *dam* mutants may be due to the indirect effect of *dam* methylation on activity of other genes.

DNA Replication

Methionine auxotrophs of *E. coli* do not replicate DNA after one round of replication, if grown in the presence of ethionine (Lark, 1968b). This result was interpreted to mean that fully methylated DNA was required for replication. This interpretation cannot be correct, however, since a *dam*$^-$ *dcm*$^-$ *hsd*$^-$ strain, which is devoid of methylation, is viable (Marinus *et al.* 1984). Alternatively, some evidence favors a model in which DNA replication proteins containing ethionine cannot function (Lark, 1979).

The origin of replication *(ori)* of the *E. coli* K-12 chromosome contains 10 times as many -GATC- sequences than expected on a random basis (Sugimoto *et al.* 1979; Meijer *et al.* 1979); this is a general feature of the *ori* region of Enterobacteria (Cleary *et al.* 1982). Methylation of these sites is not required for initiation of chromosome replication. One line of evidence is that *dam* methylase-deficient strains are viable (Marinus *et al.* 1984). Further evidence comes from a study in which origin DNA was hybridized to *Dpn*I-digested DNA from a *dam-3 dcm-6* strain (Szyf *et al.* 1982). The *ori* region of the mutant strain was resistant to *Dpn*I digestion, indicating that no fully methylated sites were present. Since methylation of *dam* sites at the *ori* region is not necessary for replication, it is not clear why there is an abundance of such sites. It is a possibility that these sites might ensure successful mismatch repair in case of a mutation.

Gomez-Eichelmann and Lark (1977) studied the methylation of Okazaki pieces in *E. coli* in an *in vitro* system. Fragments were normally joined, regardless of whether or not they were methylated. When DNA containing methylated Okazaki pieces was treated with *Dpn*I, the size reduction was less than expected. It was concluded that *Dpn*I was cutting at sites close to, or at, the end of Okazaki pieces, and that the ends of such pieces must occur near *dam* sites. Okazaki pieces are normally formed in *dam* strains *in vivo* (Marinus and Morris, 1974) and *in vitro* (Gomez-Eichelmann and Lark, 1977). If the enzyme cannot methylate *dam* sites at the ends of molecules (Uriel-Shoval *et al.* 1983) and if *dam* sites occur with high frequency at the ends of Okazaki pieces, it might be expected that such pieces would be partially methylated. This result was obtained by Marinus (1976) in a DNA ligase mutant strain that accumulates Okazaki pieces. If Okazaki pieces are ligated rapidly in wild-type strains, then such newly synthesized DNA would appear fully methylated, because there would not be any unmethylated ends. This may reconcile the results of Szyf *et al.* (1982), who found that newly synthesized DNA in *E. coli* is fully methylated, and the results of Marinus (1976) in which Okazaki pieces are partially methylated.

Protection of DNA

Marinus and Morris (1974) suggested that one function of *dam* methylation was to protect DNA from endonuclease cleavage. The studies on mismatch repair of phage lambda heteroduplexes have shown that fully methylated molecules are not repaired, presumably because they are not cut by an endonuclease; whereas unmethylated or hemimethylated molecules are cut and repaired (Pukkila *et al.* 1982). Szyf *et al.* (1982) concluded from their studies of *E. coli* that methylated bases protect DNA from nuclease activity, and that unmethylated sites on a high methylation background could provide signals for carrying out biological functions. The recent finding of Lu *et al.* (1983) that mismatch repair initiates at unmethylated *dam* sites *in vitro* could be integrated here to provide a mechanism for generating single-strand breaks at *dam* sites in DNA *in vivo.* Such single-strand breaks would be controlled by methylation and could serve as signals or initiation sites for repair and recombination (Marinus and Morris, 1974; Szyf *et al.* 1982).

How could hemimethylated *dam* sites be generated *in vivo* at regions away from the replication fork? DNA repair and recombination could produce such sites, because both involve cutting and resynthesis of one or both strands of duplex DNA. Hemimethylated sites would be created if repair and recombination enzymes progressed through –GATC– sites. Since these sites occur once every 256 base pairs, the probability of producing them should be high in cells undergoing repair and/or recombination. Hemimethylated *dam* sites in DNA regions that are away from the replication fork are known to be subject to methylation (Billen, 1968; Lark, 1968a) and, therefore, are transient. Hemimethylated sites produced by repair might lead to competition between the methylase, mismatch, and excision repair enzymes and DNA polymerase 1 that causes interference in completing normal repair and recombination. These events might explain why *dam*$^-$ *recA*$^-$ strains are lethal and degrade their DNA, as well as some other *dam* phenotypes such as hyper-recombination and sensitivity to agents known to induce DNA repair mechanisms (see below).

Restriction and Modification

It is a possibility that *dam* sites are methylated to protect DNA from plasmid- or phage-encoded restriction enzymes that cleave at unmethylated sites. No such restriction endonuclease activity has yet been found on plasmids or phages infecting *E. coli.*

Phenotypes of *dam* Mutants

The phenotypes associated with *dam* mutants are: 1) increased spontaneous mutability; 2) increased sensitivity to certain chemicals and ultraviolet (UV) light; 3) a hyper-rec phenotype; 4) increased induction of lysogenic bacteriophages; 5) inviability of *dam⁻recA⁻*, *dam⁻recB⁻*, *dam⁻recC⁻*, *dam⁻recJ⁻*, *dam⁻lexA⁻*, and *dam⁻polA⁻* double mutants; 6) increased single-strand breaks in a *dam⁻lig⁻* (ligase) strain; 7) suppression of most *dam* phenotypes by *mutS⁻* and *mutL⁻*; 8) increased transposition of Tn*10* (Roberts and Kleckner, personal communication); and 9) increased precision excision of Tn*10* (Lundblad and Kleckner, 1982) and Tn*5* (Marinus and Morris, 1974; 1975; Marinus and Konrad, 1976; Zieg *et al.* 1978; Bale *et al.* 1979; McGraw and Marinus, 1980; Marinus, 1981; Arraj and Marinus, 1983). The phenotypes above are suggestive of defective DNA repair (Marinus and Morris, 1974). Ultraviolet light-induced excision and postreplication repair was found to be normal in *dam* mutants (Marinus and Morris, 1975), suggesting that some other repair process is responsible for the phenotypes. Aberrant mismatch repair is a definite possibility. Some of the phenotypes are discussed in more detail below.

Spontaneous Mutability

Strains that are mutant in the *dam* gene show an 8–250-fold increase in spontaneous mutability compared to wild type; the variation depends on the marker being tested. Reversion of missense, *amber, ochre* (Marinus and Morris, 1974; 1975), and frameshift (Glickman, 1979; Marinus, 1981) mutations was stimulated. Both + and − frameshifts have been tested, one of which *(trpA21)* gives the greatest increase of any marker tested (Marinus, 1981). The frequency of *trp-tonA* deletion formation is increased about 10-fold in *dam* mutants (Arraj and Marinus, 1983). No difference in the frequency of Mu insertion mutations between *dam⁺* and *dam⁻* strains have been found (Marinus, unpublished data).

To determine the nature of increased spontaneous mutagenesis, Glickman (1979) used the *lacI* system to demonstrate that G/C to A/T transitions were stimulated 140-fold. The *lacI* system, as used above, relies on obtaining information from spontaneous nonsense mutations. No information regarding A/T to G/C transitions or frameshift, deletion, or insertion mutations can be gained. To overcome these limitations and to determine changes at the DNA level, we have been sequencing *dam*-induced mutations in the *mnt* repressor gene of phage P22. (Marinus, Carraway, Frey and Youderian, unpublished data).

Half the mutations identified so far are frameshifts which occur in runs of identical bases. The other half are transition mutations. Frameshift mutations are thought to occur by the slippage of DNA polymerase by one of more nucleotide bases during the replication of a run of identical base pairs. Presumably, unpaired bases in such runs are not excised by the proofreading 3′ to 5′ exonuclease activity of DNA polymerases. We speculate that such lesions are recognized and repaired by the post-replicative mismatch repair system that distinguishes between a methylated parental strand of DNA and daughter strains in *dam* mutants, repair of a parental strand lesion after polymerase incorporates one less base pair into the daughter strand would generate −1 frameshift mutations. Repair of a daughter strand lesion would result in +1 frameshift mutations. Note that polymerase associated exonucleases are not capable of repairing unpaired bases within a chain and therefore the mismatch repair system may be the only one capable of removing potential frameshift mutations.

A variety of plasmids are available that contain the *dam* gene and overproduce the *dam* methylase. Most of these produce approximately 10 times more enzyme than a normal *E. coli* strain (Arraj and Marinus, unpublished data). An exception is pGG503, which overproduces *dam* methylase about 40-fold and confers increased mutability on cells carrying it (Herman and Modrich, 1981). These results suggest that cells overproducing *dam* methylase above a greater threshold level than <20X cause the production of mutations. Additional evidence for this comes from a plasmid that we have recently constructed, which contains the *tac* promoter in front of the *dam* gene. In the absence of inducer, the cell produces about 10-fold more *dam* methylase than a normal *E. coli,* and it shows normal mutability. In the presence of inducer (to allow transcription from the *tac* promoter), *dam* methylase levels increase to 200–300-fold and the spontaneous mutation frequency increases 30–50-fold (Marinus, Poteete, and Arraj, unpublished data,).

The increased mutability in overproducing cells is independent of *recA* (Herman and Modrich, 1981), as is mutability in *dam*$^-$ bacteria (McGraw and Marinus, 1980). Increased mutability occurs in *dam*$^-$*umuC*$^-$ strains (Marinus, unpublished data), confirming that *dam* mutagenesis does not occur via the *recA*/*lexA* damage-inducible pathway (Little and Mount, 1982).

The spontaneous mutability of *dam* mutants and overproducers can be explained by the mismatch repair model. This assumes that when both strains of DNA are unmethylated, mismatch repair can occur on either strand. Repair on the parental DNA strand in *dam* strains would result in mutation after the next round of replication. In normal *E. coli,* repair occurs only on the daughter strand. In *dam*-overproducing strains, DNA is methylated before it can be repaired; since repair cannot occur when both strands are fully methylated, base pair mismatches will cause mutations after the next round of replication. It will be interesting to find out if the same spectrum of mutations are obtained in *dam*$^-$ and mismatch repair-deficient strains as in *dam* overproducers.

Sensitivity to Chemical and Physical Agents

Mutation in the *dam* gene results in increased sensitivity (compared to wild type) to a variety of agents, including UV light, alkylating agents, certain antibiotics, and base analogs (Marinus and Morris, 1974, 1975; Glickman *et al.* 1978). This sensitivity is abolished in *dam*$^-$*mutL*$^-$ or *dam*$^-$*mutS*$^-$ strains in which mismatch repair is absent (McGraw and Marinus, 1980; Glickman and Radman, 1980; Karran and Marinus, 1982). This suggests that mismatch repair provoked by these agents in *dam*$^-$ strains may be contributing to cell death. In the case of N-methyl-N-nitro-N-nitrosoguanidine, the mismatches must involve O^6-methylguanine (Karran and Marinus, 1982).

Sensitivity to base analogs (and presumably to other agents) was proposed as being due to the formation of double-strand breaks in DNA resulting from overlapping mismatch repair tracts (Glickman and Radman, 1980). It is consistent with the model that *dam*$^-$*mut*$^-$ strains are resistant to 2-aminopurine (2-AP) (Glickman and Radman, 1980). Also consistent with the model was the fact that *dam* mutants appeared to be hypermutable by 2-AP and ethylmethane sulfonate (EMS; Glickman *et al.* 1978). This latter study has a flaw, however, in that the authors did not test the genotype of the mutagenized survivors. When this was done in our laboratory by R. Craig, the survivors were found to be *dam*$^-$*mut*$^-$ (Table 5.2) and not *dam*$^-$*mut*$^+$, as assumed by Glick-

Table 5.2 Frequency to Rifampicin (Rif) Resistance of *dam*$^+$ and *dam-3* Bacteria Exposed to 2-AP and the Genotype of Rifampicin-resistant Survivors[a]

	dam$^+$		*dam-3*	
2-Ap Conc. (μg/ml)	Mutation Frequency	*mut*$^-$ Among Rif^R Survivors (%)	Mutation Frequency	*mut*$^-$ Among Rif^R Survivors (%)
0	2	0(2)	24	0(47)
10	8	0(5)	530	100(48)
30	25	0(5)	2,700	100(40)
100	30	0(5)	13,000	100(47)

[a]Bacterial strains were grown in the presence of 2-AP as described by Glickman *et al.* (1979), and portions of the cultures were plated to determine the frequency of rifampicin-resistant cells/10^8 that bacteria plated. The rifampicin-resistant cells were purified and tested for the *mut* phenotype as described by McGraw and Marinus (1980). The number of survivors tested is shown in parentheses.

man *et al.* (1978). This indicates that *dam* mutants are not hypermutable by 2-AP (and probably not by EMS either).

R. Craig and M. G. Marinus have also looked for DNA strand breaks in *dam-3* cells exposed to 2-AP. Figure 5.1 shows that no DNA breaks above the background level could be detected in *dam* bacteria exposed to 400 μg 2-AP/ml when DNA from cells was analyzed on alkaline and neutral sucrose gradients. This means either that 2-AP does not cause the production of double-strand breaks or that so few are produced as to be undetectable, or that 2-AP works through some other mechanism.

In the presence of 2-AP, we were able to show that the DNA damage-inducible *(din)* repair system (SOS) was fully induced in *dam* mutants, but not in *dam*$^-$*mut*$^-$ or *dam*$^+$ cells (Craig, Arraj, and Marinus, unpublished data). There was no significant induction of the *din* system in *dam*$^-$ cells in the absence of 2-AP. It is not known what the inducing signal is, although *dam* mutants exposed to 2-AP cannot replicate DNA. It is significant that 2-AP results in full induction of the *recA/lexA* error-prone repair system in *dam*$^-$ cells, because it allows an easy assay for environmental mutagens that may have weak activity in other test systems. We are currently testing if other agents elicit the *din* response in *dam*$^-$ strains.

The *dam* mutants may be quite useful, therefore, in determining the mechanisms of drug action and in screening for potential mutagens/carcinogens with low activity. The *dam* mutants show increased mutability to certain frameshift mutagens (Glickman, 1982).

Inviability of *dam* with Other Mutations

Mutant strains bearing mutations in *dam* and *recA, recB, recC, lexA,* or *polA* are inviable (Marinus and Morris, 1974;1975). Presumably, these gene products are required for repair of lesions that occur spontaneously in *dam* mutants (Marinus and Morris, 1974;1975; Glickman and Radman, 1980). Incubation of *dam-3* *recA*$_{ts}$ strains at the nonpermissive temperature results in DNA degradation (Marinus and Morris, 1975). Incubation of *dam-3* *recB*$_{ts}$ or *dam-3* *recC*$_{ts}$ strains at the nonpermissive temperature results in an immediate halt in cell division, but no apparent DNA damage (Marinus, unpublished data,). The inviability of *dam*$^-$*recB*$^-$,*C*$^-$ mutants can be suppressed by *sbcA*$^-$ or *sbcB*$^-$ mutations (McGraw and Marinus, 1980).

The inviability of *dam recA* mutants can be suppressed by mutation in *mutL, mutS, sinA* (McGraw and Marinus, 1980), or *mutH* (Glickman and Radman, 1980). These data tend to favor a role for methylation in mismatch repair, since the *mut*$^-$ strains are deficient in such repair (Rydberg, 1977). Another suppressor, *sin*A, has not yet been identified; however, it may be important, since it is not defective in mismatch repair (McGraw and Marinus, 1980).

It may be necessary for *dam* mutants to be able to induce DNA damage-inducible (SOS) repair at some stage during the cell cycle; this may explain

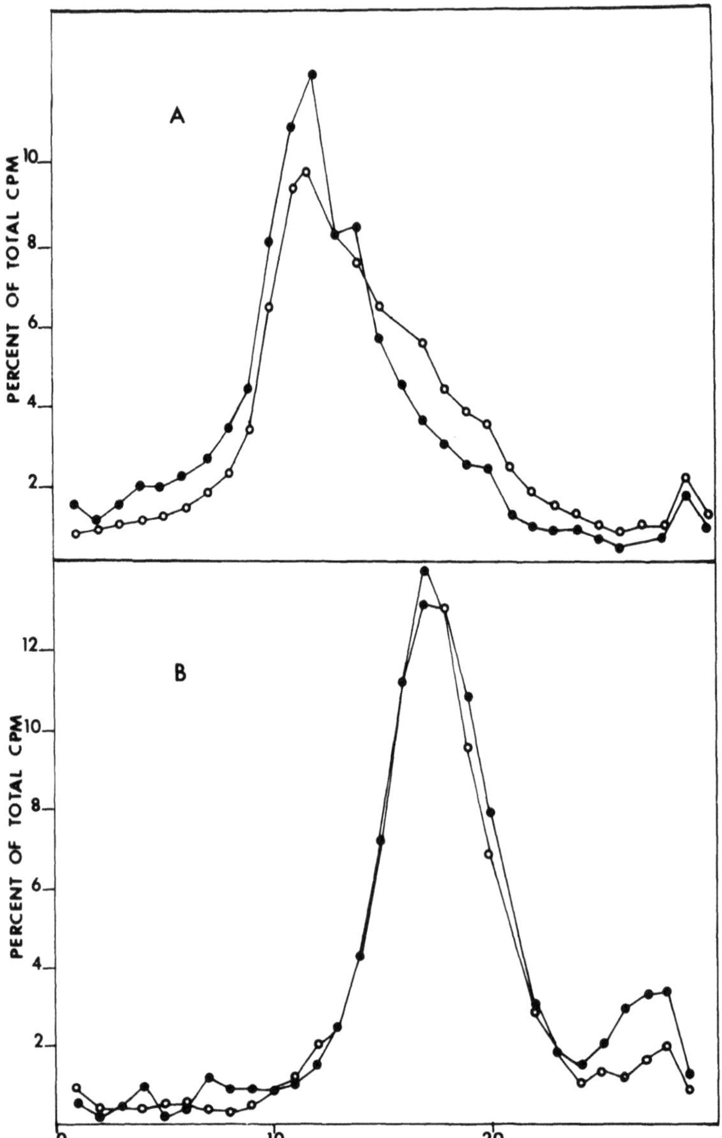

Figure 5.1 Sedimentation of DNA in alkaline (A) and neutral (B) sucrose gradients. *Dam-3* cultures were grown with either ^{3}H-thymidine or ^{14}C-thymidine, and one of these was exposed to 400 μg 2-AP/ml for 2 hours. The cells were harvested, combined, and the DNA was analyzed on sucrose gradients as described by Marinus and Morris (1974). Closed symbols are from cells exposed to 2-AP. Reversing the isotope labeling did not later the sedimentation profiles.

the inviability of *dam* with *recA* or *lexA*. These latter gene products are absolute requirements for inducible repair (Little and Mount, 1982). We have recently shown that *dam* mutations combined with a variety of mutant alleles that are associated with the damage-inducible response are viable (Table 5.3).

The SOS system regulates a variety of genes (*din, sfi, uvr,* and so on; see Little and Mount, 1982) through the action of *recA* and *lexA* gene products. The inviability with *dam* is not due to the action by any of the secondary genes tested. That the $recA^{-}$ derivative of a $dam^{-}lexA^{-}spr^{-}tif^{-}sfiA^{-}$ strain is inviable indicates that the lethal effect is mediated through *recA*. One interpretation of these results is the one given above (i.e., *dam* mutants need to be able to induce the SOS system to survive). It will be of interest both to determine the identity of the *sin*A suppressor (which allows viability of $dam^{-}recA^{-}$ bacteria) and to test the inviability of a $dam^{-}lexA^{-}$ strain containing an operator mutation in *recA*. Such a strain should be viable, because it would be producing *recA* protein.

Goze and Sedgewick (1978) measured some parameters of the damage-inducible repair (SOS) response in $uvrA^{-}$ derivatives of *dam* mutants exposed to UV light. They found increased induction, reactivation, and mutagenesis of phage lambda and increased His^{+} reversion. Although these increases were only 2–4-fold, the results suggested that the effects of UV light and the *dam* mutation potentiate induction of repair.

Methylation of *dam* and *dcm* Sites in Phage and Plasmid DNA by *E. coli* Enzymes

Phages and plasmids can be divided into three groups on the basis of methylation of *dam* and *dcm* sites: 1) those in which all sites are methylated; 2) those in which none are methylated; and 3) those that have some sites methylated. Group 1 included pBR322, pBR325, the F plasmid, single-stranded phage RF (fd, fl), and lambda *dv*. Group 2's most famous member is phage T7, and that of group 3 is phage lambda (Pirrota, 1975; Lacks and Greenberg, 1977; Vovis and Lacks, 1977; Sutcliffe, 1978; Dreiseikelmann *et al.* 1979; Razin *et al.*

Table 5.3 Viability of dam^{-} Derivatives with Mutant Alleles of the Damage-inducible (SOS) Repair System

Inviable combinations	Viable combinations
$recA^{-}dam^{-}$	$recF^{-}dam^{-}$
$lexA^{-}dam^{-}$	$lexA^{-}spr^{-}tif^{-}sfiA^{-}dam^{-}$
$lexA^{-}spr^{-}recA^{-}sfiA^{-}dam^{-}$	$uvrA^{-}dam^{-}$, $uvrB^{-}dam^{-}$, ($uvrC^{-}dam^{-}$), $sfiA^{-}dam^{-}$, $dinA^{-}dam^{-}$, $dinB^{-}dam^{-}$, $dinD^{-}dam^{-}$, $dinF^{-}dam^{-}$, $dam^{-}lexB^{-}$, $dam^{-}tif^{-}zab^{-}$, $dam^{-}umuC^{-}$

1980). The reduction in methylation of phages T7 and lambda is thought to be due to inability of DNA methylation to keep up with replication (see Chapter 7). Growth of phages and plasmids in groups 1 and 3 in a *dam*⁻*dcm*⁻ strain results in the absence of methylated bases from DNA. The fact that unmethylated lambda and fd phages could grow in *dam* mutants showed that *dam* methylation was not essential for their life-style (Marinus and Morris, 1975; Bale *et al.* 1979). No instance has been found in which a phage or plasmid that lacks *dam* or *dcm* methylation is unable to reproduce normally in methylation-deficient strains. Methylation of phage lambda in a strain containing the *dam*-overproducing plasmid pTP166 occurs at all *dam* sites (Marinus, unpublished data).

Phage lambda is unusual because its *dam* sites are partially methylated. It also is the system in which the most convincing evidence for mismatch repair exists. It will be interesting to see if this correlation holds in the future in other systems. Directed mismatch repair also can occur in phage T7, which contains few, if any, methylated bases in DNA (Bauer *et al.* 1981). This directed mismatch repair requires the *mutH, S,* and *L* gene products, but clearly cannot involve *dam* methylation; therefore, it indicates that there may exist a *dam*-independent mismatch repair system.

Phages T2 and T4 have their own *dam* methylase, which recognizes -GATC- sequences, but its structure is quite different from that of the *E. coli* enzyme (Schlagman and Hattman, 1983). It is clear that the T2 and T4 *dam* methylase is not essential, and that the enzyme is not involved in mismatch repair (Schlagman and Hattman, 1983). Further studies on this system may uncover roles for *dam* methylation other than mismatch repair and also may allow comparative studies to be carried out on the effect of the T4 methylase in *E. coli.*

That cytosine methylation is not required for replication of phage ϕX174 has been noted above.

DNA Methylation in Bacteria Other than *E. coli*

Other members of the Enterobacteria species have a *dam* gene (Brooks *et al.* 1983), and one might suspect that the biological function of *dam* is the same as that for *E. coli.* Most bacterial species examined so far contain either one or both methylated bases (Hall, 1971), although a notable exception is *Micrococcus radiodurans,* which has none (Schein *et al.* 1972). There are no methylated *dam* or *dcm* sites in *Bacillus subtilis* or *Staphylococcus aureus* (Driseikelmann and Wackernagel, 1981). It would be worthwhile to use newer methods (see Hattman, 1981) to survey a wider range of organisms. Little is known about the function of methylated bases in DNA in organisms other than *E. coli.* The following examples, which do not represent a thorough survey of the available literature, may yield significant information in the future.

DNA methylation in pneumococci is particularly interesting, because restriction enzymes *Dpn*I and *Dpn*II were isolated from this organism (Lacks and Greenberg, 1975); also, a well-defined mismatch repair system has been described (e.g., Claverys *et al.* 1981; Lacks *et al.* 1982). These studies show that DNA methylation clearly has no role in the strand selection process in mismatch repair in pneumococci.

Strains of *Streptococcus pneumoniae* isolated from nature contain either *Dpn*I or *Dpn*II (Muckerman *et al.* 1982). These enzymes probably are part of a restriction-modification system, since a restrictable phage has been isolated. A strain (Rx) containing neither *Dpn*I or *Dpn*II has been known for sometime, and it was isolated spontaneously in the laboratory. Strain Rx can be transformed to *Dpn*I$^+$ or *Dpn*II$^+$, but DNA from *Dpn*I$^+$ strains fail to transform Rx to *Dpn*II$^+$ and vice versa. This does not support a direct role of methylation in "differentiation" in pneumococci (Lacks and Greenberg, 1977), although methylation still could be involved in more complicated models. However, this will be an interesting area in which to study complementary restriction-modification systems, especially when the *Dpn* genes are cloned.

5-Methylcytosine is present in virulent, but not avirulent, *Mycobacterium tuberculosis* strains (Srivastava *et al.* 1981). The source of 5-meCyt is a plasmid-borne enzyme. It would be useful to determine the virulence of normally avirulent strains containing the 5-meCyt-producing plasmid.

Bacillus subtilis contains a cytosine methylase that is induced in competent cells (Ganesan, 1982). Competent cells show a 40-fold induction of methylase and a 700-fold enrichment for transformation. This methylase is part of a restriction system, because incubation of the purified methylase with DNA results in cleavage. Studies with mutant cells that are deficient in methylase or in transforming ability might uncover a realtionship between them.

Haemophilus species contain DNA that hybridize to the *E. coli dam* gene; *Haemophilus* DNA is resistant to cleavage by *Mbo*I (Brooks *et al.* 1983). This suggests the *Haemophilus* species may produce a methylase recognizing -GATC- sequences. It is surprising that homology between *Haemophilus* species and *E. coli dam* genes should exist, since these organisms are not related. Further investigation in this area should yield information significant for methylase function.

Caulobacter bacteroides, which displays a series of developmental changes in its life cycle, contains 5-meCyt and 6-meAde. A difference in specific activity of DNA methylases was detected between swarmer and stalked forms (Degnen and Morris, 1973). It would be worthwhile to extend these observations by using restriction enzymes as a probe for methylation status. This might allow a difference to be detected that correlates with differential gene expression in the different cell types.

Neisseria gonorrhoeae produces six cytosine and one adenine methylase (Korch *et al.* 1983). Strains deficient in the adenine methylase that recognizes -GATC- sequences have been isolated, but they are not hypermutable. This indicates that the function of the gonococcal methylase is not the same as that for *dam* of *E. coli.* Although several methylases are produced by the *N. gon-*

orrhoeae strain examined, no restriction enzymes were found; this suggests that the methylases are not part of a restriction-modification system. Korch *et al.* (1983) suggest that the modifications may provide the cells with a mechanism to distinguish gonococcal DNA from other DNA. The properties of cytosine methylase-deficient strains are keenly awaited.

Summary

DNA methylation is not essential for viability of *E. coli*. Methylated bases in promoter and other regulatory regions can influence the rate of transcription of genes. The frequency of spontaneous mutations is influenced by DNA methylation. DNA adenine methylation allows a repair process to discriminate between DNA strands containing base pair mismatches. DNA methylation-deficient mutants are useful in recombinant DNA research and in mutagenicity studies.

Acknowledgments

Studies on DNA methylation in this laboratory are supported by United States Public Health Service grant GM30330. I wish to thank Lena DeSantis for her patience during the preparation of this manuscript.

References

Arraj JA, Marinus MG: Phenotypic reversal in *dam* mutants of *Escherichia coli* K-12 by a recombinant plasmid containing the *dam*$^+$ gene. *J Bacteriol* 1983;153:562–565.

Bale A, d'Alarcao M, Marinus MG: Characterization of DNA adenine methylation mutants of *Escherichia coli* K-12. *Mutat Res* 1979;59:157–165.

Bauer J, Krammer G, Knippers R: Asymmetric repair of bacteriophage T7 heteroduplex DNA. *Molec Gen Genet* 1981;181:541–547.

Billen D: Methylation of the bacterial chromosome: an event at the "replication point?" *J Mol Biol* 1968;31:477–486.

Borek E, Srinivasan PR: The methylation of nucleic acids. *Ann Rev Biochem* 1966;35:275–297.

Boyer HW, Chow LT, Dugaiczyk, A, Hedgpeth J, Goodman HM: DNA substrate for the *Eco*RII restriction endonuclease and modification methylase. *Nature New Biol* 1973;244:40–43.

Brooks JE, Blumenthal RM, Gingeras TR: The isolation and characterization of the *Escherichia coli* DNA adenine methylase *(dam)* gene. *Nucl Acids Res* 1983;11:837–851.

Clarke L, Carbon J: A colony bank containing synthetic ColEl hybrid plasmids representative of the entire *E. coli* genome. *Cell* 1976;9:91–99.

Claverys JP, Mejean V, Gasc AM, Galibert F, Sicard AM: Base specificity of mismatch repair in *Streptococcus pneumoniae. Nucleic Acids Res* 1981;9:2267–2280.

Cleary JM, Smith DW, Harding NE, Zyskind JW: Primary structure of the chromosomal origins *(oriC)* of *Enterobacter aerogenes* and *Klebsiella pneumoniae:* Comparisons and evolutionary relationships. *J Bacteriol* 1982;150:1467–1471.

Coulondre C, Miller JM, Farrabaugh PJ, Gilbert W: Molecular basis of base substitution hotspots in *Escherichia coli. Nature* 1978;274:775–780.

Degnen ST, and Morris NR: Deoxyribonucleic acid methylation and development in *Caulobacter bacteroides. J Bacteriol* 1973;116:48–53.

Doskocil J, Sormova Z: The occurence of 5-methylcytosine in bacterial deoxyribonucleic acids. *Biochim Biophys Acta* 1965;95:513–515.

Dreiseikelmann B, Eichenlaub R, Wackernagel W: The effect of differential methylation by *Escherichia coli* of plasmid DNA and phage T7 and λ DNA on the cleavage by restriction endonuclease *Mbo*I from *Moraxella bovis. Biochim Biophys Acta* 1979;562:418–428.

Dreiseikelmann B, Wackernagel W: Absence in *Bacillus subtilis* and *Staphylococcus aureus* of the sequence-specific deoxyribonucleic acid methylation that is conferred in *Escherichia coli* K-12 by the *dam* and *dcm* enzymes. *J Bacteriol* 1981;147:259–261.

Duncan BK, Miller JH: Mutagenic deamination of cytosine residues in DNA. *Nature* 1980;287:560–561.

Dunn DB, Smith JD: Occurrence of a new base in the deoxyribonucleic acid of a strain of *Bacterium coli. Nature* 1955;175:336–337.

Friedman S: The effect of 5-azacytidine on *E. coli* DNA methylase. *Biochem Biophys Res Commun* 1979;89:1327–1333.

Friedman S: The inhibition of DNA (cytosine-5) methylases by 5-azacytidine. The effect of azacytosine-containing DNA. *Mol Pharmacol* 1981;19:314–320.

Friedman S: Bactericidal effect of 5-azacytidine on *Escherichia coli* carrying *Eco*RII restriction-modification enzymes. *J. Bacteriol* 1982;151:262–268.

Fujimoto D, Srinivasan PR, Borek E: On the nature of the deoxyribonucleic acid methylases. Biological evidence for the multiple nature of enzymes. *Biochemistry* 1982;4:2849–2865.

Ganesan AT: Uptake, restriction, modification and recombination of DNA molecules during transformation in *B. subtilis,* in Ganesan AT, Hoch J (eds). *Molecular Cloning and Gene Regulation in Bacilli.* New York, Academic Press, 1982.

Geier GE, Modrich P: Recognition sequence of the *dam* methylase of *Escherichia coli* K-12 and mode of cleavage of *Dpn*I endonuclease. *J Biol Chem* 1979;254:1408–1413.

Glickman BW, van den Elsen P, Radman M: Induced mutagenesis in *dam*⁻ mutants of *Escherichia coli,* a role of 6-methyladenine residues in mutation avoidance. *Mol Gen Genet* 1978;163:307–312.

Glickman BW: Spontaneous mutagenesis in *Escherichia coli* strains lacking 6-methyladenine residues in their DNA. An altered mutational spectrum in *dam*⁻ mutants. *Mutation Res* 1979;61:153–162.

Glickman BW, Radman M: *Escherichia coli* mutator mutants deficient in methylation-instructed DNA mismatch correction. *Proc Natl Acad Sci USA* 1980;77:1063–1067.

Glickman BW: Methylation instructed mismatch correction as a postreplication error avoidance mechanism in *Escherichia coli,* in Lemontt JF, Generoso M (eds): *Molecular and Cellular Mechanisms of Mutagenesis.* New York, Plenum Press, 1982, pp 65–88.

Gold M, Hurwitz J, Anders M: The enzymatic methylation of RNA and DNA II. On the species specificity of the methylation enzymes. *Proc Natl Acad Sci USA* 1963;50:164–169.

Gomez-Eichelmann MC, Lark KG: EndoR. *Dpn*I restriction of *Escherichia coli* DNA synthesized *in vitro.* Evidence that the ends of Okazaki pieces are determined by template deoxynucleotide sequence. *J Mol Biol* 1977;117:621–635.

Goze A, Sedgwick S: Increased UV inducibility of SOS functions in a *dam-3* mutant of *Escherichia coli* K-12 *uvrA. Mutat Res* 1978;52:323–331.

Hall RH: *The Modified Nucleosides in Nucleic Acids.* New York, Columbia University Press, 1971.

Hall CV, Yanofsky C: Cloning and characterization of the gene for *Escherichia coli* tryptophanyl transfer ribonucleic acid synthase. *J Bacteriol* 1981;148:941–949.

Hattman S, Schlagman S, Cousens L: Isolation of a mutant of *Escherichia coli* defective in cytosine specific deoxyribonucleic acid methylase activity and partial protection of bacteriophage λ against restriction by cells containing the N-3 drug resistance factor. *J Bacteriol* 1973;115:1130–1107.

Hattman S: Partial purification of the *Escherichia coli* K-12 *mec*$^{+}$ deoxyribonucleic acid-cytosine methylase. In vitro methylation completely protects bacteriophage lambda deoxyribonucleic acid against cleavage by R. *Eco*RII. *J Bacteriol* 1977;129:1330–1334.

Hattman S, Brooks JE, Masurekar M: Sequence specificity of the Pl-modification methylase (M. *Eco*Pl) and the DNA methylase (M. *Eco* dam) controlled by the *E. coli dam*-gene. *J Mol Biol* 1978;126:367–380.

Hattman S, Gribbin C, Hutchinson CA III: In vivo methylation of bacteriophage ϕ X174. *J Virology* 1979;32:845.

Hattman S: DNA methylation, in Boyer PD (ed): *The Enzymes.* New York, Academic Press, 1981, pp 15A, 517–548.

Hattman S: DNA methyltransferase-dependent transcription of the phage Mu *mom* gene. *Proc Natl Acad Sci USA* 1982;79:5518–5521.

Hays JB, Korba BE: DNA from recombinogenic λ bacteriophages generated by *arl* mutants of *Escherichia coli* is cleaved by single-strand specific endonuclease Sl. *Proc Natl Acad Sci USA* 1979;76:6066–6070.

Herman GE, Modrich P: *Escherichia coli* K-12 clones that overproduce *dam* methylase are hypermutable. *J Bacteriol* 1981;145:644–646.

Herman GE, Modrich P: *Escherichia coli dam* methylase. Physical and catalytic properties of the homogeneous enzymes. *J Biol Chem* 1982;257:2605–2612.

Karran P, Marinus MG: Mismatch correction at 06-methylguanine residues in *E. coli* DNA. *Nature* 1982;296:868–869.

Kondoh H, Ozeki H: Deletion and amber mutants of *fla* loci in *Escherichia coli* K-12. *Genetics* 1976;84:403–421.

Konrad EB: Method for the isolation of *Escherichia coli* mutants with enhanced recombination between chromosomal duplications. *J Bacteriol* 1977;130:167–172.

Korba BE, Hayes JB: Partially deficient methylation of cytosine in DNA at CCA/TGG sites stimulates genetic recombination of bacteriophage lambda. *Cell* 1982a;28:531–541.

Korba BE, Hayes JB: Novel mutations of *Escherichia coli* that produce recombinogenic lesions in DNA. V. Recombinogenic plasmids from *arl* mutants of *Escherichia coli* are unusually sensitive to nuclease S_1 and partially deficient in cytosine methylation at CC(A/T)GG sequence. *J Mol Biol* 1982b;157:213–235.

Korch C, Hagblom P, Normark S: Sequence-specific DNA modification in *Neisseria gonorrhoeae. J Bacteriol* 1983;155:1324–1332.

Kramer W, Schughart K, Fritz HJ: Directed mutagenesis of DNA cloned in filamentous phage: influence of hemimethylated GATC sites on marker recovery from restriction fragments. *Nucl Acids Res* 1982;10:6475–6485.

Lacks S, Greenberg B: A deoxyribonuclease of *Diplococcus pneumoniae* specific for methylated DNA. *J Biol Chem* 1975;250:4060–4066.

Lacks S, Greenberg B: Complementary specificity of restriction endonuclease of *Diplococcus pneumoniae* with respect to DNA methylation. *J Mol Biol* 1977;114:153–168.

Lacks SA, Dunn JJ, Greenberg B: Identification of base mismatches recognized by the heteroduplex-DNA-repair system of *Streptococcus pneumoniae. Cell* 1982;31:327–336.

Lark C: Studies on *in vivo* methylation of DNA in *Escherichia coli* $15T^-$. *J Mol Biol* 1968a;31:389–399.

Lark C: Effect of the methionine analogs, ethionine and nor-leucine, on DNA synthesis in *Escherichia coli* $15T^-$ DNA synthesis in *Escherichia coli* $15T^-$. *J Mol Biol* 1968b;31:401–414.

Lark C: Methylation-dependent DNA synthesis in *Escherichia coli* mediated by DNA polymerase I. *J Bacteriol* 1979;137:44–50.

Lieb M: A fine structure map of spontaneous and induced mutations in the lambda repressor gene including insertions of IS elements. *Mol Gen* Genet 1981;184:364–371.

Lieb M: Specific mismatch correction in bacteriophage lambda crosses by very short patch repair. *Molec Gen Genet* 1983;191:118–125

Lindahl T: DNA repair enzymes. *Ann Rev Biochem* 1982;51:61–87.

Little JW, Mount DW: The SOS regulatory system of *Escherichia coli. Cell* 1982;29:11–22.

Lu A-L, Clark S, Modrich P: Methyl-directed repair of DNA base pair mismatches in vitro. *Proc Natl Acad Sci USA* 1983;80:4639–4643.

Lundblad V, Kleckner N: Mutants of *Escherichia coli* K-12 which affect excision of transposon Tn*10*, in Lemontt JF, Generoso M (eds): *Molecular and Cellular Mechanisms of Mutagenesis.* New York, Plenum Press, 1982, pp 245–258.

Marinus MG: Location of DNA methylation genes on the *Escherichia coli* K-12 genetic map. *Mol Gen Genet* 1973;127:47–55.

Marinus MG, Morris NR: Isolation of deoxyribonucleic acid methylase mutants of *Escherichia coli* K-12. *J Bacteriol* 1973;114:1143–1150.

Marinus MG, Morris NR: Biological function for 6-methyladenine residues in the DNA of *Escherichia coli* K-12. *J Mol Biol* 1974;85:309–322.

Marinus MG, Morris NR: Pleiotropic effects of a DNA adenine methylation mutation *(dam-3)* in *Escherichia coli* K-12. *Mutat Res* 1975;28:15–26.

Marinus MG: Adenine methylation of Okazaki fragments in *Escherichia coli*. *J Bacteriol* 1976;128:853–854.

Marinus MG, Konrad EB: Hyper-recombination in *dam* mutants of *Escherichia coli* K-12. *Mol Gen Genet* (1976);149:273–277.

Marinus MG: The function of methylated bases in DNA of *Escherichia coli*, in Seeberg E, Kleppe K (eds): in *Chromosome Damage and Repair*. New York, Plenum Press, 1981, pp 469–473.

Marinus MG, Carraway M, Frey AZ, Brown L, Arraj JA: Insertion mutations in the *dam* gene of *Escherichia coli* K-12. *Molec Gen Genet* 1983;288–289.

May MS, Hattman S: Deoxyribonucleic acid cytosine methylation by host and plasmid controlled enzymes. *J Bacteriol* 1975a;122:129–138.

May MS, Hattman S: Analysis of bacteriophage deoxyribonucleic acid sequences methylated by host and R-factor controlled enzymes. *J Bacteriol* 1975b;123:768–770.

McGraw BR, Marinus MG: Isolation and characterization of Dam^+ revertants and suppressor mutations that modify secondary phenotypes of *dam-3* strains of *Escherichia coli* K-12. *Mol Gen Genet* 1980;178:309–315.

Meijer M, Beck E, Hansen G, Bergman HEN, Messer W, von Meyenburg K, Scholler H: Nucleotide sequence of the origin of replication of the *Escherichia coli* K-12 chromosome. *Proc Natl Acad Sci USA* 1979;76:580–584.

Muckerman CC, Springhorn SS, Greenberg B, Lacks SA: Transformation of restriction endonuclease phenotype in *Streptococcus pneumoniae*. *J Bacteriol* 1982;152:183–190.

Nene V, Glass RE: Genetic studies on the β subunit of *Escherichia coli* RNA polymerase. I. The effect of known, single amino acid substitutions in an essential protein. *Mol Gen Genet* 1982;188:399–404.

Ohmori H, Tomizawa JI, Maxam A: Detection of 5-methylcytosine in DNA sequences. *Nucl Acids Res* 1978;5:1479–1486.

Pirrotta V: Two restriction endonucleases from *Bacillus globiggi*. *Nucl Acids Res* 1976;3:1747–1760.

Plasterk RHA, Vrieling H, van de Putte P: Transcription initiation of Mu *mom* depends on methylation of the promoter region and a phage-coded transactivator. *Nature* 1983;301:344–347.

Pukkila P, Peterson J, Herman G, Modrich P, Meselson M: Effects of high levels of DNA adenine methylation on methyl directed mismatch repair in *E. coli*. *Genetics* 1983;104:571–582.

Razin A, Goren D, Friedman J: Studies on the biological role of DNA methylation: inhibition of methylation and maturation of the bacteriophage ϕX174 by nicotinamide. *Nucl Acids Res* 1975;2:1967–1974.

Razin A, Riggs AD: DNA methylation and gene function. *Science* 1980;210:604–610.

Razin A, Urieli S, Pollack Y, Greenbaum Y, Glazer G: Studies on the biological role of DNA methylation; IV. Mode of methylation in *E. coli*. *Nucl Acids Res* 1980;8:1783–1792.

Razin A, Friedman J: DNA methylation and its possible biological roles. *Prog Nucleic Acid Res Mol Biol* 1981;25:33–52.

Roberts RJ: Restriction and modification enzymes and their recognition sequences. *Nucl Acids Res* 1982;10:r117–r144.

Rydberg, B: Bromouracil mutagenesis and mismatch repair in mutator strains of *Escherichia coli. Mutat Res* 1978;52:11–24.

Ryokowski M, Meselson M: Methyl-directed DNA mismatch repair in *E coli. Proc Natl Acad Sci USA* (in press, 1984).

Santi DV, Garrett CE, Barr PJ: On the mechanism of inhibition of DNA-cytosine methyltransferases by cytosine analogs. *Cell* 1983;33:9–10.

Schein A, Bardahl BJ, Low M, Borek E: Deficiency of the DNA of *Micrococcus radiodurons* in methyladenine and methylcytosine. *Biochem Biophys Acta* 1972;272:481–485.

Schlagman S, Hattman S, May MS, Berger L: In vivo methylation by *Escherichia coli mec*$^+$ deoxyribonucleic acid-cytosine methylase protects λ against *in vitro* cleavage by the RII restriction endonuclease (R. *Eco*RII). *J Bacteriol* 1976;126:990–996.

Schlagman SL, Hattman S: Molecular cloning of a functional *dam*$^+$ gene coding for phage T4 DNA adenine methylase. *Gene* 1983;22:139–156.

Srivastava R, Gopinthan KP, Ramakrishnan T: Deoxyribonucleic acid methylation in Mycobacteria. *J Bacteriol* 1981;148:716–719.

Sugimoto K, Oka A, Sugisaki H, Takanami M, Nishimura A, Yasuda S, Hirota Y: Nucleotide sequence of *Escherichia coli* K-12 replication origin. *Proc Natl Acad Sci USA* 1979;76:575–579.

Sutcliffe JG: Complete nucleotide sequence of the *Escherichia coli* plasmid pBR322. *Cold Spring Harbor Symp Quant Biol* 1979;43:77–90.

Szyf M, Greenbaum Y, Urieli-Shoval S, Razin A: Studies on the biological role of DNA methylation. V. The pattern of *E. coli* DNA methylation. *Nucl Acids Res* 1982;10:7247–7259.

Toussaint A: DNA modification of bacteriophage Mu-1 requires both host and bacteriophage functions. *J Virol* 1977;23:825–826.

Urieli-Shoval S, Greenbaum Y, Razin A: Sequence and substrate specificity of isolated DNA methylases from *Escherichia coli* C. *J Bacteriol* 1983;153:274–280.

Vovis GF, Lacks S: Complementary action of restriction enzymes Endo R.*Dpn*I and Endo R.*Dpn*II on bacteriophage f1 DNA. *J Mol Biol* 1977;155:525–538.

Wagner RW, Meselson M: Repair tracts in mismatched DNA heteroduplexes *Proc Natl Acad Sci USA* 1976;73:4135–4139.

Zieg J, Maples VF, Kushner SR: Recombination levels of *Escherichia coli* K-12 mutants deficient in various replication, recombination or repair genes. *J Bacteriol* 1978;134:958–966.

6

DNA Methylation: DNA Replication and Repair

Robert H. Grafstrom*
Daniel L. Hamilton*
Robert Yuan*

Ever since the discovery of methylated bases in DNA, investigators have been intensely interested in correlating DNA base modification with biological functions. DNA has a linear informational dimension composed of both coding and control sequences. In addition, there may be a signal dimension represented by base modification. Consequently, in both prokaryotes and eukaryotes, DNA methylation has been postulated to have functional roles in DNA replication, repair, recombination, restriction-modification, mutation, transcriptional regulation, DNA packaging and chromosome structure, and differentiation (Razin and Riggs, 1980). Although the evidence for the involvement of methylation in restriction-modification in bacteria is well documented, its involvement in the remaining processes is, at best, circumstantial. Nevertheless, the heritable pattern of methylation anticipates its probable importance in some of the processes of DNA metabolism listed above. For this reason, the data relating DNA methylation and DNA replication should be reviewed and analyzed in light of recent advances in our understanding of DNA replication and repair.

Many organisms have a postreplicative modification of DNA at their cytosine or adenine residues, or at both. Both of these modifications are in the major groove and, thus, have the potential for affecting protein-DNA interactions.

Recently, work has intensified on the possible role of methylation of eukaryotic $-_5CpG_3-$ sequences in gene regulation. Although a causal relationship for methylation and gene expression has yet to be proven, evidence to date does show a correspondence between changes in methylation patterns and in gene expression. As a result, methylation also has been implicated in differentiation

*NCI-Frederick Cancer Research Facility, P.O. Box B, Frederick, Maryland 21701

in eukaryotic cells. There are difficulties with models linking undermethylation of DNA with gene expression. Among these are the lack of detectable DNA methylation in *Drosophila* (Urieli-Shoval *et al.* 1982) and the existence of genes that are fully expressed in spite of being methylated (Gerber-Huber *et al.* 1983). A way to reconcile these findings with proposed models for eukaryotic gene regulation is not immediately obvious.

To date, there is no evidence that CpG methylation has a role in either DNA mismatch repair or restriction-modification in eukaryotic cells. However, methylation may play a role in the phenomenon of maternal inheritance in chloroplasts (Sager and Kitchin, 1975). Given the heritable nature of DNA methylation, one might expect that DNA methylation should be integrally linked to DNA replication. In fact, Holliday (1979) has proposed a model, whereby DNA replication—if uncoupled from DNA methylation—could result in changes in the methylation pattern of the gene and subsequent gene expression after two rounds of DNA replication (Figure 6.1). Similarly, DNA-damaging agents also could apparently turn on genes, if the methylation pattern is not restored after DNA repair. Hence, this model provides a testable framework with which to investigate the relationship between DNA methylation and DNA replication and repair, and it suggests both necessary and sufficient properties for the relevant DNA methylases. To this end, we will examine the nature of DNA methylation as it relates to DNA replication from a temporal point of view, since a lag in methylation following replication might lead to changes in gene expression, increased mutation frequency, or DNA degradation. We also will examine the data suggesting that DNA methylation is required for replication in light of what is now known about the various types of restriction-modification systems. Because DNA repair might also lead to

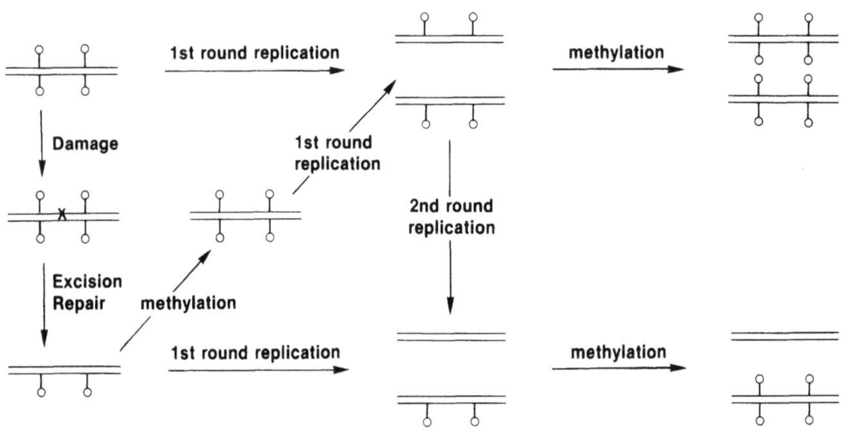

Figure 6.1 A model for the maintenance and/or alteration of DNA methylation patterns during DNA replication and repair, as proposed by Holliday (1979). The symbol (⚲) represents the methylated bases.

heritable changes of DNA methylation patterns, we will review the data relating DNA methylation and DNA repair. Finally, we will examine some of the properties of the DNA methylases to establish how they meet the conditions imposed by the model presented in Figure 6.1. A review of all the pertinent data should enable us to identify those facts of which we can be certain, those that require further investigation, and those that raise new questions.

Is Nascent DNA Methylated?

The semiconservative nature of DNA synthesis predicts that if methylation of DNA is heritable, then DNA methylation will occur only on the newly replicated strand. This problem was first studied in prokaryotes by Lark (1968a) and Billen (1968). They both demonstrated that labeling of *Escherichia coli* DNA in the presence of bromodeoxyuridine (BUdR) and [^{3}H]methionine for periods longer than 15 minutes (Billen, 1968) or 30 minutes (Lark, 1968a) resulted in methylation of DNA of hybrid density. When the resulting hybrid-density DNA was denatured, all of the [^{3}H]methyl was incorporated into the heavy strand. Since there was no detectable methylation of parent DNA, these results showed that all of the DNA methylation occurred in the newly synthesized strand.

The methylation of nascent DNA in eukaryotes was demonstrated by Kappler (1970), Burdon and Adams (1969), and Adams (1971) by using mouse cells. Both Burdon and Adams (1969) and Kappler (1970) showed that after the addition of hydroxyurea, which inhibits DNA replication by ~97%, DNA methylation is lowered by ~85%; this suggests that newly synthesized DNA is methylated. By density labeling the DNA with BUdR and isotopically labeling the DNA with methionine, they also were able to demonstrate that most of the methyl incorporated into the DNA was in the heavy strand. These results were repeated in cultured *Xenopus* cells by Bird (1978). However, all three investigators reported a significant proportion of methylation into DNA of light density, which could indicate: 1) methylation of daughter strands synthesized before BUdR was added; 2) methylation of the parent strand; or 3) methylation of DNA repair tracts. These results are discussed in more detail in further sections in this chapter. Nonetheless, in both prokaryotic and eukaryotic cells, most of the DNA methylation occurs on the newly replicated strand.

An important consideration in all of these experiments is the effect of BUdR on DNA methylation. Lark (1968a) has shown that there is no effect of BUdR on DNA methylation in *E. coli*. Similar results also have been reported for mammalian DNA methylases (Singer *et al.* 1977). An additional important point is the equilibration of size of the methionine pools in these types of experiments. Both Kappler (1970) and Lark (1968a) showed that these pools were equilibrated very rapidly.

Is Methylation Required for Replication?

Based on the model in Figure 6.1, one could predict that DNA methylation and replication would be tightly linked to prevent the loss of the methylation pattern, resulting in gene expression. However, similar logic might also predict that DNA replication would not proceed too far in the absence of DNA methylation. Both Lark (1968a) and Billen (1968) were able to uncouple DNA methylation from replication by starving *E. coli* methionine auxotrophs for methionine. Lark (1968b) tried to show that unmethylated DNA would not be replicated in a second round of DNA synthesis. However, these experiments (as well as those of Billen, 1968) were complicated by the use of the strain *E. coli* 15T$^-$, which contained both type I and type III restriction enzymes that cleave unmethylated DNA (Lark and Arber, 1970). Because type III restriction sites can be methylated on only one strand, the first round of replication in the absence of methylation results in degradation of 50% of the DNA. Type I restriction sites become available for cleavage after two rounds of replication in the absence of methionine. Thus, Lark's (1968b) conclusion that methylation is required for DNA replication is sharply biased by the presence of these restriction enzymes. In fact, her later experiments convincingly demonstrated that unmethylated DNA can be replicated (Lark and Arber, 1970) and that DNA synthesis stops not because of the lack of DNA methylation, but because of the lack of protein synthesis (Lark, 1979). Many of the other phenomena involved in her early papers also can be related to the different specificities of type I and type III restriction enzymes. However, these experiments did demonstrate that subsequent DNA methylation will take place on DNA that is replicated in the absence of methylation. Nonetheless, the fact that *dam*$^-$ strains appear to grow normally in spite of the problems with DNA repair also suggests that DNA methylation is not required for DNA replication. It could be argued that these *dam*$^-$ mutants are leaky and that only a low level of DNA methylation is required for DNA synthesis; however, the best available data suggests that this is unlikely.

Similar experiments also have been performed in mammalian cells. Adams (1974) showed that methionine starvation for up to 4 hours has no effect on DNA synthesis, and that methionine starvation for 8 hours reduces DNA replication by only 25%. In a different experiment, Culp and Black (1971) showed that an ~50% drop in [^{14}C]-BUdR incorporation occurs during 8 hours of methionine deprivation. When considered together, it appears that DNA methylation is not an absolute prerequisite for DNA synthesis; however, they may, in fact, be coupled *in vivo* in both prokaryotes and eukaryotes.

How Soon After Replication Does Methylation Occur?

Because DNA synthesis is required before DNA methylation can occur (parent DNA is not methylated), it is important to know how soon after replication that DNA methlation occurs. If the lag is too long, the probability of losing the methylation pattern increases. In mammalian cells in which DNA is synthesized in S phase, DNA methylation also has been shown to occur during S phase via several experiments (Burdon and Adams, 1969; Adams and Hogarth, 1973; and Kappler, 1970). Other reports indicate that DNA methylation might occur during mitosis (Evans and Evans, 1970; Bugler *et al.* 1980); however, both of these reports suffer from a lack of documentation on the stringency of cell synchrony and from the use of drugs that may have induced DNA repair.

An interesting aspect of DNA methylation and replication in S phase is that several investigators have shown a 1-hour lag between the maximum rate of DNA replication and that of DNA methylation (Burdon and Adams, 1969; Kappler, 1970; Geraci *et al.* 1974). As Kappler (1970) pointed out, this 1-hour lag could be due to replication of DNA with different ratios of 5-methylcytosine to thymine (e.g., heterochromatin). Adams (1971) showed that in phytohemagglutinin-stimulated lymphocytes, replication precedes DNA methylation by as much as 10 hours. However, in synchronized mouse fibroblasts, he found no discernible lag between the onset of DNA methylation and DNA synthesis (Adams, 1971).

In experiments with BUdR, both Kappler (1970) and Adams (1971) were able to detect ^{3}H methyl labeling in light/light DNA after 1 hour and 3 hours, respectively. These results were interpreted to mean that DNA methylation of nascent DNA lagged behind DNA synthesis by about 1 hour. Using uniformly labeled [^{14}C]deoxycytidine, Kappler detected only a 4-minute lag between replication and methylation—with 4.4% of the deoxycytidine being methylated. Since this value agrees with the 4.3% level of 5-methyldeoxycytidine that is detected when cells are continuously labeled and assayed for total DNA methylation, these results suggest that methylation occurs as soon as the DNA is replicated. The contradiction between the results with [^{3}H]methionine and those with [^{14}C]deoxycytidine could be partly explained by the probable contamination of the uniformly labeled deoxycytidine with 5-methyldeoxycytidine, which could be directly incorporated into the DNA.

By examining DNA methylation in purified nuclei that were labeled with BUdR at different times before harvest, Adams and Hogarth (1973) were able to show methylation of DNA that was synthesized within 10 minutes of harvesting the cells. They also were able to detect methylation of DNA that had been synthesized more than 4 hours before harvesting. They stated that they can detect 14% of the methylation after 2 minutes of DNA synthesis, calculating that this represents methylation of only 0.4% of the total DNA synthesized in one generation. These calculations suggest an overmethylation of DNA at the growing fork and subsequent demethylation. However, this conclusion disagrees with earlier data demonstrating the stability of methylated DNA *in vivo* (Burdon and Adams, 1969). There has been a preliminary report of a DNA 5-methyldeoxycytidine demethylase, but the actual identification of this activity is dependent on its further purification (Gjerset and Martin, 1982). Nonetheless, these results do seem to show that DNA methylation is not complete for several hours after DNA synthesis. More recently, Gruenbaum *et al.* (1983) suggested that, based on their studies with mouse L cells, the lag between replication and methylation may be as short as 75 seconds. These investigators permeabilized the cells in hypotonic buffer, labeled them with [α^{32}P]dGTP, and performed nearest neighbor analysis.

In *E. coli,* Lark (1968a) showed that only a small percentage of the DNA labeled *in vivo* with [^{3}H]methionine and BUdR banded in the light/light region of the CsCl gradient with DNA that had been synthesized before the 30-minute pulse. In another experiment, she also showed that most of the DNA is methylated soon after it is synthesized. These results agree with those of Billen (1968), who demonstrated that DNA synthesized in the presence of BUdR and [^{3}H]methionine for 2 minutes was methylated and banded in a broad heavy/light region on native CsCl gradients. When this DNA was denatured and analyzed again with CsCl gradients, no methylated DNA was detected with light DNA, which suggests that methylation occurred as the DNA was being synthesized.

Szyf *et al.* (1982) confirmed these earlier findings by pulse labeling DNA with ^{32}Pi, isolating the DNA, and assaying for methylation via cleavage either with *Dpn*I (a restriction enzyme that cleaves only methylated DNA) or *Mbo*I (which cleaves only unmethylated DNA) at -GATC- sequences. Although lacking in quantitation, their results—which show that DNA is sensitive to cleavage by *Dpn*I but not *Mbo*I—suggest that DNA is methylated soon after it is replicated (within 30 seconds).

Hence, in both mammalian cells and *E. coli,* most of the DNA methylation appears to occur as the DNA is being replicated. However, in eukaryotic cells, a significant proportion of DNA is methylated after replication; this is not the case for prokaryotic cells. The size of the replicated and methylated DNA was not determined in any of these experiments.

Does DNA Methylation Occur at Okazaki Fragments?

The experiments described in the previous section suggested a very close temporal coupling of DNA methylation to DNA replication. Consequently, to determine the earliest point in replication at which methylation can occur, one should ask whether or not Okazaki fragments are methylated. Although, at first glance, these studies might appear to be an esoteric exercise, it may provide important information in light of the fact that undermethylation of daughter strands is the basis for the model of DNA mismatch repair (Wagner and Meselson, 1975). If methylation and replication occur simultaneously, then the daughter strand would be fully methylated and some other mechanism for recognizing the misincorporated bases would have to be found. DNA methylation of Okazaki fragments would not *a priori* preclude its involvement in DNA mismatch repair. The only problem is the mechanism by which mismatched bases would be recognized.

The first experiments in attempting to answer this problem in eukaryotes were those of Adams (1974). Although he was able to identify putative Okazaki fragments in mouse cells with *in vivo* [^{3}H]thymidine labeling of the DNA, he was unable to detect methylation of [^{14}C]deoxycytidine in these fragments. However, because he only incorporated 424 cpm of [^{14}C] deoxycy tidine, his experiments were too insensitive to detect methylation with a frequency of 3%. Drahovsky and Wacker (1975) were also unable to detect any methylation of 9S DNA in Ehrlich ascities cells *in vivo* by using either [^{3}H]methionine or [^{14}C]deoxycytidine and by analyzing for 5-methylcytosine on thin-layer chromatography. The only other published attempts to measure methylation of Okazaki fragments in mammalian cells were those of Vanyushin's group (Demidkina *et al.* 1979; Kiryanov *et al.* 1980). They too were able to identify presumptive Okazaki fragments in mouse fibroblasts. After labeling the cell *in vivo* with [3H]methionine, they detected incorporation of the ^{3}H label (80–100% of which is in 5-methylcytosine) into these small DNA pieces and showed limited elongation of these fragments into higher molecular weight (mol wt) DNA after a 3-hour incubation. Unfortunately, the size pattern is dependent on cell density and the fragments are elongated into two to three times their original mol wt after a 3-hour chase—a result for which there is no adequate explanation. However, the validity of these findings is jeopardized by the long incubation periods needed to label the Okazaki fragments (5–10 minutes versus 30 seconds) (Tseng and Goulian, 1975) and by the lack of documentation of whether these fragments were authentic intermediates of DNA replication or were the products of DNA repair. Still, these experiments are the only reports of methylation of Okazaki fragments in mammalian cells.

Escherichia coli studies also have been attempted to determine whether or not Okazaki fragments were methylated (Marinus, 1976; Gomez-Eichelman and Lark, 1977). The latter experiments used the restriction enzyme *Dpn*I to detect methylation, because (as mentioned previously) *Dpn*I cuts DNA at -GATC- sequences only if both strands are methylated. Using an *in vitro* DNA-synthesizing system, they were able to demonstrate that *E. coli* DNA, when cut with *Dpn*I, was the same size as Okazaki fragments; they hypothesized that -GATC- sequences may play a role in DNA replication as a signal for discontinuous synthesis. Their results showed that Okazaki fragments that were synthesized *in vitro* in the presence of nicotinamide mononucleotide (NMN) to inhibit ligation were resistant to cutting with *Dpn*I, whether they were synthesized in the presence or absence of AdoMet. These results were interpreted to mean that Okazaki fragments contain -GATC- sequences that are not randomly located, but are near the termini; therefore, shifts in their mol wt were not detected after enzyme digestion. There are four major problems with this study. First, the limit digests of DNA in this crude lysate system yielded DNA fragments that were the same size as Okazaki fragments, indicating a frequency of -GATC- sequences of 1 per kilobase (kb); however, Szyf *et al.* (1982) reported limit digests of purified *E. coli* DNA with *Dpn*I that yielded fragments of approximately 0.3 kb. These data suggest that the experiment of Gomez-Eichelmann and Lark did not result in a limit digest. Second, Razin *et al.* (1975) had shown that nicotinamide was an effective inhibitor of

Dam methylase; therefore, the use of NMN as a ligase inhibitor also may have reduced the ability of the Dam methylase to methylate Okazaki fragments. In fact, methylation of Okazaki fragments was detected only when the fragments were allowed to elongate in the presence of β-diphosphopyridine nucleotide after NMN had been removed. Whether this was related to removal of the NMN or to the ligation of the fragment into higher mol wt DNA is not known. Third, the contention that Okazaki fragments would be smaller than chromosomal DNA after *Dpn*I digestion assumed both the random distribution of -GATC- sequences, as well as the nonrandom initiation of Okazaki fragments. However, if -GATC- sequences in the chromosome are random, and if discontinuous synthesis initiates at random, then one would expect the same size distribution for digests of both Okazaki fragments and the entire chromosome. Finally, the direct experiment of labeling the Okazaki fragment with [^{3}H]AdoMet *in vitro*—which would have detected both adenine and cytosine methylation and would have been conclusive—was never performed. Nonetheless, their demonstration that the synthesis of Okazaki fragments in this *in vitro* system was the same for both *dam*$^-$ and wild-type *E. coli* suggests that *dam* methylation is not necessary for discontinuous DNA synthesis.

In the earlier experiments of Marinus (1976), the methylation of Okazaki fragments *in vivo* was detected by the combined use of *E. coli* temperature-sensitive ligase mutants *(lig-4)* and labeling with [^{3}H]adenine. Okazaki fragments (10S DNA) were isolated from alkaline sucrose sedimentation velocity gradients and were analyzed for the percentage of 6-methyladenine. His results showed that 0.96 mol% of the adenines were methylated in the Okazaki fragments and that 1.47 mol% of adenines in bulk DNA were methylated. These results demonstrated that Okazaki fragments were methylated *in vivo* (although at a lower rate than chromosomal DNA) and suggested that there may indeed be a lag after replication before the newly synthesized DNA is fully methylated. Similar results (i.e., undermethylation of Okazaki fragments) also were reported by Kiryanov *et al.* (1980) for mouse fibroblasts; he detected that 2.7% of the cytosine was methylated in 5S DNA compared to 4.4% in chromosomal DNA.

In conclusion, methylation of adenine does occur in Okazaki fragments during replication of *E. coli* DNA; but, the evidence for methylation of cytosine in Okazaki fragments during replication of eukaryotic DNA is equivocal. However, in both cases, the model for DNA mismatch repair involving undermethylation of the daughter strand may still be valid in lower—as well as in higher—organisms, because Okazaki fragments appear to be undermethylated.

Is the Pattern of DNA Methylation Restored After Repair?

The model (Figure 6.1) proposes that DNA methylation must occur after the repair of DNA damage to restore the heritable pattern of methylation. This feature presupposes at least two classes of DNA methylases: one linked to

DNA replication and one linked to restoration of the informational integrity of the chromosome that is linked to DNA repair. Although the mechanism for DNA mismatch repair is not fully understood, overwhelming genetic evidence supports its existence in *E. coli* and other prokaryotes. Work is now focusing on the biochemical mechanism by which DNA mismatch repair is directed to the daughter strand through the purification of the putative repair proteins, MutH, MutL, and MutS (Grafstrom and Hoess, 1983; Lu *et al.* 1983). The study of DNA mismatch repair in eukaryotic cells is only in its infancy, and comparisons here would be premature. However, *E. coli* possesses a heritable pattern of DNA methylation, and comparisons can be made between the tendency of *E. coli* to maintain a pattern of DNA methylation and a similar process in eukaryotes. Recent experiments with eukaryotic cells have focused on the effect of chemical carcinogens on DNA methylation and gene expression, as well as on the DNA methylases themselves. Thus, it is important to understand both the mechanism by which DNA methylation protects DNA before DNA damage (mismatch repair) and the manner in which the correct methylation pattern is restored after DNA repair.

The earliest studies of DNA methylation and repair in prokaryotes are those of Ryan and Borek (1971) and Whitfield and Billen (1972). Both studies deal with the nature of methylation after DNA damage by ultraviolet (UV) radiation and address the problem of both adenine and cytosine methylation. Both studies also demonstrated the existence of damage-dependent DNA methylation, although in one case, the DNA is hypermethylated (Ryan and Borek, 1971) and in the other (Whitfield and Billen, 1972), it is hypomethylated. The ratio of 6-methyladenine to 5-methylcytosine is unchanged in experiments in which DNA synthesis still occurred (Whitfield and Billen, 1972). However, the ratio of 6-methyladenine to 5-methylcytosine was significantly higher in the absence of repair replication due to the unequal loss of adenine and cytosine methylation (Ryan and Borek, 1971). Furthermore, DNA methylation occurred only in the repaired region and not in the undamaged regions of the parent strands (Whitfield and Billen, 1972). The use of *E coli* $15T^-$ by Billen probably did not affect these results, because he did not starve the cells for methionine, which would have resulted in DNA restriction in this strain (Lark and Arber, 1970). The major problem in interpreting any of these results, however, is the lack of quantitation of the dose of UV radiation. Neither report states the effect of the dose on cell survival; although Ryan used a dose that was an order of magnitude higher than Whitfield's, the actual effective doses might be similar. The complicated cascade of events (SOS repair) that follows UV irradiation of *E. coli* might adversely affect studies of DNA methylation and repair, and these problems should be reinvestigated. For example, Ryan's results suggest that the Dam methylase was suppressed by UV radiation, but other investigators (Dunn and Smith, 1958; Yudelevich and Gold, 1969) have shown just the opposite. DNA methylation, in fact, may be linked to SOS repair, because *dam* mutants are slightly sensitive to UV radiation and because *dam recA, dam recB, dam recC,* and *dam lexA* double mutants are lethal to the cell (Marinus and Morris, 1974,1975). This linkage of DNA methylation to DNA repair in *E. coli* also might have been demonstrated by Lark (1979). In these experiments, treatment of *E. coli* methionine auxotrophs *in vivo* with ethionine induced a *polI*-dependent DNA synthesis, which suggests that ethionine was acting as either a direct or indirect DNA-damaging agent. This DNA synthesis was dependent on both AdoMet and ATP; although the role for AdoMet and/or DNA methylation in this repair synthesis remains unresolved, these provocative results warrant further study.

As mentioned previously, Adams (1971), Kappler (1970), and Bird (1978) detected methylation of parent DNA in mouse cells. Further investigation of this phenomenon by Hilliard and Sneider (1975) demonstrated that the methylation of this preexisting DNA was due to methylation of DNA repair patches. Using [^{3}H]deoxycytidine, they found that these DNA repair tracts were overmethylated. They also found that the DNA repair appeared to be caused by the method of cell synchrony, because cells synchronized by using a mitotic-G1 arrest method showed no DNA repair-directed DNA methylation, whereas cells synchronized with an S phase block did. After damaging peripheral human lymphocytes with nitrogen mustard, Drahovsky *et al.* (1976) showed that DNA repair-directed DNA methylation occurred in both repetitive and unique sequences; however, repetitive sequences were undermethylated, whereas unique sequences appeared overmethylated. Boehm and Drahovsky (1981) showed that N-methyl-N-nitrosourea (MNU) was able to inhibit the extent of methylation in Raji cells. Treatment with levels of MNU that were said to have only slight effects on cell proliferation caused a nearly 25% decrease in the level of DNA methylation. This decrease persisted for 7 days after treatment, as measured by the conversion of [^{14}C]deoxycytidine to 5-methyl-[^{14}C]deoxycytidine. Analysis of the bulk DNA with the restriction enzymes *Hpa*II and *Msp*I indicated that this decrease in 5-methylcytosine content was caused by a mechanism that invoked a hemimethylated DNA intermediate, which was completely unmethylated after replication. More recently, Lieberman's group (Kastan *et al.* 1982) also showed that confluent human diploid fibroblasts were able to methylate DNA after excision repair in cells that had been damaged by UV radiation, N-acetoxy-2-acetylaminofluorene (NAAAF), or MNU. This damage-directed DNA methylation occurred at a reduced rate and to a lesser extent, thus yielding only 1.5% methylation of DNA repair patches as compared to 3.4% in parent DNA. Even in logarithmically growing cells, these investigators observed only 2.5% methylation in repair patches, with no skewing of either repair or methylation in different classes of repetitive DNA. Interestingly, the undermethylated repair patches could be further methylated after a round of DNA replication. Although none of these studies measured the effect of the damaging agents on cell survival, the data suggest that DNA damage and repair might lead to abnormal methylation patterns, which could result in heritable changes of gene expression at the damaged site.

Although all of the studies mentioned above deal with the methylation of DNA after DNA repair, recent experiments have focused on the direct effect of the carcinogens on the DNA methyltransferase itself. Salas *et al.* (1979)

showed that purified DNA damaged with N-acetoxy-2-acetylaminofluorene, which forms C8 adducts with guanine, was a poor substrate for a partially purified DNA methyltransferase from rat brain. These studies were extended to show that DNA containing 2-acetylaminofluorene acts as an irreversible inhibitor of the DNA methylase reaction. These results were investigated in more detail by Wilson and Jones (1983), using a hemimethylated DNA substrate that is prepared by treating cells *in vivo* with azacytidine and a crude preparation of DNA methyltransferase. They were able to detect significant inhibition of the methylase activity when the hemimethylated substrate was treated with various DNA-damaging agents. This effect was dose-dependent and appeared to be caused by adduct formation—not simply by single- or double-strand breaks in the DNA, although these breaks also inhibited activity of the enzyme. Thymine dimers had no effect on the enzyme activity. Treatment of the enzyme with the carcinogens themselves (N-methyl-N-nitro-N-nitrosoguanidine [MNNG], ethylnitrosourea [ENU], or nitrogen mustard [HN2]) also substantially reduced the activity of the enzyme. These experiments suffer from the use of a large molar ratio of carcinogen to DNA and the lack of quantitation of DNA damage, except for DNA adducts of benzo-(α)pyrene diolepoxide and apurinic lesions. Many of the DNA lesions produced by the carcinogens are well characterized. It should have been possible to determine which lesion caused the inhibition of the enzyme. In addition, the effect of single- or double-strand breaks of DNA on enzyme activity could have been better defined by the use of enzymatic digests of DNA, instead of UV irradiation and sonication of BUdR-containing DNA. It is well known that treatment of DNA by lowering the pH to 1.6 not only introduces apurinic lesions in the DNA, but also denatures it. This means that single-stranded DNA, not apurinic sites, could be responsible for the loss of enzymatic activity. Finally, in the *in vivo* experiments with benzo(α)pyrene, they detected a drop in DNA methylation similar to that seen by Boehm and Drahovsky (1981). However, no quantitation of cell damage was reported.

In conclusion, in both prokaryotic cells and eukaryotic cells, methylation of DNA repair tracts does occur after DNA damage excision repair, and abnormal methylation may cause heritable changes in gene expression. However, the problems that investigators have confronted in studying DNA repair (e.g., quantitation of damage, multiple adduct formation, and demonstration of causal relationships between DNA damage and lethality or mutagenicity) cannot be ignored when studying DNA methylation. They only add to the complexity of the problem.

Conclusions

DNA methylation appears to be an important function for most cells, because they expend a considerable amount of energy preserving their methylation pattern and passing it on to their daughter cells. The model presented in Figure

6.1 has provided a useful framework with which to examine DNA methylation and its relationship to DNA replication and repair. In both prokaryotes and eukaryotes, an uncoupling of these processes could lead to DNA degradation, increased mutation frequency, or heritable changes in gene expession. Therefore, the cell must be able to regulate the amount, type, and timing of DNA methylation.

To maintain the heritable pattern of methylation requires two events: 1) methylation of the newly synthesized DNA strand after semiconservative replication; and 2) methylation of repair tracts. Repair methylation probably proceeds by a continuous scanning of the DNA for hemimethylated repair patches. Methylation of newly replicated DNA could be carried out by different mechanisms: 1) direct coupling of DNA methylation to DNA replication; 2) DNA synthesis requiring DNA methylation; and 3) sequence-specific methylation of DNA. As shown in the preceding sections, DNA methylation is linked to DNA replication, but they can be uncoupled; DNA methylation requires DNA replication, but DNA replicated in the absence of methylation can be methylated later on. Thus, DNA replication does not appear to require DNA methylation. However, it must be noted that the physiologic state produced by attempts to inhibit DNA methylation would interfere with protein synthesis, which in turn could result in the arrest of DNA replication. As for the third possiblity, the purified Dam methylase acts equally well on the hemimethylated or unmethylated -GATC- sequences and the partially purified eukaryotic methylases do not require hemimethylated DNA, nor do they show any strong sequence specificity. Furthermore, unlike the bacterial methylases, which only methylated heterologous DNA substrates, the mammalian DNA methylases methylated homologous DNA. This fact is consistent with both the methylation of Okazaki fragments at the replication fork in prokaryotic cells and the existence of a lag between DNA methylation and replication (or at least an undermethylation of newly replicated DNA) in eukaryotic cells.

In prokaryotes, there are two important reasons for maintaining the correct methylation pattern. One is protection from any endogenous restriction enzyme. The other is DNA repair, for which there is ample genetic and biochemical evidence to support a mechanism of methyl-directed mismatch correction. Since the *in vitro* fidelity of bacterial DNA polymerases is much lower than the incidence of mutation, the cells have developed a postreplicative mechanism for correcting misincorporated nucleotides. It has been proposed that this repair process discriminates between the undermethylated daughter strand and the fully methylated parent strand to direct the repair machinery to the incorrect base. In *E. coli,* the lack of homologous DNA methylation by the purified methylase does not preclude the existence of methyl-directed mismatch repair, but it may be indicative of the low level of undermethylation necessary for a mismatch sytem to operate. This system must work rapidly and efficiently in prokaryotes, since the best evidence supports DNA methylation at the replication fork in the form of methylation of Okazaki fragments. Therefore, the predominant form of methylation in prokaryotes appears to be related to a housekeeping function; that is, maintaining the informational integrity of the DNA.

Based on the knowledge of prokaryotic systems, we could assume that most of the eukaryotic DNA methylation is also associated with housekeeping functions, such as DNA repair. Although there is no direct evidence for mismatch repair in eukaryotes, the rate of nucleotide misincorporation may be increased by the lack of a proofreading exonuclease associated with the DNA polymerase. This fact argues for the existence of a very efficient mismatch repair system to insure the accuracy of the information encoded by the DNA. The need for a methyl-directed mismatch repair system could explain the undermethylation of the chromosome after DNA replication, as is evidenced by the ability of eukaryotic methylases to methylate homologous DNA. After DNA replication and repair, the DNA methylase must act to restore the methylation pattern in the DNA repair tracts. Subsequently, the daughter chromosomes are separated, folded, and packaged as a prelude to mitosis. Methylation might also have an important role in all of these processes. Thus, different (or the same) methylation patterns of limited specificity could be used as signals for DNA repair and determination of chromosome structure.

Superimposed on the generalized methylation pattern could be a subset of CpG methylation that is involved in the regulation of gene expession. By necessity, these "regulatory" methylases (or the corresponding demethylases) would have to be highly specific to turn on or off specific genes. So far, there has been no indication that eukaryotic cells contain a large number of different methylases (or demethylases); *in vitro* experiments with partially purified DNA methylases give no evidence of the type of seqeunce specificity seen with the bacterial methylases. This puzzling lack of specificity can be explained in several different ways: 1) the regulatory methylases represent a very small fraction of the total methylase activity and cannot be detected against the high background of generalized methylase activity; 2) there is only one catalytic subunit, and its specificity is defined by its interaction with other protein subunits and/or with membrane components; 3) the specificity is defined by the substrate that might not be simply DNA, but a protein-DNA complex (e.g., a nucleosome structure); and 4) DNA methylation (or its absence) is necessary, but not sufficient, for control of gene expression (i.e., a second enzyme system would sense the methylation pattern as a signal for interacting with the gene). In a sense, this is what presumably occurs with mismatch repair; the enzyme system recognizes the new strand by its undermethylation, but then interacts specifically with the incorrect base. Consequently, whereas housekeeping methylases would prefer hemimethylated DNA, regulatory methylases might recognize an unmethylated DNA sequence. Once the regulatory methylase modifies a gene, the site is transferred to the domain of the housekeeping methylase to preserve the methylation pattern for subsequent generations.

Our understanding of the nature and timing of DNA methylation in prokaryotes has required a combination of biochemical and genetic techniques; similar ones also will be needed to fully understand the eukaryotic methylases.

Thus far, these enzymes have not been purified to homogeneity nor subjected to a detailed analysis of their reaction mechanisms. Certainly, questions pertaining to both the substrate and the sequence specificity of eukaryotic methylases will require just as sophisticated techniques as those used in the studies of the prokaryotic enzymes. Recombinant DNA technologoy has given us a glimpse of the complexity of this problem. It is this technology that will provide us with the cloned genes, the collections of mutants, and the complex DNA substrates with which to better understand the biological roles of eukaryotic DNA methylation.

Acknowledgments

The authors would like to thank Nat Sternberg, Ph.D. and Cheeptip Benyajati, Ph.D. for their helpful comments on this manuscript.

Research was sponsored by the National Cancer Institute, Department of Health and Human Services, under Contract No. NO1-CO-23909 with Litton Bionetics, Inc. The contents of this publication do not necessarily reflect the views or policies of the Department of Health and Human Services, nor does mention of trade names, commercial products, or organizations imply endorsement by the United States Government.

References

Adams RLP: The relationship between synthesis and methylation of DNA in mouse fibroblasts. *Biochim Biophys Acta* 1971;254:205–212.

Adams RLP, Hogarth C: DNA methylation in isolated nuclei: old and new DNA's are methylated. *Biochim Biophys Acta* 1973;331:214–220.

Adams RLP: Newly synthesized DNA is not methylated. *Biochim Biophys Acta* 1974;335:365–373.

Billen D: Methylation of the bacterial chromosome: an event at the "replication point?" *J Mol Biol* 1968;31:477–486.

Bird AP: Use of restriction enzymes to study eukaryotic DNA methylation: II. The symmetry of methylated sites supports semiconservative copying of the methylation pattern. *J Mol Biol* 1978;118:49–60.

Boehm TLJ, Drahovsky D: Hypomethylation of DNA in Raji cells after treatment with N-methyl-N-nitrosourea. *Carcinogenesis* 1981;2:39–42.

Bugler B, Bertaux O, Valencia R: Nucleic acids methylation of synchronized BHK 21 HS 5 fibroblasts during the mitotic phase. *J Cell Physiol* 1980;103:149–157.

Burdon RH, Adams RLP; The in vivo methylation of DNA in mouse fibroblasts. *Biochim Biophys Acta* 1969;174:322–329.

Culp LA, Black PH: DNA synthesis in normal and virus-transformed mammalian cells after methionine deprivation. *Biochim Biophys Acta* 1971;247:220–232.

Demidkina NP, Kiryanov GI, Vanyushin BF: Methylation of newly synthesized DNA in culture of mouse fibroblasts. *Biochem USSR* 1979;44:1115–1123.

Drahovsky D, Wacker A: Enzymatic methylation of replication-DNA intermediates in Ehrlich ascites tumor. *Naturwissenschaften* 1975;62:189–190.

Drahovsky D, Lacko I, Wacker A: Enzymatic DNA methylation during repair synthesis in non-proliferating human peripheral lymphocytes. *Biochim Biophys Acta* 1976;447:139–143.

Dunn DB, Smith JD: The occurrence of 6-methylaminopurine in deoxyribonucleic acids. *Biochem J* 1958;68:627–636.

Evans HH, Evans TE: Methylation of the deoxyribonucleic acid of physarum polycephalum at various periods during the mitotic cycle. *J Biol Chem* 1970;245:6436–6441.

Geraci D, Eremenko T, Cocchiara R, Granieri A, Scarano E, Volpe P: Correlation between synthesis and methylation of DNA in Hela cells. *Biochem Biophys Res Comm* 1974;57:353–358.

Gerber-Huber S, May FEB, Westley BR, Felber BK, Hosbach HA, Andres A-C, Ryffel GU: In contrast to other Xenopus genes the estrogen-inducible vitellogenin genes are expressed when totally methylated. *Cell* 1983;33:45–51.

Gjerset RA, Martin DW Jr: Presence of a DNA demethylating activity in the nucleus of Murine erythroleukemic cells. *J Biol Chem* 1982;257:8581–8583.

Gomez-Eichelmann MC, Lark KG: Endo R *Dpn*I restriction of Escherichia coli DNA synthesized in vitro. Evidence that the ends of Okazaki pieces are determined by template deoxynucleotide sequence. *J Mol Biol* 1977;117:621–635.

Grafstrom RH, Hoess RH: Cloning of *mutH* and identification of the gene product. *Gene* 1983;22:245–253.

Gruenbaum Y, Cedar H, Razin A: Substrate and sequence specificity of a eukaryotic DNA methylase. *Nature* 1982;295:620–622.

Gruenbaum Y, Szyf M, Cedar H, Razin A. Methylation of replicating and post-replicated mouse L-cell DNA. *Proc Natl Acad Sci USA* 1983;80:4919–4921.

Hilliard JK, Sneider TW: Repair methylation of parental DNA in synchronized cultures of Novikoff hepatoma cells. *Nucl Acid Res* 1975;2:809–819.

Holliday, R: A new theory of carcinogenesis. *Br J Cancer* 1979;40:513–522.

Kappler JW: The kinetics of DNA methylation in cultures of a mouse adrenal cell line. *J Cell Physiol* 1970;75:21–32.

Kastan MB, Gowans BJ, Lieberman MW: Methylation of deoxycytidine incorporated by excision-repair synthesis of DNA. *Cell* 1982;30:509–516.

Kiryanov GI, Kirnos MD, Demidkina NP, Alexandrushkina NI, Vanyushin BF: Methylation of DNA in L cells on replication. *FEBS Letters* 1980;112:225–228.

Lark C: Studies on the in vivo methylation of DNA in Escherichia coli $15T^-$. *J Mol Biol* 1968a;31:389–399.

Lark C: Effect of the methionine analogs, ethionine and norleucine, on DNA synthesis in Escherichia coli $15T^-$. *J Mol Biol* 1968b;31:401–414.

Lark C, Arber W: Host specificity of DNA produced by Escherichia coli. XIII. Breakdown of cellular DNA upon growth in methionine of strains with $r15^+$, $rp1^+$ or $rn3^+$ restriction phenotypes. *J Mol Biol* 1979;52:337–348.

Lark C: Methylation-dependent DNA synthesis in Escherichia coli mediated by DNA polymerase I. *J Bacteriol* 1979;137:44–50.

Lu A-L, Clack S, Modrich P: Methyl-directed repair of DNA base-pair mismatches in vitro. *Proc Natl Acad Sci USA* 1983;80:4639–4643.

Marinus MG, Morris NR: Biological function for 6-methyl-adenine residues in the DNA of Escherichia coli K12. *J Mol Biol* 1974;85:309–322.

Marinus MG, Morris NR: Pleiotropic effects of a DNA adenine methylation mutation (dam-3) in Escherichia coli K12. *Mut Res* 1975;28:15–26.

Marinus MG: Adenine methylation of Okazaki fragments in Escherichia coli. *J Bacteriol* 1976;122:853–854.

Razin A, Goren D, Friedman J: Studies on the biological role of DNA methylation: Inhibition of methylation and maturation of the bacteriophage ΦX174 by nicotinamide. *Nuc Acids Res* 1975;2:1967–1973.

Razin A, Riggs A: DNA methylation and gene function. *Science* 1989;210:604–610.

Ryan AM, Borek E: Methylation of DNA in ultraviolet-irradiated bacteria. *Biochim Biophys Acta* 1971;240:203–214.

Sager R, Kitchin R: Selective silencing of eukaryotic DNA. *Science* 1975;189:426–433.

Salas CE, Pfohl-Leszkowicz A, Lang MC, Dirheimer G: Effect of modification by N-acetoxy-N-2-acetylaminofluorene on the level of DNA methylation. *Nature* 1979;278:71–72.

Singer J, Stellwagen RH, Roberts-Ems J, Riggs AD: 5-methylcytosine content of rat hepatoma DNA substituted with bromodeoxyuridine. *J Biol Chem* 1977;252:5509–5513.

Szyf S, Gruenbaum Y, Urieli-Shoval S, Razin A: Studies in the biological role of DNA methylation: V. The pattern of E. coli DNA methylation. *Nucl Acids Res* 1982;10:7242–7259.

Tseng BY, Goulian M: DNA synthesis in human lymphocytes: Intermediates in DNA synthesis, in vitro and in vivo. *J Mol Biol* 1975;99:317–337.

Urieli-Shoval S, Gruenbaum Y, Sedat J, Razin A: The absence of detectable methylated bases in Drosophila melanogaster DNA. *FEBS Letters* 1982;146:148–152.

Wagner RW, Meselson M: Repair tracts in mismatched DNA heteroduplexes. *Proc Natl Acad Sci USA* 1976;73:4135–4139.

Whitfield BL, Billen D: In vivo methylation of Escherichia coli DNA following ultraviolet and X-irradiation. *J Mol Biol* 1972;63:363–372.

Wilson VL, Jones PA: Inhibition of DNA methylation by chemical carcinogens in vitro. *Cell* 1983;32:239–246.

Yudelevich A, Gold M: A specific DNA methylase induced by bacteriophage 15. *J Mol Biol* 1969;40:77–91.

7

DNA Methylation Patterns: Formation and Biological Functions

Aharon Razin*

It is now an established fact that the distribution of methylated bases along DNA form a distinct pattern that is characteristic of the cell in which this DNA is harbored. It is also becoming clear that the biological functions that are fulfilled by DNA methylation are dependent on these patterns. The relationship between DNA methylation patterns and the various biological functions that are associated with these patterns is discussed in other chapters in this volume. However, it is appropriate to mention one familiar, clear-cut example of how a pattern of methylation exerts its biological function. The biological function of the restriction-modification systems in bacteria is to protect the cell from invasion of foreign genetic material. This protection is achieved via degradation of foreign DNA by restriction enzymes that are sequence-specific endonucleases. The DNA cleavage activity of these endonucleases is refractory to methylation at their recognition sites (see Chapter 5). The methylation is carried out by sequence-specific modification enzymes, which can methylate efficiently the restriction enzyme recognition sites of the host DNA. Since this methylation must provide protection to all sites in the host DNA, the pattern of methylation obtained in this case is determined by the distribution of the restriction sites along the DNA.

It will become clear to readers of this chapter that different methylation patterns are formed by different methylases to serve the various biological functions. Therefore, it stands to reason that the key elements in forming patterns of methylation are the enzymatic activities that catalyze the site-specific methylations. Two such activities have been suggested to be operating in the eukaryotic cell, depending on the state of methylation of the DNA (Riggs,

*Department of Cellular Biochemistry The Hebrew University, Hadassah Medical School, Jerusalem, Israel

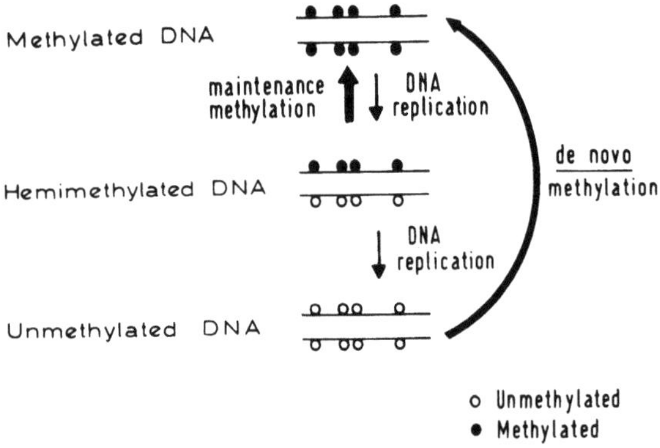

Figure 7.1 Interconversions between the various states of DNA methylation. Maintenance methylase activity methylates hemimethylated DNA. This activity is responsible for the maintenance of the methylation pattern. If two consecutive rounds of replication take place in the absence of DNA methylation, a complete loss of methyl groups occurs in 50% of the molecules. To remethylate these unmethylated sites, *de novo* methylating activity is required.

1975; Holliday and Pugh, 1975): *de novo* methylation of completely unmethylated sites and maintenance methylation of hemimethylated sites (Figure 7.1). The maintenance methylation takes place during DNA synthesis, when hemimethylated heteroduplexes are formed in which the parental strand is methylated and the newly synthesized DNA strand is not. Presumably, *de novo* methylation can occur at any time, provided conditions in the cell (e.g., enzyme level and rate of replication) favor this activity. The ratio of maintenance/*de novo* activity probably is another factor in determining the methylation pattern. Modification enzymes must have a high maintenance/*de novo* ratio to ensure efficient methylation of host DNA and to avoid methylation of the foreign DNA.

A major part of this chapter will be devoted to a detailed discussion of present knowledge concerning the biochemical mechanisms involved in the establishment of a methylation pattern and its clonal inheritance. The catalytic characteristics of the methylases, the methylation capacity of a given cell, concomitant replication, and gene activity all may have an effect on the formation and stable inheritance of the methylation pattern.

The Methylation Patterns of Various DNAs

The pattern of DNA methylation may differ in many respects among the various organisms (Table 7.1). While 5-methylcytosine (m^5Cyt) is the only methylated base found in vertebrate and plant DNA (Razin and Riggs, 1980; Gruenbaum *et al.* 1981a), unicellular organisms such as bacteria (Razin and Friedman, 1981) or the eukaryotic unicellular organism *Tetrahymena* (Gar-

Table 7.1 Characteristics of DNA Methylation in Various Organisms

Feature	Vertebrates	Plants	E. coli
Methylated base	m^{5}Cyt	m^{5}Cyt	m^{5}Cyt, m^{6}Ade
Methylated sequences	CpG	CpG, CpXpG	CC$^{A}_{T}$GG, GATC
Extent of methylation	$\leq$ 90% of all sites	$\leq$ 90% of all sites	$\sim$ 100% of the sites
Symmetry of the site	+	+	+
Number of methylases	Probably one	?	Several
DNA substrate specificity	Hemimethylated preferred		Hemimethylated and unmethylated

ovsky *et al.* 1973; Rae and Steele, 1978) may contain N^6-methyladenine (m^{6}Ade) in addition to m^{5}Cyt or as the only methylated base. Although the DNA of most organisms is methylated, in DNA of insects such as *Drosophila,* no methylated bases have been detected (Urieli-Shoval *et al.* 1982). However, with few exceptions, the rule is that the DNA of eukaryotic organisms is methylated at cytosine residues located at CpG-containing sequences. In plant DNA, cytosine residues at CpXpG sequences are methylated in addition to those at CpGs (Gruenbaum *et al.* 1981a). It should be noted that no complete methylation of all methylatable sequences has been found in eukaryotic DNA. In contrast, the DNA of *Escherichia coli* is methylated both at adenine residues (m^{6}Ade) of essentially all GATC sequences and at the inner cytosine residue (m^{5}Cyt) of all CC$^{A}_{T}$GG sequences (Razin *et al.* 1980). It should be emphasized that these methylations represent most of the methylated sites in *E. coli* and are not associated with restriction-modification. Modifications associated with restriction-modification systems of types 1 and 3 are two orders of magnitude less frequent (see Chapters 3 and 6). Examination of the methylatable sites in eukaryotic DNA (CpG; CpXpG) and in *E. coli* DNA (GATC; CC$^{A}_{T}$GG) reveals that all are palindromic sequences with 180° rotational symmetry. Thus, every methylatable cytosine or adenine residue on one strand of the DNA is symmetrically complemented by a methylated base on the opposite strand. It is now well established that methylation of methylatable sites is symmetric on both strands (Bird and Southern, 1978; Cedar *et al.* 1979; Szyf *et al.* 1982). The biochemical significance of this fact will be clarified below.

A key element in the elucidation of the pattern of DNA methylation is the determination of the distribution of the methyl groups along the DNA. It was relatively easy to investigate the distribution of m^{5}Cyt and m^{6}Ade in *E. coli.* Since those bases appear in CC$^{A}_{T}$GG and GATC sequences, respectively, and as all these sites in the *E. coli* DNA are essentially methylated (Razin *et al.* 1980), the distribution of these sites along the DNA reflects the distribution of the methyl groups. Analysis of the distribution of the sites using the appropriate restriction enzymes (Sau3AI to cleave at GATC and BstNI to cleave at CC$^{A}_{T}$GG) revealed a random distribution (Szyf *et al.* 1982). This random distribution of methyl groups suggests that DNA methylation in *E. coli* plays a general role, such as protecting the DNA from endonucleolytic cleavage rather than functioning at the gene level. In fact, *dam* mutants of *E. coli* that are defective in the methylation of GATC sites, show high mutability, and their DNA shows a high frequency of single-stranded breaks (Marinus and Morris, 1975). For a more detailed discussion, see Chapter 6 in this volume. Biological processes (e.g., genetic recombination, DNA replication, and repair) must involve a controlled production of single-strand breaks. A pattern of methylation that is composed of a few specific sites that remain transiently hemimethylated on a high background of methylated sites may direct endonucleolytic activity to produce single-stranded breaks. It should be noted, however, that the random distribution of the methyl groups along the entire *E. coli* chromosome does not necessarily mean that there are not some specific sequences that are more methylated than others. In fact, the *E. coli* origin of replication is very rich in methylated GATC sites. It recently has been shown that absence of methylation at the origin has no effect on replication (Szyf *et al.* 1982). However, the significance of this cluster of methylated sites can be understood on different grounds. Such a cluster will increase the probability of having transiently hemimethylated sites at this region, which might be important for proper initiation of replication or (alternately) faithful conservation of this important sequence.

It has been fairly well established that m^5Cyt in eukaryotic DNA is nonrandomly distributed. For example, the methylated base m^5Cyt appears in repetitive sequences in a several-fold higher abundance than in unique sequences (Salomon *et al.* 1969; Miller *et al.* 1974; Harbers *et al.* 1975; Solage and Cedar, 1978; Ehrlich *et al.* 1982). The methylated cytosine residues are also nonrandomly distributed with respect to the nucleosomal structure of the eukaryotic chromosome. DNA associated with the nucleosomes is significantly more methylated, and very little methylation is observed in spacer regions between nucleosome cores (Razin and Cedar, 1977; Solage and Cedar, 1978). A recent study demonstrates that the highest abundance of m^5Cyt is found in DNA associated with nucleosomes that contain histone H1 (Ball *et al.* 1983).

A closer insight into the nonrandom pattern of methylation in certain eukaryotes has been obtained by the use of restriction enzymes that recognize CpG-containing sequences (see Chapter 8). On the level of genomic DNA, two compartments have been observed. One compartment contains a high percentage of methylated sites, while the other compartment carries either very little or no m^5Cyt. The most striking division into two such compartments has been observed in sea urchin (Bird *et al.* 1979). Similar methylation compartments have been observed in other organisms across the phylogenetic spectrum in DNA from invertebrates (Bird and Taggart, 1980), slime molds (Whittaker *et al.* 1981), and vertebrates (Naveh-Many and Cedar, 1982; Cooper *et al.* 1983).

The most detailed information on the pattern of DNA methylation came from the analysis of methylation at specific gene sequences via the use of CpG enzymes and Southern blot hybridization with specific probes. By this analysis, methylation is probed at specific sites in a single gene region. In many cases, certain sites in these sequences were found to be hypomethylated when the gene is actively expressed. The same gene sequences were found to be heavily methylated in tissues where the gene is not expressed (see Chapter 16). Genes that code for housekeeping functions have been shown to be undermethylated in the 5′ region of all tissues (Table 7.2). It has been suggested that the methylatable sites may be classified into three groups: those that are methylated in all tissues (m^+), sites that are not methylated in any of the tissues (m^-), and the variable group of sites (m^v) that are methylated in some tissues, but not in others (Mandel and Chambon, 1979). Included in the last group are sites that were found to be partially methylated presumably because these sites are not methylated in all cells of the same tissue.

Both the nature of this kind of methylation patterns and the strong correlation between the methylation pattern of specific genes with their activity (Razin and Riggs, 1980) suggest that the methylation pattern itself may determine the capability of a gene to be transcribed (see Chapter 8). Tissue specificity of the methylation pattern suggests that differentiation may involve a divergence of the methylation pattern during development. This type of mechanism would satisfy the requirements of a differentiated state, since DNA methylation patterns fulfill two basic principles: (1) a methylation pattern can be clonally inherited in somatic cells; and (2) the methylation pattern is prone to changes during embryogenesis and may be reversible under certain conditions. The clonal inheritance of a methylation pattern and the processes leading to changes in the methylation pattern will be discussed in subsequent sections of this chapter.

Table 7.2 Methylation Patterns of "Housekeeping" Genes

Species	Gene	Pattern of Methylation	References
Hamster	aprt	5′ region hypomethylated in all tissues	Stein *et al.* (1983)
Mouse	dhfr	5′ region hypomethylated in all tissues	Stein *et al.* (1983)
	rRNA genes	Hypomethylated in all tissues	Bird *et al.* (1981b)
Rat	rRNA genes	Hypomethylated in all tissues	Reilly *et al.* (1982) Kunnath and Locker (1982)
	Amplified rRNA genes	Hypomethylated in rat hepatoma cell line	Tantravahi *et al.* (1981)
Xenopus	rRNA genes	60-bp repeated spacer hypomethylated in somatic tissues methylated in sperm	Bird *et al.* (1981a)
	Somatic 5S RNA genes	Hypomethylated in liver	Sims *et al.* (1983)

DNA Methyl Transferases (Methylases) and the Methylation Reaction

The source for methyl groups in the DNA is the universal methyl donor S-adenosylmethionine (SAM). The activated methyl group in SAM is transferred to specific sites in the DNA by a postreplication methyl transfer process catalyzed by a specific methyl transferase (methylase). The reaction does not require magnesium ions and, in many cases, is inhibited by Mg^{++}. The preferred DNA substrate for the methylase has been an enigma for a long time. While the *E. coli* methylases (the first to be isolated and characterized) prefer duplex-unmethylated DNA as substrate (Nesterenko *et al.* 1979), mammalian methylases show better activity *in vitro* with single-stranded DNA as substrate (Drahovsky and Morris, 1971; Simon *et al.* 1978; Adams *et al.* 1979; Roy and Weissbach, 1975). However, it is clear that in most of the *in vivo* methylation reactions neither of these substrates is likely to be the substrate for the methylase. In light of experimental data indicating that methylation of DNA occurs during replication (Lark, 1968; Billen, 1968; Szyf *et al.* 1982; Kappler, 1971; Gruenbaum *et al.* 1983), it is clear that most of the methylase activity within the cell is on hemimethylated DNA (see Figure 7.1). Attempts have been made in the past to obtain hemimethylated DNA to test this assumption. Sneider *et al.* (1975) prepared DNA from ethionine-treated cells, Turnbull and Adams (1976) used DNA after methionine starvation, and Adams *et al.* (1979) prepared hemimethylated DNA by nick translation. In all of these methods, homogeneously hemimethylated DNA could not be guaranteed. It only recently has become possible to obtain a proper hemimethylated substrate (Gruenbaum *et al.* 1981b; Taylor and Jones, 1982). The most effective substrate for *in vitro* methylation by a eukaryotic methylase proved to be an *in vitro*-synthesized heteroduplex that is methylated at all cytosine moieties on one strand (Figure 7.2). The methylase activity using this substrate can be compared to that with an homologous, nonmethylated duplex DNA that is syn-

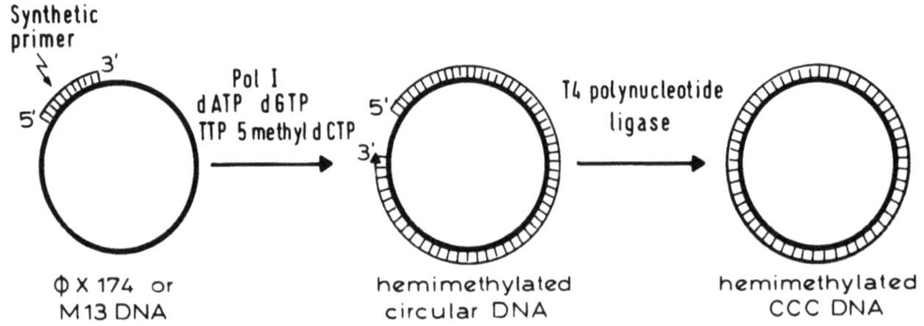

Figure 7.2 Schematic illustration of the *in vitro* synthesis of hemimethylated DNA. The newly synthesized strand is methylated at all cytosine residues. (From Gruenbaum *et al.* 1981b.)

thesized *in vitro* by the same enzymatic synthesis. Using these substrates, the mammalian methylase preparations show activity between one and two orders of magnitude higher with the hemimethylated DNA than with unmethylated *in vitro*-synthesized substrate (Table 7.3). The methylation observed with the nonmethylated duplex DNA represents *de novo* methylation of unmethylated sites, whereas activity measured with the hemimethylated substrate is considered to be representative of maintenance methylase activity (see Figure 7.1). This activity functions *in vivo* to maintain the pattern of methylation during DNA synthesis.

It should be noted that no evidence exists for the presence of two different methylases in the eukaryotic cell. Although it presently cannot be ruled out that a *de novo* methylase and a maintenance methylase both operate in the cell side by side, it is more likely that—under suitable conditions—the maintenance methylase shows *de novo* activity as well. This presumption is based on several observations. The ratio *de novo*/maintenance is not changed during purification. A homogeneously purified methylase performs both activities at the same ratio, as observed with a crude enzyme preparation (Bestor and Ingram, 1983). Both activities are inhibited to the same extent by an antimethylase monoclonal antibody (Drahovsky, personal communication). In any event, *de novo* methylation seems to be very inefficient in eukaryotic cells. In DNA-mediated gene transfer experiments, *de novo* methylation has been shown to be a very rare event (Pollack *et al.* 1980; Wigler *et al.* 1981). In other

Table 7.3 Substrate Specificity of DNA Methylases

	Relative Activity		
Source of Methylase	Hemimethylated	Unmethylated	References
Prokaryotic methylases			
E. coli dam	1	1	Urieli-Shoval *et al.* (1983)
E. coli mec	1	1	Urieli-Shoval *et al.* (1983)
Modification type I	50–100	1	Vovis *et al.* (1974)
Modification type II	1	1	Rubin and Modrich (1977)
Eukaryotic methylases			
Mouse ascites	100–200	1	Gruenbaum *et al.* (1982)
Mouse spleen	10	1	Jones and Taylor (1981)
Rat liver	10–20	1	Kaplan *et al.* (unpublished results)
Mouse T-cell line	10	1	Szyf *et al.* (1984)
Mouse L cells	10–20	1	Szyf *et al.* (1984)
Mouse F9 teratocarcinoma cells	10	1	Szyf *et al.* (1984)
Mouse placenta	25–50	1	Pfeifer *et al.* (1983)
Mouse erythroleukemia cells metaseI	35	1	Bestor and Ingram (1983)
metaseII	50	1	

experiments, genes that were injected into zygotes underwent methylation by a very slow *de novo* methylation process (Niwa *et al.* 1983). Although a very inefficient *de novo* activity has been demonstrated in the experiments discussed above, it is also clear that *de novo* methylation takes place in vertebrates. Bovine satellite DNA is undermethylated in sperm in comparison with somatic tissues (Sturm and Taylor, 1981), and viral genomes acquire a methylation pattern on integration (Sutter and Doerfler, 1980). A better understanding of the enzymatic mechanisms that underly DNA methylation in eukaryotes requires characterization of methylases purified to homogeneity (Simon *et al.* 1983; Pfeifer *et al.* 1983; Bestor and Ingram, 1983) and the use of both antimethylase monoclonal antibodies and the cloned methylase gene.

The *E. coli dam* and *mec* methylases show similar activity with unmethylated and hemimethylated DNA (Urieli-Shoval *et al.* 1983). The *dam* methylase also has been shown to require two independent binding events to methylate a GATC site symmetrically on both strands. This feature distinguishes the *E. coli* DNA methylases from eukaryotic DNA methylases. While *E. coli* DNA methylases seem to be insensitive to the state of methylation of the specific site on the opposite strand, the eukaryotic methylases are clearly more efficient when the site is already methylated on one strand. The methylases of restriction-modification systems fall into two classes: 1) type I modification enzymes, which have a substrate specificity similar to eukaryotic methylases—namely, hemimethylated DNA being the preferred substrate; and 2) type II modification enzymes, which show equal activity with both substrates (see Table 7.3). If an analogy can be drawn between eukaryotic methylases and the modification enzymes type I and type II, it would be predicted that the eukaryotic maintenance methylase is an oligomer the same as the type I modification enzymes. Both prefer hemimethylated DNA as substrate. The oligomeric form may be required to provide the symmetry that is needed to carry out mainte-

Table 7.4 Sequence Specificity of L Cells Methylase: *In Vivo* Experiment*

		Cleavage by Restriction Enzyme	
Enzyme	Sequence	Before Transfection	After Transfection
HpaII	$C^{m}CGG$	−	−
MspI	$^{m}CCGG$	−	+
HhaI	$G^{m}CGC$	−	−
SmaI	$CC^{m}CGGG$	−	−
HaeIII	$GG^{m}CC$	−	+
AluI	$AG^{m}CT$	−	+
EcoRII	$C^{m}C(^{A}_{T})GG$	−	+

**In vitro* synthesized, hemimethylated ϕX DNA (Gruenbaum *et al.* 1981b) was introduced by DNA-mediated gene transfer into L cells (Stein *et al.* 1982). DNA from isolated clones was subjected to analysis of ϕX sequences for methylation at CpC, CpA, CpG, and CpT sites. Since all cytosines in the *in vitro* newly synthesized strand of the ϕX DNA were methylated, none of the enzymes cleaved before transfection. After transfection, the only resistant sites were CpG-containing sites suggesting that faithful maintenance methylation and subsequent replication eliminate all nonrelevant (non-CpG) methylations.

nance methylation. Another prediction that can be made is that the monomeric form of the eukaryotic methylase carries out *de novo* methylation the same as the type II modification enzymes that are active as monomers (Modrich, 1983).

Sequence specificity of the eukaryotic DNA methylases varies from a very high level of specificity to CpG sequences exerted by the maintenance methylase activity (Gruenbaum *et al.* 1982; Stein *et al.* 1982) to a very low level of specificity of *de novo* methylation activity as expressed by methylation of CpA and CpT sequences, as well as CpG sequences (Simon *et al.* 1983). The high specificity of the maintenance activity provides the mechanism to eliminate methylation by the *de novo* activity at sequences other than CpG, provided that DNA replication continues. A direct proof of the high specificity of the maintenance methylase activity *in vivo* and its ability to eliminate methylation at sites other than CpG appears in the results of an *in vivo* experiment presented in Table 7.4.

Methylation with Respect to Replication

The fact that DNA is methylated not as a result of incorporation of methylated nucleotides into DNA during its synthesis, but rather as a result of a site-specific methylation of postreplicating DNA, raises the question of how the methylation process relates to the replication of the DNA. It has been shown that the methylation of *E. coli* DNA occurs at/or near the replication fork (Lark, 1968; Billen, 1968). Since these original studies have been published, efforts have been made to determine the precise timing of DNA methylation in eukaryotes. These studies led to conflicting results, probably because different measures were taken to synchronize the cell cultures used. For example, arresting the cell cycle at various points in S phase will result in artificial lag periods between replication and methylation. One study that avoided the use of synchronized cells (Kappler, 1970) arrived at the conclusion that DNA methylation in a eukaryotic cell occurs (as in *E. coli*) very close to the fork (1–2 minutes after DNA is replicated, it undergoes methylation). This early observation recently received strong support from a kinetic study performed in permeabilized cells (Gruenbaum *et al.* 1983). In this study, [α^{32}P]dGTP was introduced into mouse L cells, and the methylation of the newly synthesized DNA was followed by monitoring the level of labeled m^{5}dCMP by nearest neighbor analysis. Via this analysis, a 75-second lag was observed between the synthesis of the DNA and methylation. It is now clear that this lag is also very short in *E. coli* (Szyf *et al.* 1982). In view of the fact that methylation occurs at the replication fork, it was of interest to determine whether the methylation of DNA in eukaryotic cells is restricted to S phase or whether it can be observed throughout the cell cycle. Methylation has been observed during G2 (Adams, 1971), as well as during mitosis (Bugler *et al.* 1980); recently, it has been reported that delayed methylation can occur after the S phase (Woodcock *et al.* 1983). This probably reflects methylation of sites that are particularly resistant and, therefore, are methylated slowly. If these sites remain unmethylated until cell division, this will lead to a heritable loss of methylation at some sites. The implication of such a loss of methyl groups in the DNA will be discussed in the next section.

How Is a Methylation Pattern Formed?

In *E. coli* in which the methylation pattern is actually a reflection of the distribution of methylatable sites, the complete methylation of all sites forms the pattern (Szyf *et al.* 1982). It is still unclear whether some sites remain transiently unmethylated during replication. A recent study of the kinetics of methylation during DNA replication in *E. coli* (Szyf *et al.* 1984) reveals that the methylase exhibits differential methylation efficiencies with respect to the various sites on the DNA (Figure 7.3). A limiting intracellular level of the methylase, in conjunction with different affinities of the enzyme to specific sites, results in a sequential hypomethylation of specific sites in rapidly replicating DNA (e.g., λ phage or pBR322 DNA). These sites undergo methylation in a reversed sequence of events when DNA replication comes to a halt.

In eukaryotes, partial undermethylation of CpG-containing sequences forms a discrete tissue-specific pattern of methylation. However, the mechanism by which such a methylation pattern is formed is still obscure. Since the methylation pattern is tissue-specific, it was tempting to speculate that the pattern is established at an early embryonic stage, and, subsequently, is clonally inherited

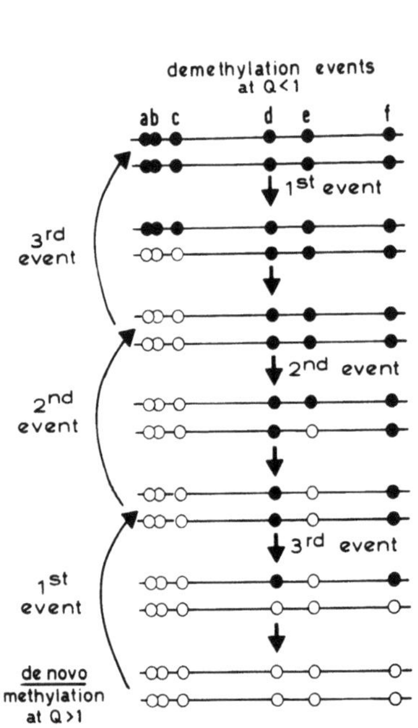

Figure 7.3 Effect of the methylation quotient (Q) on the methylation pattern. The methylation quotient (Q) designates the ratio of methylation capacity (number of methylase molecules times turnover number) over newly replicated methylated sites. At $Q < 1$ (low methylase level or very rapid rate of DNA replication), sequential demethylation occurs. Presumably, sites (a, b, c) with low affinity towards the methylase undergo demethylation first and high affinity sites (f) last. At $Q > 1$ (very high methylase level or slow replication), *de novo* methylation takes place in an inverse sequence of events. The schematic physical map shown here represents the location of CC$^{A}_{T}$GG sites in pBR322 DNA. ○ = ummethylated; ● = methylated. (From Szyf *et al.* 1984.)

(Figure 7.4). It was further suggested that the formation of the pattern involves changes in the methylation pattern by *de novo* methylation of unmethylated sites and a loss of methyl groups by a passive demethylation. It was suggested that this passive demethylation can be achieved by active replication of the DNA with no concomitant methylation. The question was how is hypomethylation obtained at specific sites?

Several possible biochemical mechanisms have been proposed in the past for specific demethylation (Razin and Riggs, 1980). One such mechanism postulated the existence of site-specific determinator proteins that bind to the DNA at the time of methylation and prevent specific sites from being methylated by the maintenance methylase. A second and trivial mechanism might involve a specific demethylase. Although demethylase activity has recently been claimed (Gjerset and Martin, 1982), it should await more convincing data before such a mechanism for demethylation can be claimed. A third possibility has been discussed above that is based on the assumption that the intracellular methylase level is limiting and that the methylase has different affinities to the various methylatable sites. According to this mechanism, an interplay between the intracellular level of the methylase and the rate of replication can determine

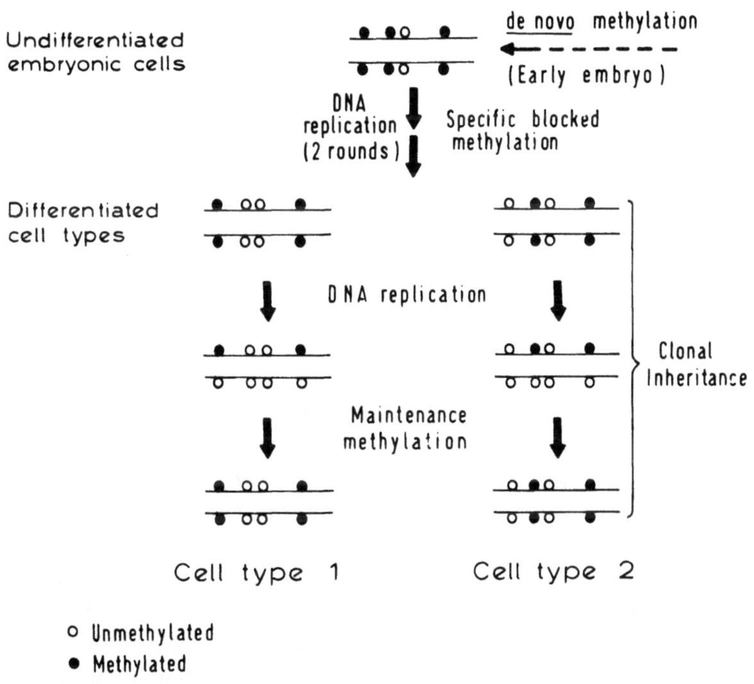

Figure 7.4. Establishment and clonal inheritance of tissue-specific methylation patterns. Changes in the methylation pattern occur early in development (by *de novo* methylation and specific demethylation). In the somatic tissue, the pattern is conserved and clonally inherited.

whether *de novo* methylation or demethylation will occur (see Figure 7.3). In conjunction with determinator proteins, this type of mechanism may result in the formation of differentiated patterns of DNA methylation (see Figure 7.4). This type of mechanism alone may explain the massive demethylation that takes place at the stage of differentiation to extraembryonic tissues (see Table 7.5).

Supporting evidence for this type of mechanism in eukaryotes already exists. First, the substantial inhibition (30–100%) of intracellular methylase activity by a very modest replacement of the cytosine residues in the DNA by azacytidine (Creusot *et al.* 1982; Taylor and Jones, 1982) implies that the intracellular level of the enzyme is limiting. It has been shown that partial hepatectomy in the rat induces a several-fold increase in the intracellular level of methylase activity. This induction presumably is the mechanism employed by the somatic cells to prevent a change in the methylation pattern caused by rapid proliferation (Kaplan, Szyf and Razin, unpublished results). It is also a well-known fact that the extrachromosomal copies of rRNA genes (Bird and Southern, 1978) and various viral genomes (Doerfler, 1983) are not methylated, whereas the integrated counterparts of these genes are methylated. Although it cannot be ruled out that the eukaryotic methylases are not active in trans (Wettstein and Stevens, 1983; Saenundsen *et al.* 1983), it is more likely that the intracellular methylase level is not sufficient to methylate the rapidly replicating extrachromosomal DNA. This suggestion is now supported by experimental data. Polyoma virus DNA (Subramanian, 1982) or SV40 DNA (Yisraeli, personal communication) were methylated *in vitro* at HpaII sites and analyzed posttransfection. A loss of the methyl groups has been observed as a result of the DNA replication *in vivo.*

If the establishment of the DNA pattern of methylation in the various tissues is indeed taking place during early stages of embryogenesis, as was suggested before (Razin and Riggs, 1980), changes in the pattern of methylation must be observed at these stages of development. The study of methylation patterns at early stages of mammalian embryogenesis is not an easy task. However, changes in the extent of DNA methylation have been observed in various embryonic tissues, as well as in mouse teratocarcinoma cells that were induced to differentiate (Razin *et al.* 1983). The extraembryonic tissues, such as yolk sac and placenta, were found to undergo undermethylation in comparison to the DNA of the cells of the embryo proper (Table 7.5). In addition, the teratocarcinoma cell DNA undergoes demethylation in response to treatment with retinoic acid. The decrease in the number of methyl moieties occurs at a large number of CpG sites spread out over the entire genome, as indicated by a restriction enzyme analysis of several mouse genes that include dhfr, β-globin and the H2K major histocompatibility gene family. Although these changes in DNA methylation seem to be associated with the differentiation process, it is not clear whether tissue-specific methylation patterns are the result of these changes.

Table 7.5 Extent of CpG Methylation in Various Cell Types of the Mouse

	Tissue	% CpG Methylation
Adult somatic tissues	Liver	78
	Kidney	82
	Brain	75
	Lung	76
	Spleen	82
Germ line and embryonic cells	Testes	69
	Sperm	67
	Placenta	49
	Yolk sac	53
	Embryo (9–10 days)	77
Cell lines	L cells	65
	3T3 cells	70
	F9 teratocarcinoma	65
	T-cell line	70
	Ascites tumor cells	67

DNA methylation analysis was according to Gruenbaum *et al.* (1981b).

Clonal Inheritance of the Methylation Pattern

The discovery of cell-specific methylation patterns suggested that a specific enzymatic mechanism faithfully replicates the methylation pattern for many cell generations. As mentioned above, methylation of DNA occurs primarily at the replication fork. Based on that notion, and on the fact that methylatable sites have a palindromic structure and (therefore) are methylated symmetrically on both DNA strands (see Table 7.1), it has been suggested that a maintenance methylase might carry out this important biological function of clonal inheritance of the methylation pattern. This methylase, of course, would be expected to preferentially use hemimethylated DNA as substrate (see Figure 7.1).

In a cell where all methylatable sites in its DNA are methylated, as is the case in *E. coli,* clonal inheritance of the methylation pattern might be achieved by a simple semiconservative process. The methylase recognizes a sequence, and it methylates opposite to the methylated base on the parental strand. This can be considered to be a replication of the methylation pattern by a semiconservative mechanism. Conversely, when the pattern of methylation is composed of unmethylated sites as well as methylated sites (as in vertebrates), clonal inheritance of the methylation pattern is more complicated and should depend on a highly specific maintenance methylase that is capable of discriminating between relevant (CpG) and nonrelevant (CpC, CpA, and CpT) hemimethylated sites (see Table 7.4). Methyl groups at nonrelevant sites probably do not serve as templates for methylation by the maintenance methylase. In this manner, the nonrelevant methylations will be lost during subsequent rounds of replication. Such high-fidelity methylation is an absolute requirement if a cell-specific methylation pattern in multicellular organisms is indeed associated with stable differentiation, as was proposed in 1975 (Riggs, 1975; Holliday and Pugh, 1975). DNA-mediated gene transfer experiments were used recently by several research groups to demonstrate clonal inheritance of a methylation pattern *in vivo* (Pollack *et al.* 1980; Wigler *et al.* 1981; Stein *et al.* 1982; Harland, 1982). In one type of experiment, nonmethylated and *in vitro*-methylated gene sequences were introduced into Ltk^- cells by cotransfection with the *Herpes simplex* virus tk gene. Genomic DNA was isolated from Ltk^+ clones after 25, 100, and 200 cell generations. The DNA was restricted with various CpG restriction enzymes, and the restriction patterns were analyzed by blot hybridization. The results of these experiments clearly demonstrate clonal inheritance of the pattern of methylation that was introduced into the foreign DNA by *in vitro* methylation. The methylation pattern was stabilized *in vivo* at an early stage after the integration of the foreign DNA into the cell genome (Pollack *et al.* 1980), and it was faithfully conserved thereafter for hundreds of cell generations (Stein *et al.* 1982). To test the sequence specificity of this maintenance methylation, hemimethylated DNA was synthesized *in vitro* (see Figure 7.2) and was used for DNA-mediated gene transfer experiments (Stein *et al.* 1982). The genomic DNA isolated from the Ltk^+ colonies was analyzed, as described above, with various restriction enzymes. The results of this analysis indicated that the maintenance methylation *in vivo* is specific to –CpG– sequences (see Table 7.4). Methyl groups at other sequences (e.g., CpC, CpA, or CpT) were eliminated by dilution during multiple cell divisions, and the sites became sensitive to the respective restriction enzymes.

These properties of maintenance methylation have been verified *in vitro* by using crude preparations of methylase from various mammalian sources. In a kinetic study using crude methylase from mouse ascites tumor cells, the activity of the enzyme was found to be two orders of magnitude higher with *in vitro*-synthesized, hemimethylated ϕX 174 or M13 DNA than with the respective unmethylated DNA. This result indicates that the somatic cell methylase primarily carries out maintenance methylation. Further analysis of the characteristics of this enzymatic activity revealed that the enzyme methylates exclusively at hemimethylated CpG sites (Gruenbaum *et al.* 1982). In a recent study using a crude methylase preparation from regenerating rat liver, the enzyme showed a 10–20-fold higher activity using *in vitro*-synthesized, hemimethylated M13 DNA as substrate than with *in vitro*-synthesized, unmethylated M13 duplex DNA. While the product of the methylation reaction with the hemimethylated DNA showed methylation exclusively at CpG, the unmethylated substrate underwent methylation at –CpA– and –CpT– sequences as well (Simon *et al.* 1983). The result of this analysis clearly demonstrates that while maintenance methylation is CpG-specific, *de novo* methylation might occur at any cytosine residue. Although this has not yet been validated by *in vivo* experiments, it is not unexpected, since nonspecific *de novo* methylation *in vivo* may be corrected by highly specific maintenance methylation under conditions of ongoing replication (see Table 7.4).

Conclusions

Cell-specific patterns of DNA methylation seem to add a level of information that complements the genetic information encoded in the DNA sequence. This information is clonally inherited in somatic cells by a maintenance methylase activity that replicates the methylation pattern with high fidelity following DNA synthesis. A close relationship can be drawn between the nature of a given methylation pattern and the biological role played by this pattern. In many prokaryotes, the DNA methylation patterns obtained by modification enzymes and other DNA methylases can provide resistance to nucleolytic activity. Therefore, it comes as no surprise that all methylatable sites in bacterial DNA are found to be fully methylated. However, transiently, hemimethylated sites may serve as hot spots for nucleolytic activity that results in single-strand breaks that presumably are needed for biological functions (e.g., recombination and DNA replication and repair).

In contrast, methylation patterns of eukaryotic DNA are more complex. In general, the –CpG– sequences are partially methylated (see Table 7.1). The incomplete methylation of –CpG– sequences in eukaryotic DNA allows changes in the methylation pattern, both by *de novo* methylation of unmethylated sites and by loss of methyl groups via decreased methylation during DNA synthesis. These changes, if they occur during embryogenesis and are directed by determinator proteins, may result in the differentiation of a master methylation pattern. Based on detailed analysis of methylation patterns of a variety of specific genes (see Chapters 8 and 16) and on direct experiments using *in vitro*-methylated genes (Vardimon *et al.* 1982; Fradin *et al.* 1982; Jahner *et al.* 1982; Cedar *et al.* 1983), it can be concluded that methylation of critical sites in eukaryotic gene sequences block expression of the corresponding genes. Since the general rule is that undermethylation of a few critical sites on a high background of methylation in the region of gene sequences correlates with activity of the genes, it is conceivable that changes at the time of differentiation must involve demethylation. At least formation of extraembryonic tissues recently has been shown to be associated with demethylation (Razin *et al.* 1984). One question yet unsolved concerns the biochemical mechanism that is responsible for the specific loss of methyl groups. A passive mechanism for the loss of methyl groups is generally accepted. Failure of the maintenance methylase to methylate at specific sites during two rounds of replication will result in the loss of methyl groups at these sites. The failure of the methylase to properly methylate all relevant sites can result via several different events: 1) a decrease in the intracellular level of the maintenance methylase or an increase in the rate of DNA replication; 2) inhibition of the methylase by specific proteins that bind to the DNA at the methylatable site; and 3) active transcription that is coupled to DNA replication. Changes of the methylation pattern by *de novo* methylation and loss of methyl groups can form new methylation patterns. Once a methylation pattern is established (perhaps around blastula), it is clonally inherited in the somatic tissue (see Figure 7.4).

Acknowledgments

Support from the National Institutes of Health Grant No. GM 20483 is acknowledged. I wish to thank my colleagues Moshe Szyf and Yaakov Pollack for helpful suggestions. I am especially grateful to Caroline Gopin for her careful and patient preparation of the manuscript.

References

Adams RLP: The relationship between synthesis and methylation of DNA in mouse fibroblasts. *Biochim Biophys Acta* 1971;254:205–212.

Adams RLP, McKay EL, Craig LM, Burdon RH: Mouse DNA methylase: Methylation of native DNA. *Biochim Biophys Acta* 1979;561:345–357.

Ball DJ, Gross DS, Garrard WT: 5-methylcytosine is localized in nucleosomes that contain histone H1. *Proc Natl Acad Sci (USA)* 1983;80:5490–5494.

Bestor TH, Ingram VM: Two DNA methyltransferases from murine erythroleukemia cells: Purification sequence specificity and mode of interaction with DNA. *Proc Natl Acad Sci (USA)* 1983;80:5559–5563.

Billen D: Methylation of the bacterial chromosome: an event at the "replication point?" *J Mol Biol* 1968;31:477–486.

Bird AP: Use of restriction enzymes to study eukaryotic DNA methylation. 2. The symmetry of methylated sites supports semi-conservative copying of the methylation pattern. *J Mol Biol* 1978;118:49–60.

Bird AP, Southern EM: Use of restriction enzymes to study eukaryotic DNA methylation. 1. The methylation pattern in rDNA from Xenopus laevis. *J Mol Biol* 1978;118:27–47.

Bird AP, Taggart MH, Smith BA: Methylated and unmethylated DNA compartments in the sea urchin genome. *Cell* 1979;17:889–901.

Bird AP, Taggart MH: Variable patterns of total DNA and rDNA methylation in animals. *Nucleic Acids Res* 1980;8:1485–1497.

Bird AP, Taggart MH, Macleod D: Loss of rDNA methylation accompanying the onset of ribosomal gene activity in early development of X. laevis. *Cell* 1981a;26:381–390.

Bird AP, Taggart MH, Gehring CA: Methylated and unmethylated rRNA genes in the mouse. *J Mol Biol* 1981b;152:1–17.

Bugler B, Bertaux O, Valencia R: Nucleic acids methylation of synchronized BHK21 HS5 fibroblasts during the mitotic phase. *J Cell Phys* 1980;103:149–157.

Cedar H, Solage A, Glaser G, Razin A: Direct detection of methylated cytosine in DNA by use of the restriction enzyme MspI. *Nucleic Acids Res* 1979;6:2125–2132.

Cedar H, Stein R, Gruenbaum Y, Naveh-Many T, Sciaky-Gallili N, Razin A: Effect of DNA methylation on gene expression. *Cold Spring Harbor Symposia Quant Biol* 1982;47:605–609.

Cooper DN, Taggart MH, Bird AP: Unmethylated domains in vertebrate DNA. *Nucleic Acids Res* 1983;11:647–657.

Creusot F, Acs G, Christman JK: Inhibition of DNA methyltransferase and induction of Friend erythroleukemia cell differentiation by 5-azacytidine and 5-aza-2′-deoxycytidine. *J Biol Chem* 1982;257:2041–2048.

Doerfler W: DNA methylation and gene activity. *Ann Rev Biochem* 1983;52:93–124.

Drahovsky D, Morris NR: Mechanism of action of rat liver DNA methylase. 2 Interactions with single-stranded methyl-acceptor DNA. *J Mol Biol* 1971;61:343–356.

Ehrlich M, Gama-Sosa MA, Huang LH, Midgett RM, Kuo KC, McCune RA, Gehrke C: Amount and distribution of 5-methylcytosine in human DNA from different types of tissues or cells. *Nucleic Acids Res* 1982;10:2709–2721.

Fradin A, Manley JL, Prives CL: Methylation of simian virus 40 HpaII site affects late, but not early, viral gene expression. *Proc Natl Acad Sci (USA)* 1982;79:5142–5146.

Garovsky MA, Hattman S, Pleger GL: [^{6}N] methyladenine in the nuclear DNA of a eucaryote. Tetrahymena pyriformis. *J Cell Biol* 1973;56:697–701.

Gjerset RA, Martin DW: Presence of a DNA demethylating activity in the nucleus of murine erythroleukemia cells. *J Biol Chem* 1982;257:8581–8583.

Gruenbaum Y, Naveh-Many T, Cedar H, Razin, A: Sequence specificity of methylations in higher plant DNA. *Nature* 1981a;292:860–862.

Gruenbaum Y, Cedar H, Razin A: Restriction enzyme digestion of hemimethylated DNA. *Nucleic Acids Res* 1981b;11:2509–2515.

Gruenbaum Y, Cedar H, Razin A: Substrate and sequence specificity of a eukaryotic DNA methylase *Nature* 1982;295:620–622.

Gruenbaum Y, Szyf M, Cedar H, Razin, A: Methylation of replicating and post-replicated mouse L-cell DNA. *Proc Natl Acad Sci (USA)* 1983;80:4919–4921.

Harbers K, Harbers B, Spencer JH: Nucleotide clusters in deoxyribonucleic acids. 12 The distribution of 5-methylcytosine in pyrimidine oligonucleotides of mouse L-cells satellite DNA and main band DNA. *Biochem Biophys Res Commun* 1975;66:738–746.

Harland RM: Inheritance of DNA methylation in microinjected eggs of Xenopus laevis. *Proc Natl Acad Sci (USA)* 1982;79:2323–2327.

Holliday R, Pugh JE: DNA modification mechanisms and gene activity during development. *Science* 1975;187:226–232.

Jähner D, Stuhlmann H, Stewart CL, Harbers K, Loehler J, Simon I, Jaenisch R: De novo methylation and expression of retroviral genomes during mouse embryogenesis. *Nature* 1982;298:623–628.

Jones PA, Taylor SM: Hemimethylated duplex DNAs prepared from 5-azacytidine-treated cells. *Nucleic Acids Res* 1981;9:2933–2947.

Kappler JW: The kinetics of DNA methylation in cultures of a mouse adrenal cell line. *J Cell Physiol* 1970;75:21–32.

Kappler JW: The 5-methylcytosine content of DNA: tissue specificity. *J Cell Physiol* 1971;78:33–36.

Kunnath L, Locker J: Variable methylation of the rRNA genes of the rat. *Nucleic Acids Res* 1982;10:3877–3892.

Lark C: Studies on the *in vivo* methylation of DNA in *Escherichia coli* $15T^-$. *J Mol Biol* 1968;31:389–399.

Mandel JL, Chambon P: DNA methylation: organ specific variations in the methylation pattern within and around ovalbumin and other chicken genes. *Nucleic Acids Res* 1979;7:2081–2090.

Marinus MG, Morris NR: Pleiotropic effects of a DNA adenine methylation mutation (dam-3) in *Escherichia coli* K-12. *Mutat Res* 1975;28:15–26.

Miller OJ, Schnedl W, Allen J, Erlanger BF: 5-methylcytosine localized in mammalian constitutive heterochromatin. *Nature* 1974;251:636–637.

Modrich P: Studies of sequence and recognition by type 2 restriction and modification enzymes. *CRC Crit Rev Biochem* 1983;13:287–323.

Naveh-Many T, Cedar H: Topographical distribution of 5-methylcytosine in animal and plant DNA. *Mol Cell Biol* 1982;2:758–762.

Nesterenko VF, Buryanov YaI, Baev AA: Isolation and properties of DNA-cytosine methylase 1 from Escherichia coli MRE600. *Biochimiya* 1979;44:130–141.

Niwa O, Yokota Y, Ishida H, Sugahara, T: Independent mechanisms involved in suppression of the Moloney Leukemia virus genome during differentiation of murine teratocarcinoma cells. *Cell* 1983;32:1105–1113.

Pfeifer GP, Grunwald S, Bohem TLS, Drahovsky, D: Isolation and characterization of DNA cytosine 5-methyltransferase from human placenta. *Biochim Biophys. Acta* 1983;740:323–330.

Pollack Y, Stein R, Razin A, Cedar, H: Methylation of foreign DNA sequences in eukaryotic cells. *Proc Natl Acad Sci (USA)* 1980;77:6463–6467.

Rae PMM, Steele RE: Modified bases in the DNAs of unicellular eukaryotes: an examination of distributions and possible roles with emphasis on OH methyl uracil in Dinoflagellates. *Biosystems* 1978;10:37–53.

Razin A, Cedar H: Distribution of 5-methylcytosine in chromatin. *Proc Natl Acad Sci (USA)* 1977;74:2725–2728.

Razin A, Riggs AD: DNA methylation and gene function. *Science* 1980;210:604–610.

Razin A, Urieli S, Pollack Y, Gruenbaum Y, Glaser, G: Studies on the biological role of DNA methylation: 4. Mode of methylation of DNA in E. coli cells. *Nucleic Acids Res* 1980;8:1783–1792.

Razin A, Friedman J: DNA methylation and its possible biological roles. *Prog Nucleic Acids Res Mol Biol* 1981;25:33–52.

Razin A, Webb C, Szyf M, Yisraeli J, Rosenthal A, Naveh-Many T, Sciaky-Gallili N, Cedar H: Variations in DNA methylation during mouse cell differentiation in vivo and in vitro. *Proc Natl Acad Sci (USA)* 1984;81:2275–2279.

Reilly JG, Thomas CA Jr, Lundell MJ: Methylation of mouse ribosomal DNA genes. *DNA* 1982;1:259–266.

Riggs AD: X inactivation, differentiation and DNA methylation. *Cytogenet Cell Genet* 1975;14:9–11.

Roy PH, Weissbach A: DNA methylases from Hela cells nuclei. *Nucleic Acids Res* 1975;2:1669–1684.

Rubin RA, Modrich P: EcoRI methylase: physical and catalytic properties of the homogeneous enzyme. *J Biol Chem* 1977;252:7265–7272.

Saenundsen AK, Perlmann C, Klein G: Intracellular Epstein-Barr Virus DNA is methylated in and around the EcoRI-J fragment in both producer and non-producer cell lines. *Virology* 1983;126:701–706.

Salomon R, Kaye AM, Hertzberg M: Mouse nuclear satellite DNA: 5-methylcytosine content, pyrimidine isoplith distribution and electron microscopic appearance. *J Mol Biol* 1969;43:581–592.

Simon D, Grunert F, von Acken U, Döring HP, Kröger, H: DNA methylase from regenerating rat liver: purification and characterization. *Nucleic Acids Res* 1978;5:2153–2167.

Simon D, Stuhlmann A, Jahner D, Wagner H, Werner E, Jaenisch R: Retrovirus genomes methylated by mammalian but not bacterial methylase are noninfectious. *Nature* 1983;304:275–277.

Sims MA, Doering JL, Hoyle HD: DNA methylation patterns in the 5S DNAs of Xenopus laevis. *Nucleic Acids Res* 1983;11:277–290.

Sneider TW, Teague WM, Rogachevsky LM: S-adenosylmethionine dependent DNA-cytosine 5-methyltransferase from a Novikoff hepatoma cell line. *Nucleic Acids Res* 1975;2:1685–1700.

Solage A, Cedar H: Organization of 5-methylcytosine in chromosome DNA. *Biochemistry* 1978;17:2934–2938.

Stein R, Gruenbaum Y, Pollack Y, Razin A, Cedar H: Clonal inheritance of the pattern of DNA methylation in mouse cells. *Proc Natl Acad Sci (USA)* 1982;79:61–65.

Stein R, Sciaky-Gallili N, Razin A, Cedar H: Pattern of methylation of two genes coding for housekeeping functions. *Proc Natl Acad Sci (USA)* 1983;80:2422–2426.

Sturm KS, Taylor JH: Distribution of 5-methylcytosine in the DNA of somatic and germ line cells from bovine tissues. *Nucleic Acids Res* 1981;9:4537–4546.

Subramanian KN: Sites on gene expression in a genome functioning autonomously in a vertebrate host. *Nucleic Acids Res* 1982;10:3475–3486.

Sutter D, Doerfler W: Methylation of integrated adenovirus type 12 DNA sequences in transformed cells is inversely correlated with viral gene expression. *Proc Natl Acad Sci* 1980;77:253–256.

Szyf M, Gruenbaum Y, Urieli-Shoval S, Razin A: Studies on the biological role of DNA methylation: 5. The pattern of E. coli DNA methylation. *Nucleic Acids Res* 1982;10:7247–7259.

Szyf M, Avraham-Haezni K, Reifman A, Shlomai J, Kaplan F, Oppenheim A, Razin A: A DNA methylation pattern is determined by the intracellular level of the methylase. *Proc Nat Acad Sci (USA)* 1984;81:3278–3282.

Tantravahi U, Guntaka R, Erlanger BF, Miller OJ: Amplified ribosomal RNA genes in a rat hepatoma cell line are enriched in 5-methylcytosine. *Proc Natl Acad Sci (USA)* 1981;78:489–493.

Taylor SM, Jones PA: Mechanism of action of eukaryotic DNA methyltransferase. Use of 5-azacytosine-containing DNA. *J Mol Biol* 1982;162:679–692.

Turnbull JF, Adams RLP: DNA methylase: purification from ascites cells and the effect of various DNA substrates on its activity. *Nucleic Acids Res* 1976;3:677–695.

Urieli-Shoval S, Gruenbaum Y, Sedat J, Razin A: The absence of detectable methylated bases in Drosophila melanogaster DNA. *FEBS Lett* 1982;146:148–152.

Urieli-Shoval S, Gruenbaum Y, Razin A: Sequence and substrate specificity of isolated DNA methylases from *Escherichia coli* C. *J Bacteriol* 1983;153:274–280.

Vardimon L, Kressmann A, Cedar H, Maechler M, Doerfler W: The expression of a cloned adenovirus gene is inhibited by in vitro methylation. *Proc Natl Acad Sci (USA)* 1982;79:1075–1077.

Vovis GF, Horiuchi K, Hartman N, Zinder ND: Restriction endonuclease B and f1 heteroduplex DNA. *Nature New Biol* 1973;246:13–16.

Vovis GF, Horiuchi K, Zinder D: Kinetics of methylation by a restriction endonuclease from E. coli B. *Proc Natl Acad Sci (USA)* 1974;71:3810–3813.

Wettstein FO, Stevens JG: Shope Papilloma Virus DNA is extensively methylated in Non-Virus-Producing Neoplasms. *Virology* 1983;126:493–504.

Whittaker PA, McLachlan A, Hardman N: Sequence organization in nuclear DNA from Physarum polycephalum: methylation of repetitive sequences. *Nucleic Acids Res* 1981;9:801–814.

Wigler M, Levy D, Perucho M: The somatic replication of DNA methylation. *Cell* 1981;24:33–40.

Wilson VL, Jones PA: DNA methylation decreases in aging but not in immortal cells. *Science* 1983;220:1055–1057.

Woodcock DM, Adams JK, Allan RG, Cooper IA: Effect of several inhibitors of enzymatic DNA methylation on the in vivo methylation of different classes of DNA sequences in a cultured cell line. *Nucleic Acids Res* 1983;11:489–499.

8

DNA Methylation and Gene Expression[1]

Howard Cedar*

Understanding the mechanisms involved in the regulation of gene expression in eukaryotic organisms is one of the major challenges facing the modern molecular biologist. Despite recent advances that permit the dissection of individual genes at the nucleotide level, the factors involved in the orchestration of gene activity during development are not well understood. Two general models usually are proposed for explaining differential gene expression. Since the genetic information seems to be faithfully inherited by every somatic cell and germ line, it has been suggested that chromatin conformation plays a role in the selective expression of individual genes in specific cell types. There certainly is a large body of evidence that convincingly shows that active gene regions have a unique chromosomal conformation characterized by a generally increased accessibility to a variety of DNA probes (Chapter 15). A second model that has been put forward raises the possibility that changes in the DNA methylation pattern of specific genes may be responsible for the observed tissue-specific gene expression. This article will provide some of the evidence that links DNA methylation to the regulation of gene activity and will attempt to probe the mechanism by which this modification may act to modulate gene expression in specific cell types.

Many tissue-specific genes have been mapped and their methylation pattern has been established by using various methyl-sensitive, CpG-containing enzymes. The study of this pattern in different tissue DNAs shows a striking correlation between gene activity and undermethylation in the general region

[1]This chapter is dedicated to the memory of my father, Morris Cedar, a gentle man of great integrity.

*Department of Cellular Biochemistry, The Hebrew University, Hadassah Medical School, Jerusalem, Israel

surrounding the regulatory and structural portion of these genes (Chapter 16). The globin family of genes is the most carefully studied of this class of gene. In this case, the adult genes are undermethylated in the hemoglobin-producing reticulocytes, but they are heavily methylated in other tissue DNAs. Conversely, the fetal genes are found to be undermethylated in the fetal liver, which contains the cells actively producing fetal hemoglobin (Chapter 12). This same tissue-specific pattern also has been observed for many other genes and gene families (Chapter 16).

It should be noted that these correlations do not, in every case, encompass all of the methylatable sites in the region of the specific gene. Thus, while the chicken α-globin gene shows a clear-cut difference in the level of methylation in many sites covering the entire transcribed region (Weintraub *et al.* 1981), in the case of the rabbit globin gene, there are only three sites that show any correlation with gene activity; also, the changes are of a partial nature (Shen and Maniatis, 1980). In general, the correlation recorded for globin from human and rabbit sources is rather weak. It has been argued that the interpretation of these results is difficult indeed, since it is hard to obtain pure populations of reticulocytes in these organisms. The strong correlation seen in the chicken system is testimony to the fact that it is easy to purify the hemoglobin-producing cells in this organism. It also should be noted that there are tissue-specific variations in the methylation pattern of many genes that are not correlated with differences in gene expression. Thus, not every relative undermethylation necessarily correlates with gene activation. There also are several cases in which no correlation can be observed between gene expression and the DNA methylation pattern (Chapter 16). In these cases, it should be noted that a correlation cannot be strictly ruled out, since restriction enzymes do not asasy every available CpG residue.

The relationship between methylation and RNA transcription also has been analyzed at the biochemical level. Mouse liver or kidney DNA has a constant level of DNA methylation, so that about 75 percent of the CpG residues are in a modified form. When this total DNA is hybridized to poly A-containing RNA from the same tissues and the complementary DNA is isolated by column chromatography, this fraction was found to be only 40 percent methylated; this indicates that, in general, actively transcribed sequences are undermethylated, but are not totally unmethylated relative to the rest of the genome (Naveh-Many and Cedar, 1981).

The above discussion concentrated on methyl moieties in cells from the animal kingdom, which are modified exclusively at CpG residues. This dinucleotide presumably represents the only methylatable site in these organisms, since only this pair contains symmetric cytosines in the two complementary DNA strands. All of these sites probably are methylated at both cytosine residues; this duplication serves as a marker during replication, when the newly synthesized DNA becomes methylated at the exact sites where the nascent strand is methylated. This mechanism provides a neat way of conserving the pattern of modification from generation to generation. In addition to CpG methylation, plant DNA contains methyl moieties in the prototype trinucleotide sequence

CXG, where X can be A, C, or T (Gruenbaum *et al.* 1981). The inheritance of this site is made possible by the trinucleotide cytosine symmetry. It has been shown that the correlation between gene expression and DNA undermethylation is also true for plant CpG residues; recent experiments show that this correlation extends to CXG residues as well (Naveh-Many and Cedar, unpublished results). Although this correlation has not yet been demonstrated for a specific plant gene, there are restriction enzymes available that recognize such sites and are sensitive to DNA methylation (Gruenbaum *et al.* 1981). This result suggests that the correlation between gene activity and DNA methylation is a basic one, since it is observed at a variety of different methyl-containing sites.

DNA Methylation and Differentiation

The very presence of a tissue-specific pattern of methylation indicates that such a pattern can be stably inherited in somatic cells. In fact, this hypothesis has been proven by experiments using DNA-mediated gene transfer in which it was shown that foreign DNA inherits its own methylation fingerprint (Pollack *et al.* 1980; Stein *et al.* 1982; Wigler *et al.* 1981). It is also clear from the tissue-specific nature of the methylation distribution that changes in this pattern must take place during normal animal cell development. One hint as to the nature of these changes can be obtained by studying the tissue-specific methylation pattern of sperm DNA, which represents part of the starting material for the differentiation process and which contains the exact methylation pattern that is inherited from the male organism. These studies show that almost all of the tissue-specific differences observed *in vivo* must be due to some process of selective demethylation (Chapter 16). Thus, for example, all of the globin genes—as well as other tissue-specific gene sequences—are maximally methylated in sperm DNA. How and when these specific methylation changes occur is not known. In addition to those changes that occur in the region of tissue-specific abundant gene sequences, it has been shown that certain satellite sequences actually undergo *de novo* methylation during normal development (Sturm and Taylor, 1981; Ehrlich *et al.* 1982; Pages and Roizes, 1982). In several organisms, certain satellite sequences are more highly methylated in some somatic cells, while being relatively undermethylated in sperm DNA. While this observation is convincing, the exact mechanism for this change has not been elucidated. Although it is reasonable that this is due to the direct *de novo* methylation of these repeated sequences, it is also possible that changes in the copy number of methylated or unmethylated sequences may explain these results. None of these studies attempted to quantitate the number of these specific satellite sequences in various tissues and in sperm. It also has been suggested that *de novo* methylation may be involved in the process of X inactivation, which also occurs during the early developmental events of the animal embryo (Chapter 13). Partial evidence that this, indeed, is the case was shown by the fact that X-linked Hprt genes in sperm DNA are active in a transfection assay, while the Hprt gene on the inactive X in native somatic cells is clearly not competent in this same assay (Kratzer *et al.* 1983). It has been hypothesized that this inactivation at the level of the DNA itself is due to methylation; numerous experiments showing that 5-aza-C can activate various X chromosomal genes support this hypothesis (Mohandes *et al.* 1981; Jones *et al.* 1982; Lester *et al.* 1982). Currently, however, *de novo* methylation has not been demonstrated for any specific, cloned X chromsome-linked gene. In addition to those sequences that probably undergo *de novo* methylation during the normal process of development, it has also been shown that genes inserted into mouse zygotes by microinjection get methylated at some stage of the differentiation process (Palmiter *et al.* 1982; Jahner *et al.* 1982). This same phenomenon is also mimicked after the transfection of gene sequences into totipotential teratocarcinoma cell lines (Cedar *et al.* 1982; Stewart *et al.* 1982; Gautsch and Wilson, 1983).

In addition to the various gene-sequences that undergo either demethylation or *de novo* methylation during embryogenesis, several genes maintain identical methylation patterns during this process. Most notable are the aprt gene, the dhfr gene, and various ribosomal RNA genes. All of these gene sequences are highly methylated over large regions of the structural gene, but they are almost devoid of methylation at their 5′ ends (Chapter 16). Furthermore, this pattern is found in every somatic tissue and in germ-line sperm DNA as well. Since these genes have housekeeping functions and probably are expressed in every tissue and in every cell during the differentiation process, it could be argued that these genes might never be in an inactive state. Thus, the fixed methylation pattern may be a mechanism for marking these genes for constitutive activity. Since most expressed genes probably fall into this category, one might suggest that this constitutive pattern is actually written into the genetic information by way of DNA modification and that only a small number of tissue-specific genes actually undergo changes in their methylation pattern during development. One interesting gene family in this regard are the collagen genes. Whereas this gene is heavily methylated in the structural gene and is unmethylated in its 5′ region in all tissue sources, its activity is limited to fibroblast cells (McKeon *et al.* 1982). By way of comparison with other genes with a similar methylation pattern, it might be expected to be active in many tissue types. The fact that this gene plays a prominent role in the early stages of differentiation may have something to do with this interesting methylation pattern.

DNA Methylation Affects Gene Expression

The consistent correlation between gene activity and undermethylation strongly suggests a role for DNA modification in the regulation of gene expression during animal development. Alternatively, it is possible that changes in gene expression are mediated through other mechanisms and that demethylation or *de novo* methylation are posttranscriptional events that occur after the change in gene activity. Several lines of evidence suggest that DNA methylation can, indeed, inhibit specific gene expression in animal cells; however, there is no clear-cut data showing that DNA modification is the causal factor in the orchestration of gene regulation that occurs during differentiation.

Two types of experimental approaches have been used to test the effect of DNA methylation on gene activity. One approach is to insert methylated DNA into animal cells either by DNA-mediated gene transfer or by injection techniques and to observe the effect of this modification on the transcription of the genes. An alternate approach is to cause DNA hypomethylation by treatment of cells with 5-azacytidine (5-aza-C) and to follow the effects of this treatment on specific gene expression. The results of both of these methods support the model that DNA methylation can inhibit gene expression, probably at the level of RNA transcription.

Several methods have been employed for inserting gene sequences into animal cells. Using DNA-mediated gene transfer by means of calcium phosphate precipitation and by direct application onto growing cells, DNA is introduced into a small number of competent cells. Using one of several gene markers, transformed cells can be selected and grown into mass culture. Using this method, many types of gene sequences have been introduced into fibroblast cell lines; these genes generally are active in this foreign environment. However, several types of experiments have suggested that certain gene sequences methylated *in vitro* are not active in such cells. When the plasmids containing the HSV tk gene or the chicken tk gene were methylated by using the HpaII methylase, the ability of these vectors to produce tk^+ colonies was 5–10-fold impaired (Pollack *et al.* 1980; Wigler *et al.* 1981). The expression of other gene sequences also has been found to be inhibited by *in vitro* methylation. To show that methylation affects transcription and not the process of transfection, methylated and unmethylated aprt genes were inserted into tk^- $aprt^-$ L cells by DNA-mediated gene cotransfer, by using selection for the tk^{+-} phenotype. Almost all of the resulting clones, indeed, contained the aprt gene, but only the unmodified gene was expressed in these cells (Stein *et al.* 1982).

The fact that methylated DNA is not active provides the basis of a biological assay for DNA modification. Whereas methyl-sensitive restriction enzymes show that mouse-endogenous viral sequences are heavily methylated, the fact that these sequences are inactive in a DNA transformation assay suggests that these methyl moieties, indeed, inhibit the expression of these genes in the cells in which they are found. When these genes are activated, however, they become efficient in the transformation assay (Harbers *et al.* 1981). This assay also has been used as evidence that DNA modification plays a role in X inactivation. The activity of the X-linked hgprt gene can be assayed by DNA-mediated gene transfer into $hgprt^-$ cells and selection in HAT medium (Liskay and Evans, 1980). While DNA from the active X chromosome is active in this assay, the DNA from the inactive X is unable to produce hgprt colonies (Chapter 11). In a sense, this assay for DNA methylation—although not site-specific—is more informative than restriction enzyme assays, since one can directly ascertain the effect of DNA modification on gene activity.

It should be noted that by using this form of DNA-mediated gene transfer, almost all inserted genes are expressed to some extent; some are even hormone-inducible when tested in cells containing the proper receptive molecules. *In vivo,* however, these genes are expressed in a tissue-specific manner, and as a result, the endogenous counterparts of these genes are not expressed in the experimental cell system (i.e., L cells). Thus, the basic factors that inhibit the expression of these genes in L cells do not seem to affect the expression of the introduced genes. Since the activity state of these genes probably has been predetermined by the developmental history of the cells, one can speculate that DNA-mediated gene transfer bypasses this developmental process, resulting in a partially active gene. It is interesting that the only factor to date that inhibits the expression of stably introduced genes is DNA methylation. Furthermore, in some instances, selection for the nonexpression of the tk gene has led to its *in vivo, de novo* methylation (Ostrander *et al.* 1982; Clough *et al.* 1982). This suggests that this modification may also be involved in the inhibition of tissue-specific genes in cells in which they are not expressed (i.e., L cells).

Other assay techniques, in addition to DNA-mediated gene transfer, have been used to study the effects of DNA methylation on gene expression. Gene transcription, for example, may be measured by injection of circular plasmids into Xenopus oocytes. This system is characterized by transcription fidelity, and it probably mimics transcription *in vivo.* When the adenovirus E2a gene was methylated *in vitro* at the CCGG sites, transcription of this gene was reduced by one to two orders of magnitude in this sytem (Vardimon *et al.* 1982). Similar results were obtained when SV40 DNA was injected into these eggs (Fradin *et al.* 1982). While early transcription was not affected by DNA methylation (Graesmann *et al.* 1983), the transcription of the late functions was clearly inhibited by modification. This correlates with the methylation of a unique CCGG site located in the leader portion of the late region RNA.

One word of caution should be noted with reference to these experiments in which DNA is methylated artificially *in vitro.* It is not clear whether the changes introduced *in vitro* have a parallel significance *in vivo.* In the case of the aprt gene, it is quite clear that this gene is heavily methylated *in vivo,* but it lacks methylation at its 5′ end (Stein *et al.* 1983). Since this gene probably is expressed in every cell type and at every stage of differentiation, and since it has a constant methylation pattern throughout development, changes in methylation probably do not play a role in its regulation; therefore, *in vitro* methylation is quite artificial with no parallel *in vivo.* This same criticism applies to the tk gene and to the other viral genes assayed for the effect of methylation. In the case of the globin gene, it is known that there is a correlation between gene expression and hypomethylation; since its methylation pattern changes during development, it may be assumed that the *in vitro* methylation mimics the situation *in vivo.* In any event, it should be remembered that the introduction of methyl groups *in vitro* is equivalent to a mutational change in the DNA sequence of the gene, and in each case, it must be verified that this is representative of what occurs *in vivo.*

An alternate approach for showing that DNA methylation inhibits gene activity is to use substances that cause demethylation and assay for the activation of specific genes. Both 5-aza-C and its deoxy analog 5-aza-deoxycytidine can be used to efficiently inhibit DNA maintenance methylation in dividing cells (Chapter 9). After a short treatment with these drugs that spans one or two cell cycles, one can obtain a considerable degree of undermethylation, which then remains inheritable. Using this method, it has been shown that this drug can cause undifferentiated cells to develop various differentiated phenotypes; and, it can cause the activation of previously inactive genes, including those on the inactive X chromosome. In several instances, it has been shown that activation is accompanied by demethylation within the region of the specific gene (Groudine *et al.* 1981; Ley *et al.* 1982). Since 5-aza-C is an analog of 5-methylcytosine, and since it appears to inhibit the action of the cellular methylase following its incorporation into the DNA, it has been assumed that this drug leads to gene activatin by causing local changes in methylation. The implications of this result are that many genes in a particular cell type are locked into an inactive state by cytosine methylation and that this block can only be released by removing their methyl moieties. While this is a reasonable assumption, it should be noted that one cannot rigorously rule out the possibility that 5-aza-C leads to gene activation by some unknown mechanism and that the changes in the methylation pattern of a specific activated gene is an effect of this activation.

One of the gray areas in the interpretation of 5-aza-C results is the apparent specificity of this compound in activating some genes, but not others. In the experiment of Groudine *et al.* (1981), chicken-endogenous virus genes were activated by treatment with 5-aza-C. Although virus production and hypomethylation of the endogenous gene were observed in over 80% of the cells, the chicken globin genes in these same cells remained fully methylated and were not activated by this treatment. A quick survey of the various gene systems that undergo induction after 5-aza-C treatment shows that each gene is sensitive to this drug to different degrees. There could be various reasons for these differences in sensitivity. Although it is reasonable to assume that methyl moieties inhibit gene expression *in vivo,* it is not known for each gene how many methyl groups in the proper location are responsible for inhibiting gene activity. Thus, while the activation of certain genes may be achieved by hypomethylation at one or two selective sites, the activation of other genes may require more extensive changes in the methylation pattern. Thus, even if demethylation occurs in a random manner, it is still possible that each gene will undergo activation with different kinetics. Another possibility is that different CpG sequences undergo demethylation at varying rates, depending on their flanking DNA sequences. Finally, it is reasonable to assume that there may be multiple mechanisms for gene inactivation and that some genes, in fact, may not be under methylation control. All of these unknown factors make it difficult to interpret experiments employing 5-aza-C treatment; however, all of the data is, indeed, consistent with the notion that this drug causes a generalized demethylation that can activate individual genes to an extent based on the degree to which the expression of each gene is controlled by DNA methylation.

Mechanism of Action of DNA Methylation

Since the data supporting a role for DNA modification in gene regulation is certainly convincing, it might be worthwhile to try to understand the modification mechanisms by which changes in gene expression are caused. The first question is which regions of the gene environment are, indeed, influenced by DNA methylation. A great deal of evidence suggests that the major determining factor in the expression of any gene is the nature of its promoter region, which includes various control sequences placed at fairly defined places at the 5′ end of each eukaryotic gene. In fact, many experiments show that even when a particular gene promoter is linked to different structural genes, transcription control depends on the characteristics of the promoter. Thus, it is reasonable to assume that DNA methylation acts mainly on 5′ upstream sequences; a good deal of experimental data supports this contention. Although there is a general correlation between gene activity in various tissues and DNA hypomethylation, this correlation is best seen in the 5′ regions of these genes (Chapter 11 and 16). Other genes in eukaryotic cells, which are generally in an active state, have a distinct DNA methylation pattern in which the 3′ end of the gene is heavily methylated, while the 5′ end lacks 5 methylcytosine. This also suggests that methylation in the structural gene may have little effect on gene activity, while changes in the methylation state of the 5′ regions may be critical for determining gene expression. It is interesting that when the HSV tk gene was methylated with the HpaII methylase and its activity was assayed by DNA-mediated gene transfer, the effects of modification were found to be marginal (Pollack *et al.* 1980). Conversely, total CpG methylation of this gene leads to a 500-fold reduction in its ability to transfect L cells (Cedar *et al.* 1982). The best explanation for this effect is that the promoter region of this gene (including 120 nucleotides upstream from the mRNA start site) lacks CCGG sites, but it contains numerous other CpG moieties (Wagner *et al.* 1981). This concept has recently been tested in a more direct manner. The human globin gene was inserted into M13 and was hemimethylated by primer extension by using dmCTP as one of the nucleotide substrates. When inserted into animal cells growing in culture, this gene remains fully methylated at all CpG moieties, but it loses other cytosine methylations via dilution after many generations of growth. By using various different primers in this reaction, one can prepare DNA molecules that are selectively methylated in specific regions of the gene. Using this direct approach, Busslinger *et al.* (1983) showed that methylation in the 5′ region of the gene was sufficient to inhibit transcription of this gene *in vivo,* and methylation exclusively in the 3′ region was without effect. These results are particularly significant in view of the fact that there is a clear correlation *in vivo* between the expression of this gene in fetal liver and hypomethylation of one detectable CCGG site located −50 from the mRNA transcription start (van der Ploeg and Flavell, 1980). Similar experiments on the effect of DNA methylation on the HSV tk gene and the hamster aprt gene also show that the 5′ region of these genes is the major site at which methylation affects gene expression (Keshet and Cedar, preliminary results). Other experimental approaches have been employed to demonstrate that 5′ methylation is sufficient to inactivate certain viral genes (Fradin *et al.* 1982; Kruczek and Doerfler, 1983). If gene expression is primarily controlled by DNA methylation in the promoter region, it is hard to understand how human (Lawn *et al.* 1980) or mouse β-globin genes could be controlled by DNA modification, since these genes lack CpG residues for 200 nucleotides upstream from the transcription start site.

Assuming that methylation acts mainly at the 5′ regions of tissue-specific genes, it would be interesting to know how this modification actually affects gene transcription. Several laboratories have attempted to show a direct effect on transcription by assaying the effect of methylation on RNA synthesis *in vitro* via the use of purified enzymes or cell extracts. None of these efforts have proved successful to date, either because these *in vitro* systems are lacking factors necessary for faithful transcription or because DNA methylation affects RNA synthesis at a higher order level (probably affecting protein-DNA interactions). In this regard, it is interesting that in all of the systems in which an effect has been demonstrated, injected DNA is packaged in a nucleosomal array. In the case of DNA-mediated gene transfer by calcium phosphate precipitation, it is also clear that active genes maintain the nuclease sensitivity hot spots that characterize endogenous active genes (Weintraub, 1983; Sweet *et al.* 1982). The suggestion, of course, is that methylation probably inhibits gene expression by changing DNA-protein interactions.

Several lines of evidence support—but do not prove—this hypothesis. Although nucleosomes appear to be randomly distributed on the DNA, it is evident that these subunits may have a more defined placement in specific regions of the genome, including the region surrounding active genes. Several studies show that the relationship between these nucleosomes and methylated CpG sites is not random, since over 75% of these moieties are concentrated within nucleosomes, which leaves a lower density of methylation in the micrococcal nuclease-sensitive linker regions (Razin and Cedar, 1977; Solage and Cedar, 1978; Ball *et al.* 1983). Other experiments also show a correlation between DNAse hypersensitivity, which is characteristic of active genes, and hypomethylation in the same regions. It is interesting that HSV tk-containing clones could be reverted to tk^- by selection in BudR, and a fraction of the resulting clones were found to have undergone *de novo* methylation in the region of this gene (Sweet *et al.* 1982). This change in the methylation pattern was accompanied by the abolishment of the nuclease-hypersensitive region at the 5′ end of the tk gene. Once again, these results demonstrate a clear correlation between chromatin structure and methylation, but they do not prove a cause-and-effect relationship. In any event, it is certainly true that changes in chromatin conformation can be accomplished in the absence of variations in the methylation pattern.

Several more direct lines of evidence also support the idea that DNA methylation may affect protein-DNA interactions. In prokaryotes, it has been shown that changes in methylation, in fact, alter the binding of certain regulatory proteins (Fisher and Caruthers, 1979). Furthermore, all restriction enzyme interactions are strongly influenced by the presence of methyl moieties in their recognition sites (Roberts, 1983; McClelland, 1981). It is not hard to imagine that the presence of this modification can serve as a marker and affect the interaction of the DNA in this region with regulatory proteins. Furthermore, it is quite reasonable that DNA methylation may alter the tertiary structure of DNA itself, which (in turn) may affect the direct interaction of other sequences with RNA polymerase or may alter its interaction with other chromatin structural or regulatory proteins. Since the melting temperature of the 5mC=G base pair is considerably higher than that of the normal C=G hydrogen bonds, it is conceivable that this modification affects local DNA melting (Gill *et al.* 1974; Ehrlich *et al.* 1975). In addition, it may have either a positive or negative effect on the formation of non-B form DNA, which may be found at critical sites in the genome. It is already well documented that CpG methylation increases the rate of formation of Z-form DNA (Behe and Felsenfeld, 1981) and that this unusual negatively coiled DNA is not a good substrate for the placement of nucleosomes (Nickol *et al.* 1982). The significance of Z-form DNA is discussed in another chapter in this volume.

Postexpression Variations in DNA Methylation

As described in this chapter, the animal organism shows a dynamic movement of methylation during the process of development. Thus, we have seen examples of tissue-specific genes that are heavily methylated in germ line DNA, but become hypomethylated in the tissue of expression. There are certain housekeeping genes that have a constant methylation pattern during differentiation and there are other sequences, including satellites, that undergo *de novo* methylation. Since *in vitro* DNA methylation has been shown to inhibit gene expression, the implication of these studies is that genes are turned on and off during differentiation by a careful orchestration of methylation changes. While this scheme certainly is reasonable, there is presently no hard data supporting this theory.

One important criterion of this concept is that the changes in methylation should proceed the alterations in gene expression. There is no system in which this has been investigated, probably because the determination steps in animal development occur in a very small population of cells that would not be amenable to blot hybridization analysis. While Groudine and Weintraub (1981) suggested that hypomethylation preceded the activation of the chicken globin gene in 35-hour chicken embryos, this was not tested experimentally, since the formation of erythrocytes is a nonsynchronized process; thus, it makes it difficult to study individual cell types.

An alternate possibility that is consistent with all of the current data is that *changes* in gene expression are mediated by other factors and that gene hypomethylation or *de novo* methylation are posttranscriptional events. The methylation pattern of several liver-specific genes provides an excellent model for this hypothesis. Both the vitellogenin gene and the albumin gene are very undermethylated in the adult liver. Although both genes are turned on in the fetal liver at approximately 18 days, both genes are still heavily methylated at this stage, which suggests that the hypomethylation observed in this system follows the changes in gene expression (Vedel *et al.* 1983; Andrews *et al.* 1982; Gerber-Haber *et al.* 1983). In the case of the chicken vitellogenin gene, one CCGG site at the 5′ end of the gene was clearly shown to be methylated, despite the fact that this gene was being actively expressed after hormone treatment of the liver (Wilks *et al.* 1982). Furthermore, these same methylated molecules were shown to become DNAseI-sensitive at about the same time that activation occurred. This suggests that chromatin conformation—but not DNA methylation—are correlated with gene activity (Burch and Weintraub, 1983). This same site, however, was found to be completely undermethylated in the adult liver. A similar story has been observed for the rat liver PEPCK gene (Benvenisty *et al.* 1983). While this gene is only activated at the time of birth, it is inducible via cAMP after about 17–18 days of development, although it is fully methylated at this stage. The Xenopus vitellogenin gene is even more unusual, because it remains fully methylated even in the adult liver. (Gerber-Haber *et al.* 1983). Other nonliver genes appear to go through a similar process, as shown by the fact that crystallin undergoes hypomethylation *following* its activation in lens tissue (Grainger *et al.* 1983). It should be noted that in all of these studies, only a limited number of restriction enzyme sites were assayed for DNA methylation. Thus, if there were more decisive, timely changes in the methylation pattern at other sites, these would go undetected in these experiments. It is certainly conceivable that some critical CpG sites are involved in gene determination, while other sites undergo hypomethylation after activation of the gene. In the albumin gene, there is some evidence that this, indeed, may be the case. One 5′ CCGG site that is consistently undermethylated in albumin-producing hepatoma cell lines was partially undermethylated in fetal liver; this suggests that changes at this site might actually precede the activation of the gene (Ott *et al.* 1982). Difficulties in isolating pure populations of fetal hepatocytes also complicate the picture by masking possibly unmethylated sites.

A similar mode of thinking may be applied to events in either early embryos or teratocarcinoma cells in which gene inactivation is associated with *de novo* methylation (Chapter 10). When MuLV is inserted into undifferentiated teratocarcinoma cells, it is rapidly inactivated, although the *de novo* methylation of these sequences occurs only after 3 weeks post infection; this suggests that other unknown factors cause inactivation, while DNA methylation is only a secondary event (Gautsch and Wilson, 1983). Conversely, this methylation may play an important role in the inhibition of these exogenous genes at later stages of development. When teratocarcinoma cells containing the MuLV gene are induced to differentiate, these integrated genes remain silent, despite the fact that these differentiated cells are now permissive hosts for the MuLV injection. Thus, although DNA modification may not be involved in gene regulation at on stage of development, it may play an important role in other cell types.

The same type of phenomenon may also be true of X chromosome inactivation. It is possible that inactivation takes place by some unknown mechanism, but that the inactive state is maintained by *de novo* methylation, which then keeps these genes off in all subsequent cell generations. A hint that this is a reasonable mechanism comes from studies of the inactivation of the X chromosome during the formation of the extraembryonic membranes. In this case, it appears that the inactivation process takes place without any associated DNA modification, showing that DNA methylation may not be necessary to bring about the actual inactivation (Kratzer *et al.* 1983).

A model that is consistent with all of the above facts is that DNA methylation plays a role in inhibiting gene expression, so that if a particular gene is methylated in a critical spot, this modification would prevent its transcription. Conversely, it is possible that *de novo* DNA methylation or hypomethylation may not be the primary factor involved in changing the activity state of a gene. Once these changes are made, however, DNA modification or the lack of it may act to perpetuate the new pattern of gene expression.

The following example might give an indication of how this system may function. A particular tissue-specific gene is fully methylated in the germ line DNA and, as a result, is inactive in all cells during early development. At one actual point in time, a specific factor may appear that can *override* the effect of this gene-specific methylation and, thus, turn on this gene. As a result of either certain protein-DNA interactions or transcription itself, this gene may now undergo demethylation; this, in turn, may perpetuate the new activity state of the gene, even in the absence of the initial triggering reaction. While this is an alternative to the idea that *changes* in methylation direct changes in gene activity, there is no strong experimental support for either hypothesis. One interesting and suggestive experiment has been provided by Macleod and Bird (1983). While the Xenopus ribosomal genes are fully methylated in every tissue (except for a small area in the spacer region), this area is also heavily methylated in sperm DNA. When sperm ribosomal DNA and somatic cell ribosomal DNA were assayed for transcriptional activity by injection into unfertilized oocytes, both templates were highly active. The implication of this experiment is that either this methylation difference plays absolutely no role in gene expression or that factors in the oocyte are able to override the effect of this additional methylation.

Conclusion

On careful inspection, one is almost forced to the conclusion that methylation changes alone are not the driving force in controlling gene regulation. Since the observed variations in the methylation patterns are quite tissue-specific, it is hard to imagine how specific demethylation would occur without the participation of some other protein or nucleic acid factor. Thus, even if gene activation is dependent on the removal of certain methyl groups, it is obvious that other factors must first recognize this methylated gene and promote its demethylation. It has been assumed that hypomethylation is caused by the specific inhibition of maintenance methylation during replication. Since this methylase activity is present throughout the cell cycle and appears to act quite efficiently, it is quite possible that some protein factor must bind to the specific gene region to prevent methylation until the next round of DNA replication, when a totally unmethylated gene molecule can segregate out. Thus, both the determination event and the actual demethylation event probably would require the presence of specific proteins. Without decreasing the importance of methylation as a possible mediator of gene regulation, it is important to point out that the key events in differentiation almost certainly require other factors.

Although there are numerous examples of genes that do not appear to undergo hypomethylation concordant with its activation, one must be careful in interpreting these experiments. These studies all employ restriction enzymes to assay specific methylation sites within the region of a particular gene. Since these restriction enzymes are inhibited by both fully methylated and hemimethylated sites, it is certainly possible that these genes become hemimethylated—either preceeding or together with the activation of the gene—and that this change would go undetected by the assay system. Thus, it has been noted that the globin gene in Friend erythroleukemia cells undergo activation after treatment with specific inducers, but that the gene itself does not undergo hypomethylation at three restriction sites in the gene domain (Sheffery *et al.* 1982). Since there is a decrease and eventual stoppage of cell division after induction, it is possible that many of these active cells undergo specific hypomethylation during one replication cycle and then remain in a hemimethylated state. The same lack of hypomethylation has been observed for the muscle-specific actin and myosin genes in the differentiating rat myoblast system. In this case, it is worth noting that the methylation pattern of these genes is the same, both in skeletal muscle cells as well as in other rat tissues (M. Shani, personal communication). One simple explanation of this is that once muscle is formed by the fusion of myoblasts during development, the nuclei of these formed cells do not undergo more than one round of replication. Thus, even if demethylation occurred during the activation process, it would not be observed, since there are no further division cycles to segregate out hypomethylated DNA. If this type of hypomethylation occurred during or following fusion, the muscle cell DNA may indeed remain hemimethylated for the life of the organism. This should be amenable to experimentation. In one experiment in which the metalothenine gene was activated by treatment of cells with 5-aza-C, only one cycle of DNA replication in the presence of the drug was required to observe active clones after selection in cadmium-containing medium (Compere and Palmiter, 1981). This suggests that this gene already may be active in its hemimethylated state. This may not be a general phenomenon, since the activation of the Hprt gene seems to require two rounds of DNA replication before viable clones can be detected.

There certainly is enough evidence to date to suggest that DNA methylation may play a role in the regulation of gene expression, although the exact details of this process have not been worked out. Conversely, it is quite clear that, in many cases, gene control is carried out in the absence of a DNA modification process. It has been well documented that *Drosophila* species and probably most other insects lack even low levels of CpG modification (Urieli-Shoval *et al.* 1982). Although it has not been ruled out that a small number of central methyl sites do not play a role in this organism, it is probable that gene regulation occurs without the aid of DNA methylation. Even within the animal world, there are experimental examples of tissue-specific regulation that does not require DNA modification. The chicken crystallin gene, for example, has been injected into various mouse primary cell types, and it has been shown to be active only in lens-derived cells (Kondoh *et al.* 1983). In a similar vein, the chick lysozyme promoter is also active only in primary oviduct cells (Renkawitz *et al.* 1982). Furthermore, enhancing elements also have been shown to provide tissue specificity. The IgG enhancer, for example, clearly provides a means of activating genes in IgG-producing myeloma cells (Banerji *et al.* 1983; Gillies *et al.* 1983), and other tissue-specific enhancers are now the subject of experimental interest. While these enhancing elements may not be the only factors involved in the differentiation of tissue specificity, it is clear that this mechanism works even in the absence of the inhibiting effect of DNA methylation.

On another level, gene induction also appears to occur in the absence of DNA methylation changes. Thus, when various gene sequences were inserted into mouse erythroleukemia cells via DNA-mediated gene transfer, only globin-like genes were activated during the induction procedure, despite the fact that none of these inserted genes were methylated (Chao *et al.* 1983; Wright *et al.* 1983). All of these examples demonstrate that gene regulation probably occurs at many independent levels, with DNA modification being one link in this chain.

It has been pointed out that while the difference between the minimally active and maximally active state of a gene in prokaryotes may be about 1,000-fold, this difference is much larger in eukaryotes in which it has been suggested that tissue-specific differences of certain genes reach 10^8–10^9 (Ivarie *et al.* 1983). The task of maintaining these differences probably requires many mechanisms working in parallel. DNA modification most certainly plays a role in this process, together with other biological mechanisms.

References

Andrews GK, Dziadek M, Tamaoki T: Expression and methylation of the mouse α-Fetoprotein gene in eukaryotic, adult and neoplastic tissues. *J Biol Chem* 1982;257:5148–5153.

Ball DT, Gross DS, Garrard WT: 5-methylcytosine is localized in nucleosomes that contain histone H1. *Proc Natl Acad Sci (USA)* 1983;80:5490–5494.

Banjeri J, Olson L, Schaffner W: A lymphocyte specific cellular enhancer is located downstream of the joining region in immunoglobin heavy chain genes. *Cell* 1983;33:729–740.

Behe M, Felsenfeld G: Effects of methylation on a synthetic polynucleotide: The B-2 transition in poly (dG-m^{5}dc) poly (dG-m^{5}dC). *Proc Natl Acad Sci (USA)* 1981;78:1619–1623.

Benvenisty N, Mencher D, Meyuhas O, Razin A, Reshef L: Methylation of rat cytosolic phosphoenol pyruvate carboxykinase gene: Pattern associated with tissue specificity and development. *Proc Natl Acad Sci (USA)* 1984; in press.

Burch JBE, Weintraub H: Temporal order of chromatin structural changes associated with activation of the major chicken vetellogenin gene. *Cell* 1983;33:65–76.

Busslinger M, Hurst J, Flavell RA: DNA methylation and the regulation of globin gene expression. *Cell* 1983;34:197–206.

Cedar H, Stein R, Gruenbaum Y, Naveh-Many T, Sciaky-Gallili N, Razin A: Effect of DNA methylation on gene expression. *Cold Spring Harbor Symposia Quant Biol* 1982;47:605–609.

Chao MV, Mellon P, Charmay P, Maniatis T, Axel R: The regulated expression of β -globin genes introduced into mouse erythroleukemia cells. *Cell* 1983;32:483–493.

Clough DW, Kunkel LM, Davidson RL: 5-azacytidine induced reactivation of a Herpes Simplex thymidine kinase gene. *Science* 1982;216:70–73.

Compere SJ, Palmiter RD: DNA methylation controls the inducibility of the mouse metallothionein-1 gene in lymphoid cells. *Cell* 1981;25:233–240.

Ehrlich M, Ehrlich K, Mayo JA: Unusual properties of the DNA from Xanthomonas phage XP-12 in which 5-methylcytosine completely replaces cytosine. *Biochim Biophys Acta* 1975;395:109–119.

Ehrlich M, Gama-Sosa MA, Huang LH, Midgett RM, Kuo KC, McCune RA, Gehrke, C: Amount and distribution of 5-methylcytosine in human DNA from different types of tissues or cells. *Nucleic Acids Res* 1982;10:2709–2721.

Fisher EF, Caruthers MH: Studies on gene control regions. 12. The functional significance of a Lac operator constitutive mutations. *Nucleic Acids Res* 1979;7:401–416.

Fradin A, Manley JL, Prives CL: Methylation of simian virus 40 Hpa II site affects late but not early viral gene expression. *Proc Natl Acad Sci (USA)* 1982;79:5142–5146.

Gautsch JW, Wilson M: Delayed *de novo* methylation in teratocarcinoma suggests additional tissue-specific mechanisms for controlling gene expression. *Nature* 1983;301:32–37.

Gerber-Haber S, May FEB, Westlay BR, Felber BK, Hosbach HA, Andres A-C,

Ryffel GV: In contrast to other Xenopus genes the estrogen-inducible vitellogenin genes are expressed when totally methylated. *Cell* 1983;33:43–51.

Gill JE, Mazrimas JA, Bishop CC Jr: Physical studies on synthetic DNAs containing 5-methylcytosine. *Biochim Biophys Acta* 1974;335:330–348.

Gillies SD, Morrison SL, Oi VT, Tonegawa S: A tissue-specific transcription enhancer element is located in the major intron of a rearranged immunoglobin heavy chain gene. *Cell* 1983;33:717–728.

Graessmann M, Graessmann A, Wagner H, Werner E, Simon D: Complete DNA methylation does not prevent polyoma and simian virus 40 virus gene expression. *Proc Natl Acad Sci (USA)* 1983;80:6470–6474.

Grainger RM, Hazard-Leonards RM, Samaha F, Hougan LM, Lesk MR, Thomson GH: Is hypomethylation linked to activation of crystallin genes during development? *Nature* 1983;306:88–91.

Groudine M, Eisenmann R, Weintraub H: Chromatin structure of endogenous retroviral genes and activation by an inhibitor of DNA methylation. *Nature* 1981;292:311–317.

Groudine M, Weintraub H: Activation of globin genes during chicken development. *Cell* 1981;24:393–401.

Gruenbaum Y, Naveh-Many T, Cedar H, Razin A: Sequence specificity of methylation in higher plant DNA. *Nature* 1981;292:860–862.

Harbers K, Schnieke A, Stuhlman H, Jahner D, Jaenish R: DNA methylation and gene expression: endogenous retroviral genome becomes infectious after molecular cloning. *Proc Natl Acad Sci (USA)* 1981;78:7609–7613.

Ivarie RD, Schacter BS, O'Farrell PH: The level of expression of the rat growth hormone gene in liver tumor cells is at least eight orders of magnitude less than that in anterior pituitary cells. *Mol Cell Biol* 1983;3:1460–1467.

Jahner D, Stuhlmann H, Stewart CL, Harbers K, Loehler J, Simon J, Jaenish R: De novo methylation and expression of retroviral genomes during mouse embryogenesis. *Nature* 1982;298:623–628.

Jones PA, Taylor SM, Mohandas T, Shapiro LJ: Cell cycle specific reactivation of an inactive X-chromosome locus by 5-azadeoxycytidine. *Proc Natl Acad Sci (USA)* 1982;79:1215–1219.

Kondoh H, Yasuda K, Okada TS: Tissue specific expression of a cloned chick δ-crystallin gene in mouse cells. *Nature* 1983;301:440–442.

Kratzer PG, Chapman VM, Lambert H, Evans RE, Liskay RM: Differences in the DNA of the inactive X chromosomes of fetal and extra-embryonic tissues of mice. *Cell* 1983;33:37–42.

Kruczek I, Doerfler W: Expression of the chloramphenicol acetyltransferase gene in mammalian cells under the control of adenovirus type 12 promoters: effect of promoter methylation on gene expression. *Proc Natl Acad Sci (USA)* 1983;80:7586–7590.

Lawn RM, Efstratiadis A, O'Connell C, Maniatis T: The nucleotide sequence of the human β-globin gene. *Cell* 1980;21:647–651.

Lester SC, Korn NJ, Demars R: Depression of genes on the human inactive X-chromosome: evidence for differences in locus-specific rates of depression and rates of transfer of active and inactive genes after DNA-mediated transformation. *Somatic Cell Genet* 1982;8:265–284.

Ley TJ, Decimone J, Anagnou NP, Kelter GH, Humphries RK, Turner PH, Young NS, Helter P, Nienhuis AW: 5-azacytidine selectively increases α-globin synthesis in a patient with β^+ thalassemia. *N Engl J Med* 1982;307:1469–1475.

Liskay RM, Evans RJ: Inactive X chromosome DNA does not function in DNA-mediated cell transformation for the hypoxanthine phosphribosyl transferase gene. *Proc Natl Acad Sci (USA)* 1980;77:4895–4898.

McClelland M: The effect of sequence specific DNA methylation on restriction endonuclease cleavage. *Nucleic Acids Res* 1981;9:5859–5866.

McCleod D, Bird A: Transcription in oocytes of highly methylated rDNA from Xenopus laevis sperm. *Nature* 1983;306:200–203.

McKeon C, Ohkabo H, Pastan J, Crombrugghe B: Unusual methylation pattern of the $\alpha 2(1)$ collagen gene. *Cell* 1982;29:203–210.

Mohandes T, Sparkes PS, Shapiro LJ: Reactivation of an inactive human X-chromosome: evidence for X-inactivation by DNA methylation. *Science* 1981;211:393–396.

Naveh-Many T, Cedar H: Active gene sequences are undermethylated. *Proc Natl Acad Sci (USA)* 1981;78:4246–4250.

Nickol J, Behe M, Felsenfeld G: Effect of the B-Z transition in poly (dG-m^{5}dC)·poly (dG-m^{5}dC) on nucleosome formation. *Proc Natl Acad Sci (USA)* 1982;79:1771–1775.

Ostrander M, Vogel S, Silverstein S: Phenotypic switching in cells transformed with the Herpes Simplex virus thymidine kinase gene. *Mol Cell Biol* 1982;2:708–714.

Ott M-O, Sperling L, Cassio D, Levilliers T, Sala-Trepat J, Weiss, MC: Undermethylation at the 5′ end of the albumin gene is necessary but not sufficient for albumin production by rat hepatoma cells in culture. *Cell* 1982;30:825–833.

Pages M, Roizes G: Tissue specificity and organization of CpG methylation in calf satellite DNA 1. *Nucleic Acids Res* 1982;10:565–576.

Palmiter RD, Chen HY, Brinster RL: Differential regulation of metallothionein-thymidine kinase fusion genes in transgenic mice and their offspring. *Cell* 1982;29:701–710.

Pollack Y, Stein R, Razin A, Cedar H: Methylation of foreign DNA sequences in eukaryotic cells. *Proc Natl Acad Sci (USA)* 1980;77:6463–6467.

Razin A, Cedar H: Distribution of 5-methylcytosine in chromatin. *Proc Natl Acad Sci (USA)* 1977;74:2725–2728.

Renkawitz R, Beng H, Graf T, Matthias P, Grez M, Schutz G: Expression of a chicken lysozyme recombinant gene is regulated by progesterone and dexamethasone after microinjection into oviduct cells. *Cell* 1982;31:167–176.

Roberts RJ: Restriction and modification enzymes and their recognition sequences. *Nucleic Acid Res* 1983;11:r135–r173.

Sheffery M, Rifkind RA, Marks P: Murine erythroleukemia cell differentiation: DNAse 1 hypersensitivity and DNA methylation near the globin genes. *Proc Natl Acad Sci (USA)* 1982;79:1180–1184.

Shen CKJ, Maniatis T: Tissue-specific DNA methylation in a cluster of rabbit β-like globin genes. *Proc Natl Acad Sci (USA)* 1980;77:6634–6638.

Solage A, Cedar H: Organization of 5-methylcytosine in chromosomal DNA. *Biochemistry* 1978;17:2934–2938.

Stein R, Gruenbaum Y, Pollack Y, Razin A, Cedar H: Clonal inheritance of the pattern of DNA methylation in mouse cells. *Proc Natl Acad Sci (USA)* 1982;79:61–65.

Stein R, Razin A and Cedar H: In vitro methylation of the hamster adenine phosphoribosyltransferase gene inhibits its expression in mouse L-cells. *Proc Natl Acad Sci (USA)* 1982;79:3418–3422.

Stein R, Sciaky-Gallili N, Razin A, Cedar H: Pattern of methylation of two genes coding for housekeeping functions. *Proc Natl Acad Sci (USA)* 1983;80:2422–2426.

Stewart CL, Stuhlman H, Jahner D, Jaenisch R: De novo methylation, expression and infectivity of retroviral genomes introduced into embryonal carcinoma cells. *Proc Natl Acad Sci (USA)* 1982;79:4098–4102.

Sturm KS, Taylor JH: Distribution of 5-methylcytosine in the DNA of somatic and germline cells from bovine tissues. *Nucleic Acids Res* 1981;9:4537–4546.

Sweet RW, Chao MV, Axel R: The structure of the thymidine kinase gene promoter: nuclease hypersensitivity correlates with expression. *Cell* 1982;31:347–353.

Urieli-Shoval S, Gruenbaum Y, Sedat J, Razin A: The absence of detectable methylated bases in Drosophila melanogaster DNA. *FEBS Lett* 1982;146:148–152.

van der Ploeg LHT, Flavell RA: DNA methylation in the human $\gamma\delta\beta$-globin locus in erythroid and non-erythroid tissues. *Cell* 1980;19:947–958.

Vardimon L, Kressman A, Cedar H, Maechler M, Doerfler W: Expression of a cloned adenovirus gene is inhibited by in vitro methylation. *Proc Natl Acad Sci (USA)* 1982;79:1073–1077.

Vedel M, Gomez-Garcia M, Sala M, Sala-Trepat JM: Changes in methylation pattern of albumin and α-fetoprotein genes in developing rat liver and neoplasma. *Nucleic Acids Res* 1983;11:4335–4354.

Wagner MJ, Sharp JA, Summers WC: Nucleotide sequence of the thymidine kinase gene of herpes simplex virus type 1. *Proc Natl Acad Sci (USA)* 1981;78:1441–1445.

Weintraub H, Larsen A, Groudine M: α-globin gene switching during the development of chicken embryos: expression and chromosome structure. *Cell* 1981;24:333–344.

Weintraub H: A dominant role for DNA secondary structure informing hypersensitive structures in chromatin. *Cell* 1983;32:1191–1203.

Wigler M, Levy D, Perucho M: The somatic replication of DNA methylation. *Cell* 1981;24:33–40.

Wilks AF, Cozens PJ, Mattaj JW, Jost JP: Estrogen induces a demethylation at the 5′ end region of the chicken vitellogenin gene. *Proc Natl Acad Sci (USA)* 1982;79:4252–4255.

Wright S, de Boer E, Grosveld FG, Flavell RA: Regulated expression of the human β-globin gene family in murine erythroleukemia cells. *Nature* 1983;305:333–336.

9

Gene Activation by 5-Azacytidine

Peter A. Jones*

The nucleoside analog 5-azacytidine (5-aza-C) was first synthesized in Czechoslovakia in 1963 (Piskala and Sorm, 1964); it also has been isolated from streptoverticillium (*Streptoverticillus ladakanus,* Hanka *et al.* 1966; Bergy and Herr, 1966). 5-Aza-C differs from cytidine only by the inclusion of a nitrogen atom in the 5 position of the pyrimidine ring (Figure 9.1). It was originally developed for use as a cancer chemotherapeutic agent and is still used in the treatment of certain types of acute myelogenous leukemia. However, recent interest in the drug has been directed toward its remarkable ability to induce the expression of repressed genes in eukaryotic cells and to act as an inhibitor of DNA methylation.

Chemistry

Pithova *et al.* (1965) first proposed that 5-aza-C underwent a facile ring cleavage to yield an unstable N-formylguanylribosylurea intermediate in neutral aqueous solutions, which was followed by the loss of formate to form ribosylguanylurea in an irreversible reaction (Figure 9.2). In strong acid solutions, the initial reaction is cleavage of the glycosidic linkage to yield 5-azacytosine, 5-azauracil, and ribose (Figure 9.2). This reaction scheme was subsequently confirmed by later studies by Notari and DeYoung (1975), Israili *et al.* (1976), Beisler (1978), and more recently by Chan *et al.* (1979). The hydrolysis in neutral solutions, therefore, is a two-step reaction in which the initial ring opening to the N-formylguanylribosylurea intermediate occurs with an equilibrium

*University of Southern California Comprehensive Cancer Center, 2025 Zonal Avenue, Los Angeles, California 90033.

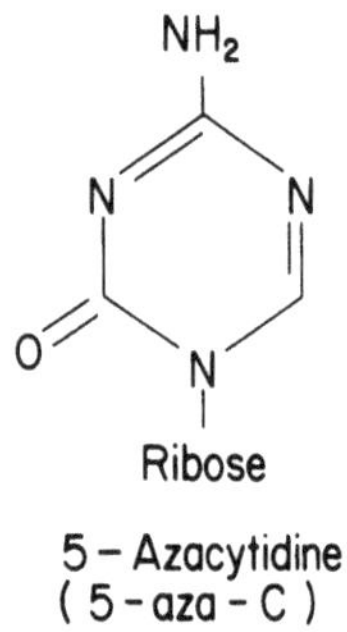

Figure 9.1 Structure of 5-azacytidine (5-Aza-C).

constant of 0.58 ± 0.03 at pH 5.6 (Chan *et al.* 1979). The drug appears to be maximally stable at pH 7 (Chan *et al.* 1979), and the half-life in phosphate-buffered saline (pH 7.4) at 37°C is approximately 6 hours (Constantinides, 1977).

More recent studies by Lin *et al.* (1981) have demonstrated that the deoxy analog 5-aza-2'-deoxycytidine (5-aza-CdR) decomposed in a similar manner to 5-aza-C in alkaline solutions. However, in neutral solutions or water, there were marked differences in the decompositions of the two compounds. In neutral solution, 5-aza-CdR decomposed to both the formylguanylribosylurea

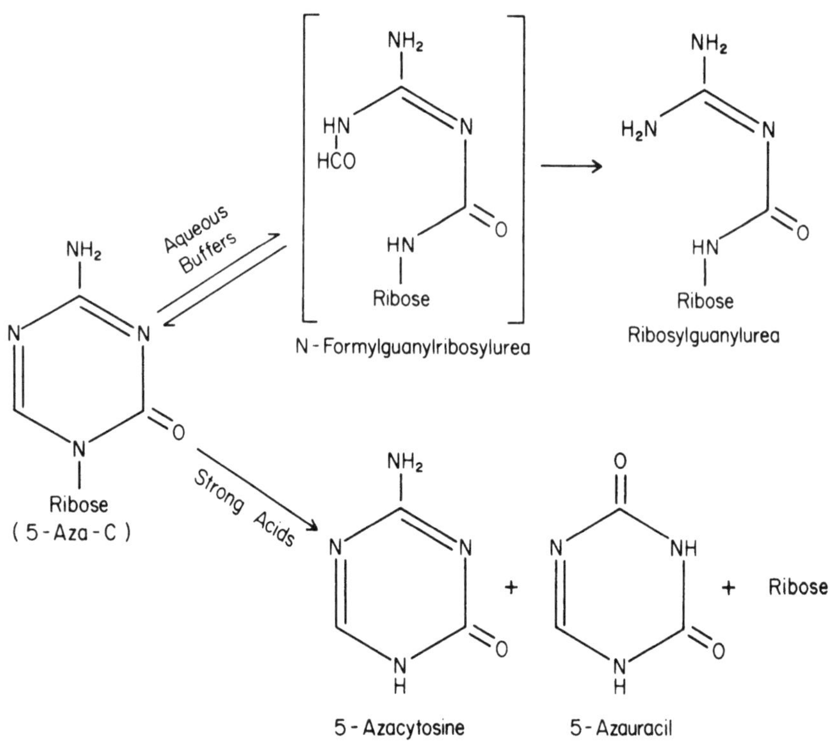

Figure 9.2 Decomposition of 5-azacytidine under neutral and acidic conditions.

intermediate and three unknown compounds that were chromophoric at 254 nm. The instability of 5-aza-C and 5-aza-CdR in neutral solutions has complicated their clinical use, and it has encouraged the development of other analogs such as 5,6-dihydro-5-azacytidine (Beisler *et al.* 1976) and pseudoisocytidine (Chou *et al.* 1979), which possess more stable ring systems.

Phosphorylation of Cytidine Analogs

The enzyme uridine-cytidine kinase catalyzes the phosphorylation of uridine, cytidine, and a number of pyrimidine analogs to the monophosphate level in the presence of ATP and Mg^{++} (Caputto, 1962; Anderson, 1973). This enzyme is important in the salvage of pyrimidines and is especially active in rapidly proliferating cells, since phosphorylation of pyrimidine nucleosides is the rate-limiting step in the production of nucleotides (Anderson, 1973). The phosphorylation of 5-aza-C by uridine-cytidine kinase has been studied extensively (see Vesely and Cihak, 1978 for review). It appears that 5-aza-C, cytidine, and uridine are phosphorylated by a single enzyme and not by separate, closely related enzymes (Lee *et al.* 1974; Liacouras and Anderson, 1979). The Km of the purified kinase from hepatoma cells for 5-aza-C is 20–120-fold higher than that for uridine and cytidine, respectively (Liacouras and Anderson, 1979). 5-Aza-C is a relatively weak, competitive inhibitor of cytidine phosphorylation, whereas both cytidine and uridine are strong competitive inhibitors of 5-aza-C phosphorylation (Liacouras and Anderson, 1979). After

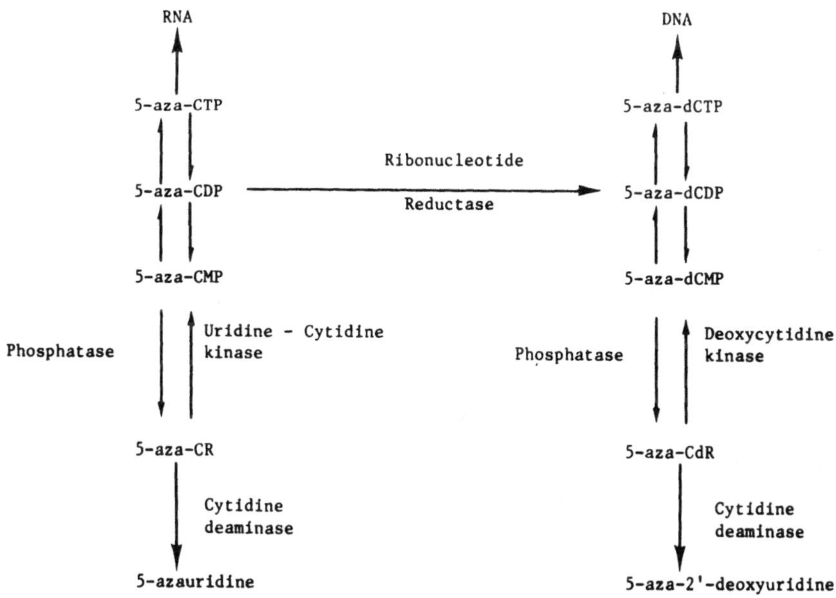

Figure 9.3 Metabolism of 5-azacytidine and 5-aza-2'-deoxycytidine.

conversion to the nucleotide level, 5-aza-C may be incorporated into the RNA and DNA of treated cells (Figure 9.3; Li *et al.* 1970).

In contrast, 5-aza-CdR is phosphorylated by a deoxycytidine kinase (Figure 9.3) with an apparent Km that is 5-fold higher than that for deoxycytidine (Momparler and Derse, 1979). Deoxycytidine acts as a potent competitive inhibitor of the phosphorylation of 5-aza-CdR, and it can completely eliminate the effects of the analog in several systems (Momparler and Goodman, 1977). Thus, 5-aza-CdR and 5-aza-C are phosphorylated by different enzymes; also, 5-aza-C, but not 5-aza-CdR, is incorporated into RNA and may produce defects in RNA function.

Incorporation of high levels of the 5-azacytosine ring into DNA in *Escherichia coli* results in helix instability and disruption of secondary structure (Zadrazil *et al.* 1965). It has been suggested that the lability of incorporated 5-azacytosine may be the cause of this helix instability; the possibility of ring cleavage in the polynucleotide cannot be completely ruled out. However, we have found that greater than 85% of the incorporated 5-azacytosine in eukaryotic DNA can be recovered intact from treated cells (Jones and Taylor, 1981).

Both 5-aza-C and 5-aza-CdR can be deaminated by cytidine deaminase to form the uridine and deoxyuridine derivatives, respectively (Chabot *et al.* 1983). Both of these products are biologically inactive and are unlikely to be responsible for the profound effects of the drugs on DNA methylation and cellular differentiation.

Inhibition of Nucleic Acid Methyltransferases by 5-Azacytidine

Prokaryotic Systems

Friedman (1979) was the first to report that 5-aza-C inhibited the DNA (cytosine − 5) methylase in *E. coli* K12 grown in the presence of the drug without inhibiting the DNA (adenine − N6) methylase. These results were extended in a more comprehensive study that showed that DNA extracted from 5-aza-C-treated *E. coli* could inhibit the methyltransferase extracted from the same cells in an *in vitro* reaction (Friedman, 1981). The inhibition was time-dependent and could be decreased by adding excess substrate DNA. However, the inhibitory capacity of DNA that contained 5-azacytosine was destroyed by incubation with DNAse, although the inhibited enzyme was not reactivated by treatment with nucleases. The DNA containing 5-azacytosine also inhibited the EcoRII and HpaII modification methylases. The results were the first to suggest that DNA containing 5-azacytosine irreversibly inhibited DNA methylases. These studies are the only ones in which purified enzymes have been used, and they also were important in showing that the inhibition by 5-azacytosine was specific for DNA cytosine methyltransferases.

Inhibition of RNA Methylation

Incorporation of 5-aza-C into nuclear RNA substantially impairs ribosomal RNA processing in cultured Novikoff hepatoma cells (Weiss and Pitot, 1974, 1975). The drug causes a substantial inhibition of RNA methylation within the cells; the methylation of 45S RNA was 74% of control values, whereas methylation of 32S RNA was almost the same as the control (Weiss and Pitot, 1974). 5-Aza-C also impairs the amino acid acceptor function of tRNA (Lee and Karon, 1976), perhaps by inhibiting the modification of cytosine residues (Lu *et al.* 1976). Transfer RNA isolated from 5-aza-C-treated cells, therefore, can inhibit cell-free protein synthesis (Momparler *et al.* 1976).

Treatment of mice with 5-aza-C leads to a marked reduction in the 5mC content of liver tRNA (Lu *et al.* 1976). This decrease in 5mC content was shown to be due to a specific and rapid decrease in the activity of the tRNA 5mC methylase (Lu and Randerath, 1979). Importantly, the results not only were found with 5-aza-C, but with several other analogs modified in the 5 position, including 5-fluorocytidine (Lu *et al.* 1979) and 5-fluorouracil (Tseng *et al.* 1978).

The mechanism of inhibition of tRNA methylation was investigated in more detail by Lu and Randerath (1980). Their data suggested that new RNA synthesis was a prerequisite for the inhibition of the enzyme by 5-aza-C *in vivo.* A slowly sedimenting RNA (4 − 7S) from the livers of mice treated with 5-aza-C specifically decreased the activity of tRNA cytosine 5-methyltransferase activity when present in an *in vitro* tRNA methyltransferase assay. Thus, administration of 5-aza-C to mice led to the synthesis of a low molecular weight (mol wt) RNA fraction, which inactivated the tRNA methyltransferase both *in vivo* and *in vitro.* The data were consistent with the idea that the tRNA methyltransferase was strongly inhibited by the presence of the 5-azacytosine ring in the RNA molecule in a manner similar to that found for the DNA methyltransferases.

Inhibition of Eukaryotic DNA Methyltransferases

5-Aza-C and 5-aza-CdR inhibited the methylation of newly synthesized DNA in C3H 10T1/2 clone 8 cells (Jones and Taylor, 1980) and in L1210 cells growing in the peritoneal cavity of the mouse (Wilson *et al.* 1983). The DNA strand containing 5-azacytosine remained undermethylated in the round of DNA synthesis after analog incorporation; the extent of inhibition of DNA modification was dependent on the 5-aza-C concentration over a narrow concentration range. Woodcock *et al.* (1983) showed that 5-aza-C was more effective in inhibiting methylation of DNA that was modified immediately after synthesis and was a less potent inhibitor of delayed methylation.

Our initial experiments (Jones and Taylor, 1980) clearly showed that the degree of substitution of 5-azacytosine for cytosine was not sufficient to account for all of the inhibition (approximately 85%) of cytosine methylation that was found at high analog concentrations. Thus, we suggested that the occurrence of a 5-azacytosine residue at a modification site might severely impede the progress of the enzyme along the DNA helix, so that low levels of the fraudulent base could cause a considerable inhibition of DNA methylation. Therefore, the drug would cause more than a point inhibition of methylation at a substituted cytosine. This would result in a gross undermethylation of the DNA in treated cells (Figure 9.4).

The incorporation of 5-azacytosine into DNA does, indeed, lead to the formation of hemimethylated DNA in treated cells (Jones and Taylor, 1981). Figure 9.4 also shows that symmetrically demethylated DNA would be formed during the second division after 5-aza-C treatment. This would result if the

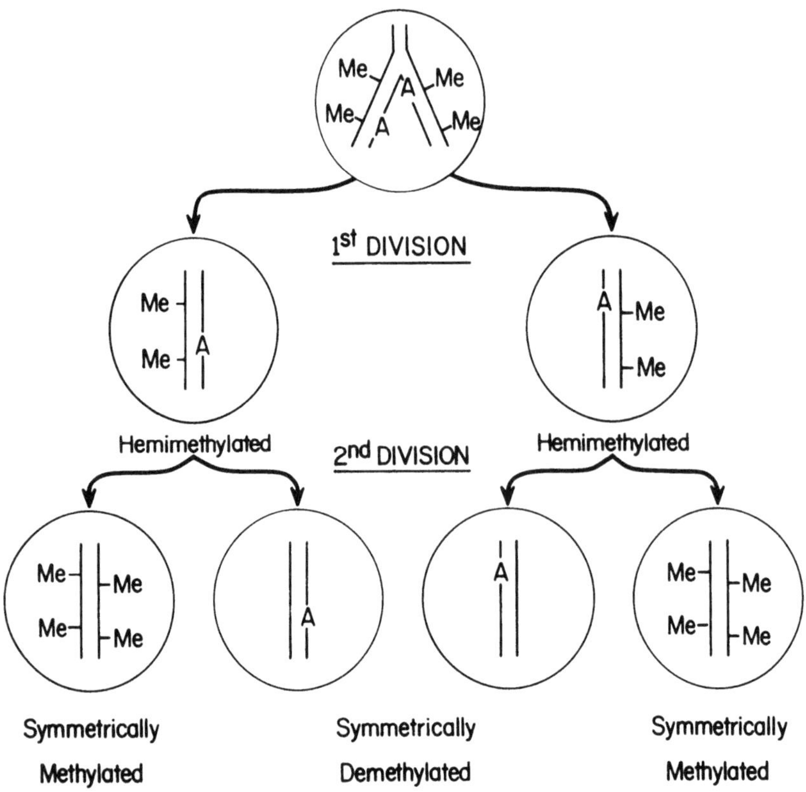

Figure 9.4 Possible mechanism of action of 5-azacytidine. The replication of DNA containing 5-methylcytosine (Me) is shown at the top of the diagram. If 5-azacytosine (A) were incorporated into newly synthesized DNA and inhibited DNA methylation, the daughter cells of the first division would contain hemimethylated DNA. Symmetrically demethylated DNA in this DNA would result after the second division after treatment, because of the specificity of the DNA methyltransferase.

"maintenance methylase" only applied methyl groups to hemimethylated, newly synthesized duplex DNA.

The hypomethylation induced by 5-aza-C, therefore, is heritable in the absence of further drug treatment, although it is clear that methylation levels can be restored after drug removal to an extent that varies with the cell type. Creusot *et al.* (1982) found considerable delayed methylation of cytosine residues incorporated during 5-aza-CdR treatment after the fraudulent nucleoside had been removed. Gasson *et al.* (1983) have reported *de novo* methylation that occurred over several weeks after cessation of 5-aza-C treatment in a T-lymphoid cell line. However, Adams *et al.* (1982) obtained derivatives of L cells with much decreased 5mC levels after 5-aza-CdR exposure in which the decreased methylation was maintained over six months. We have recently obtained 10T1/2 derivatives, which are hypomethylated after many consecutive treatments with 5-aza-CdR and which show slowly progressive increases in methylation over many population doublings (Flatau *et al*, 1984). The general tendency to increase methylation after treatment may be related to the finding that DNA modification levels are more variable in culture than previously thought (Wilson and Jones, 1983).

Mechanism of Inhibition of DNA Methyltransferases

Tanaka *et al.* (1980) demonstrated that 5-aza-C inhibited DNA methylation in Ehrlich ascites cells, and they clearly showed that the administration of the drug resulted in a marked reduction in the level of DNA methylase that could be extracted from the treated cells. Tanaka *et al.* (1980) also showed that 5-aza-C itself did not inhibit the methylation reaction when added to the assay, suggesting that the analog must be incorporated into DNA to inhibit DNA methylation.

It was subsequently found that DNA extracted from cells treated with 5-aza-C was hemimethylated and acted as an efficient acceptor of methyl groups from S-adenosylmethionine in the presence of a crude mouse spleen DNA methyltransferase (Jones and Tyalor, 1981). However, increasing levels of incorporated 5-azacytosine in the DNA inhibited the action of the methyltransferase, suggesting that incorporation of the analog into DNA might be responsible for the inhibitory effect of 5-aza-C on DNA methylation. Creusot *et al.* (1981) extended the results of Tanaka *et al.* (1980) and showed that 5-aza-C and 5-aza-CdR led to rapid time- and dose-dependent decreases in DNA methyltransferase activity in treated Friend erythroleukemia cells. They also obtained evidence that the analogs must be incorporated into DNA to mediate their effects on the enzyme.

The DNA methyltransferase apparently forms a tight binding complex with hemimethylated duplex DNA that contains high levels of 5-azacytosine, which cannot be dissociated by high concentrations of salt (Creusot *et al.* 1981; Taylor and Jones, 1982b). Treatment of cultured cells with biologically effective concentrations of 5-aza-C and other cytidine analogs modified in the the 5 position (Figure 9.5) also results in a loss of extractable active enzyme (Tanaka *et al.* 1980; Creusot *et al.* 1982; Taylor and Jones, 1982b). The formation of an enzyme-azacytosine complex might be irreversible, so that new protein synthesis is required to reverse the block of DNA methylation within treated cells.

We were unable to detect any inhibition of DNA methylation in the presence of either the nucleoside 5-aza-C or the penultimate deoxynucleotide 5-aza-2'-deoxycytidine triphosphate (Taylor and Jones, 1982b). This finding also provided indirect evidence that the drug must be incorporated into DNA to inhibit DNA methylation. However, Bouchard and Momparler (1983) have now directly shown that 5-aza-2'-deoxycytidine triphosphate must be incorporated into newly synthesized DNA to inhibit DNA methyltransferase. Purified mammalian DNA polymerase catalyzed the incorporation of 5-aza-dCTP into DNA with an apparent Km value of 3 μmol, whereas the Km for dCTP was 2 μmol. The apparent V_{max} for 5-aza-dCTP incorporation was slightly lower than that for dCTP. However, template studies showed that the fraudulent nucleotide analog was incorporated into poly dIC, but not into poly dAT; this suggests that incorporation followed Watson-Crick base pairing. 5-Azacytosine also was incorporated into *Micrococcus luteus* DNA by nick translation; the presence of low levels of 5-azacytosine residues in the nick-translated DNA provided a significant inhibition of the template acceptor activity for DNA methylase (Bouchard and Momparler, 1983). Thus, incorporation of 5-aza-dCTP into DNA directly inhibited the DNA methylase action.

Therefore, 5-Aza-C, shows a broad ability to inhibit cytosine methylation in DNA, rRNA, and tRNA, and it inhibits modification of DNA by prokaryotic and eukaryotic enzymes. The common feature of all these inhibitory activities appears to be the formation of a tight binding complex that leads to enzyme inactivation. Further studies on the formation of such complexes may prove useful in our understanding of the chemistry of the transmethylation reaction. Indeed, Santi *et al.* (1983) have recently suggested that the methyltransferases

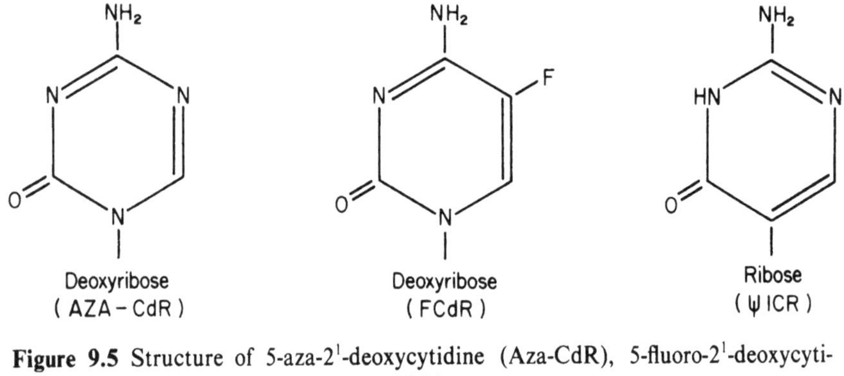

Figure 9.5 Structure of 5-aza-2'-deoxycytidine (Aza-CdR), 5-fluoro-2'-deoxycytidine (FCdR), and pseudoisocytidine (ψICR).

may form a covalent bond with the 6 position of the pyrimidine ring in a manner similar to the interaction of thymidilate synthetase with deoxyuridine monophosphate. The five substituted pyrimidines, therefore, may act as mechanism-based inhibitors of this reaction, which leads to the formation of irreversible covalent binding of the methylase to DNA (Figure 9.6).

Mesenchymal Cell Differentiation and 5-Azacytidine

5-Aza-C has remarkable effects on the stability of the differentiated state in cultured cells. Micromolar concentrations of the analog induce the formation of functional muscle, fat, and cartilage cells several weeks after exposure of the C3H 10T1/2 clone 8 line (Reznikoff *et al.* 1973) of mouse embryo cells to the drug (Constantinides *et al.* 1977; 1978; Taylor and Jones, 1979). All of these phenotypes were shown to be fully differentiated by several morphologic and biochemical criteria.

The striated muscle cells contained many nuclei (up to 2,000) in a single tubular cytoplasm, and they resulted from the fusion of mononucleated precursors, as in those found with normal muscle cells in culture. They had elevated levels of myosin ATPase activity and developed fully functional acetylcholine receptors on their surfaces that were capable of eliciting a twitching response to added neurotransmitters. The cells also twitched spontaneously in culture and contained regular striations with a repeating unit length of 2.7 μmol, which was similar to the values reported for muscle cells in culture. They represented fully differentiated muscle cells that are never seen in untreated control cultures of 10T1/2 cells. In addition, the 10T1/2 cells have been

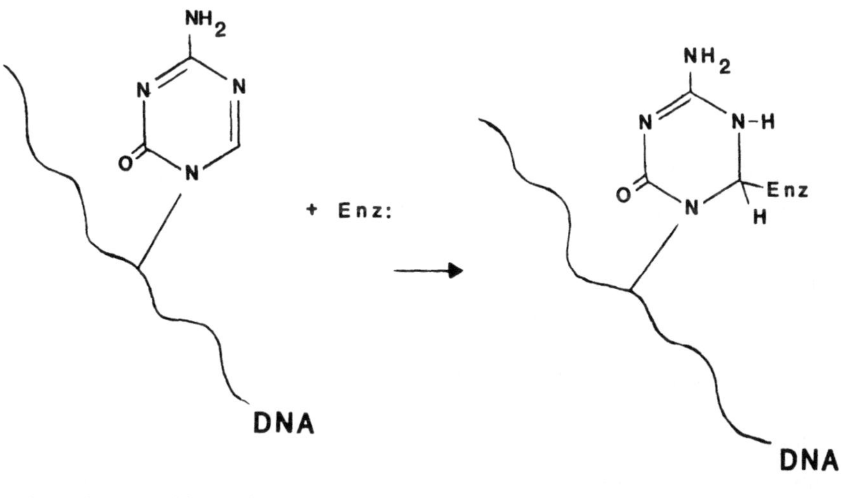

Figure 9.6 Possible mechanism for the formation of a tight binding complex between DNA methyltransferases and 5-azacytosine-containing DNA.

treated with a wide variety of agents over the last 10 years; yet, multinucleated myotubes have never been observed. The effects of 5-aza-C on differentiation, therefore, are specific, and the multinucleated cells develop 12 days after exposure to the analog in the absence of further drug treatment.

Fully differentiated adipocytes and chondrocytes are also induced by 5-aza-C, so that the changes in differentiation are not restricted to the muscle phenotype (Taylor and Jones, 1979). Such changes also are not restricted to cells of the 10T1/2 line, since the same three mesenchymal phenotypes were found in treated 3T3 cultures (Taylor and Jones, 1979). 5-Aza-C can also change the differentiated state of cells lines derived from adult mice. Thus, cells of the CVP3SC6 line (Nesnow and Heidelberger, 1976) form muscle cells after exposure to the analog (Taylor and Jones, 1982a). It is also possible to induce the formation of muscle cells from cloned lines of oncogenically transformed 10T1/2 cells (Taylor and Jones, 1979). Therefore, the changes in differentiation induced by 5-aza-C have been found in at least four different mouse cell lines.

The mesenchymal differentiation induced by 5-aza-C in these systems is not due to a selection of cells with a predisposition to differentiate. A high proportion of subclones (Taylor and Jones, 1979) or single 10T1/2 cells (Taylor and Jones, 1982a) formed muscle and fat colonies after drug treatment. All three of the new phenotypes (muscle, fat, and cartilage) were also displayed, in some cases, from the progeny of single 10T1/2 cells exposed to 5-aza-C. These facts established that the mechanism of action was not due to generalized cytotoxic activity; it also should be remembered that a wide variety of other cytotoxic agents (e.g., cytosine arabinoside) fail to induce the differentiation of myogenic cells in treated cultures.

There is an absolute requirement of cell division for the new phenotypes to develop after 5-aza-C treatment. More than two divisions are required for the expression of the muscle phenotype (Taylor and Jones, 1982a). The maximum number of muscle cells arises approximately 8–10 divisions after drug treatment. Since these mitoses occur in the absence of the analog, it is unlikely that 5-azacytosine has to be physically present in the DNA of the cells actively expressing a new phenotype. This fact distinguishes the effects of 5-aza-C on differentiation from those found with other analogs for the extinction of phenotypes (Rutter *et al.* 1973). In these latter cases, there often is a requirement for the continued presence of the fraudulent molecule in DNA, so that the effects on differentiation are apparent.

5-Aza-C shows a remarkable cell cycle specificity for its effects on differentiation. Synchronized 10T1/2 cells are maximally sensitive to phenotypic change when treated in the early part of the S phase of the cell cycle (Constantinides *et al.* 1978; Taylor and Jones, 1982a), and virtually no muscle is induced in cultures treated in G1 phase. There is also a substantial decrease in sensitivity to the drug in cells treated in the latter part of the S phase. The differences between responses in early and late S phase cannot be accounted for by an enhanced incorporation of 5-azacytosine into DNA that is replicated early in S phase (Taylor and Jones, 1982a). Thus, 5-azacytosine may be incorporated into DNA, which replicates early in the S phase of the cell cycle and which is important in the control of mesenchymal differentiation, to change cellular phenotype.

Induction of Differentiation in Other Systems

Cells derived from several different species respond to 5-aza-C treatment by the expression of previously suppressed genetic information (Table 9.1). Particular excitement was generated by the finding that the analog could induce the expression of genes located on inactive X chromosomes in a variety of hybrid systems. X inactivation has been found to be highly stable and refractory to reactivation by a large number of different stimuli. However, Mohandas

Table 9.1 Induction of Gene Expression by 5-Azacytidine

System	Species	References
New Mesenchymal Phenotypes	Mouse	Constantinides *et al.* (1977, 1978)
(muscle, fat, cartilage)		Taylor and Jones (1979)
		Mondal and Heidelberger (1980)
	Chinese hamster	Sager and Kovac (1982)
X chromosome reactivation	Human and Mouse	Mohandas *et al.* (1981)
		Jones *et al.* (1982)
		Graves (1982)
		Lester *et al.* (1982)
Metallothionein-1 inducibility	Mouse	Compere and Palmiter (1981)
Emetine resistance	Chinese hamster	Worton *et al.* (1983)
Thymidine kinase	Mouse	Clough *et al.* (1982)
(cellular and HSV)	Chinese hamster	Harris (1982)
Blood cell differentiation	Human	Bodner *et al.* (1981)
		Christman *et al.* (1983)
	Mouse	Boyd and Schrader (1982)
		Creusot *et al.* (1982)
		MacLeod *et al.* (1983)
Fetal hemoglobin	Baboon	DeSimone *et al.* (1982)
	Human	Ley *et al.* (1982)
		Charache *et al.* (1982)
Melanocyte differentiation	Mouse	Silagi and Graf (1981)
Prolactin	Rat	Ivarie and Morris (1982)
New tumor antigens	Mouse	Frost and Kerbel (1984)
Glucocorticoid receptor	Mouse	Gasson *et al.* (1983)
Endogenous and exogenous viruses	Hamster	Altanerova (1972)
	Chicken	Groudine *et al.* (1981)
	Human	Ben-Sasson and Klein (1981)
	Mouse	Niwa and Sugahara (1981)
		Tennant *et al.* (1982)
		McGeady *et al.* (1982)
		Hoffman *et al.* (1982)

et al. (1981) demonstrated that a high percentage of hybrid cells (approximately 1%) surviving treatment with 5-aza-C expressed human hypoxanthine-guanine phoshoribosyl transferase located on a structurally normal, inactive X chromosome that was present within the mouse-human hybrid cells. These studies were subsequently confirmed by Lester *et al.* (1982) and Graves (1982), who obtained similar results using a *Mus musculus* × *Mus Caroli* cell line. Induction of gene expression in the hybrid cells was also highly cell cycle-dependent, and the highest frequency of derepression was found in cells treated in the late S phase, when the inactive X chromosome was replicating (Jones *et al.* 1982).

Considerable interest has recently been generated by the finding that 5-aza-C can induce the expression of globin genes in baboons (DeSimone *et al.* 1982) and in humans suffering from either thalassemia (Ley *et al.* 1982) or sickle cell anemia (Charache *et al.* 1982). Thus, the effects of 5-aza-C on differentiation are not confined only to tissue culture cells, but they can also be duplicated with normal cells *in vivo*. These studies are exciting, since they suggest that it might be possible to treat certain human diseases by the induction of gene expression to compensate for the lack of a particular gene product. Conversely, it is important to remember that 5-aza-C may induce the expression of other genes within the patient and also may be capable of inducing oncogenic changes (see below). Therefore, the use of this analog should only be contemplated in those patients in whom other lines of treatment have been demonstrated to be totally ineffective.

5-Aza-C induces the expression of several different endogenous and exogenous viruses from many different species (Table 9.1). The first demonstration of this was reported by Altanerova (1972), who found that the fraudulent nucleoside induced the expression of Rous sarcoma virus from transformed hamster cells. Although Altanerova suggested that this was due to the mutagenic effects of the analog, it is more likely that the results were due to the more general ability of 5-aza-C to induce gene expression.

Table 9.1 thus demonstrates that a large variety of genes can be induced by transient exposure of cells to 5-aza-C. The effects of the drug are not only found in tissue culture cells, but also in normal diploid cells in living animals (DeSimone *et al.* 1982; Ley *et al.* 1982). Conversely, it is clear that 5-aza-C is not effective at inducing differentiation or visible changes in phenotype under all conditions.

Wolf and Migeon (1982) failed to induce the expression of X-linked genes from diploid human cells with 5-aza-C treatment, and the drug inhibits differentiation in mammary gland explants (Bolander, 1983). Therefore, there may be restrictions on developmental potential that prevent 5-aza-C treatment from inducing new phenotypes in all cases. Evidence has recently been obtained that 5-aza-C treatment may be a necessary, but not sufficient, condition for gene expression in the globin system (Ley *et al.* 1983). Murine erythroleukemia cells containing human chromosome 11 express human β and δ globins, but not the human γ gene, when induced to differentiate with hexamethylene bisacetamide

(HMBA). However, if the cells are treated with 5-aza-C before differentiation is induced with HMBA, then there is a 40-fold increase in the level of human γ-globin expression. Thus, 5-aza-C treatment may set the gene in an expressable state, but further changes in chromatin structure (induced by HMBA in this case) may be essential for expression.

Unpublished results in our laboratory suggest that low-passage human fibroblasts or bovine endothelial cells do not undergo visible changes in morphology after 5-aza-C treatment. This may indicate that the differentiated state is more tightly controlled in diploid cells than in the hybrid or aneuploid cells used in many of the studies reported in Table 9.1. Nevertheless, the drug can induce gene expression in living animals, so that the cells that respond may be primed for the expression of the particular gene in question.

Evidence that 5-Azacytidine Activates Genes by Inhibition of DNA Methylation

It has been important to establish whether the remarkable effects of 5-aza-C on cell differentiation are mediated by perturbation of methylation patterns within DNA. This is particularly important, because the drug is toxic and has many other effects on treated cells that might be equally capable of inducing changes in the differentiated state. Some of the evidence that suggests that 5-aza-C acts through a DNA modification mechanism is summarized in Table 9.2.

Concentrations of the drug that induce the muscle phenotype in 10T1/2 cells also are highly effective inhibitors of the methylation of newly incorporated cytosine residues (Jones and Taylor, 1980). Although 5-aza-C has other effects on cultured cells, it should be remembered that some of these effects (outlined in Table 9.3) are only seen with higher drug concentrations. The effects of the drug on differentiation appear to be specific for the 5 position of the cytosine ring; only analogs that contain a modified 5 position (i.e., 5-aza-2^{1}-deoxycytidine, 5-fluoro-2^{1}-deoxycytidine, and pseudoisocytidine; see Figure 9.5) induce the muscle phenotype in treated cultures (Jones and Taylor, 1980). Conversely, analogs such as 6-azacytidine or cytosine arabinoside are totally ineffective in inducing muscle differentiation within the cells. Thus, at the very least, the 5 position of the cytosine ring appears to be involved in the control of mesenchymal differentiation.

Studies with these analogs also established that the lability of the 5-azacytosine ring in aqueous solution (see Figure 9.2) did not appear to be important in eliciting this response. 5-Fluoro-2^{1}-deoxycytidine and pseudoisocytidine, which have very stable ring structures, also inhibited DNA methylation and perturbed the differentiated state. Since 5-aza-CdR and 5-fluoro-2^{1}-deoxycytidine cannot enter RNA due to the irreversible nature of the reaction catalyzed by the ribonucleotide reductase enzyme (see Figure 9.3), this data also supports a DNA-linked mechanism of action.

Table 9.2 Evidence that 5-Azacytidine Activates Genes by Inhibition of DNA Methylation

Evidence	References
Biologically active concentrations inhibit DNA methylation	Jones and Taylor (1980)
Differentiation effect is specific for 5 position of cytosine ring	Jones and Taylor (1980)
5-Aza-CdR is active at 1/10th the concentration of 5-Aza-C	Constantinides *et al.* (1978) Groudine *et al.* (1981)
Change in expression is heritable after many generations in the absence of further 5-aza-C or selective conditions	Taylor and Jones (1979) Mohandas *et al.* (1981) Ivarie and Morris (1982) Harris (1982)
Hypomethylation is induced at specific sites within expressed genes	Compere and Palmiter (1981) Groudine *et al.* (1981) Clough *et al.* (1982) Ley *et al.* (1982)
5-Aza-C-reactivated DNA functions in DNA-mediated transfection of hypoxanthine phosphoribosyl transferase	Venolia *et al.* (1982) Lester *et al.* (1982)
Activation frequencies are several orders of magnitude higher than mutation frequencies	Groudine *et al.* (1981) Jones *et al.* (1982) Harris (1982)

Table 9.3 Induction of Chromosome Damage, Oncogenic Transformation and Differentiation by 5-Azacytidine

Concentration	Toxicity[a] (% Survivors)	Chromosome[b] Breaks	Sister[c] Chromatid Exchanges	Mutations[d]	Oncogenic[e] Transformation	Differentiation[f]
None	100%	1%	5.2	9	0/148	0
0.1 μmol	100%	0%	5.4	—	0/10	—
0.3 μmol	70%	—	—	—	0/11	—
1.0 μmol	40%	0%	7.2	—	28/20	344
2 μmol	—	—	—	—	—	12,289
3 μmol	25%	—	—	3	2/10	10,184
4 μmol	—	—	10.8	—	—	—
10 μmol	5%	5%	18.0	2	1/22	1,750

[a]Determined as percent 10T1/2 cells surviving 24-hr exposure (Benedict *et al.* 1977).

[b]Percentage of $A(T_1)Cl$ 3 hamster fibrosarcoma cells with chromosomal damage (Benedict *et al.* 1977).

[c]Average number of sister chromatid exchanges/metaphase in $A(T_1)Cl$ 3 cells (Banerjee and Benedict, 1979).

[d]Number of ouabain-resistant 10T1/2 cells ($\times 10^{-6}$) (Landolph and Jones, 1982).

[e]Total number of morphologically transformed 10T1/2 foci per number of dishes treated (Benedict *et al.* 1977).

[f]Total number of muscle nuclei per 60-mm dish of treated 10T1/2 cells (Jones and Taylor, 1980).

The changes induced by 5-aza-C are heritable in the absence of further drug treatment, so that even though the presence of the fraudulent ring may directly affect protein-DNA interactions (Adams *et al.* 1982) or lead to the covalent binding of the methyltransferase to DNA (Santi *et al.* 1983; see Figure 9.6.), a heritable change is induced in the absence of further analog treatment. Cells can be induced to change their phenotype after a 5-minute exposure to 5-aza-C during the sensitive part of the S phase (Taylor and Jones, 1982a). Thus, a switching mechanism appears to be activated after a very short treatment time.

Many of the studies in Tables 9.1 and 9.2 also have directly shown that changes in methylation are elicited within the genes whose expression is induced by analog treatment. However, it was clear from the original studies of Groudine *et al.* (1981) that an error-correcting mechanism seems to exist that is capable of remethylating regions of DNA in which a hypomethylation was induced by 5-aza-C. Thus, although stable undermethylation of endogenous viral genomes was found in chicken cells treated with 5-aza-C, the same was not found for globin sequences whose expression was not induced by the drug.

The above studies all provide very strong correlative evidence that the effects of 5-aza-C are linked to effects on methylation. More direct evidence that the drug acts by inducing a change at the DNA level came from the important studies of Venolia *et al.* (1982) and Lester *et al.* (1982). These workers extracted DNAs from hybrid cell lines containing 5-aza-C-reactivated X chromosomes, and they clearly demonstrated the ability of this DNA to transform $HPRT^-$ recipient cells. Thus, the drug had altered the DNA structure of the X chromosome DNA within the hybrids. This provided strong support that the effects of the analog were mediated at the level of DNA and were stably inherited.

Taken together, the information in Table 9.2 provides a strong case that 5-aza-C induces a structural change in DNA that alters the expressibility of genes. This change is most probably related to the ability of the analog to change methylation patterns.

Carcinogenicity of 5-Azacytidine

Animal Studies

Several studies have indicated that 5-aza-C may be carcinogenic in mice. Stoner *et al.* (1973) treated A/He mice 6–8 weeks of age with intraperitoneal injections of the analog three times weekly for 8 weeks at total dose levels of 90, 62, and 33 mg/kg. The animals were sacrificed 24 weeks after the first injection and were scanned for the number of lung tumors. The number of tumors per mouse at the highest dose was 0.73, which was significantly higher ($p < 0.05$) than in untreated animals (tumor incidence was approximately

0.20%). The increase in the number of lung tumors was not statistically significant at lower doses of 5-aza-C.

In a National Cancer Institute study (1978), groups of 35 male and 35 female B6C3F1 mice 38 days of age were given intraperitoneal injections of 5-aza-C three times weekly for 52 weeks. Lymphatic and granulocytic tumors were observed in 17 of the 29 surviving female mice of the low-dose group (2.2 mg/kg) at a highly significant incidence ($p < 0.001$) when compared with the control group (tumor incidence was 0 out of 14). Of the 17 animals, 10 had granolocytic tumors; however, no significant increase of tumors was observed in male mice.

The analog has also been shown to induce a high frequency of mouse leukemia in AKR strain mice (Vesely and Cihak, 1973). This strain of mouse shows an extremely high frequency of spontaneous leukemia in female mice, and it ranges from 71–85% of all mice showing signs of the disease between 4–9 months of age in untreated controls. 5-Aza-C and IUdR were both shown to decrease significantly the time at which leukemia was first seen in the susceptible strain.

However, in all these studies, the animals showed significant tumor incidence in untreated controls, so that it is not clear whether the drug acted as a direct inducer of the malignant disease or simply hastened its appearance in highly susceptible strains. Dose responses to 5-Aza-C also are complex and often show peak responses at intermediate concentrations (Benedict *et al.* 1977; Jones and Taylor, 1980). Therefore, it may be important to repeat some of these studies over wider dose ranges.

Cell Culture Studies

5-Aza-CR and several other chemotherapeutic agents induce morphologic transformation in mouse C3H 10T1/2 clone 8 cells (Benedict *et al.* 1977). Although the transformation was scored purely on a morphologic basis in this study, tumors have been subsequently obtained from foci derived from 5-aza-C-treated cultures (Benedict and Jones; Taylor and Jones, unpublished data). Additionally, we have found that the drug can induce oncogenic transformation in Balb/3T3 cells (Wilson and Jones, unpublished observations), so that the chemical is capable of inducing oncogenic changes in at least two culture systems.

Mutagenesis

The data regarding the mutagenic activity of 5-aza-C are difficult to interpret, since the analog, at best, appears to be only weakly mutagenic in bacterial and eukaryotic systems. Fucik *et al.* (1965) initially reported that the chemical was weakly mutagenic for one of two strains of *E. coli*. Marquardt and Marquardt

(1977) also have reported that 5-aza-C is very weakly mutagenic in the TA100 strain of *Salmonella typhimurium*. The analog is mutagenic for arboviruses (Halle, 1968) and is weakly mutagenic in the RNA vesicular stomatitus virus (Pringle, 1970). Thus, the agent is very weakly mutagenic for RNA viruses and, possibly, also for bacteria.

Marquardt and Marquardt (1977) reported that 5-aza-C was weakly mutagenic in eukaryotes, and it reverted V79 Chinese hamster cells to 8-azaguanine resistance at dose levels ranging from 0.4 μmol to approximately 10 μmol. In contrast to these results, we were unable to obtain consistent dose dependency curves for the induction of mutation in V79 cells to 8-azaguanine resistance or in 10T1/2 cells to ouabain resistance (Landolph and Jones, 1982). These results, which were obtained in the same cell line used for differentiation studies, do not support the idea that 5-aza-C-induced changes in differentiation have a mutational basis. It is also clear that the frequencies of gene activation (Groudine *et al.* 1981; Jones *et al.* 1982; Harris, 1982) often are several orders of magnitude higher than those expected on a purely mutational basis.

Chromosome Damage by 5-Azacytidine

5-Aza-C is a toxic agent at high concentrations, and it has direct effects on eukaryotic chromosomes. Table 9.3 summarizes this data, so that the effects of equal concentrations of the analog on the property being studied can be appreciated.

High concentrations of 5-aza-C (40 μmol) were first shown to be capable of inducing chromatid breaks in the hamster fibrosarcoma cell line Don-C by Karon and Benedict (1972). Subsequently, Benedict *et al.* (1977) showed that 10 μmol of 5-aza-C induced chromosomal aberrations in 5% of hamster fibrosarcoma $A(T_1)CL$-3 cells that survived treatment. Conversely, no significant increases in chromosomal aberrations were seen in cells surviving 1 μmol analog treatment. The production of sister chromatid exchanges by 5-aza-C has also been measured by Banerjee and Benedict (1979), who found significant increases in the number of sister chromatid exchanges in the $A(T_1)Cl_3$ line treated with 4 μmol or 10 μmol of 5-aza-C. This result was confirmed by Chambers and Taylor (1982), who showed that 5-fluorodeoxycytidine and 5-aza-C at concentrations of 10 μmol increased by 50% the background level of sister chromatid exchanges in Chinese hamster ovary cells. Since significant numbers of chromosomal aberrations or sister chromatid exchanges are not induced by the drug at concentrations that are maximally effective at inducing oncogenic transformation or differentiation in 10T1/2 cells, it is unlikely that the effects of 5-aza-C on chromosomal morphology and sister chromatid exchange are linked to the profound effects of the analog on differentiation.

Very high concentrations of 5-aza-C (50–100 μmol) affect the condensation of G bands in cultures of human lymphocytes (Viegas-Pequignot and Dutrillaux, 1976). Lower concentrations of the drug (1–10 μmol) prevent the condensation of R band-positive heterochromatin (G-C-rich segments) in bovine or primate cells (Viegas-Pequignot and Dutrillaux, 1981). Therefore, 5-Aza-C may be useful in detecting GC-rich heterochromatin in certain types of mammalian species, although (once again) the concentrations used were higher than those that induce significant changes in the differentiated state of cells.

Table 9.3 also summarizes present data on the ability of 5-aza-C to induce ouabain-resistant mutants in the 10T1/2 cell system (Landolph and Jones, 1982) and morphologic transformation in the same cell system (Benedict *et al.* 1977). Once again, it is apparent that the induction of mutations by the analog does not correlate with its ability to induce muscle cells in treated cultures, since concentrations that are highly efficient inducers of muscle formation do not increase the background mutation frequency. Interestingly, the response of oncogenic transformation and production of muscle cells appear to be similar, suggesting that the induction of these two phenotypic changes may be linked in some way.

5-Aza-C at very high concentrations (40 μmol) causes severe fragmentation of the DNA of *E. coli* cells grown in the presence of the analog. At higher analog concentrations, there also appeared to be a decrease in the Tm of the extracted DNA and a decrease in hyperchromicity (Zadrazil *et al.* 1965). However studies in our laboratory have failed to show differences in the Tm of DNA extracted from eukaryotic cells treated with 10^{-5} M of 5-aza-C (Jones, unpublished observations). Thus, while very high drug concentrations may cause instability of the DNA helix, it is unlikely that the concentration of the analog used to induce differentiation (1–3 μmol) causes substantial changes in the physical properties of DNA in treated eukaryotic cells.

Acknowledgment

Supported by National Institutes of Health Grants No. GM30892 and CA33592.

References

Adams RLP, Fulton J, Kirk D: The effect of 5-azadeoxycytidine on cell growth and DNA methylation. *Biochim Biophys Acta* 1982;697:286–294.

Altanerova V: Virus production induced by various chemical carcinogens in a virogenic hamster cell line transformed by Rous sarcoma virus. *J Natl Cancer Inst* 1972;49:1375–1380.

Anderson EP: Nucleoside and nucleotide kinases; in Boyer PD (ed): *The Enzymes.* New York, Academic Press, 1973, vol 9 pp 49–96.

Banerjee A, Benedict WF: Production of sister chromatid exchanges by various cancer chemotherapeutic agents. *Cancer Res* 1979;39:797–799.

Beisler JA: Isolation, characterization and properties of a labile hydrolysis product of the antitumor nucleoside, 5-azacytidine. *J Med Chem* 1978;21:204–208.

Beisler JA, Abbasi MM, Driscoll JS: Dihydro-5-azacytidine hydrochloride, a biologically active and chemically stable analog of 5-azacytidine. *Cancer Treatment Rep* 1976;60:1671–1674.

Benedict WF, Banerjee A, Gardner A, Jones PA: Induction of morphological transformation in mouse C3H 101/2 clone 8 cells and chromosomal damage in hamster A(T_1)Cl-3 cells by cancer chemotherapeutic agents. *Cancer Res* 1977;37:2202–2208.

Ben-Sasson SA, Klein G: Activation of the Epstein-Barr virus genome by 5-azacytidine in latently infected human lymphoid lines. *Int J Cancer* 1981;28:131–135.

Bergy ME, Herr RR: Microbial production of 5-azacytidine. II. Isolation and chemical structure. *Antimicrob Ag Chemother* 1966, pp 625–630.

Bodner AJ, Ting RC, Gallo RC: Induction of differentiation of human promyelocytic leukemia cells (HL-60) by nucleosides and methotrexate. *J Natl Cancer Inst* 1981;67:1025–1030.

Bolander FF: The effect of 5-azacytidine on mammary gland differentiation *in vitro*. *Biochem Biophys Res Commun* 1983;111:150–155.

Bouchard J, Momparler RL: Incorporation of 5-Aza-2^1-deoxycytidine-5^1-triphosphate into DNA: interactions with mammalian DNA polymerase and DNA methylase. *Mol Pharmacol* 1983;24:109–114.

Boyd AW, Schrader JW: Derivation of macrophage-like lines from the pre-B lymphoma ABLS 8.1 using 5-azacytidine. *Nature* 1982;297:691–694.

Caputto R: Nucleotide kinases, in Boyer PD, Lardy H, Myrback K (eds): *The Enzymes*. New York, Academic Press, 1962, vol 6 pp 133–138.

Chabot GG, Bouchard J, Momparler RL: Kinetics of deamination of 5-aza-2^{1}deoxycytidine and cytosine arabinoside by human liver cytidine deaminase and its inhibition by 3-deazauridine, thymidine or uracil arabinoside. *Biochem. Pharmacol* 1983;32:1327–1328.

Chambers JC, Taylor JH: Induction of sister chromatid exchanges by 5-fluorodeoxycytidine: correlation with DNA methylation. *Chromosome* 1982;85:603–609.

Chan KK, Giannini DD, Staroscik JA, Sadee W: 5-Azacytidine hydrolysis kinetics measured by high pressure liquid chromatography and ^{13}C-NMR spectroscopy. *J Pharm Sci* 1979;68:807–812.

Charache S, Dover GJ, Smith KD, et al: Treatment of sickle cell anemia with 5-azacytidine. *Blood* 1982;60(suppl 1):449.

Chou T-C, Burchenal JH, Fox JJ, Watanabe KA, Chu CK, Philips FS: Metabolism and effects of 5-(β-D-ribofuranosyl) isocytosine in P815 cells. *Cancer Res* 1979;39:720–728.

Christman JK, Mendelsohn N, Herzog D, Schniederman N: Effect of 5-azacytidine on differentiation and DNA methylation in human promyelocytic leukemia cells (HL-60). *Cancer Res* 1983;43:763–769.

Clough DW, Kunkel LM, Davidson RL: 5-Azacytidine induced reactivation of a Herpes Simplex thymidine kinase gene. *Science* 1982;216:70–73.

Compere SJ, Palmiter RD: DNA methylation controls the inducibility of the mouse metallothionein-1 gene in lymphoid cells. *Cell* 1981;25:233–240.

Constantinides PG: *Myogenic Conversion of Cultured Cells by Azapyrimidines*. M.Sc. Thesis, University of Stellenbosch Medical School, South Africa, 1977.

Constantinides PG, Jones PA, Gevers W: Functional striated muscle cells from non-myoblast precursors following 5-azacytidine treatment. *Nature* 1977;267:364–366.

Constantinides PG, Taylor SM, Jones PA: Phenotypic conversion of cultured mouse embryo cells by aza pyrimidine nucleosides. *Devel Biol* 1979;66:57–71.

Creusot F, Acs G, Christman JK: Inhibition of DNA methyltransferase and induction of Friend erythroleukemia cell differentiation by 5-azacytidine and 5-aza-2'-deoxy-cytidine. *J Biol Chem* 1982;257:2041–2948.

DeSimone J, Heller P, Hall L, Zwiers D: 5-Azacytidine stimulates fetal hemoglobin synthesis in anemic baboons. *Proc Natl Acad Sci (USA)* 1982;79:4428–4431.

Flatau E, Jones PA: Production of 5-azacytidine resistant cell lines (submitted).

Friedman S: The effect of azacytidine on *E. Coli* DNA methylase. *Biochem Biophys Res Commun* 1979;89:1324–1333.

Friedman S: The inhibition of DNA (Cytosine-5) Methylases by 5-Azacytidine—The effect of azacytosine-containing DNA. *Mol Pharmacol* 1981;19:314–320.

Frost P, Kerbel RS: The selection of strongly immunogenic "Tum-" variants from tumors at high frequency using 5-azacytidine. *J Exptl Med* (1984); (in press).

Fucik V, Zandrael S, Sormova Z, Sorm F: Mutagenic effect of 5-azacytidine in bacteria. *Coll Czech Chem Comm* 1965;30:2883–2886.

Gasson JC, Ryden T, Bourgeois S: Role of *de novo* DNA methylation in the glucocorticoid resistance of a T-lymphoid cell line. *Nature* 1983;302:621–632.

Graves JAM: 5-Azacytidine-induced re-expression of alleles on the inactive X-chromosome in a *Mus musculus* × *M. caroli* cell line. *Exp Cell Res* 1982;141:99–105.

Groudine M, Eisenman R, Weintraub H: Chromatin structure of endogenous retroviral genes and activation by an inhibitor of DNA methylation. *Nature* 1981;292:311–317.

Halle S: 5-Azacytidine as a mutagen for arboviruses. *J Virol* 1968;2:1228–1229.

Hanka LJ, Evans JS, Mason DJ, Dietz A: Microbial production of 5-azacytidine. Production and biological activity. *Antimicrob Ag Chemother* 1966, pp 619–624.

Harris M: Induction of thymidine kinase in enzyme-deficient Chinese hamster cells. *Cell* 1982;29:483–492.

Hoffmann JW, Steffen D, Gusella J, Tabin C, Bird S, Cowing D, Weinberg RA: DNA methylation affecting the expression of murine leukemia proviruses. *J Virol* 1982;44:144–157.

Israili ZH, Volger WR, Mingioli ES, Pirkle JL, Smithwick RW, Goldstein JH: The disposition and pharmacokinetics in humans of 5-azacytidine administered intravenously as a bolus or by continuous infusion. *Cancer Res* 1976;36:1453–1461.

Ivarie RD, Morris JA: Induction of prolactin deficient variants of CH_3 rat pituitary tumor cells by ethyl methanesulfonate: reversion by 5-azacytidine a DNA methylation inhibitor. *Proc Natl Acad Sci (USA)* 1982;79:2967–2970.

Jones PA, Taylor SM: Hemimethylated duplex DNAs prepared from 5-azacytidine treated cells. *Nucl Acids Res* 1981;9:2933–2947.

Jones PA, Taylor SM: Cellular differentiation, cytidine analogs and DNA methylation. *Cell* 1980;20:85–93.

Jones PA, Taylor SM, Mohandas T, Shapiro LJ: Cell cycle specific reactivation of an inactive X-chromosome locus by 5-azadeoxycytidine. *Proc Natl Acad Sci (USA)* 1982;79:1215–1219.

Karon M, Benedict WF: Chromatid breakage: Differential effect of inhibitors of DNA synthesis during G_2 phase. *Science* 1972;178:62–63.

Landolph JR, Jones PA: Mutagenicity of 5-azacytidine and related nucleosides in C3H/10T1/2 C18 and V79 cells. *Cancer Res* 1982;42:817–823.

Lee TT, Karon M: Inhibition of protein synthesis in 5-azacytidine-treated HeLa cells. *Biochem Pharmacol* 1976;25:1737–1742.

Lee T, Karon M, Momparler RL: Kinetic studies on phosphorylation of 5-azacytidine with the purified uridine-cytidine kinase from calf thymus. *Cancer Res* 1974;34:2482–2488.

Lester SC, Korn NJ, DeMars R: Derepression of genes on the human inactive X-chromosome: evidence for differences in locus-specific rates of derepression and rates of transfer of active and inactive genes after DNA-mediated transformation. *Somatic Cell Genet,* 1982;8:265–284.

Ley TJ, DeSimone J, Anagnou NP, Heller GH, Humphries RK, Turner PH, Young NS, Heller P, Nienhius AW: 5-Azacytidine selectively increases β-globin synthesis in a patient with $\beta+$ thalassemia. *N Engl J Med* 1982;307:1469–1475.

Ley T, et al: Personal communication, 1983.

Liacouras AS, Anderson EP: Uridine-cytidine kinase IV. Kinetics of the competition between 5-azacytidine and the 2 natural substrates. *Mol Pharmacol* 1979;15:331–340.

Li LH, Olin EJ, Buskirk HH, Rineke LM: Cytotoxicity and mode of action of 5-azacytidine on L1210 leukemia. *Cancer Res* 1970;2760–2769.

Lin K-T, Momparler RL, Rivard GE: High performance liquid chromatographic analysis of chemical stability of 5-aza-2^{1}-deoxycytidine. *J Pharm Sci* 1981;70:1228–1232.

Lu L-JW, Chiang GH, Medina D, Randerath K: Drug effects on nucleic acid modification. 1. Specific effect of 5-azacytidine on mammalian transfer RNA modification *in vivo. Biochem Biophys Res Commun* 1976;68:1094–1101.

Lu L-JW, Randerath K: Effects of 5-azacytidine on transfer RNA methyltransferases. *Cancer Res* 1979;39:940–948.

Lu L-JW, Randerath K: Mechanism of 5-azacytidine induced transfer RNA cytosine-5-methyltransferase deficiency. *Cancer Res* 1980;40:2701–2705.

Lu L-J, Tseng WC, Randerath K: Effects of 5-fluorocytidine on mammalian transfer RNA and transfer RNA methyltransferases. *Biochem Pharmacol* 1979;28:489–495.

Macleod CL, Hyman R, Bourgeois S, Hays E: *Differentiation of a T Lymphoma Line in vitro.* UCLA Symposium, 1983.

Marquardt H, Marquardt M: Induction of malignant transformation and mutagenesis in cell cultures by cancer chemotherapeutic agents. *Cancer (Phil)* 1977;40:1930–1934.

McGeady ML, Jhappan C, Ascocione R, Van de Woude GF: In vitro methylation of specific regions of the cloned moloney sarcoma virus genone inhibits its transforming activity. *Cell Mol Biol* 1982;3:305–314.

Mohandas T, Sparkes RS, Shapiro LJ: Reactivation of an inactive human X chromosome: Evidence for X inactivation by DNA methylation. *Science* 1981;211:393–396.

Momparler RL, Derse D: Kinetics of phosphorylation of 5-aza-2'-deoxycytidine by deoxycytidine kinase. *Biochem Pharmacol* 1979;28:1443–1444.

Momparler RL, Goodman J: *In vitro* cytotoxic and biochemical effects of 5-aza-2'-deoxycytidine. *Cancer Res* 1977;37:1636–1639.

Momparler RL, Siegel S, Avila F, Lee T, Karon, M: Effect of tRNA from 5-azacytidine treated hamster fibrosarcoma cells on protein synthesis *in vitro* in a cell free system. *Biochem Pharmacol* 1976;25:389–392.

Mondal S, Heidelberger C: Inhibition of induced differentiation of C3H/10T1/2 Clone 8 mouse embryo cells by tumor promoters. *Cancer Res* 1980;40:334–338.

National Cancer Institute: Bioassay of 5-azacytidine for possible carcinogenicity (Technical report series No. 42, DHEW Publ. No. [NIH] 78–842). Washington, DC, United States Government Printing Office, 1978.

Nesnow S, Heidelberger C: The effects of modifiers of microsomal enzymes on chemical oncogenesis in cultures of C3H mouse cell lines. *Cancer Res* 1976;36:1801–1808.

Niwa O, Sugahara T: 5-Azacytidine induction of mouse endogenous type C virus and suppression of DNA methylation. *Proc Natl Acad Sci (USA)* 1981;78:6290–6294.

Notari RE, DeYoung JL: Kinetics and mechanism of degradation of the antileukemic agent 5-azacytidine in aqueous solution. *J Pharm Sci* 1975;64:1148–1157.

Piskala A, Sorm F: Nucleic acids components and their analogues. Synthesis of 1-glycosyl derivatives of 5-azauracil and 5-azacytosine. *Coll Czech Chem Commun* 1964;29:2060–2067.

Pithova P, Piskala A, Pitha J, Sorm F: Nucleic acids components and their analogues LXVI. Hydrolysis of 5-azacytidine and its connection with biological activity. *Coll Czech Commun* 1965;30:2801–2811.

Pringle CR: Genetic characteristics of conditional lethal mutants of vesicular stomatitus virus induced by 5-fluorouracil, 5-aza-cytidine and ethyl methane sulfonate. *J Virol* 1979;5:559–567.

Reznikoff CA, Brankow DW, Heidelberger C: Establishment and characterization of a cloned line of C3H mouse embryo cells sensitive to postconfluence inhibition of cell division. *Cancer Res* 1973;33:3231–3238.

Rutter WJ, Pictet RL, Morris W: Towards molecular mechanisms of developmental processes. *Ann Rev Biochem* 1973;42:601–646.

Sager R, Kovac P: Pre-adipocyte determination either by insulin or by 5-azacytidine. *Proc Natl Acad Sci (USA)* 1982;79:480–484.

Santi DV, Garrett CE, Barr PJ: On the mechanisms of inhibition of DNA-cytosine methyltransferases by cytosine analogs. *Cell* 1983;33:9–10.

Silagi S, Graf LH: Induction of melanin and plasminogen activator by 5-azacytidine (abstract) *J Cell Biol* 1981;91:1023.

Stoner GD, Shimkin MB, Kniazeff AJ, Weisberger JH, Weisburger EK, Gori GB: Test for carcinogenicity of food additives and chemotherapeutic agents by pulmonary tumor response in Strain A mice. *Cancer Res* 1973;33:3069–3085.

Tanaka M, Hibasami H, Nagai J, Ikeda T: Effects of 5-azacytidine on DNA methylation in Ehrlichs' ascites tumor cells. *Aust J Exp Biol Med Sci* 1980;58:391–396.

Taylor SM, Jones PA: Multiple new phenotypes induced in 10T1/2 and 3T3 cells treated with 5-azacytidine. *Cell* 1979;17:771–779.

Taylor SM, Jones PA: Changes in phenotypic expression in embryonic and adult cells treated with 5-azacytidine. *J Cell Physiol* 1982a;111:187–194.

Taylor SM, Jones PA: Mechanisms of action of eukaryotic DNA methyltransferase: use of 5-azacytosine containing DNA. *J Mol Biol* 1982b;162:679–692.

Tennant RW, Olten JA, Myer FE, Rascati RJ: Induction of retrovirus gene expression in mouse cells by some chemical mutagens. *Cancer Res* 1982;42:3050–3055.

Tseng W-C, Medina D, Randerath K: Specific inhibition of transfer RNA methylation and modification in tissues of mice treated with 5-fluorouracil. *Cancer Res* 1978;38:1250–1257.

Venolia L, Gartler SM, Wassman ER, Yen P, Mohandas T, Shapiro LJ: Transformation with DNA from 5-azacytidine-reactivated X chromosomes. *Proc Natl Acad Sci (USA)* 1982;79:2352–2354.

Vesely J, Cihak A: High frequency induction *in vivo* of mouse leukemia in the AKR strain by 5-azacytidine. *Experimentia* 1973;29:1132–1133.

Vesely J, Cihak A: 5-Azacytidine: mechanism of action and biological effects in mammalian cells. *Pharmac. Ther* 1978;2:813–840.

Viegas-Pequignot E, Dutrillaux B: Segmentation of human chromosomes induced by 5-ACR (5-azacytidine). *Human Genet* 1975;34:247–254.

Viegas-Pequignot E, Dutrillaux B: Detection of G-C rich heterochromatin by 5-azacytidine in mammals. *Human Genet* 1981;57:134–137.

Weiss JW, Pitot HC: Alteration of ribosomal precursor RNA in Novikoff hepatoma cells by 5-azacytidine. Studies on methylation of 45S and 32S RNA. *Arch Biochem Biophys* 1974;165:588–596.

Weiss JW, Pitot HC: Effects of 5-azacytidine on nucleolar RNA and preribosomal particles in Novikoff hepatoma cells. *Biochemistry* 1975;14:316–326.

Wilson VL, Jones PA: DNA methylation decreases in aging but not in immortal cells. *Science* 1983;220:1055–1057.

Wilson VL, Momparler RL, Jones PA: Inhibiton of DNA methylation of L1210 leukemic cells by 5-aza-2'-deoxycytidine: A possible mechanism of chemotherapeutic action. *Cancer Res* 1983;43:3493–3496.

Wolf SF, Migeon BR: Studies on X chromosome DNA methylation in normal human cells. *Nature* 1982;295:667–672.

Woodcock DM, Adam JK, Allan RG, Cooper IA: Effect of several inhibitors of enzymatic DNA methylation on the *in vivo* methylation of different classes of DNA sequences in a cultured human cell line. *Nucl Acids Res* 1983;11:489–499.

Worton RG, Grant SA, Duff C: Gene inactivation and reactivation at the EMT locus in Chinese hamster cells, in Sternberg NL, Pearson ML, (eds): *Gene Transfer and Cancer*. New York, Raven Press, (in press, 1983).

Zadrazil S, Fucik V, Bartl P, Sormova Z, Sorm F: The structure of DNA from Escherichia Coli cultured in the presence of 5-azacytidine. *Biochim Biophys Acta* 1965;108:701–703.

10

DNA Methylation in Early Mammalian Development

Detlev Jähner*
Rudolf Jaenisch*

The genomic DNAs of vertebrates and many invertebrates contain 5-methylcytosine (5-meCyt) as the only modified base (Table 10.1). Considerable interest in DNA methylation has been created by increasing evidence that links methylation patterns to patterns of gene expression in differentiation. An inverse correlation between methylation and transcriptional activity has been found for a number of developmentally regulated genes; it will be reviewed elsewhere (see Chapter 8).

If DNA methylation can modulate gene expression, it is important to understand how methylation patterns are established and maintained during the life cycle of animals. A given methylation pattern can be stably inherited in somatic cells by the action of a "maintenance" methylase (Wigler, 1981). This enzyme recognizes hemimethylated sites on the parental strand of newly replicated DNA, and it copies each methylation site on the daughter strand. However, the establishment of a new methylation pattern must involve the action of a *de novo* methylase acting on unmodified DNA, and it has been postulated to occur during early embryogenesis (Riggs, 1975; Holliday and Pugh, 1975).

In this chapter, we will first summarize the data that established the constancy of methylation in animals. We will then focus on retroviruses that were instrumental in demonstrating the presence of *de novo* methylation in embryonic cells. Finally, we will discuss the possible significance of *de novo* methylation for gene expression in mammalian embryogenesis.

---

*Heinrich-Pette-Institut für Experimentelle Virologie und Immunologie an der Universität Hamburg, Martinistrasse 52, 2000 Hamburg 20, Federal Republic of Germany

Table 10.1 Presence of 5-meCyt in Somatic and Germ Cells of Animals

Phylum	Germ Cells (Sperm)	Somatic Tissues	References[b]
Porifera	na	+	1
Coelenterata	na	+	1
Mollusca	+	+	1,2
Arthropoda	na	−	3
Echinodermata	+[a]	+	1,4–7
Tunicata	na	+	2
Vertebrata, classes			
Cyclostomata	+	na	8
Chondrychtheis	+	na	8
Ostheichtheis	+	+	1,4,8
Amphibia	+[a]	+	1,2,8,9
Reptilia	na	+	1,8
Aves	+	+	8,10
Mammalia, species			
Mouse	+	+	1,2
Rat	+	+	5,7,8,11,12
Rabbit	+	+	1,13
Pig	+	+	1
Sheep	+	+	1,7
Calf	+	+	7,11
Human	+	+	14,15

na = not analyzed.

[a]Also eggs were analyzed.

[b]Reference 1, Vanyushin *et al.* (1970); 2, Bird *et al.* (1980); 3, Rae and Steele (1979); 4, Wyatt (1951); 5, Grippo *et al.* (1968); 6, Pollock *et al.* (1978); 7, Bird *et al.* (1979); 8, Vanyushin *et al.* (1973); 9, Bird *et al.* (1981); 10, Kappler (1971); 11, Dockočil and Šorm (1962); 12, Kunnath and Locker (1982); 13, Waalwijk and Flavell (1978); 14, van der Ploeg and Flavell (1980); 15, Ehrlich *et al.* (1982).

Embryonal Development and Methylation

Constancy of Methylation Patterns

The presence of 5-meCyt has been demonstrated in the DNA of all vertebrates and many invertebrates, with the notable exception of insects (Table 10.1). The level of methylation is constant in a given species, but it varies between different species (Table 10.2). The maintenance methylase is postulated to passively maintain a given methylation pattern and to be responsible for its stable somatic inheritance (Pollack *et al.* 1980; Wigler *et al.* 1981; Stein *et al.* 1982a). Gene activation during differentiation is frequently associated with demethylation; it is believed that sequence-specific proteins inhibit the maintenance methylase during replication, which leads to methylation patterns that are specific for each tissue (Razin and Riggs, 1980). Once methylation at a given site is inhibited during a "critical" mitosis, the changed pattern of methylation would be clonally inherited as a result of the passive maintenance methylase system.

Demethylation of DNA during animal development, however, was not observed when the concentration of methyl groups was analyzed. The levels of DNA methylation did not vary signficantly between the somatic tissues in a given species. However, sperm DNA in some species appeared to be hypomethylated when compared with somatic DNA (Table 10.2). It was shown that this difference could be accounted for by hypomethylation of highly repetitive sequences in sperm relative to somatic tissues, whereas the methylation of unique DNA was not different (Sturm and Taylor, 1981; Pages and Roizes, 1982; Ehrlich *et al.* 1982; Chapman *et al.*, 1984). This suggested that highly repetitive—but not unique—sequences may be *de novo* methylated during development of some mammalian species.

A comparison of total methylation in DNA at different development stages of sea urchin, mouse, and rabbit is summarized in Table 10.3. No significant difference in methylation was observed in total DNA from sperm or ovum and later embryos of the sea urchin. Similarly, DNA from mouse embryos, from embryonal carcinoma cells, and from adult tissues had equal levels of methylation. In rabbits, a higher level of methylation was seen in DNA from early cleavage embryos and in the embryoblast of blastulas, as compared to DNA from the trophoblast cells; the significance of this difference is not known at present.

The data summarized in Tables 10.2 and 10.3 indicate that the overall level of methylation in unique DNA stays constant during development. To clearly

Table 10.2 The Level of 5-meCyt in DNA of Sperm and Somatic Tissues

	Sperm		Somatic Tissues		
Species	Mol % of 5-meCyt[a]	Fraction of Methylated HpaII sites[b]	Mol % of 5-meCyt[a]	Fraction of Methylated HpaII sites[b]	References[c]
Salmon	1.64		1.80–2.13		1
Carp	1.56		1.19–1.65		2
Chicken	0.95		0.92–1.04		2
Sheep	0.76		1.07–1.25		1
Cattle	0.75		1.01–1.40		1
Cattle	0.95		1.40		3
Rat		0.77		0.71–0.79	4
Human	0.84		0.76–1.00		5
Human		0.71		0.57–0.62	6

[a]The mol % of 5-meCyt is defined as the fraction of 5-meCyt of all bases × 100.

[b]The fraction of methylated HpaII sites was determined by HpaII/MspI digestions.

[c]Reference 1, Vanyushin *et al.* (1970); 2, Vanyushin *et al.* (1973); 3, Sturm and Taylor (1981); 4, Kunnath and Locker (1982); 5, Ehrlich *et al.* (1982); 6, van der Ploeg and Flavell (1980).

Table 10.3 The Level of 5-meCyt in Animal DNA at Different Stages of Development

		Sea Urchin		Mouse	Rabbit
		Mol% of 5-meCyt[a] (1[c])	Fraction of[b] Methylated HpaII Sites (2[c])	Fraction of[b] Methylated HpaII Sites (3[c])	Fraction of[b] Methylated HpaII Sites (4[c])
Germ line	Sperm	0.8–0.9	~0.4		
	Egg		~0.4		
	Early cleavage stages	0.9	~0.4	0.7–0.8	0.7
	Blastula	0.8	~0.4	0.7–0.8	0.5 extraembryonic 0.7 embryonic cells
Embryonic stages	Gastrula	0.9	~0.4		
	Embryonic carcinoma cells			0.7–0.8	
Adult tissues				0.7–0.8	0.5(3[c])

[a]The mol% of 5-meCyt is defined as the fraction of 5-meCyt of all bases × 100.

[b]The fraction of methylated HpaII sites was determined by HpaII/MspI digestions.

[c]Reference 1, Pollock *et al.* (1978); 2, Bird *et al.* (1979); 3, Singer *et al.* (1979); 4, Manes and Menzel (1981).

assess the possible role of methylation during development, individual genes had to be analyzed. Using restriction enzymes that are sensitive to DNA methylation, the pattern of methylation of developmentally regulated genes and housekeeping genes was compared in the DNA of sperm and somatic tissues (see Chapter 8). No site in any gene within all species that were analyzed was less methylated in sperm than in somatic cells. For housekeeping genes, a nearly identical pattern was found in all cells that were analyzed (see Chapter 7), indicating a faithful maintenance of methylation patterns during development by a maintenance methylase activity present in the cells of higher animals.

In addition to the maintenance methylase in somatic cells, a *de novo* methylase was postulated to be active in early embryos (Razin and Riggs, 1980). The action of such an enzyme may be required during the life cycle of an animal for two reasons. First, demethylation introduced by an error of the maintenance methylase in germ cells would be genetically fixed and would lead to a gradual decrease in the overall methylation level of a species, unless corrected by a *de novo* methylase. Second, if methylation has a role in determining the activity of a gene, the repression of genes that are active during oogenesis—but not in the embryo—may involve *de novo* methylation after fertilization.

Tools for Studying Development in Mammals

Thus far, the role of methylation for gene expression has not been analyzed in mammalian embryos due to the following experimental limitations. First, at present, no molecular probes are available for genes whose expression is specific for oogenesis or for early stages of development. Second, it is difficult or impossible to obtain sufficient biological material from oocytes and early cleavage stage embryos to perform biochemical analyses. Retroviruses and embryonal carcinoma (EC) cells represent two partial solutions for overcoming these limitations, thus allowing the analysis of gene regulation in early mammalian development. Mouse EC cells have in common with preimplantation embryos many characteristics and they can be grown *in vitro* as homogeneous populations. Variation in culture conditions, as well as chemical compounds, can be used to induce differentiation of EC cell lines into a variety of cell types (Martin, 1980).

Upon infection, retroviruses stably integrate into the cellular genome, and they can become germ-line determinants as endogenous viruses. (Jaenisch, 1983). These integrated proviral genomes resemble cellular genes in many aspects; molecular probes of the provirus can be easily prepared. The main experimental advantage of retroviruses as tools for studying mammalian development becomes apparent when contrasted with DNA microinjection as an alternative means of introducing genes into animals. Microinjection of DNA is limited to single cells of the preimplantation stage embryo, whereas retroviral genomes—due to the high efficiency of viral infection—can be introduced into cells of embryos at early—as well as late—stages of development. The presence and expression of the proviral genome can be studied directly after infection, as well as later on in the development in somatic and germ cells of the resulting animal. These experiments have defined a block in proviral expression in preimplantation embryos, which is not seen at later stages. A block in virus expression was also seen in EC cells grown in tissue culture, but not in cells that were induced to differentiate.

In the following sections, we will review the interaction of retroviruses with mouse embryos and EC cells. The molecular mechanisms responsible for the block in retrovirus expression involves *de novo* methylation. This may be signficant for our understanding of gene expression in early mammalian development, as discussed in the final section of this chapter.

Retroviruses, Methylation, and Mouse Development

Endogenous Retroviruses

Endogenous retroviruses are present in the genomes of many vertebrates. The replication of these viruses involves the close interaction of the virus and the host. The transcription of the integrated viral genome, termed "provirus," is dependent on the synthetic machinery of the host, wheras the transcription start site and termination signal are encoded within the viral genome. Upon infection, the viral genome becomes reverse-transcribed into double-stranded DNA, which can integrate at multiple sites into the host genome.

The genome of the laboratory mouse carries three main classes of endogenous retroviruses: A type, B type, and C type viruses, all in multiple copies (Table 10.4). The expression of these genomes is associated with different stages of mouse development. Thus, type A particles are highly expressed at the two-cell stage and in EC cells, but not in the blastocyst. However, no production of B and C type virus particles is found in preimplantation mouse embryos and EC cells; yet, it occurs at later stages of development in some strains of mice, but not in other strains. Variations in the developmental stages (when viral genomes become activated) and in the type of tissues that express C type genomes have been observed (for review, see Jaenisch and Berns, 1977; Kelly and Condamine, 1982). The analysis of the structure and the transcriptional regulation of individual endogenous retroviral genomes is difficult due to the presence of multiple viral genomes (see Table 10.4).

The analysis of viral genome structure and expression during mouse development has been possible in mouse substrains that carry a single copy of the Moloney murine leukemia virus (M-MuLV) in their germ line. These mouse strains, termed Mov substrains, have been derived either by exposing early mouse embryos to infectious virus (Jaenisch, 1976; Jähner and Jaenisch, 1980; Jaenisch *et al.* 1981) or by microinjecting cloned proviral genomes into the mouse zygote (Harbers *et al.* 1981a; Stewart *et al.* 1983a). Thus far, 14 substrains have been established, with each carrying a proviral genome at a different locus (= Mov locus). Some of these genomes have been shown to be defective in the production of infectious virus (Schnieke *et al.* 1983) and were not analyzed further. The expression of all nondefective M-MuLV genomes that were analyzed in Mov substrains of mice is dependent on the stage of development (see Table 10.4). No virus replication was found in preimplantation embryos, but it occurred regularly at later stages. Individual genomes, which are present at different loci, become activated at different stages of development. This indicates that the chromosomal position of the provirus influences proviral genome expression. In all Mov substrains, virus activation is restricted to a minor population of cells, and the type of cell involved has not yet been identified (Jaenisch *et al.* 1981). However, virus activation is not random with respect to cell type, as indicated by an absolute block to spontaneous virus activation in hematopoietic and fibroblastic cells of the Mov-1 substrain (Fiedler *et al.* 1982). A direct demonstration of the importance of the viral integration site for virus activation is the muscle-specific expression of M-MuLV genomes in a mouse derived from a zygote that was microinjected with cloned proviral DNA (Harbers *et al.* 1981a). Thus, the position of M-MuLV in the host genome has a strong influence on both the developmental stage of virus activation and the tissue of virus expression. It is likely that the regulatory

Table 10.4 Expression of Endogenous Retroviruses in Mouse Development and in EC Cells[a]

Type of Virus		Viral Copies per Haploid Genome Equivalent	Expression of Proviral Genomes in: EC Cell Lines	Preimplantation Embryos	Postimplantation Embryos	Adult Mice
Endogenous murine retroviruses (1)	A	~500	+(2)	+(5)		
	B	3–8	−(3)	−(6)		+(1)
	C	25–50	−(4)	−(6)	+(7)	+(1)
M-MuLV experimentally introduced into the germ line	Mov-1	1		−(8)	−(9)	+(9)
	Mov-3,-9,-13,-14	1		−(8)	+(8–10)	+(9, 11)

+ = Endogenous viruses are usually activated in a minor cell population; any infectious particle formed will lead to virus spread, which results in high titers of virus in the animal.

[a] Numbers in parentheses refer to the following references: 1, Coffin (1982); 2, Hojman-Montes de Oca *et al.* (1983); 3, Crépin and Gros (1979); 4, Teich *et al.* (1977); Stewart *et al.* (1982); Gautsch and Wilson (1983); 5, Yotsuyanagi and Szöllösi (1981); 6, for review, see Kelly and Condamine (1982); 7, Lerner *et al.* (1976); Strand *et al.* (1977); 8, Jaenisch, unpublished results; 9, Jaenisch *et al.* (1981); 10, Stuhlmann *et al.* (1981); 11, Stewart *et al.* (1983a).

mechanisms used to control retrovirus expression are similar to those involved in the regulation of cellular genes, indicating that retorviruses can be used to study aspects of gene control during differentiation.

Early Embryonic Cells Are Nonpermissive for Expression of M-MuLV

As shown in Table 10.4, endogenous murine C type viruses, as well as M-MuLV experimentally introduced into the germ line, do not replicate in preimplantation embryos. This block of expression may have been established directly after infection of the embryo or it may be a consequence of passages through the germ line. To address this question, preimplantation embryos were exposed to exogenous retroviruses at the 4–16-cell stage *in vitro*. Virus integration into the cellular genome occurred shortly after infection of the embryos, as indicated by the analysis of the proviral genomes in animals derived from the embryos (Jähner *et al.* 1982). No virus particles were found in infected embryos by using a sensitive assay that would have detected a single-infectious virus particle (Jaenisch *et al.* 1975). Viral genomes, however, became activated in most animals derived from infected embryos (Table 10.5). The activation of the proviral sequences was restricted to a small population of cells, as no indication of virus spread was seen in the postimplantation embryo. However, efficient replication followed when virus was introduced via direct injection into the postimplantation embryo or the newborn mouse (Jähner *et al.* 1982). These observations defined a block to virus replication that is characteristic for preimplantation mouse embryos. This block is maintained in postimplantation embryos and adult mice for viral copies introduced at the preimplantation stage. Cells of the postimplantation embryo, however, are competent to replicate virus when newly infected. The process of provirus activation has not been analyzed at the molecular level. This review will concentrate on the mechanisms that establish and maintain the block of C type retorvirus replication at different stages of mouse development.

The molecular events following infection cannot be analyzed in embryos due to the limited material available. Therefore, infection of EC cells was used to analyze the integration and expression of viral genomes. Unintegrated viral genomes were present in F9 and PCC4 cells early after exposure to M-MuLV, all viral genomes were stably integrated by 2–3 days postinfection (Stewart *et al.* 1982; Gautsch and Wilson, 1983; Niwa *et al.* 1983). As observed earlier with preimplantation embryos, no virus production was found in a variety of EC cell lines (Table 10.6). No stable viral RNA (Teich *et al.* 1977; Stewart *et al.* 1982) nor *in vitro* transcription in isolated nuclei (Gautsch and Wilson, 1983) were detected. This indicated that the block of virus expression in EC cells occurred at the transcriptional level. When infected EC cells were induced to differentiate, virus activation occurred at a low frequency only, and no detectable viral RNA was induced by differentiation of previously infected F9 cells (Stewart *et al.* 1982). However, virus replicated efficiently when the EC

Table 10.5 The Expression of the Exogenous Moloney Leukemia Retrovirus in Mouse Development

Stage of Development at Infection		Mode of Infection	Expression of M-MuLV in: Preimplantation Embryos	Postimplantation Embryos	Adult Mice	Reference[d]
Preimplantation embryos	Zygote	Microinjection of cloned DNA	—[c]	±[b]	+[a]	1,2
	4–16-cell embryos	Infection with virus	—	±	+	1–4
Postimplantation embryos	Day 8 of gestation	Infection with virus		+	+	2,5
	Newborn mice	Infection with virus			+	3,5

[a] + = Viral RNA or infectious virus detected in all mice.

[b] ± = The expression of M-MuLV introduced into preimplantation mouse embryos presumably occurs in a minor population of cells dependent on the site of proviral integration.

[c] — = No viral proteins or virus particles detected.

[d] Reference 1, Harbers *et al.* (1981a); 2, Jähner *et al.* (1982); 3, Jaenisch *et al.* (1975); 4, Jaenisch and Berns (1977); 5, Jaenisch (1980).

Table 10.6 Replication of Exogenous Retroviruses in Mouse EC Cells and Differentiated Derivatives

Stage of Differentiation at Infection	Stage of Differentiation When Analyzed	Replication of Virus[a]	Cell Line	Reference[b]
EC (undifferentiated)	Same as at infection	−	PSA4,S2,Nulli	1
		−	F9,PC13	2
		−	PCC4	3
		−	PCC6	4
		±	PSA4,S2	1
	Induced to differentiate	±	F9	2
		±	PCC4	3
EC cell, differentiated prior to infection	Same as at infection	+	PSA4,S2	1
		+	PCC4	3
Differentiated cell lines derived from EC cells	Same as at infection	+	PCD1,PCD3	5
		+	EB22/20	2
		+	SSCM	6

[a]The replication of virus was tested biologically; + = high titer of virus found; ± = inefficient activation of virus, viral titer several orders of magnitude lower than producing cells; − = no virus detected.

[b]Reference 1, Teich *et al.* (1977); 2, Stewart *et al.* (1983b); 3, Speers *et al.* (1980); 4, cited in Kelly and Condamine (1982); 5, Périès *et al.* (1977); 6, Gautsch and Wilson (1983).

cells were differentiated before infection or when differentiated derivatives of EC cell lines were exposed to virus (Table 10.6). This indicated that proviral genomes carried in cells of postimplantation mouse embryos, as well as in differentiated EC cells, can be either biologically active or inactive. Only viral genomes introduced into differentiated cells were expressed regularly, whereas viral genomes introduced at an early embryonic stage remained inactive in most differentiated cells. These results suggest that the transcriptional block of proviral genomes is established in embryonic cells and is maintained during differentiation by a cis-acting mechanism.

Methylation, Expression, and Infectivity

Restricted expression is a characteristic of endogenous viruses, as seen in species as different as mice and chickens. This prompted the analysis of the methylation levels of the proviral genomes The level of methylation of proviral genomes that is present in different organs or in cultured cells was determined by restriction enzymes that are sensitive to DNA methylation. Two classes of proviral genomes were analyzed: 1) endogenous genomes present in every cell of an animal at a given genetic locus; and 2) exogenous proviral genomes integrated at many different sites. The exogenous genomes are derived from either endogenous viruses via spontaneous activation followed by virus replication and superinfection of susceptible cells or experimental infection of an animal with virus. An inverse correlation between the transcriptional activity and the level of provirus methylation was found in all analyzed cases (see Table 10.7). Endogenous retroviruses of the mouse were methylated at all testable sites (Stuhlmann *et al.* 1981), and viral RNA was not detectable in cells carrying only the endogenous virus (Nobis and Jaenisch, 1980). Exogenous proviral genomes, however, were transcriptionally active and undermethylated in all analysed organs. In chickens, the same endogenous provirus has been found to be either active or inactive in different cell lines; only the active form was partially demethylated (see Table 10.7). The strong correlation between proviral methylation and biological activity suggested a role for methylation in suppression of endogenous retroviral genomes.

The importance of DNA methylation in the establishment of the inactive state of a retorviral genome was demonstrated by transfection assays. High molecular weight DNA from tissues of Mov substrains of mice was used to determine the biological activity of proviral genomes when transfected onto NIH 3T3 mouse fibroblasts that are permissive for virus replication (Stuhlmann *et al.* 1981). The methylated, endogenous proviral copies were not infectious when DNA from several organs of different Mov substrains was tested. However, unmethylated exogenous copies derived from endogenous proviral genomes by spontaneous activation and subsequent superinfection of the same cells were shown to be infectious (Table 10.8).

A direct role for methylation in repressing biological activity was demonstrated by experimental removal both of methyl groups from endogenous retroviral genomes and of the 5–8-Kb flanking mouse sequences. Cloning of methylated eukaryotic DNA in *Escherichia coli* removes the 5-meCyt residues, because these bacteria lack the appropriate methylases. Therefore, the biological activity of the same endogenous proviral genomes was compared before and after molecular cloning by using the transfection assay. Restriction enzyme analysis revealed that the only detectable difference between the genomic and the cloned proviral DNA was the pattern of DNA methylation. The uptake of foreign DNA by cells upon transfection has been shown to be independent of the level of methylation of the foreign DNA (Wigler *et al.* 1981; Stein *et al.* 1982b). As shown in Table 10.8, no infectious virus could be detected with the uncloned proviral DNA, whereas the same provirus was at least 70-fold more infectious after molecular cloning (Harbers *et al.* 1981b).

The view that methylation has a direct effect on gene expression was supported by the reverse experiment; that is, *in vitro* methylation of unmethylated, cloned proviral DNA and subsequent transfection (Table 10.9). Bacterial methylases specific for the sequences CCGG or GCGC reduced the infectivity of either cloned Moloney sarcoma (McGeady *et al.* 1983) or M-MuLV proviral genomes (Hoffmann *et al.*, 1982) 2–10-fold; whereas, a purified methylase from rat liver that methylates all CpG dinucleotides inhibited the infectivity of M-MuLV DNA by more than three orders of magnitude (Simon *et al.*

Table 10.7 Expression and Methylation of Endogenous and Exogenous Retroviruses in Tissues and Differentiated Cells of Mouse and Chicken

			Endogenous Virus		Exogenous Virus		
Species	Virus	Organ or Cell Type	Expression[a]	Methylation[b]	Expression[a]	Methylation[b]	Reference[c]
Mouse	M-MuLV Mov-1 (Mov-3,-13)	Spleen, liver	−	+	+	−	1
	AKR	Fibroblast	−	+	+	−	2
	MMTV	Mammary gland	−	+	+	−	3
Chicken	RAV-0	Fibroblast	−	+	+	−	4
	ev3	Embryonic red blood cell	+	±			5
	ev1		−	+			5
	ev1	Fibroblast	+	±			6

[a]The presence (+) or absence (−) of expression of viral genomes.

[b]The methylation of proviral genomes was determined by restriction enzyme analysis; + = all sites highly methylated; ± = some sites unmethylated; − = no methylation.

[c]Reference 1, Stuhlmann *et al.* (1981); 2, Hoffmann *et al.* (1982); 3, Cohen (1980); 4, Humphries *et al.* (1979); 5, Groudine *et al.* (1981); 6, Conklin *et al.* (1982).

Table 10.8 Methylation and Infectivity of Exogenous and Endogenous, and of Cloned M-MuLV Proviral Genomes

Proviral Genome	Source of DNA	Level of Methylation[a]	Specific Infectivity[b] ($\times 10^{-6}$)	Reference[c]
Exogenous viruses	Spleen, liver, brain, kidney, fibroblast cells	−	0.7–1.4	1,2
Endogenous viruses				
Mov-1,3,9,13	Spleen, liver, embryo DNA	+	<0.1	1,3,4
Cloned Mov-3 from liver	Plasmid pMov-3	−	8	3
Cloned Mov-9 from liver	Plasmid pMov-9	−	7	4

[a]The level of methylation was determined by restriction enzyme analysis; − = no methylation detectable; + = all sites tested are highly methylated.

[b]Specific infectivity defined as plaque-forming units per genome on transfection onto 3T3 cells.

[c]Reference 1, Stuhlmann *et al.* (1981); 2, Jähner *et al.* (1982); 3, Harbers *et al.* (1981b); 4, Chumakov *et al.* (1982).

1983). Thus, the suppression of the biological activity of genomic proviral DNA in the transfection assay, when compared to the cloned unmethylated provirus, was mimicked by the *in vitro* methylation with the rat liver enzyme, but not with the bacterial methylases. This result indicated that the quantitative transfection assay could be used to probe for gene activity that is dependent on DNA methylation at sites that may not be detectable for restriction enzymes.

Table 10.9 In Vitro Methylation Inhibits Biological Activity of Retroviral Genomes

Proviral Genome	Methylase	Specific Infectivity[a] ($\times 10^{-6}$)	Reference[b]
Moloney sarcoma virus (MSV)	None	0.016	1
	HpaII	0.0015	
Moloney leukemia virus (M-MuLV)	None	0.50	2
	HhaI	0.32	
	HpaII	0.30	
	None	0.60	3
	HpaII	0.12	
	Rat liver methylase	<0.0006	

[a]Specific infectivity defined as focus-forming units (for MSV) or plaque-forming units (for M-MuLV) per genome on transfection onto 3T3 cells.

[b]Reference 1, McGeady *et al.* (1983); 2, Hoffmann *et al.* (1982); 3, Simon *et al.* (1983).

Further support for the importance of methylation for expression of viral genomes came from experiments in which retroviral genomes were activated *in vivo* by treatment with the drug 5-Azacytidine (5-aza-C). This drug is believed to inactivate the maintenance methylase (for review, see Santi *et al.* 1983), which leads to a postreplicative demethylation of cellular DNA. In tissue culture, silent endogenous viruses of mouse (Niwa and Sugahara, 1981; Hoffmann *et al.* 1982) and chicken (Groudine *et al.* 1981; Conklin *et al.* 1982) were shown to be induced by treating the cells with 5-Aza-C. Similarly, treatment of animals with this drug was shown to efficiently activate M-MuLV proviruses carried in the germ line of nonviremic Mov substrains of mice (Jaenisch, unpublished).

In conclusion, the experiments summarized in Tables 10.8 and 10.9 show that the biological activity of retroviral genomes transfected into 3T3 cells can be influenced by the level of methylation present in the viral and flanking sequences. An inactive state seems to involve a higher level of methylation than an active state. Based on the assumption that gene expression of transfected DNA is controlled by mechanisms relevant for gene expression in the animal, the results indicate that DNA methylation has a role in the maintenance of the cis-acting block of retrovirus expression.

De Novo Methylation of M-MuLV in Early Embryonic Cells

Endogenous retroviral genomes in mice are highly methylated (Cohen, 1980; Stuhlmann *et al.* 1981). Since the retroviral genomes introduced into mouse embryos to derive the Mov substrains of mice were unmethylated, *de novo* methylation of the proviral genomes must have occurred at some point either during the development of the infected embryo and/or as a consequence of transmission through the germ line. To address this problem, mouse embryos were infected at different stages of embryonic development, and the level of methylation was determined in the resulting adult mice. The biological activity of these genomes was quantitatively determined by transfection of DNA from adult mice onto mouse NIH 3T3 cells. As shown in Table 10.10, no detectable *de novo* methylation occurred when postimplantation embryos or newborn mice were infected and analyzed 2–4 months later. In contrast, viral genomes that had integrated into the DNA of preimplantation embryos became methylated at all sites tested when analyzed in liver, brain, and kidney of the adult (Jähner *et al.* 1982). Thus, an efficient *de novo* methylation activity was active in the pluripotent cells of the preimplantation embryo, which was not detected in cells of the postimplantation embryo or the newborn mouse. The infectivity of the viral genomes was inversely correlated to the degree of methylation; thus, *de novo* methylation of exogenous viral genomes correlated with nonpermissivity of preimplantation mouse embryos to virus expression (compare with Table 10.5). As shown by the transfection assay, this block was cis acting and was maintained during differentiation. The continuous block of virus expression

Table 10.10 De Novo Methylation and Infectivity of Exogenous M-MuLV Genomes in Mouse Embryogenesis

Stage of Differentiation at Virus Infection		Analyses of DNA from Tissues of Adult Mice: Level of Methylation at Individual Sites (%)	Number of Sites Analysed	Specific Infectivity[a] (× 10^6)
Preimplantation stage	Zygote	>70	44	<0.01
	Morula	>70	44	<0.1
Postimplantation stage	Day 8 of gestation	<20	11	0.7–1.4
	Newborn mice	<20	4	0.9

SOURCE: All data from Jähner *et al.* (1982) and Stuhlmann *et al.* (1981).

[a]The specific infectivity is defined as the number of infectious centers induced per transfected proviral genome on mouse NIH 3T3 cells.

throughout development correlated with the maintenance of DNA methylation of the priviral genomes.

To address the question of whether *de novo* methylation was causally involved in the establishment of the inactive state of proviral genomes in early embryonic mouse cells, the temporal order of molecular events following virus infection of EC cells was analyzed. For this analysis, the kinetics of viral DNA integration and viral genome transcription was determined after virus infection. The involvement of *de novo* methylation was analyzed in two ways: 1) the kinetics of *de novo* methylation was determined by restriction enzyme analysis; and 2) the infectivity of viral DNA was determined at different intervals after the infection to follow the establishment of the cis-acting suppression of proviral activity due to *de novo* methylation. The transfection assay was crucial in this analysis, since it functionally detects the presence of methylated sites that cannot be analyzed by restriction enzymes (see previous section; see Table 10.9). For comparison, the same experiments were performed with the NIH 3T3 mouse fibroblast cell line, which exerts no block to virus expression. The results with the two EC cell lines, F9 and PCC4, and the 3T3 cell line are summarized in Table 10.11 (Stewart *et al.* 1982; Gautsch and Wilson, 1983; Niwa *et al.* 1983).

Unintegrated viral copies were present 2 days after infection in F9 and 3T3 cells; at 3 days, most of the DNA was stably integrated. In F9 cells, unmethylated viral DNA and infectivity were detected as long as unintegrated viral copies were present; but, all proviral DNA was highly methylated and was rendered noninfectious within 3 days postinfection. Thus, the kinetics of all three parameters tested (i.e., virus integration, *de novo* methylation, and loss of infectivity) changed in parallel, which indicated immediate *de novo* methylation and inactivation of the infecting DNA on genomic integration. No *de*

Table 10.11 The Kinetics of Integration, Expression, Methylation, and Infectivity of Proviral Genomes in EC Cell Lines and Fibroblasts Infected with M-MuLV

Time after Infection (Days)	Relative[a] Amount of Viral RNA			Relative[a] Amount of Unintegrated DNA			Relative[a] Amount of Unmethylated DNA			Relative[a] Infectivity of Total DNA		
	F9	PCC4	3T3	F9	PCC4	3T3	F9	PCC4	3T3	F9	PCC4	3T3
0.5		+		+++	+++ (1,2)	+++	+++	+++	++	+++		+
1		+		+++	+ (1,2)	+++	+++	+++	+++	+++		+++
1.5		–		++		++	++	+++	+++	++		+++
2		–		+	+ (1)	+	+	+++	+++	+	± (2)	++
3		–		–		+	–		+++	–		
5								+++			±	
7		–				–	–	+++	+++			
>16	–	–	+++	–	–	–	–	–	+++	–	–	+++

[a]The absolute values of parameters listed varied considerably between the different cell lines. ± = transfection assays with PCC4 cells were not quantitative; detected infectivity may have been due to a single activation event per transfected culture scoring positive in the assay.

SOURCE: All data for F9 and 3T3 cells were from Stewart *et al.* (1982). Data for PCC4 are from Gautsch and Wilson (1983) except reference 1, D'Auriol *et al.* (1981); 2, Niwa *et al.* (1983) (reference numbers in parentheses).

novo methylation and no loss of infectivity was seen when 3T3 cells were infected with M-MuLV. The noninfectious proviral genomes carried in F9 cells were potentially infectious, as shown by treatment of transfected 3T3 cells with 5-Aza-C (Stewart *et al.* 1982). This indicated that noninfectivity of the proviral genomes was due to methylation and not to some other modification of the DNA.

Similarly, as in F9 cells, the block to virus expression in PCC4 cells was established shortly after infection. Less than 10 viral RNA molecules per cell were detected during the first 24 hours postinfection, and no expression was found at later times (Gautsch and Wilson, 1983). In contrast to F9 cells, *de novo* methylation was not detected up to 14 days after infection of PCC4 cells. The crucial question of whether the undermethylated M-MuLV DNA present in PCC4 cells soon after infection was infectious cannot be answered from the published data, because no quantitative transfection assays were performed. The infectivity found in a single transfection assay 5 days postinfection could have been due to a single activation event in the transfected cells. The kinetics of viral DNA integration was not convincingly established and gave contradictory results in three laboratories working with the same cell line (compare D'Auriol *et al.* 1981; Gautsch and Wilson, 1983; Niwa *et al.* 1983). It appears possible that either a few unintegrated M-MuLV copies that were not detected by the methods employed or a rare demethylation event in the transfected cells gave a positive singal in the transfection assay (compare with Table 10.11). Therefore, it cannot be excluded that *de novo* methylation at sites other than the tested SmaI sites occurred soon after integration and was involved in blocking virus transcription.

It is clear from the results summarized in Table 10.11 that the two EC cell lines, F9 and PCC4, differ in the efficiency of *de novo*-methylating retroviral genomes. It is possible that efficient *de novo* methylation is a marker of early embryonic development and that the two cell lines correspond to different stages of embryogenesis. The important question of whether a cis- or trans-acting mechanism initiates the block of viral transcription in embryonic cells cannot be answered. In an analogy to the results obtained with transfected DNA, the *de novo* methylation may fix the transcriptionally inactive state that is inherited during subsequent development due to the action of the maintenance methylase. Activation of proviral genomes during later development is restricted to a small population of cells, suggesting that changes from an inactive to an active state may depend on activation of the chromosomal region of integration. In most cells, activation does not occur, indicating that the presence of the methyl groups has a strong influence on the maintenance of the inactive state during the life of the animal.

The low amount of viral RNA detected during the first day after infection (see Table 10.11) may be derived from unintegrated viral DNA that remained unmethylated. In fact, it has been shown that plasmid DNA microinjected into mouse zygotes is expressed before genomic integration (Harbers *et al.* 1981a; Brinster *et al.* 1982). Thus, foreign DNA introduced into embryonic cells is blocked in expression and is *de novo* methylated only following genomic integration.

Methylation, Chromatin Structure, and Regulatory Sequences

The influence of other factors that are important for gene regulation has to be considered in defining the exact role of *de novo* methylation in early embryonic cells. Chromatin structure, methylation, and more recently tissue-specific regulatory sequences have been proposed as elements that are important in expression of eukaryotic genes. The most consistent parameter revealing the active state of a gene is an opened chromatin structure, as shown for a large number of genes (Elgin, 1981) that include M-MuLV genomes integrated in tissues of mice (Breindl *et al.* 1982, 1984; van der Putten *et al.* 1982). Changes in chromatin structure, as well as changes in the pattern of DNA methylation, are cis-acting mechanisms that can act in concert during gene activation, as most convincingly shown for globin genes (Conklin and Groudine, Chapter 14) and endogenous retroviral genomes (Groudine *et al.* 1981). The inverse correlation between methylation and gene activity of several developmentally regulated genes pointed to the importance of DNA methylation for gene expression (see Chapter 8, this volume). However, for a number of genes such as the $\alpha 2$ (I) collagen gene (McKeon *et al.* 1982) and the vitellogenin genes (Gerber-Huber *et al.* 1983; Burch and Weintraub, 1983), this correlation has not been found; that is, no changes of the methylation pattern have been observed during gene activation. These studies, however, do not exclude that methylation of a subset of sites not detected by restriction enzymes is correlated with transcriptional activity. In fact, a direct role of methylation in gene transcription has been demonstrated by gene transfer of *in vitro*-methylated viral, housekeeping or or developmentally regulated genes into eukaryotic cells (compare with Table 10.9; Wigler *et al.* 1981; Vardimon *et al.* 1982; Cedar *et al.* 1983; Flavell *et al.* 1983; Simon *et al.* 1983; Busslinger *et al.* 1983). These experiments suggested that DNA methylation may be involved in the establishment and maintenance of a state of gene inactivity in the animal.

The evidence summarized above is consistent with a simple model for gene regulation during development. During one or a few critical cell cycles, trans-acting molecules may trigger a gene to switch from an inactive to a competent state that precedes gene activation. This switch may involve alterations of chromatin structure, as well as demethylation. Demethylation may be due to inhibition of the maintenance methylase at specific sites. The altered pattern of methylation may be used to establish and maintain the gene in the new state of competence during subsequent cell replication. Depending on physiologic stimuli, transcription may be initiated by interaction of specific trans-acting molecules with regulatory sequences of the gene. Such "enhancer" sequences, which control gene activity in cis, have been defined recently in viral (Khoury and Gruss, 1983) and cellular genes (Gillies *et al.* 1983; Banerji *et al.* 1983).

It is likely that these enhancers represent the targets for differentiation-specific regulatory molecules, because their effect is restricted to specific cell types. This model of gene control during development predicts that demethylation can be dissociated from actual gene transcription—as has been seen with the $\alpha 2$ (I) collagen gene (McKeon *et al.* 1982) or immunoglobulin genes (Mather and Perry, 1983). Transcriptional inactivation of the X chromosome in mouse trophoblast tissue is another example of hypomethylation being dissociated from gene expression (Kratzer *et al.* 1983).

The expression of genes experimentally introduced into the germ line of animals is consistent with the model of gene expression outlined above. Although the enhancer sequences present in the M-MuLV genome (Khoury and Gruss, 1983) are active in many differentiated cell types of the postimplantation embryo (compare with Table 10.5), activation of the highly methylated genomes in Mov substrains of mice was observed in a developmentally regulated manner. This suggested that virus activation depends on the transcriptional competence of the chromosomal region surrounding the viral integration site. Demethylation of M-MuLV regulatory sequences has recently been found in developing Mov mice. This process was governed both by the state of differentiation and by the chromosomal position, indicating that it might be involved in the activation of the proviral genomes (Jähner and Jaenisch, in preparation).

Likewise, differential expression of the rabbit β-globin gene (Lacy *et al.* 1983) and metallothionein TK fusion genes (Palmiter *et al.* 1982) in transgenic mice appeared to depend on their chromosomal position. In contrast, transgenic mice carrying multiple copies of a rearranged immunoglobin gene (Brinster *et al.* 1983) appeared to express the foreign gene independently of their site of integration. It is not clear whether this particular enhancer or whether the high number of integrated copies present in these mice caused the immunoglobulin gene to be expressed regardless of where it was integrated. The chicken transferrin gene is another example of a gene that appears to be expressed independently of its chromosomal position when inserted into the germ line of mice (McKnight *et al.* 1983).

The relationship between methylation and expression has not been established for these exogenously introduced genes. It is possible, however, that regulatory sequences of some genes are not always *de novo* methylated and inactivated in embryonic cells or are activated independently of their chromosomal position in a tissue-specific manner. This may involve specific demethylation and may be similar to demethylation of normal cellular genes during development.

It is possible that genes active in embryonic cells are controlled by enhancer elements that are different from those of genes active in differentiated cells. Enhancers active in EC cells have been defined recently by polyoma virus mutants (Fujimura *et al.* 1981; Levine, 1982; Dandolo *et al.* 1983). The results obtained with retroviral genomes indicated that the inactivation of DNA sequences occurs shortly after integration into the genome of the embryonic cell, and that *de novo* methylation is associated with this process. The integration into a given chromosomal region may determine whether, at what stage during development, and in which cell lineage the inserted gene is changed into a state of competence.

De Novo Methylation in Embryonic and Differentiated Cells

The *de novo* methylation activity of preimplantation mouse embryos methylated M-MuLV genomes at all 50 testable sites and in addition all testable sites of the adjacent cellular flanking sequences (Jähner *et al.* 1982; Jähner and Jaenisch, in preparation). Microinjection of DNA into the zygote pronucleus and transfection into EC cells have demonstrated that not only retroviral genomes, but also cloned cellular genes are substrates for the embryonic *de novo* methylase (Jähner *et al.* 1982; Palmiter *et al.* 1982; Cedar *et al.* 1983; F. Costantini, personal communication). This suggested that the enzyme has no apparent preference for sites that are present within specific sequences, and it has the potential to methylate all CpG sites of exogenous DNA integrated into the embryonic genone. As shown in Table 10.11, the kinetics of *de novo* methylation of proviral genomes differed in F9 and PCC4 cells, although the final extent of methylation was similar in both cell lines. Therefore, a potential site remains a substrate for the *de novo* methylase in many cell generations. The final level of methylation in exogenous DNA appears to be higher than the overall methylation of cellular DNA (compare with Table 10.3), indicating that a fraction of the sites in the cellular genome—but not in exogenous DNA—are protected against *de novo* methylation. However, in all cases examined thus far, the exogenous sequences were transcriptionally inactive. Since the protection of potential sites may be due to specific protein DNA interactions, the *de novo* methylation of DNA may depend on its chromosomal structure and position and on its transcriptional activity.

In contrast to the results obtained with the preimplantation mouse embryos, *de novo* methylation of viral or cloned DNAs was only rarely observed in cells corresponding to later stages of development (for review, see Wigler, 1981; see Table 10.12). In the two instances in which the kinetics of *de novo* methylation were analyzed without selection of cell clones (e.g., in cells infected with adenovirus type 12 and Epstein-Barr virus [Table 10.12], the methylation activity was much less efficient than the activity found in embryonic cells. A very small fraction of mouse L cells carrying *Herpes* TK sequences was selectable for the TK^- phenotype. The multiple TK sequences in these clones were highly methylated, indicating *de novo* methylation before selection (Hardies *et al.* 1983). *De novo* methylation either can be induced with a low frequency in transformed mammalian cells or acts on a very small fraction of chromosomal sites. Support for this interpretation comes from analysis of *de novo* methylation activity in postnatal mice. M-MuLV genomes introduced at multiple sites in cells of the postimplantation embryo did not become detectably methylated during several months postinfection (Jähner *et al.* 1982). Furthermore, all available evidence indicates that demethylation, but not *de novo* methylation,

Table 10.12 Inefficient De Novo Methylation of DNA in Differentiated Cells

Species	Cell Type	Virus or DNA Introduced into Cells	Derivation of Cells with Methylated Donor DNA	Reference[a]
		DNA		
Mouse	L cells, TK⁻	TK gene of HSV	Rare clone selected for TK⁻ phenotype	1
		TK gene of HSV + pBR322	Rare clone	2
		Virus		
Rat	Fibrosarcoma	RSV	Rare clone	3
Rat	NRK	M-MuLV		4
Hamster	Sarcoma	**Ad 12**		5
Monkey	Transformed lymphocyte	Herpes saimiri	Tumors or cell lines with many passages	6
Human	Transformed lymphocyte	EBV		7

[a]Reference 1, Christy and Scangos (1982); Ostrander *et al.* (1982); Clough *et al.* (1982); Hardies *et al.* (1983); 2, Pollack *et al.* (1980); 3, Guntaka *et al.* (1980); 4, Hoffmann *et al.* (1982); 5, Kuhlmann and Doerfler (1982); 6, Desrosiers (1982); 7, Kintner and Sugden (1981).

plays a role in the regulation of genes in terminally differentiated cells. This evidence indicates that the *de novo* methylation activity detected in some selected cell clones is of no or little biological relevance.

Due to the high *de novo* methylase activity, it is not possible to detect the presence of a maintenance methylase in embryonic cells by similar experiments as used in differentiated cells. However, extrachromosomal DNA, which is not *de novo* methylated, has been shown to be a substrate for a maintenance methylase in *Xenopus* eggs (Harland, 1982). Thus far, only a maintenance methylation activity has been isolated from EC cells (D. Simon, personal communication), although the same methods used to purify this activity resulted in the isolation of a *de novo* methylation activity from rat liver (Simon *et al.* 1978, 1983). This may suggest that *de novo* methylation and maintenance methylation are performed by the same enzyme whose specificity may be modified by factors that are developmentally regulated.

A Biological Role of De Novo Methylation?

The exact role of methylation in gene expression has yet to be defined. An attractive hypothesis suggests that methylation or hypomethylation is a means of stabilizing the structure of a gene in an active or inactive state. Thus, a hypomethylated state would be a prerequisite of transcriptional competence. Actual transcription, however, would depend on trans-acting factors that may interact with tissue-specific enhancers, as discussed above. The maintenance methylase would provide a means of keeping the pattern of methylation constant in somatic cell generations. The establishment of a new methylation pattern, however, requires the action of a *de novo* methylase that accepts unmethylated DNA as a substrate. In the following paragraphs, we will speculate about the functions that this enzyme may have during the life cycle of an animal.

We will consider three possible functions for *de novo* methylation in early embryogenesis—two of which have been mentioned in the second section in this chapter.

(1) The *de novo* methylase provides the embryo with a mechanism for inactivating any foreign DNA, and it may have evolved as a means of protecting the developing embryo against deleterious consequences of virus expression.

(2) The *de novo* methylase has a repair function and corrects the errors introduced by the maintenance methylase. Errors have been seen both as a shift in methylation patterns of cellular genes during growth in tissue culture (Shmookler Reis and Goldstein, 1982) and as a loss of methylation in DNA of aging fibroblast cultures (Wilson and Jones, 1983). Therefore, the maintenance of the constant level of methylation, as observed in a given animal species, would require *de novo* methylation in embryonic development before separation of somatic from germ-line cells.

(3) If methylation is involved in turning off genes that were expressed during oogenesis, a general and nonspecific *de novo* methylation may be a simple and secure strategy for inactivating all genes whose expression is either not required or would be deleterious for the development of the early embryo. In fact, an efficient repression of the genes that are active during oogenesis may be a condition for "resetting" the genetic program; no direct evidence has been obtained as yet. Some observations, however, are consistent with the hypothesis that, in mammals, a different set of genes other than that expressed during oogenesis is activated soon after fertilization.

Early embryogenesis in amphibians may represent one extreme that is controlled almost entirely by preformed maternal mRNA (Davidson, 1976). Likewise, early development of sea urchins appears to depend heavily on maternal mRNA (Davidson *et al.* 1982). The embryonic genome, however, is activated early, as seen, for example, by the expression of an embryo-specific class of histone genes at the 16-cell stage (Bryan *et al.* 1983). The other extreme is the mammal with its relatively small egg and very slow early development. Mammalian embryogenesis appears to depend on activation and expression of embryonic genes before the eight-cell stage, and possibly as early as the two-cell stage (Johnson *et al.* 1977; Sherman, 1979; Braude *et al.* 1979; Sawicki *et al.* 1981). For example, it has been shown that total RNA and mRNA in the mouse oocyte decreases after fertilization and begins to increase again at the two-cell stage (Pikó and Clegg, 1982). Similarly, the concentration of histone and actin mRNA is strongly diminished in two-cell mouse embryos when compared to oocytes; it increases again at the eight-cell stage (Giebelhaus *et al.*

1983). All of these observations are consistent with the notion that genes active in mammalian oogenesis become inactive after fertilization, and also that the embryonal genome becomes activated soon after the first cleavage.

Because the methylation of specific cellular genes that are active in oogenesis has not been analyzed in oocytes and early embryonic cells, the evidence for an efficient *de novo* methylase, which is characteristic for embryonic—but not differentiated—cells, rests on methylation of exogenously introduced DNA. The level of total methylation in DNA from oocytes, embryonic cells, and adult tissues was not signficantly different (Table 10.3), which argues against extensive *de novo* methylation of cellular DNA in early embryos. However, the significance of these measurements for our discussion is limited for the following reasons. Although it is clear from transfection assays with different genes that methylation is involved in gene repression, increasing evidence indicates that methylation of only a small number of CpG dinucleotides is relevant for the control of biological activity. Relevant CpG dinucleotides probably are located at specific sites (Vardimon *et al.* 1982; Flavell *et al.* 1983; Busslinger *et al.* 1983; Krūczek and Doerfler 1983), which may be different from those detectable via restriction enzyme analysis (Simon *et al.* 1983). This suggests that the great majority of methylated sites observed in cellular DNA are of little significance for gene expression. The very different levels of total methylation found in different animal species are in agreement with this assumption. A change in methylation of a small fraction of biologically significant sites would not be expected to be detectable as a change in the total level of methylation.

If the "oocyte-specific genes" are inactivated by a general and nonspecific *de novo* methylation, how are the embryonic genes activated? We propose two speculative solutions that are not mutually exclusive.

(1) The interaction of protein factors with specific DNA sequences and/or local chromatin conformation may protect potentially methylatable CpG sites against becoming *de novo* methylated in the embryonic cell. In this way, embryonic genes would become available and remain competent for transcription as long as the specific protein factors protect these sequences against *de novo* methylation.

(2) "Embryo-specific genes" may be regulated by control elements that permit expression even when methylated. This hypothesis would assume a different control of transcription in embryonic and somatic cells; possibly mediated by factors that modify the transcriptional machinery. The control elements may represent embryo-specific enhancers similar to the enhancers found in polyoma mutants that replicate in EC cells (Levine, 1982). Due to specific enhancers, individual gene members of multigene families may be active in either the embryo or the adult. Transcription of histone genes in sea urchin development may represent an example of this situation (Bryan *et al.* 1983). Single-copy genes, conversely, may be regulated by two alternative control elements; one is active in the preimplantation embryo and the other is active in somatic cells. The transcriptional control of the α-amylase gene is an example of two promoters that alternatively transcribe a single-copy gene in different

tissues (Schibler *et al.* 1983). This latter hypothesis postulates two sets of control elements; a somatic set sensitive to methylation and an embryonic set active independently of methylation. This mode of embryonic gene control would combine a safe-guard against expression of "inappropriate" genes with assured expression of the proper embryonic genes without having to rely on the interaction of specific protein factors to protect the embryonic control elements against *de novo* methylation.

De novo methylation is a characteristic of early embryonic cells and not of differentiated somatic cells. It is possible that repression of genes that are active during oogenesis is a prerequisite for activating a new genetic program promoting the preimplantation development of mammals. Early embryonic cells may not tolerate an accidental activation of "inappropriate" genes—an event that may be not as deleterious during later somatic development, when cell proliferation allows selection against a defective cell. To keep the genes not used in early embryogenesis safely repressed may be the essence for a cell to be totipotent. *De novo* methylation may be the means of assuring this repression.

Acknowledgments

We thank Drs. H. Arnold, W. Ostertag, A. Stacey and all of our colleagues in the Tumor Virus Department for many discussions and critical comments; and also E. Danckers for her patience and help in preparing the manuscript. Work from the authors' laboratory was supported by grants from the Stiftung Volkswagenwerk and the Deutsche Forschungsgemeinschaft. The Heinrich-Pette-Institut is financially supported by Freie und Hansestadt Hamburg and Bundesministerium für Jugend, Familie and Gesundheit.

References

Banerji J, Olson L, Schaffner W: A lymphocytespecific cellular enhancer is located downstream of the joining region in immunoglobulin heavy chain genes. *Cell* 1983;33:729–740.

Bird AP, Taggart MH, Smith BA: Methylated and unmethylated DNA compartments in the sea urchin genome. *Cell* 1979;17:889–901.

Bird, AP, Taggart MH: Variable patterns of total DNA and rDNA methylation in animals. *Nucleic Acids Res* 1980;8:1485–1497.

Bird A, Taggart M, Macleod D: Loss of rDNA methylation accompanies the onset of ribosomal gene activity in early development of X. laevis. *Cell* 1981;26:381–390.

Braude P, Pelham H, Flach G, Lobatto R: Post-transcriptional control in the early mouse embryo. *Nature* 1979;282:102–105.

Breindl M, Nath U, Jähner D, Jaenisch R: DNase I sensitivity of endogenous and exogenous proviral genome copies in M-MuLV-induced tumors of Mov-3 mice. *Virology* 1982;119:204–208.

Breindl M, Harbers K, Jaenisch R: Retrovirus-induced lethal mutation in collagen I gene of mice is associated with an altered chromatin structure. *Cell* 1984; in press.

Brinster RL, Chen HY, Warren R, Sarthy, A, Palmiter RD: Regulation of metallothionein-thymidine kinase fusion plasmids injected into mouse eggs. *Nature* 1982;296:39–42.

Brinster RL, Ritchie KA, Hammer RE, O'Brien RL, Arp B, Storb U: Expression of a microinjected immunoglobin gene in the spleen of transgenic mice. *Nature* 1983;306:332–336.

Bryan P, Olah J, Birnstiel M: Major changes in the 5′ and 3′ chromatin structure of sea urchin histone genes accompany their activation and inactivation in development. 1983;*Cell* 33:843–848.

Burch JBE, Weintraub H: Temporal order of chromatin structural changes associated with activation of the major chicken vitellogenin gene. *Cell* 1983;33:65–76.

Busslinger M, Hurst J, Flavell RA: DNA methylation and the regulation of globin gene expression. *Cell* 1983;34:197–206.

Cedar H, Stein R, Gruenbaum Y, Naveh-Manyy T, Sciaky-Gallili N, Razin A: Effect of DNA methylation on gene expression, in *Cold Spring Harbor Symposia on Quantitative Biology: Structures of DNA.* Cold Spring Harbor Laboratory, Cold Spring Harbor, New York, 1983, vol 47, pp 605–609.

Chapman V, Forrester C, Sanford J, Hastie N, Rossant J: Cell lineage specific undermethylation of mouse repetitive DNA. *Nature* 1984;307:284–286.

Christy B, Scangos G: Expression of transferred thymidine kinase genes is controlled by methylation. *Proc Natl Acad Sci* 1982;79:6299–6303.

Chumakov I, Stuhlmann H, Harbers K, Jaenisch R: Cloning of two genetically transmitted Moloney leukemia proviral genomes: correlation between biological activity of the cloned DNA and viral genome activation in the animal. *J Virol* 1982;42:1088–1098.

Clough DW, Kunkel LM, Davidson RL: 5-Azacytidine-induced reactivation of a herpes simplex thymidine kinase gene. *Science* 1982;216:70–73.

Coffin J: Endogenous viruses, in Weiss R, Teich N, Varmus H, Coffin J (eds): *RNA Tumor Viruses.* Cold Spring Harbor Laboratory, Cold Spring harbor, New York, 1982; pp 1109–1203.

Cohen JC: Methylation of milk-borne and genetically transmitted mouse mammary tumor virus proviral DNA. *Cell* 1980;19:653–662.

Conklin KF, Coffin JM, Robinson HL, Groudine M, Eisenman R: Role of methylation in the induced and spontaneous expression of the avian endogenous virus *ev*-1: DNA structure and gene products. *Mol Cell Biol* 1982;2:638–652.

Crépin M, Gros F: Regulation of mouse mammary tumor viral RNA synthesis in embryonal carcinoma cells and in teratocarcinoma derived myoblasts. *Biochem Biophys Res Comm* 1979;87:781–788.

Dandolo L, Blangy D, Kamen R: Regulation of polyoma virus transcription in murine embryonal carcinoma cells. *J Virol* 1983;47:55–64.

D'Auriol L, Yang WK, Tobaly J, Cavalieri F, Périès J, Emanoil-Ravicovitch R: Studies on the restriction of ecotropic murine retrovirus replication in mouse teratocarcinoma cells. *J Gen Virol* 1981;55:117–122.

Davidson EH: *Gene Activity in Early Development,* ed 2. New York, San Francisco, London, Academic Press, 1976.

Davidson EH, Hough-Evans BR, Britten RJ: Molecular biology of the sea urchin embryo, *Science* 1982;217:17–26.

Desrosiers RC: Specifically unmethylated cytidylic-guanylate sites in *Herpesvirus saimiri* DNA in tumor cells. *J Virol* 1982;43:427–435.

Doskočil J, Šorm F: Distribution of 5-methyl-cytosine in pyrimidine sequences of deoxyribonucleic acids. *Biochim Biophys Acta* 1962;55:953–959.

Ehrlich M, Gama-Sosa MA, Huang L-H, Midgett RM, Kuo KC, McCune RA, and Gehrke C: Amount and distribution of 5-methylcytosine in human DNA from different types of tissues or cells. *Nucleic Acids Res.* 1982;10:2709–2721.

Elgin SCR: DNAase I-hypersensitive sites of chromatin. *Cell* 1981;27:413–415.

Fiedler W, Nobis P, Jähner D, Jaenisch R: Differentiation and virus expression in BALB/Mo mice: endogenous Moloney leukemia virus is not activated in hemotopoietic cells. *Proc Natl Acad Sci (USA)* 1982;79:1874–1878.

Flavell RA, Grosveld F, Busslinger M, de Beer E, Kioussis D, Mellor AL, Golden L, Weiss E, Hurst J, Bud H, Bullman H, Simpson E, James R, Santamaria M, Atfield G, Festenstein H: Structure and expression of the human globin genes and murine histocompatability antigen genes, in *Cold Spring Harbor Symposia on Quantitative Buiology: Structures of DNA.* Cold Spring Harbor Laboratory, Cold Spring Harbor, New York, 1983,vol 47, pp 1067–1068.

Fujimura FK, Deininger PL, Friedmann T, Linney E: Mutation near the polyoma DNA replication origin permits productive infection of F9 embryonal carcinoma cells. *Cell* 1981;23:809–814.

Gautsch JW, Wilson MC: Delayed *de novo* methylation in teratocarcinoma suggests additional tissue-specific mechanisms for controlling gene expression. *Nature* 1983;301:32–37.

Gerber-Huber S, May FEB, Westley BR, Felber BK, Hosbach HA, Andres A-C, Ryffel GU: In contrast to the Xenopus genes the estrogen-inducible vitellogenin genes are expressed when totally methylated. *Cell* 1983;33:43–51.

Giebelhaus, DH, Heikkila JJ, Schultz GA: Changes in the quantity of histone and actin messenger RNA during the development of preimplantation mouse embryos. *Dev. Biol* 1983;98:148–154.

Gillies SD, Morrison SL, Oi VT, Tonegawa S: A tissue-specific transcription enhancer element is located in the major intron of a rearranged immunoglobulin heavy chain gene. *Cell* 1983;33:717–728.

Grippo P, Iaccarino M, Parisi E, Scarano E: Methylation of DNA in developing sea urchin embryos. *J Mol Biol* 1968;36:195–208.

Groudine M, Eisenman R, Weintraub H: Chromatin structure of endogenous retroviral genes and activation by an inhibitor of DNA methylaion. *Nature* 1981;292:311–317.

Guntaka RV, Rao PY, Mitsialis SA, Katz R: Modification of avian sarcoma proviral DNA sequences in nonpermissive XC cells but not in permissive chicken cells. *J Virol* 1980;34:569–572.

Harbers K, Jähner D, Jaenisch R: Microinjection of cloned retroviral genomes into mouse zygotes: integration and expression in the animal. *Nature* 1981a;293:540–542.

Harbers K, Schnieke A, Stuhlmann H, Jähner D, Jaenisch R: DNA methylation and gene expression: endogenous retorviral genome becomes infectious after molecular cloning. *Proc Natl Acad Sci (USA)* 1981b;78:7609–7613.

Hardies SC, Axelrod DE, Edgell MH, Hutchison III: Phenotypic variation associated with molecular alterations at a cluster of thymidine kinase genes. *Mol Cell Biol* 1983;3:1163–1171.

Harland RM: Inheritance of DNA methylation in microinjected eggs of *Xenopus laevis. Proc Natl Acad Sci (USA)* 1982;79:2323–2327.

Hoffmann JW, Steffen D, Gusella J, Tabin C, Bird S, Cowing D, Weinberg RA: DNA methylation affecting the expression of murine leukemia proviruses. *J. Virol* 1982;44:144–157.

Hojman-Montes de Oca F, Dianoux L, Périès J, Emanoil-Ravicovitch R: Intracisternal A particles: RNA expression and DNA methylation in murine teratocarcinoma cell lines. *J Virol* 1983;46:307–310.

Holliday R, Pugh JE: DNA modification mechanisms and gene activity during development. *Science* 1975;187:226–232.

Humphries EH, Glover C, Weiss RA, Arrand JR: Differences between the endogenous and exogenous DNA sequences of Rous-associated virus-O. *Cell* 1979;18:803–815.

Jaenisch R, Fan H, Croker B: Infection of preimplantation mouse embryos and of newborn mice with leukemia virus: tissue distribution of viral DNA and RNA and leukemogenesis in the adult animal. *Proc Natl Acad Sci (USA)* 1975;72:4008–4012.

Jaenisch R: Germ line integration and Mendelian transmission of the exogenous Moloney leukemia virus. *Proc Natl Acad Sci (USA)* 1976;73:1260–1264.

Jaenisch R, Berns A: Tumor virus expression during mammalian embryogenesis, in Sherman M, (ed): *Concepts in Embryogenesis*. Cambridge, Massachusetts, MIT Press, 1977, pp 267–314.

Jaenisch R: Retroviruses and embryogenesis: microinjection of Moloney leukemia virus into midgestation mouse embryos. *Cell* 1980;19:181–188.

Jaenisch R, Jähner D, Nobis P, Simon I, Löhler J, Harbers K, Grotkopp D: Chromosomal position and activation or retroviral genomes inserted into the germ line of mice. *Cell* 1981;24:519–529.

Jaenisch R: Endogenous retroviruses. *Cell* 1983;32:5–6

Jähner D, Jaenisch R: Integration of Moloney leukaemia virus into the germ line of mice: correlation between site of integration and virus activation. *Nature* 1980;287:456–458.

Jähner D, Stuhlmann H, Stewart CL, Harbers K, Löhler J, Simon I, Jaenisch R: *De novo* methylation and expression of retroviral genomes during mouse embryogenesis. *Nature* 1982;298:623–628.

Johnson MH, Handyside AH, Braude PR: Control mechanisms in early mammalian development, in Johnson MH (ed): *Development in Mammals*. Amsterdam, New York, Oxford, Elsevier/North-Holland Biomedical Press, 1977, vol 2, pp 67–97.

Kappler JW: The 5-methylcytosine content of DNA: tissue specificity. *J Cell Physiol* 1971;78:33–36.

Kelly F, Condamine H: Tumor viruses and early mouse embryos. *Biochim Biophys Acta* 1982;651:105–141.

Khoury G, Gruss P: Enhancer elements. 1983;*Cell* 33:3113–314.

Kintner C, Sugden G: Conservation and progressive methylation of Epstein-Barr viral DNA sequences in transformed cells. *J Virol* 1981;38:305–316.

Kratzer PG, Chapman VM, Lambert H, Evans RE, Liskay RM: Differences in the DNA of the inactive X chromosomes of fetal and extraembryonic tissues of mice. *Cell* 1983;33:37–42.

Kruczek I, Doerfler W: Expression of the chloramphenicol acetyltransferase gene in mammalian cells under the control of adenovirus type 12 promoters: effect of promotor methylation on gene expression. *Proc Natl Acad Sci (USA)* 1983;80:7586–7590.

Kuhlmann I, Doerfler W: Shifts in the extent and patterns of DNA methylation upon explanation and subcultivation of adenovirus type 12-induced hamster tumor cells. *Virology* 1982;118:169–180.

Kunnath L, Locker J: Characterization of DNA methylation in the rat. *Biochem Biophys Acta* 1982;699:264–271.

Lacy E, Roberts S, Evans EP, Burtenshaw D, Constantini FD: A foreign β-globin gene in transgenic mice: integration at abnormal chromosomal positions and expression in inappropriate tissues. *Cell* 1983;34:343–358.

Lerner RA, Wilson CB, Del Villano BC, McConahey PJ, Dixon FJ: Endogenous oncornaviral gene expression in adult and fetal mice: quantitative, histologic, and physiologic studies of the major viral glycoprotein, gp70. *J Exp Med* 1976;143:151–166.

Levine AJ: The nature of he host range restriction of SV40 and Polyoma viruses in embryonal carcinoma cells. *Curr Top Microbiol Immunol.* 1982;101:1–30.

Manes C, Menzel P: Demethylation of CpG sites in DNA of early rabbit trophoblast. *Nature* 1981;293:589–590.

Martin GR: Teratocarcinomas and mammalian embryogenesis. *Science* 1980; 209:768–776.

Mather EL, Perry RP: Methylation status and DNase I sensitivity of immunoglobulin genes: changes associated with rearrangement. *Proc Natl Acad Sci (USA)* 1983;80:4689–4693.

McGeady ML, Jhappan C, Ascione R, Vande Woude, GF: In vitro methylation of specific regions of the cloned Moloney sarcoma virus genome inhibits its transforming activity. *Mol Cell Biol* 1983;3:305–314.

McKeon C, Ohkubo H, Pastan I, de Crombrugghe B: Unusual methylation pattern of the $\alpha 2$(I) collagen gene. *Cell* 1982;29:203–210.

McKnight G, Hammer R, Kuenzel E, Brinster R: Expression of the chicken transferrin gene in transgenic mice. *Cell* 1983;34:335–341.

Niwa O, Sugahara T: 5-Azacytidine induction of mouse endogenous type C virus and suppression of DNA methylation. *Proc Natl Acad Sci (USA)* 1981;78:6290–6294.

Niwa O, Yokota Y, Ishida H, Sugahara T: Independent mechanisms involved in suppression of the Moloney leukemia virus genome during differentiation of murine teratocarcinoma cells. *Cell* 1983;32:1105–1113.

Nobis P, Jaenisch R: Passive immunotherapy prevents expression of endogenous Moloney virus and amplivication of proviral DNA in BALB/Mo mice. *Proc Natl Acad Sci (USA)* 1980;77:3677–3681.

Ostrander M, Vogel S, Silverstein S: Phenotypic switching in cells transformed with the herpes simplex virus thymidine kinase gene. *Mol Cell Biol* 1982;2:708–714.

Pages M, Roizes G: Tissue specificity and organization of CpG methylation in calf satellite DNA I. *Nucleic Acids Res* 1982;10:565–576.

Palmiter RD, Chen HY, Brinster RL: Differential regulation opf metallothionein-thymidine kinase fusion genes in transgenic mice and their offspring. *Cell* 1982;29:701–710.

Périès J, Alves-Cardoso E, Canivet M, Debons-Guillemin MC, Lasneret J: Lack of multiplication of ecotropic murine C-type viruses in mouse teratocarcinoma primitive cells: brief communication. *J Natl Cancer Inst.* 1977;59:463–465.

Pikó L, Clegg KB: Quantitative changes in total RNA, total poly(A), and ribosomes in early mouse embryos. *Dev Biol* 1982;89:362–378.

Pollock JM Jr, Swihart M, Taylor JH: Methylation of DNA in early development: 5-methyl cytosine content of DNA in sea urchin sperm and embryos. *Nucleic Acids Res* 1978;5:4855–4863.

Pollock Y, Stein R, Razin A, Cedar H: Methylation of foreign DNA sequences in eukaryotic cells. *Proc Natl Acad Sci (USA)* 1980;77:6463–6467.

Rae PMM, Steele RE: Absence of cytosine methylation at C-C-G-G and G-C-G-C sites in the rDNA coding regions and intervening sequences of Drosophila and the rDNA of other higher insects. *Nucleic Acids Res* 1979;6:2987–2995.

Razin A, Riggs AD: DNA methylation and gene function. 1980;210:604–610.

Riggs AD: X inactivation, differentiation, and DNA methylation. *Cytogenet Cell Genet* 1975;14:9–25

Santi DV, Garrett CE, Barr PJ: On the mechanism of inhibition of DNA-cytosine methyltransferases by cytosine analogs. *Cell* 1983;33:9–10.

Sawicki JA, Magnuson T, Epstein CJ: Evidence for expression of the paternal genome in the two-cell mouse embryo. *Nature* 1981;294:450–451.

Schibler U, Hagenbüchle O, Wellauer PK, Pittet AC: Two promoters of different strengths control the transcription of the mouse alpha-amylase gene *Amy*-1^a in the parotid gland and the liver. *Cell* 1983;33:501–508.

Schnieke A, Stuhlmann H, Harbers K, Chumakov I, Jaenisch R: Endogenous Moloney leukemia virus in nonviremic Mov substrains of mice carries defects in the proviral genome. *J Virol* 1983; 45:505–513.

Sherman MI: Developmental biochemistry of preimplantation mammalian embryos. *Ann Rev Biochem* 1979;48:443–470.

Shmookler Reis RJ, Goldstein S: Variability of DNA methylation patterns during serial passage of human diploid fibroblasts. *Proc Natl Acad Sci (USA)* 1982;79:3949–3953.

Simon D, Grunert F, v Acken U, Doring HP, Kröger H: DNA-methylase from regenerating rat liver: purification and characterization. *Nucleic Acids Res.* 1978;5:2153–2167.

Simon D, Stuhlmann H, Jähner D, Wagner H, Werner E, Jaenisch R: Retrovirus genomes methylated by mammalian but not bacterial methylase are non-infectious. *Nature* 1983; 304:275–277.

Singer J, Roberts-Ems J, Luthardt FW, Riggs AO: Methylation of DNA in mouse early embryos, teratocarcinoma cells and adult tissues of mouse and rabbit. *Nucleic Acids Res* 1979; 7:2369–2385.

Speers WC, Gautsch JW, Dixon FJ: Silent infection of murine embryonal carcinoma cells by Moloney murine leukemia virus. *Virology* 1980;105:241–244.

Stein R, Gruenbaum Y, Pollack Y, Razin A, Cedar H: Clonal inheritance of the pattern of DNA methylation in mouse cells. *Proc Natl Acad Sci (USA)* 1982a;79:61–65.

Stein R, Razin A, Cedar H: *In vitro* methylation of the hamster adenine phosphoribosyltransferase gene inhibits its expression in mouse L cells. *Proc Natl Acad Sci. (USA)* 1982b;79:3418–3422.

Stewart C, Harbers K, Jähner D, Jaenisch R: X-chromosome linked transmission and expression of retroviral genomes microinjected into mouse zygotes. *Science* 1983a;22:760–762.

Stewart C, Jähner D, Stuhlmann H, Jaenisch R: Retroviruses as probes for studying gene epxression in mouse embryogenesis, in *Conferences on Cell Proliferation: Teratocarcinoma Stem Cells.* Cold Spring Harbor Laboratory, Cold Spring Harbor, New York, vol 10: (1983b) 379–385.

Stewart CL, Stuhlmann H, Jähner D, Jaenisch R: *De novo* methylation, expression, and infectivity of retroviral genomes introduced into embryonal carcinoma cells. *Proc Natl Acad Sci (USA)* 1982;79:4098–4102.

Strand M, August T, Jaenisch R: Oncornavirus gene expression during embryonal development of the mouse. *Virology* 1977;76:886–890.

Stuhlmann H, Jähner D, Jaenisch R: Infectivity and methylation of retroviral genomes is correlated with expression in the animal. *Cell* 1981;26:221–232.

Sturm KS, Taylor JH: Distribution of 5-methyl-cytosine in the DNA of somatic and germline cells from bovine tissues. *Nucleic Acids Res* 1981;9:4537–4546.

Teich NM, Weiss RA, Martin GR, Lowy DR: Virus infection of murine teratocarcinoma stem cell lines. *Cell* 1977; 12:973–982.

van der Ploeg LHT, Flavell RA: DNA methylation in the human $\gamma\delta\beta$-globin locus in erythroid and nonerythroid tissues. *Cell* 1980;19:947–958.

van der Putten H, Quint W, Verma IM, Berns A: Moloney murine leukemia virus-induced tumors: recombinant proviruses in active chromatin regions. *Nucleic Acids Res* 1982;10:577–592.

Vanyushin BF, Tkacheva SG, Belozersky AN: Rare bases in animal DNA. *Nature* 1970;225:948–949.

Vanyushin BF, Mazin AL, Vasilyev VK, Belozersky AN: The content of 5-methylcytosine in animal DNA: the species and tissue specificity. *Biochim Biophys Acta* 1973;299:397–403.

Vardimon L, Kressmann A, Cedar H, Maechler M, Doerfler W: Expression of a cloned adenovirus gene is inhibited by *in vitro* methylation. *Proc Natl Acad Sci (USA)* 1982;79:1073–1077.

Waalwijk C, Flavell RA: DNA methylation at a CCGG sequence in the large intron of the rabbit β-globin gene: tissue-specific variations. *Nucleic Acids Res* 1978;5:4631–4641.

Wigler M, Levy D, Perucho M: The somatic replication of DNA methylation. *Cell* 1981;24:33–40.

Wigler MH: The inheritance of methylation patterns in vertebrates. *Cell* 1981; 24:285–286.

Wilson VL, Jones P: DNA methylation decreases in aging but not in immortal cells. *Science* 1983;220:1055–1057.

Wyatt GR: Recognition and estimation of 5-methyl-cytosine in nucleic acids. *Biochem J* 1951;48:581–590.

Yotsuyanagi Y, Szöllösi D: Early mouse embryo intracisternal particle: fourth type of retrovirus-like particle associated with the mouse. *J Natl Cancer Inst* 1981;67:677–685.

11

Specific Promoter Methylations Cause Gene Inactivation

Walter Doerfler*
Klaus-Dieter Langner*
Inge Kruczek†
Lily Vardimon‡
Doris Renz*

One of the important tasks of current research in molecular biology of eukaryotes is the elucidation of more complex coding signals, particularly of those controlling the expression of eukaryotic genes. Methylated bases (i.e., 5-methylcytosine [5-meCyt] residues) in eukaryotes constitute this type of conspicuous signal. As amply documented in this volume, the introduction of methylated nucleotides into a DNA sequence may serve different functions. In this concise review, our discussion will be restricted to the role of DNA methylation in the long-term shut-off of eukaryotic genes. It has been shown by different lines of investigations that DNA methylation at specific sites can cause gene inactivation (for reviews, see Razin and Riggs, 1980; Ehrlich and Wang, 1981; Doerfler, 1981; Razin and Friedman, 1981; Doerfler, 1983; Riggs and Jones, 1983; Doerfler, 1984a,b). It has also become apparent that the absence of DNA methylation is a necessary—but not sufficient—precondition for gene expression (van der Ploeg and Flavell, 1980; Kuhlmann and Doerfler, 1982; Ott *et al.* 1982). The biochemical mechanisms underlying the regulatory function of DNA methylation are still completely unknown. Obviously, methylated bases could modulate protein-DNA interactions, as has been clearly demonstrated for the restriction endonucleases in prokaryotes. Alternatively, methylated nucleotides could induce structural alterations of DNA (Behe and Felsenfeld, 1981; Behe *et al.* 1981).

In several recent reviews (Razin and Riggs, 1980; Drahovsky and Böhm,

*Institute of Genetics, University of Cologne, Cologne, West Germany.
†Institute of Biochemistry, Munich University, Munich, West Germany.
‡Developmental Biology Laboratory, Cummings Life Science Center, University of Chicago, Chicago, Illinois.

1980; Doerfler, 1981, 1983, 1984a,b; Ehrlich and Wang, 1981; Hattmann, 1981; Razin and Friedmann, 1981; Riggs and Jones, 1983), and in other chapters of this volume, basic facts on DNA methylation, its biochemistry, and its functional significance have been described; therefore, they will not be repeated. Previous reviews from this laboratory have dealt with functional aspects of DNA methylation (Doerfler, 1981, 1984a), the signal value of DNA methylation (Doerfler, 1983), and the possible role of DNA methylation in viral oncogenesis and persistence (Doerfler, 1984b).

In this chapter, we will briefly review the evidence supporting the concept that DNA methylation can inactivate eukaryotic genes; this inactivation probably is a long-term switch-off. In the main part of this review, the results of *in vitro* DNA methylation experiments will be discussed. Special reference will be made to viral systems that have served an important role in research on the function of DNA methylation. Integrated viral genes have many of the properties of eukaryotic genes—also with respect to DNA methylation. It has proved to be of particular interest that with some of the DNA viral genomes, one can study the state of methylation in both the integrated and the free intracellular form of viral DNA or DNA isolated from purified virions. Each of these forms of the viral genome is in a particular functional state, with respect to differential gene expression, and gene expression probably is controlled by quite different mechanisms.

In designing *in vitro* DNA methylation experiments by using specific cloned viral genes, we have been guided by the concept that methylation patterns at certain sites of these genes first have to be studied in transformed cells in culture—with the gene under investigation in both the active and inactive forms. A perfect inverse correlation was observed between the methylation of 5′–CCGG–3′ sites in the E2a region of integrated adenovirus type 2 DNA in transformed hamster cells and the expression of this region (Vardimon *et al.* 1980). Accordingly, the molecularly cloned E2a region of adenovirus type 2 DNA was *in vitro* methylated with the prokaryotic HpaII DNA-methyltransferase, and it was injected as either methylated or unmethylated plasmid into the nuclei of *Xenopus laevis* oocytes for activity determinations (Vardimon *et al.* 1982a). Without such precise correlations between *in vivo* and *in vitro* studies, it is doubtful that meaningful results can be obtained, since we do not know *a priori* the sequences responsible for functional modulations by DNA methylation.

DNA Methylation and Gene Activity: A Careful Evaluation of the Evidence

Hypotheses that correlate DNA methylation and gene activity in eukaryotes evolved both from observations of the differential states of DNA methylation in developing organisms and on the basis of theoretic considerations (Holliday and Pugh, 1975; Riggs, 1975; Sager and Kitchin, 1975). This field was fully developed after a series of experimental approaches was designed that yielded insights into the functional significance of DNA methylation. The main lines of evidence—extensively reviewed earlier (Razin and Riggs, 1980; Doerfler, 1981, 1983; Ehrlich and Wang, 1981)—are described in the following subsections.

Inverse Correlations

Inverse correlations between the extent of DNA methylation and the level of expression of a given gene have been established in different systems by analyzing a number of eukaryotic and viral genes. Since these correlations have been mainly restricted, thus far, to 5′–CCGG–3′ and 5′–GCGC–3′ sites, it could not be expected that they would apply to all the genes investigated without exceptions. Moreover, DNA methylation may not be the only decisive factor in the regulation of all genes. In general, it is risky; hence, it is inappropriate to extrapolate the findings from a given set of systems so that they will hold true for all other systems as well. Furthermore, studies are needed as to whether these inverse correlations indicate that DNA methylation is the cause or the consequence of gene inactivation.

Experimental evidence was provided by studies on DNA methylation of a certain gene either in different organs of one organism or in viral systems (Christman *et al.* 1977; Bird *et al.* 1979; Desrosiers *et al.* 1979; McGhee and Ginder, 1979; Mandel and Chambon, 1979; Sutter and Doerfler, 1979, 1980; Cohen, 1980; Guntaka *et al.* 1980; van der Ploeg and Flavell, 1980; Vardimon *et al.* 1980). In a certain gene that was actively expressed in one organ, many of the 5′–CCGG–3′ sites were found to be unmethylated. The DNA in the same gene in other organs not expressing this gene was completely methylated at the 5′–CCGG–3′ sites. In integrated viral genomes, unexpressed viral genes were hypermethylated, while actively transcribed genes showed low levels of methylation. The 5′–CCGG–3′ sites were preferentially investigated because of the availability of the isoschizomeric restriction endonuclease pair HpaII and MspI (Waalwijk and Flavell, 1978), which permitted determination of the state of methylation at all 5′–CCGG–3′ sites. In many genes, the state of methylation at these sites appears to have special significance. In this way, both inverse correlations between the extent of DNA methylation at 5′–CCGG–3′ sites in specific genes and the degree to which these genes are expressed have been originally established in several different systems (Desrosier *et al.* 1979; Mandel and Chambon, 1979; Sutter and Doerfler, 1979, 1980; Cohen, 1980; Guntaka *et al.* 1980; van der Ploeg and Flavell, 1980; Vardimon *et al.* 1980; Hynes *et al.* 1981; Weintraub *et al.* 1981; Kruczek and Doerfler, 1982; Kuhlmann and Doerfler, 1982; Smith *et al.* 1982; Jähner *et al.* 1983). In some cases, these inverse correlations have been extended to other sequences containing 5′–CpG–3′ dinucleotides (e.g., the sequence 5′–GCGC–3′, the recognition sequence of the restriction endonuclease HhaI [Kruczek and Doerfler, 1982; Kuhlmann and Doerfler, 1982]).

As expected, inverse correlations between gene activity and DNA methylation at 5′–CCGG–3′ and 5′–GCGC–3′ sites do not invariably exist. In some genes, DNA methylation at sites other than HpaII sites or HhaI sites may be important in regulation; other genes, which are not permanently inactivated, may not respond to that signal at all. Thus, unusual methylation patterns, with respect to expression, have been reported for vitellogenin genes A1 and A2 in hepatocytes (Gerber-Huber *et al.* 1983). These genes are heavily methylated irrespective of expression, while other *Xenopus laevis* genes show the usual inverse correlations. Perhaps estrogen inducibility of the vitellogenin genes could offer an explanation. Similarly, the DNA around the start site of transcripton of the α_2 (I) collagen gene in the chicken was not methylated, and the central and 3′ regions of the gene were methylated irrespective of transcription (McKeon *et al.* 1982). For the two rat insulin genes, no evidence has been adduced for a correlation between DNA methylation and gene expression (Cate *et al.*, 1983).

Studies on the Expression of *In Vitro*-Methylated Genes

Investigations directed toward the core of the problem used *in vitro*-methylated cloned genes that were microinjected into the nucleus of oocytes of *Xenopus laevis* (Vardimon *et al.* 1981; Fradin *et al.* 1982; Vardimon *et al.* 1982a,b, 1983) or into mammalian cells in culture (Stein *et al.* 1982; Waechter and Baserga, 1982; McGeady *et al.* 1983). Genes methylated at specific sites usually were not expressed in these experiments. The use of sequence-specific *de novo* DNA methyltransferases from prokaryotic organisms allowed the determination of highly specific sequences that somehow were involved in the control of gene expression. Other sequences (e.g., the sequence 5′–GGCC–3′), when methylated *in vitro,* did not exhibit any effect on gene expression (Doerfler *et al.* 1982; Vardimon *et al.* 1982b). These data, accumulated from work in different systems and in different laboratories, provided direct support for the notion of methylated sequences at highly specific sites playing an important part in the regulation of gene expression. These observations demonstrated that DNA methylation caused transcriptional inactivation (at least in the genes tested), and that it was not just a consequence of gene inactivation.

From our own contributions to this particular problem, it appeared to be important that work on the *in vitro* methylation of a certain viral gene was preceded by a thorough investigation of the state of methylation of the same gene in the integrated form in transformed cells. Since we had only two DNA methyltransferases (HpaII and HhaI) available for these studies, one had to ascertain that the 5′–CCGG–3′ and 5′–GCGC–3′ sites were actually involved in the regulation of this particular gene. Thus, *in vivo* studies proved to be an important precondition for the successful design of *in vitro* investigations. Moreover, it was essential to ascertain that *in vitro* DNA methylation of a gene was complete before its transcriptional activity was tested.

Inhibition of DNA Methyltransferases by 5-Azacytidine

Further evidence is derived from work with the cytidine analog 5-azacytidine (5-aza-C). This compound can be incorporated into replicating DNA and—due to its chemical structure—it cannot be methylated. However, the main inhibitory role of 5-azacytidine toward DNA methylation emanates from its ability to inhibit DNA-methyltransferases (Christman *et al.* 1980; Jones and Taylor, 1981; Creusot *et al.* 1982; Jones *et al.* 1982). In a number of different systems, it has been possible to activate previously dormant genes and to induce their expression (Taylor and Jones, 1979; Groudine *et al.* 1981; Conklin *et al.* 1982; Harris, 1982; Jones *et al.* 1982). The original report on the gene-activating property of 5-aza-C described the activation of a complex set of cellular functions leading to *in vitro* differentiation of mouse fibroblasts to twitching muscle cells (Constantinides *et al.* 1977).

By 5-aza-C treatment, specific genes could be turned on, even in animals (DeSimone *et al.* 1982) and humans (Ley *et al.* 1982). The inhibitory drug stimulated fetal hemoglobin synthesis in anemic baboons, and it selectively increased γ-globin synthesis in patients with β^+ thalassemia. Concomitant demethylations in the genes involved could also be shown (Ley *et al.* 1982). Since 5-aza-C is a highly toxic agent, treatment of humans certainly will be contraindicated in most cases.

It is, perhaps, not to be expected that all dormant cellular genes can be turned on by treatment of cells with 5-aza-C (Doerfler *et al.*, 1982). The absence of DNA methylation appears to be a necessary—but not sufficient—precondition for gene activation (van der Ploeg and Flavell, 1980; Kuhlmann and Doerfler, 1982; Ott *et al.* 1982). A crucial mechanism like gene activity may be subject to multifaceted regulatory mechanisms, with DNA methylation constituting one important parameter. Thus, depending on the stringency of inactivation for a given gene, 5-aza-C treatment may or may not lead to the activation of a certain gene or set of genes.

Interestingly, ultimate chemical carcinogens inhibit the transfer of methyl groups to hemimethylated DNA in *in vitro* studies that use DNA methyltransferase from mouse spleen. Perhaps, carcinogenic agents may lead to heritable changes in DNA methylation patterns by a variety of mechanisms (Wilson and Jones, 1983).

For a complete understanding of the role of DNA methylation in gene regulation, methods will be required to determine the state of methylation possibly at all the 5′–CpG–3′ sites in a gene and its adjacent sequences. Such a method has now become available (Church and Gilbert, 1984). Improved *in vitro* transcription systems also will be required to evaluate the effect of methylation in different 5′–CpG–3′ sites. Presently available *in vitro* transcription systems do not seem to respond to DNA methylation.

The results obtained by 5-aza-C induction are interesting, but they also are problematic to some degree in that a toxic analog could always have additional, poorly understood effects; thus, the interpretation of the induction phenomena may be more complicated than previously thought.

Manipulated Genes and *In Vitro* Methylation

It has been shown, in a number of different systems, that absence of DNA methylation at the 5′-ends and/or promoter sites of eukaryotic or viral genes correlates with gene expression (Flavell *et al.* 1982; Kruczek and Doerfler, 1982, 1983; Ott *et al.* 1982; Sanders Haigh *et al.* 1982; Wilks *et al.* 1982; Busslinger *et al.*, 1983). In one instance (Wilks *et al.* 1982), estrogen treatment of chickens caused demethylation of an HpaII site at the 5′-end of the vitellogenin gene. It appeared to be an attractive hypothesis that specific DNA methylations close to, and/or at, the promoter end of a gene might play a role in the switch-off of transcription. Of course, these results still left open the question of what site or sites precisely exerted the decisive effects.

The methodologic repertoire of recombinant DNA technology may help to attempt a direct approach to this problem. In our own laboratory, two sets of experiments have been initiated; these results will be discussed in detail in the overview that follows.

(1) The E2a gene of adenovirus type 2 DNA had been previously shown to be transcriptionally inactivated by HpaII methylation judged by microinjection into the nuclei of *Xenopus laevis* oocytes, whereas the unmethylated gene was transcribed (Vardimon *et al.* 1981, 1982a). The same gene methylated at 5′–GGCC–3′ sites was not inactivated (Vardimon *et al.* 1982b). We now have *in vitro* manipulated the gene and methylated the 5′–CCGG–3′ sites, either at the 5′-end or in the 3′-region of the gene (see Figure 11.2).

(2) The plasmid pSVO CAT carries both the prokaryotic chloramphenicol acetyltransferase (CAT) gene and a HindIII site just in front of that gene for the insertion of foreign promoter sequences (Gorman *et al.* 1982). Several adenovirus promoters were inserted at the HindIII site, and they proved to be active in facilitating expression of the CAT gene. Promoters that carried HpaII or HhaI sites upstream from the TATA signal could be inactivated by methylation at 5′–CCGG–3′ sites (Kruczek and Doerfler, 1983). Promoters with no HpaII sites (e.g., the early SV40 promoter in the plasmid pSV2 CAT), or with HpaII or HhaI sites downstream from the TATA signal, were insensitive to DNA methylation. These results support the concept that DNA methylation at specific sites (5′–CCGG–3′) in the promoter region, at least of adenovirus genes, causes transcriptional inactivation. Thus, for some genes, a causal relationship between DNA methylation and transcriptional inactivation has been established. It undoubtedly will be necessary to further refine these studies and to extend them to other genes as well before generalized conclusions can be drawn.

Specific DNA Methylations as Regulatory Signals

From many different lines of investigations, which were briefly reviewed in the preceding section, a general resumé can be formulated on the role of specific DNA methylations in gene regulation. DNA methylations appear to represent long-term signals in gene inactivation. Whenever a gene has to be permanently turned off, DNA methylation can be employed as a signal. Genes that are no longer active and are not required after certain stages in development have passed, perhaps, are kept inactive by methylations at specific sites. On the other hand, when certain genes will have to be activated occasionally, it would be inopportune to inactivate them by methylation. In virus-transformed cells, certain viral genes must not be activated, or else the inactive state of the viral genome may be reversed and the transformed cell possibly may be destroyed by viral replication. Thus, transformed cells will be selected in which the late viral genes involved in replication are incapacitated by DNA methylation. This reasoning may help to explain why some of the most convincing inverse correlations between the levels of DNA methylation and the extent of gene activity usually have been found in viral systems. Possibly, the methods employed to isolate stably transformed cells also select for those patterns of methylation that are advantageous for the maintenance of the transformed state.

Since high levels of DNA methylation are associated with the inactivation of the corresponding genes, and since the DNA in many tumor cells has been found to be strikingly undermethylated, it is not unreasonable to investigate whether cellular genes that are normally silent in tumor cells may have been reactivated. This, untimely reactivation of normally inactive genes may be causally related to the low levels of DNA methylation in tumor cells (Feinberg and Vogelstein, 1983). The reactivation of cellular genes also could be associated with the oncogenic phenotype of these cells. This apparently plausible way of correlating well-documented phenomena probably is still too simplistic, but it may help to design more meaningful experiments.

When reviewing inverse correlations between DNA methylation and gene activity in viral and nonviral eukaryotic genes, it becomes apparent that there are exceptions to this simple rule. First, the question will have to be asked whether the sequences investigated for the levels of methylation are always the relevant ones. For operational reasons, the sequence 5′–CCGG–3′, which can be cleaved with the HpaII/MspI isoschizomeric restriction endonuclease pair, has been most frequently studied. In the case of lacking correlations, it has to be ascertained that the sites investigated have functional significance.

More importantly, however, if one accepted the notion that specific DNA methylations were a signal for long-term gene inactivation, one would not expect genes to be inactivated by methylation that need to be occasionally reactivated or need to have a low level of activity. It is obvious that mechanisms other than DNA methylation must be intrinsically involved in gene inactivation (e.g., specific proteins binding to regulatory DNA sequences). DNA methylations, in fact, may modulate these protein-DNA interactions. For example, certain genes encoding essential functions occasionally may have to be reactivated and should not be shut off permanently. Another interesting example is offered by genes that can be activated by steroid hormones; they usually are not methylated. For other genes, we may not understand their temporal activity patterns well enough to predict the extent of DNA methylation. It is assumed that highly specialized genes remain active only in certain organs of an animal. However, can one be certain that these specific messages occasionally are not produced in other cells of the organism also? If such necessities arose, DNA methylation would not be the signal of choice to inactivate this set of genes. With this way of reasoning, a scheme can be developed that operationally subdivides genes into at least three classes:

1. Permanently inactivated genes
2. Inactive genes that occasionally are reactivated
3. Active genes

This scheme undoubtedly will have to be refined as an improved understanding of differential gene activities emerges. In this simple scheme, DNA methylation would be expected to be a signal of inactivation predominantly in class I genes.

There is increasing evidence that DNA methylations at/or close to the promoter region of a gene exert a function in gene inactivation. For some genes, DNA methylations upstream from the TATA signal have proved to be essential, while downstream methylations have not affected gene activity. Therefore, correlations between DNA methylations and the inactive state of genes will have to be sought predominantly at the promoter regions of a gene.

The possibility exists that each 5-meCyt residue in DNA may have its own functional significance that is not necessarily related only to gene activity. Thus, cell- or tissue-specific patterns of DNA methylation may be superimposed on DNA methylation patterns, which reflect the inactivated states of genes. Finally, once a gene has been permanently inactivated by specific promoter methylations, additional methylations of that gene may be related either to the organization of chromatin in that part of the genome or to the presence of DNA methyltransferases—or yet to other factors not directly responsible for the regulation of gene expression.

Studies on DNA Methylations in the Adenovirus System

Our own results in this system have recently been reviewed (Doerfler, 1981, 1983, 1984a). The most important findings shall therefore be restated briefly. Virion DNA that is isolated from purified virus particles is not detectably methylated (Günthert *et al.* 1976; Sutter *et al.* 1978; von Acken *et al.* 1979; Eick *et al.* 1983b). Trace amounts of 5-meCyt occasionally spotted in virion DNA preparations probably are due to contaminations of virion preparations with host cell DNA that contains 3.6% (KB DNA) to 4.4% (HEK DNA) of 5-meCyt (Günthert *et al.* 1976). It is still an interesting problem as to how adenovirus DNA avoids becoming methylated when replicating in the nuclei of host cells whose DNA is markedly modified. The highly efficient and rapid rate of viral DNA replication in productively infected cells, on the one hand, and the tight association of DNA methyltransferases with cellular DNA, on the other hand, may be some of the factors explaining the lack of virion DNA methylation. It is also interesting to note that the cellular DNA covalently linked to adenovirus type 12 DNA termini in symmetric recombinants (SYREC) between viral and host DNA is not methylated. The same cellular sequences, however, that still reside in the host genome are extensively methylated. The SYREC DNAs presumably originated from a viral-host DNA recombination event and are packaged into virions (Deuring *et al.* 1981; Deuring and Doerfler, 1983).

Free intracellular adenovirus DNA, both in productively infected human cells and in hamster cells abortively infected with adenovirus type 12, is not methylated either (Vardimon *et al.* 1980). In productively infected human cells, both the parental viral DNA present in the cells early after infection and the newly synthesized viral DNA accumulating late in the infection cycle are not methylated at the 5′–CCGG–3′ and the 5′–GCGC–3′ sites. The intracellular adenovirus type 2 DNA was analyzed by using the restriction endonucleases HpaII, MspI, and HhaI, and by using small restriction fragments of adenovirus type 2 DNA as hybridization probes for Southern (1975) blots of total intracellular DNA from infected cells. Similarly, by using the restriction endonucleases TaI (5′-TCGA-3′) and DpnI (5′-GATC-3′), which are N6-methyl Ade-sensitive, adenine methylations at these sites could be ruled out (Wienhues and Doerfler, manuscript in preparation). The results were correlated to the nucleotide sequence of adenovirus type 2 DNA (R. J. Roberts, U. Pettersson, F. Galibert, and J. Sussenbach, personal communication). Thus, there is no evidence that DNA methylations would play a role either in the regulation of late viral genome functions in productive infection or in abortive infection. Since the removal of methylated bases from any genome probably requires at least two replication cycles plus the concomitant inhibition of maintenance DNA methyltransferases, it would be inopportune for a viral genome to succumb to that switch-off mechanism of the host cell. It is still an open question as to whether active enzymatic demethylation (Gjerset and Martin, 1982) can occur.

Adenovirus DNA previously unmethylated was found to be extensively methylated at 5′–CCGG–3′ and 5′–GCGC–3′ sites on insertion into the genome of adenovirus-transformed cells (Sutter *et al.* 1978; Sutter and Doerfler, 1979, 1980; Vardimon *et al.* 1980; Kruczek and Doerfler, 1982). Thus, the viral DNA had to be *de novo* methylated at some time after integration into the host genome. The analysis of methylation patterns of adenovirus type 12 DNA in virus-induced tumors revealed that, in tumor cell DNA, the integrated viral DNA sequences were still undermethylated. On explantation of tumor cells into culture, the extent of methylation of viral DNA increased gradually with continued passage of the tumor cells (Kuhlmann and Doerfler, 1982). This increase was not random, but followed a certain pattern that eventually led to the hypermethylation of the integrated late viral genes (Kuhlmann and Doerfler, 1983). Similar observations were made when the cells were subcloned early after explantation from the tumor. At present, it is not understood which factors regulate the establishment of organized patterns of methylation of the viral genome after insertion into the host genome. Concomitant with—but perhaps only temporally related to—the increase in viral DNA methylation, the viral genome had a tendency to be excised again and to be lost from the host cells during early passages after explantation (Kuhlmann and Doerfler, 1983).

Inverse correlations between the extent of DNA methylation and the expression of integrated adenovirus genes in adenovirus-transformed cells have been described (Sutter and Doerfler, 1979, 1980; Vardimon *et al.* 1980; Kruczek and Doerfler, 1982). Inactive genes are hypermethylated; actively expressed genes are undermethylated. Similar inverse correlations have been described in many other viral and nonviral eukaryotic systems (for reviews, see Razin and Riggs, 1980; Doerfler, 1981, 1983).

T637 hamster cells are an adenovirus type 12-transformed cell line containing about 22 copies of adenovirus type 12 DNA per diploid genome (Sutter *et al.* 1978; Stabel *et al.* 1980). Morphologic revertants of this cell line have been isolated and characterized (Groneberg *et al.* 1978; Groneberg and Doerfler, 1979; Eick *et al.* 1980; Eick and Doerfler, 1982; Eick *et al.* 1983a). Some of these revertants contain only one remaining copy of viral DNA or fragments thereof (Eick *et al.* 1980). In comparison to the levels of methylation of viral DNA sequences in cell line T637, DNA methylation is markedly increased for viral DNA in the revertants; also, the extent of expression is strikingly diminished (Eick *et al.* 1980; Schirm and Doerfler, 1981).

Methylations at the Promoter and 5′-ends of Genes Affect Gene Expression

The inverse correlation established, in many different eukaryotic systems, between the extents of DNA methylation at 5′–CCGG–3′ sites and gene expression may not pertain to DNA methylation along the entire stretch of a gene or a group of genes. Evidence has been deduced from studies on adenovirus type 12-transformed hamster cells that 5-meCyt positioned at the 5′–CCGG–3′ and 5′–GCGC–3′ sites of the 5′-termini of viral genes, and in the promoter/leader sequences, correlate to the shut-off of integrated viral genes (Kruczek and Doerfler, 1982). The data summarized in Figure 11.1 demonstrate that some of the 5′–CCGG–3′ (MspI) sites at the 3′-termini are methylated, either in the E1 regions or in the E2a regions of integrated adenovirus type 12 genomes in cell lines T637, HA12/7, and A2497-3; although, these regions are expressed in all three cell lines (Ortin *et al.* 1976; Schirm and Doerfler, 1981; Esche and Siegmann, 1982). In contrast, in cell line HA12/7, all of the MspI sites at the 5′-terminus of the E3 region are methylated (Kruczek and Doerfler, 1982); consequently, this region fails to be expressed (Schirm and Doerfler, 1981; Esche and Siegmann, 1982; Kruczek and Doerfler, 1982).

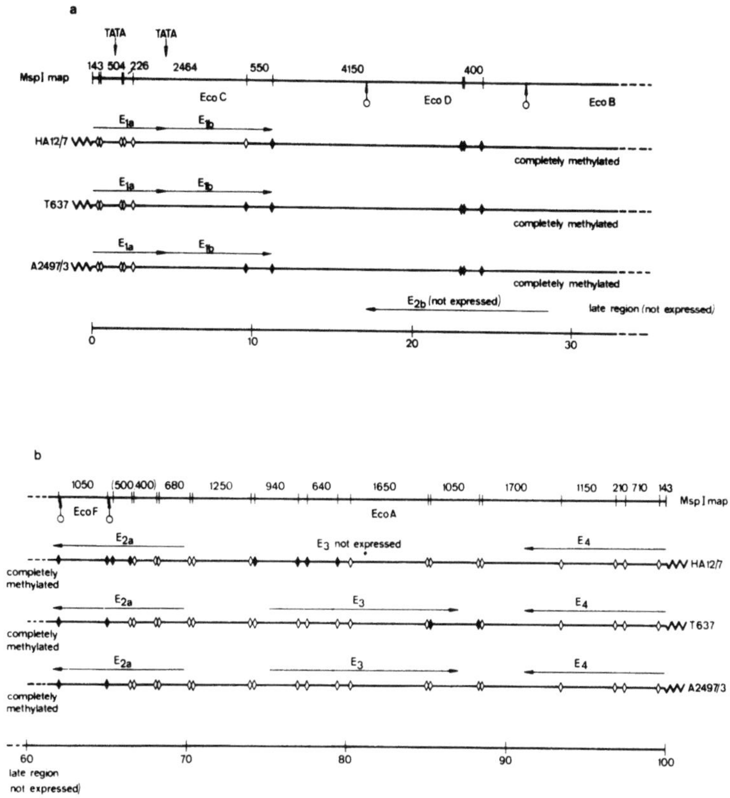

Figure 11.1 Functional maps of the left (a) and the right (b) halves of the Ad12 DNA molecules integrated in cell lines HA12/7, T637, and A2497-3. The horizontal lines represent the Ad12 genomes (————) integrated into the genomes of cell lines (∧∧∧) as indicated. The MspI maps of the left (a) and right (b) ends of Ad12 DNA are presented in the top line. Vertical bars on and figures above the horizontal lines indicate the locations of the MspI sites and the sizes of some of the MspI fragments, respectively. The arrows (♁) designate EcoRI sites. TATA marks the locations of presumptive Goldberg-Hogness signals in the E1a regions (Sugisaki *et al.* 1980; Bos *et al.* 1981). The unmethylated (◇) and methylated (◆) 5′-CCGG-3′ sites in the integrated Ad12 genomes in cell lines HA12/7, T637, and A2497-3 are designated by open or closed symbols as indicated. The horizontal arrows indicate map positions and direction of transcription of the individual early regions in each of the cell lines. The presence of an arrow indicates expression of corresponding regions of the Ad12 genome. MspI sites to the right of the EcoRI-D/B and to the left of the EcoRI-A/F junctions have not been mapped. The early regions of Ad12 DNA (Ortin *et al.* 1976; Esche and Siegmann, 1982) and a fractional length scale have also been indicated. The map of the E1 region of Ad12 DNA is as described by Bos *et al.* (1981). (From Kruczek and Doerfler, 1982.)

Similar correlations have been observed for the 5′–GCGC–3′ (HhaI) sites. There is considerable evidence that demonstrates that DNA methylations at the 5′-termini represent the decisive shut-off signals. This evidence is derived from several eukaryotic genes, the adenine phosphoribosyl-transferase and dihydrofolate reductase genes (Stein *et al.* 1983), the globin gene (Flavell *et al.*, 1982; Busslinger *et al.* 1983), the rDNA in *Xenopus laevis* (La Volpe *et al.* 1982), the albumin gene in rat hepatoma cells (Ott *et al.* 1982), the chicken α-globin gene (Sanders Haigh *et al.* 1982), and the vitellogenin gene in chickens (Wilks *et al.* 1982). The biochemical mechanisms responding to 5-meCyt in specific positions of eukaryotic DNA in the control of gene expression are unknown.

The E2a Region of Adenovirus Type 2 DNA Shows Consistent Inverse Correlations Between DNA Methylations at 5′–CCGG–3′ Sites and Gene Activity

The adenovirus type 2-transformed hamster cell lines HE1, HE2, and HE3 (Johansson *et al.* 1978; Cook and Lewis, 1979) have been used to study the function of 5-meCyt at specific sites. It has been demonstrated, both by direct immunoprecipitation on SDS-polyacrylamide gels with cell extracts (Johansson *et al.* 1978) and by *in vitro* translation of hybrid-selected messenger RNAs (Esche, 1982), that cell lines HE2 and HE3 do not express the adenovirus type 2-specific, DNA-binding protein; whereas, cell line HE1 does express this 72 Kd protein. The DNA-binding protein of adenovirus type 2 (van der Vliet and Levine, 1973) is encoded in the E2a region of the viral genome. All three hamster cell lines contain the intact E2a region in the integrated adenovirus type 2 genomes (Vardimon and Doerfler, 1981). We have been able to demonstrate: 1) that in addition to the intact E2a regions, each cell line contains the complete late promoter of the E2a region (Vardimon *et al.* 1982b); and 2) that in cell line HE1, in which the DNA-binding protein is synthesized, this late promoter is being used (Vardimon *et al.* 1983). The E2a region is controlled by a complex array of promoter sequences. Two promoters are used predominantly early, and one region is used late, in the productive infection cycle (Chow *et al.* 1979; Baker and Ziff, 1981). Thus, the failure of the DNA-binding protein to be expressed in cell lines HE2 and HE3 cannot be due simply to a defective gene or to deletions in the promoter region of the E2a segment. In cell lines HE2 and HE3, all 14 5′–CCGG–3′ sites (Figure 11.2) are methylated; in cell line HE1, all of the same sites are unmethylated; as determined by restriction enzyme analysis using HpaII and MspI endonucleases (Vardimon *et al.* 1980). Thus, with the E2a region of adenovirus type 2 DNA in three transformed hamster cell lines, we have established an example of a striking inverse correlation between DNA methylation at 5′–CCGG–3′ sites distributed over the

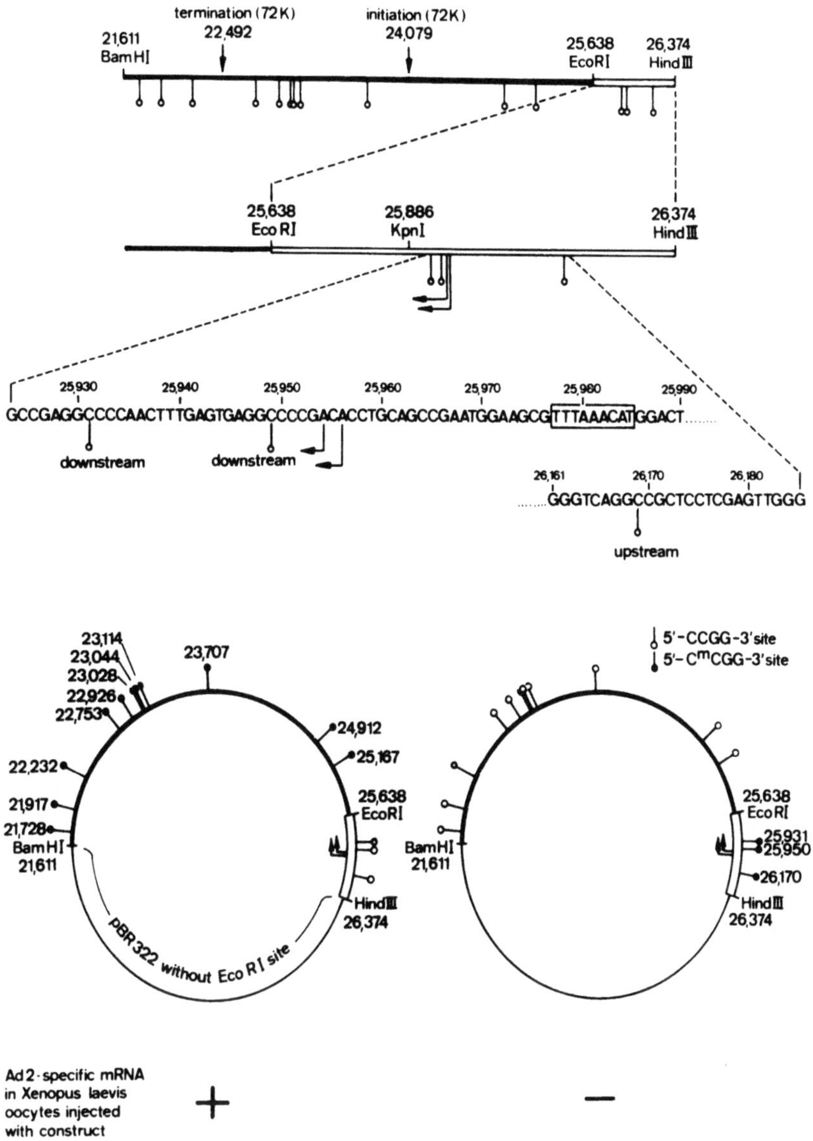

Figure 11.2 Selective methylations of 5′-CCGG-3′ sequences in the promoter or in the body of the E2a region of adenovirus type 2 DNA. The E2a region of adenovirus type 2 DNA was cloned as a BamHI-HindIII fragment. The 3′ located BamHI-EcoRI or the 5′ located EcoRI-HindIII fragment was methylated with the HpaII DNA methyltransferase at the 5′–CCGG–3′ sequences. The methylated 3′ or 5′ segment was religated with the unmethylated 5′ or 3′ fragment, respectively. Either construct was subsequently microinjected into nuclei of *Xenopus laevis* oocytes, and adenovirus type 2 E2a-specific RNA was searched by blot hybridization experiments, as described in the text. The unmethylated (⚲) and the methylated (⚲) 5′–CCGG–3′ (HpaII) sequences have been indicated. The numbers refer to the nucleotide numbers in the adenovirus type 2 sequence. Experimental details have been described elsewhere. (From Langner *et al.* 1984.)

entire length of the gene and the expression of the DNA-binding protein. It is worth noting that the 5′-GGCC-3′ (HaeIII) sites in the E2a region are not methylated (Vardimon *et al.* 1982b). It appears that the HaeIII site does not exert any effect on gene expression via DNA methylation.

The correlation between DNA methylation of specific sequences and shut-off of these genes could indicate a causal relationship between DNA methylation and gene expression; or, DNA methylation could be a consequence of the absence of gene expression. To determine more precisely, and in a more direct way, the role that DNA methylation can play in the control of gene expression, we have developed systems in which the effects of *in vitro* methylation at specific sites of certain viral segments on gene expression can be directly investigated.

In Vitro Methylation of Cloned Adenovirus Genes at Specific Sites Causes Transcriptional Inactivation

The cloned E2a region of adenovirus type 2 DNA was *in vitro* methylated by using the HpaII DNA methyltransferase isolated from *Haemophilus parainfluenzae* (Mann and Smith, 1977). This enzyme methylates the 3′-C residues in the 5′-CCGG-3′ sequences. This prokaryotic DNA methyltransferase was chosen for *in vitro* methylation experiments, because within the adenovirus type 2-transformed hamster cell lines HE2 and HE3, the E2a region in the integrated Ad2 genomes was methylated at the 14 5′-CCGG-3′ sites, and because the E2a region was not expressed. Conversely, all 14 5′-CCGG-3′ sites of the E2a region were unmodified in cell line HE1, in which the E2a region was expressed. When the *in vitro*-methylated E2a genes of adenovirus type 2 DNA were microinjected into nuclei of *Xenopus laevis* oocytes (Telford *et al.* 1979; Grosschedl and Birnstiel, 1980), the microinjected DNA remained methylated for at least 24 hours; also, its transcription into mRNA was completely blocked (Vardimon *et al.* 1981, 1982a). In contrast, unmethylated DNA that was independently microinjected into nuclei of *Xenopus laevis* oocytes was readily expressed as adenovirus type 2-specific RNA. It could be shown that the late E2a promoter of unmethylated DNA was used in *Xenopus* oocytes, although nonspecific starts of transcription were also observed. Viral RNA was not detectably initiated on the methylated E2a fragment. Unmethylated sea urchin histone genes coinjected with methylated adenovirus type 2 E2a DNA continued to be expressed in *Xenopus* oocytes (Vardimon *et al.* 1982a). Thus, the transcriptional block of the E2a region of adenovirus type 2 DNA was not artefactual, but it was causally related to methylation at the 5′-CCGG-3′ sites. We also methylated the 5′-GGCC-3′ sites in the E2a region of adenovirus type 2 DNA (i.e., the inverse sequence) by using the BsuRI methyltransferase from *Bacillus subtilis* (Günthert *et al.* 1981). After microinjection into nuclei of *Xenopus laevis*, the BsuRI-methylated E2a region continued to be expressed (Vardimon *et al.* 1982b). In cell lines HE1, HE2, and HE3, the 5′–GGCC–3′ sites in the E2a regions of integrated adenovirus type 2 DNA were not methylated. Thus, we concluded that DNA methylations had to occur at highly specific sequences to be functionally relevant. Furthermore, DNA methylations at specific sites (5′–CCGG–3′), in some way, caused gene inactivation. Similar conclusions were reached by researchers who *in vitro*-methylated SV40 DNA (Fradin *et al.* 1982), the hamster adenine phosphoribosyltransferase gene (Stein *et al.* 1982), the cloned thymidine kinase gene of herpes simplex virus (Waechter and Baserga, 1982), or specific regions of the cloned Moloney sarcoma genome (McGeady *et al.*, 1983); they also observed transcriptional inactivation of these genes. Peculiarly, in the case of the thymidine kinase gene from *Herpes simplex* virus, methylation of adenine at the EcoRI site led to gene inactivation on microinjection into hamster cells (Waechter and Baserga, 1982). In SV40 DNA, there is only one HpaII site close to the late promoter of this genome. When this site was methylated, the late genes were inactivated, and the early viral genes continued to be expressed on microinjection into *Xenopus laevis* oocytes (Fradin *et al.* 1982). It cannot be decided, at present, whether the functionally relevant sites for DNA methylation will be the same ones for all genes. It is conceivable that some genes may respond to methylation at different sites. Moreover, it will have to be determined which parts of a gene or of its regulatory sequences have to be methylated for gene inactivation to ensue (see the following section).

Methylation of Specific Sequences in the Promoter Parts of Viral Genes Leads to Transcriptional Inactivation

The question arose as to whether *in vitro* methylation of specific sequences in the promoter part of genes would lead to transcriptional inactivation. In this section, three different experimental approaches will be described.

(1) The promoter part of the cloned E2a segment of adenovirus type 2 DNA was selectively methylated and rejoined to the main portion of the gene; the construct was then microinjected into nuclei of *Xenopus laevis* oocytes to test transcriptional activity (Langner *et al.* 1984).

(2) The isolated promoters of two adenovirus type 12 genes were ligated into the plasmid pSVO CAT developed by Gorman *et al.* (1982). A viral promoter was inserted into the HindIII site just in front of the CAT gene. The activity of the construct was assayed by a chromatographic procedure in cell extracts after transfection of mouse Ltk$^-$ cells. Constructions that use the prokaryotic CAT gene have proved to be very useful in testing various eukaryotic promoters for their activity. The CAT constructs containing adenovirus promoters were specifically methylated and transfected; and, the biological activities of the methylated and the unmethylated constructs were compared.

(3) Promoter methylations of the γ-globin gene lead to inhibition of γ-globin gene expression (Busslinger *et al.* 1983).

In these *in vitro* methylation experiments involving also viral promoters, the DNA methyltransferases were used from *H. parainfluenzae* (HpaII) specific for the 3′-C residue in the sequence 5′–CCGG–3′ and from *H. haemolyticus* (HhaI) specific for the internal C in the sequence 5′–GCGC–3′. Again, these sequences were deliberately chosen for *in vitro* methylation, because inverse correlations between promoter methylations and adenoviral gene activities had been established for precisely these sequences (Kruczek and Doerfler, 1982).

Promoter Methylations in the E2a Region of Adenovirus Type 2 DNA

By using the technologic repertoire of genetic engineering, we have *in vitro* methylated the cloned E2a region of adenovirus type 2 DNA, so that there was modification of either the 5′-terminal EcoRI-HindIII fragment carrying the late promoter of the E2a region with three 5′–CCGG–3′ (HpaII) sites or the 3′-terminal EcoRI-BamHI fragment carrying the main part of the gene with 11 5′–CCGG–3′ sites. Both the anatomy of the cloned E2a segment and the locations of the HpaII sites are indicated in Figure 11.2. The isolated methylated fragments were religated to the unmodified 3′-terminal or 5′-terminal fragment, respectively, and were circularized with the plasmid DNA. Either construct combination was subsequently microinjected into nuclei of *Xenopus laevis* oocytes; the expression of the adenovirus type 2 E2a gene was monitored by Northern blot analyses (Alwine *et al.* 1977) of the total RNA extracted 24 hours after microinjection. Adenovirus type 2 DNA was [^{32}P]-labeled by nick translation (Rigby *et al.* 1977) and was used as a hybridization probe. The results have demonstrated that methylations of the three HpaII sites in the promoter and 5′ region eliminate transcription of the viral DNA fragment; whereas, methylations at the 11 HpaII sites in the main part of the gene appear not to affect transcriptional activity of the construct (Langner *et al.* 1984). It was verified by S1 mapping of the transcripts that the Ad2-specific RNA synthesized from the construct, which carried 11 methylated HpaII sites in the main part of the gene, was initiated at the same sites as in Ad2-infected human KB cells (Langner *et al.* 1984). In control experiments, the same contructs were designed in which the 5′- or 3′-terminal fragments were mock methylated in an identical reaction that lacked the methyl donor S-adenosylmethionine. Both mock-methylated constructs were fully active after microinjection into nuclei of *Xenopus laevis* oocytes (Langner *et al.* 1984). These results demonstrate that the three 5′–CCGG–3′ sites that are decisive in the shut-off of the E2a region by methylation reside in the promoter and 5′ region of the gene. It is not yet clear to what extent the other 11 such sites in the bulk of the gene have a modulating or an (as yet) unknown effect on gene expression. The fact that there is such a perfect inverse correlation between the apparent methylation of all 14 5′–CCGG–3′ sites and gene expression in the E2a region of adenovirus type 2 DNA, in the transformed hamster lines HE1, HE2, and

HE3, raises the question of what biological significance methylation may indeed have at each individual 5′–CCGG–3′ site. Further detailed work will be required to resolve this problem.

Promoter Methylations at 5′-CCGG-3′ or 5′-GCGC-3′ Sites Upstream from the TATA Signal Lead to Transcriptional Inactivation

The effect of DNA methylations, at specific promoter sites, on gene expression was tested by using a sensitive and quantitative assay system (Kruczek and Doerfler, 1983). The plasmid pSVO CAT (Gorman *et al.* 1982) contains both the prokaryotic CAT gene and a HindIII site in front of it for experimental promoter insertion. On insertion into pSVO CAT, the E1a and protein IX gene promoters from adenovirus type 12 DNA (Figure 11.3) were capable of mediating CAT expression on transfection into mouse Ltk^- cells. The CAT-promoting activity of the early simian virus 40 (SV40) promoter in plasmid pSV2 CAT (Gorman *et al.* 1982) is refractory to methylation by the HpaII or HhaI DNA methyltransferase at 5′–CCGG–3′ or 5′–GCGC–3′ sequences, respectively, because this promoter lacks such sites. The CAT sequence of this plasmid carries four HpaII and no HhaI sites. Methylations of the HpaII sites in the coding region of either the CAT gene or the plasmid do not affect expression. This experiment served as a control for establishing that methylations at these sites had no inhibitory function on CAT expression. The E1a promoter of Ad12 DNA, comprising the left-most 525 base pairs of the viral genome, carries two 5′–CCGG–3′ and three 5′–GCGC–3′ sites upstream from the left-most TATA signal. The 5′–CCGG–3′ sequence closer to the TATA signal is part of one of the enhancer signals described in the E1a promoter region of adenovirus type 12 DNA (Hearing and Shenk, 1983) (Figure 11.3). Methylation of the HpaII or HhaI sites incapacitates the E1a promoter (Figure 11.4). The promoter of protein IX gene of Ad12 DNA (Figure 11.3) contains one 5′–CCGG–3′ and one 5′–GCGC–3′ site downstream and two 5′–GCGC–3′ sites $>$ 300 base pairs upstream from the TATA motif and probably outside the promoter. The protein IX promoter is not inactivated by methylation of these sites (Kruczek and Doerfler, 1983).

The experimental findings that have been described demonstrate that critical methylations of specific sites in the promoter of certain viral genes can affect gene expression. It appears that 5′–CCGG–3′ and 5′–GCGC–3′ sites upstream from the TATA signal of the E1a promoter of Ad12 DNA have decisive regulatory functions. Methylations of sites $>$ 300 base pairs upstream from the TATA signal have no effect. Methylations of the four HpaII sites either in the sequence of the CAT gene or in the plasmid sequence in pSV2 CAT, for example, had no effect on CAT expression. Of course, the possibility cannot be excluded that specific DNA methylations at 5′–CpG–3′ sites other than HpaII or HhaI sites in either the body of the CAT gene or the E1a gene may have an effect on gene expression.

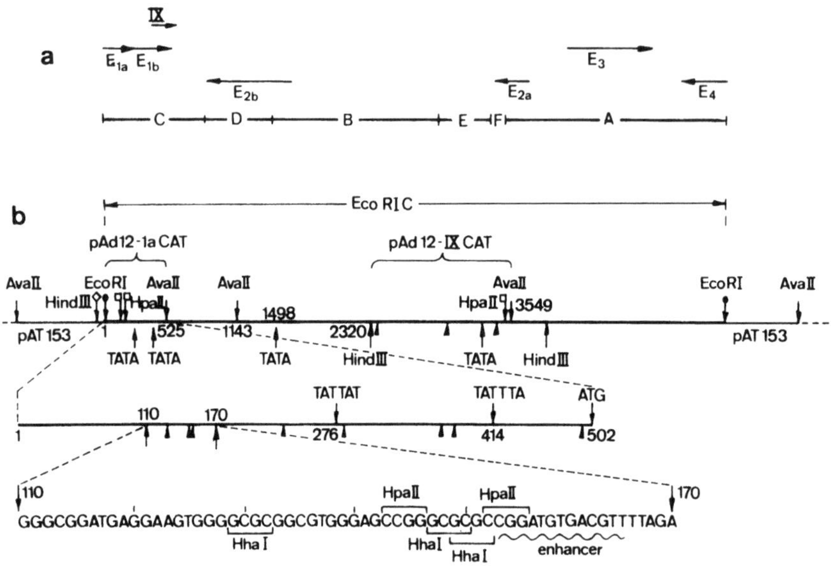

Figure 11.3 Map of the adenovirus type 12 genome and functional anatomy of its left-most EcoRI-C fragment. (a) Functional map of Ad12 DNA (Kruczek and Doerfler, 1982). E1a, E1b, E2a, E3, and E4 designate the early regions of the viral genome; IX designates the location of the protein IX gene of Ad12. Letters refer to EcoRI DNA fragments on the restriction maps. (b) Detailed map of the left terminal EcoRI-C fragment with restriction sites as indicated. Numbers refer to base pairs (bp) as counted from the left terminus of viral DNA. TATA, TATTAT, and TATTTA designate the presumptive Goldberg-Hogness boxes. The nucleotide sequence between bp 110 and bp 170 (Sugisaki *et al.* 1980; Bos *et al.* 1981) is also included. HpaII (5′-CCGG-3′) and HhaI (▲ 5′-GCGC-3′) sites are indicated. The symbol ∿∿∿ designates one of the E1a enhancers of adenovirus type 12 DNA (Hearing and Shenk, 1983). (From Kruczek and Doerfler, 1983, with minor modifications.)

It is interesting to mention that in the E1a promoter region used in the pAd12-1a CAT construct, the pentanucleotide GGGCG occurred six times. Either the same motif or the complementary motif also was found to be clustered several times in the 21-base pair repeats of SV40 DNA (Everett *et al.* 1983), in the myc oncogene (Colby *et al.* 1983), in the immediate early promoter of *Herpes simplex* virus type 1 (Cordingley *et al.* 1983), in the promoter region of the thymidine kinase gene (McKnight, 1982)—as well as in hamster and human DNA sequences linked to Ad12 DNA in the Ad12-induced tumor line CLAC1 (Stabel and Doerfler, 1982)—and in the symmetric recombinant (SYREC 2) between Ad12 and human DNAs, respectively (Deuring and Doerfler 1983). Lastly, it should be recalled that the E1a region of adenoviruses has a regulatory function in the expression of many other adenoviral genes (Jones and Shenk, 1979).

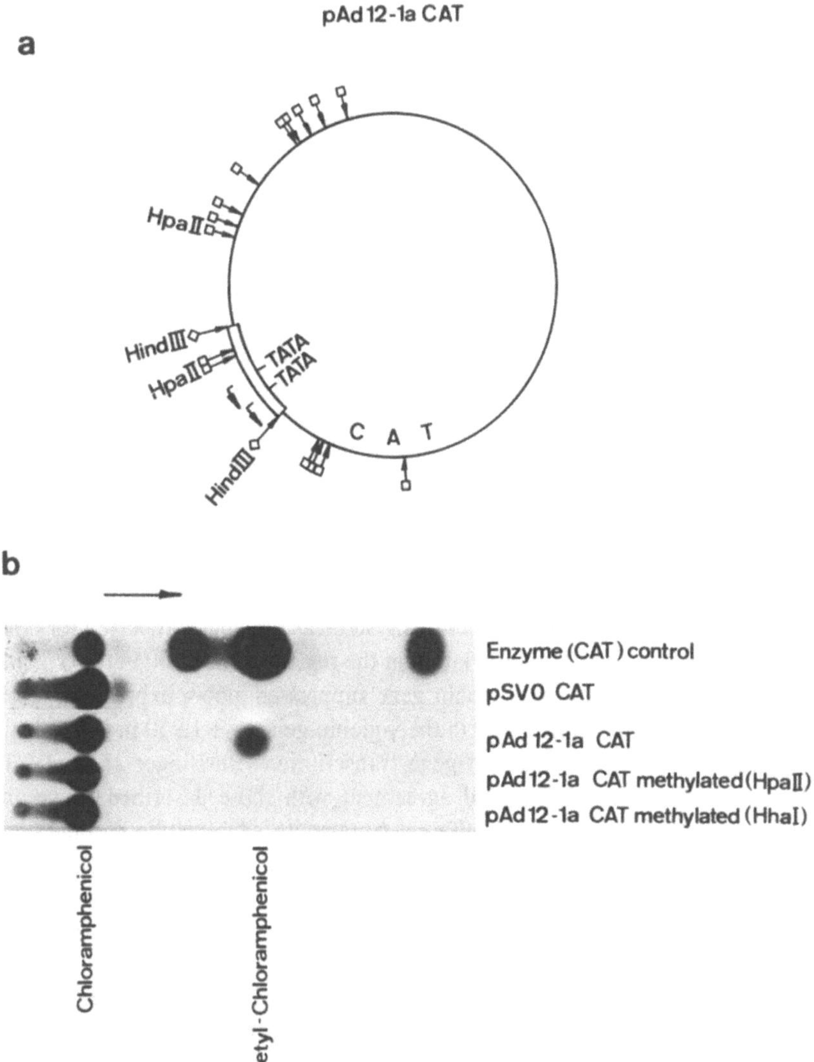

Figure 11.4 The E1a promoter of Ad12 DNA mediates CAT expression and is inhibited by methylation. (a) Map of the pAd12-1a CAT construct. ⫯ symbols designate 5′–CCGG–3′ (HpaII) sites; → designates the approximate TATA positions. (b) Experimental design and details were described elsewhere (Kruczek and Doerfler, 1983) and in the text. The pAd12-1a CAT preparation was *in vitro* methylated by the HpaII or HhaI DNA methyltransferase. Conversion of chloramphenicol to acetylated forms of chloramphenicol by extracts of transfected cells as indicated was measured by autoradiography of chromatography sheets. (From Kruczek and Doerfler, 1983.)

Specific Promoter Methylations of the γ-Globin Gene Cause Inhibition of γ-Globin Gene Transcription

It also has been demonstrated that the methylation of cytidine residues in the 5′ region of the γ-globin gene, between nucleotides −760 to +100, eliminates transcription of the gene (Busslinger *et al.* 1983). There are 11 5′–CpG–3′ dinucleotides in this stretch of 860 nucleotides. It has not yet been shown which of these nucleotides are the decisive ones in affecting the regulation of gene expression. The authors have used the *in vitro* methylation procedure (Stein *et al.* 1982) based on replicating a gene—in this case the γ-globin gene—in the single-stranded vector M13 in the presence of 5-methyl dCTP. By choosing the appropriate primer fragment for this *in vitro* replication reaction, specific segments of the gene and its flanking sequences can be methylated. In this way, all the cytidine residues in a certain segment will be methylated. Thus, this method will not permit identification of the functionally significant nucleotides. The cloned γ-globin gene, thus hemimethylated in various regions, was then cotransfected into Ltk^- mouse cells with the thymidine kinase gene of *Herpes simplex* virus. Cell clones containing both the γ-globin and thymidine kinase genes were selected. It was found that the methylation pattern of the ^{m}CpG residues was faithfully inherited in this system, as previously reported for other genes (Stein *et al.* 1982). Methylation in the region from −760 in the 5′ flanking region to +100 in the γ-globin gene suppressed globin expression; methylation of c-residues from +100 in the γ-globin gene to +1,950 in the 3′ flanking region did not affect globin gene transcription (Busslinger *et al.* 1983). These findings are in very good agreement with those described above and those obtained in a number of different systems. It is interesting to mention in this context that in 15 different mammalian genes, the 5′-flanking sequences are enriched in 5′–CpG–3′ dinucleotides, whereas the 3′-regions are rather depleted of these sequences (McClelland and Ivarie, 1982). A high frequency of 5′–CpG–3′ sequences at the left terminus of adenovirus DNA also has been described (Doerfler *et al.* 1982). This 5′–CpG–3′-rich area precedes the E1a region that is important in the regulation of the expression of most other early viral genes (Jones and Shenk, 1979).

Conclusions

Many different lines of evidence have demonstrated that the methylation of specific sites can cause transcriptional inactivation of eukaryotic genes. It has also been shown, at least for several viral and nonviral eukaryotic genes, that the specific sites most relevant for the control of expression are located in the promoter regions of these genes. We have now presented data that would point at specific sequences, perhaps involving enhancer-like elements, that are located upstream from the TATA signal of the E1a promoter of adenovirus type 12 DNA. However, it would be premature to generalize these findings.

Moreover, one has to remember that CAT expression after transfection is transient and probably is extrachromosomal.

Many other questions remain open for further investigations. These investigations will be required to elucidate the function of each methylated C residue in either the promoter or the body portions of a gene. If promoter methylations by themselves could effect gene inactivation, it remains enigmatic as to why sites distributed over the entire gene are methylated in permanently inactivated genes. Could each methyl group that is introduced exert a unique function? In this context, it will be of great importance to determine the state of methylation of all the C residues occurring in a gene or, at least, in the promoter region of this gene. Genomic sequencing methods that have been recently developed (Church and Gilbert, 1984) should make this goal attainable.

Another puzzling observation that deserves more detailed investigations is the shift in the levels of DNA methylation in adenovirus DNA sequences after the explantation of adenovirus type 12-induced tumor cells into culture (Kuhlmann and Doerfler, 1982). Does the chromatin organization at, and around, the inserted foreign (viral) DNA sequences matter? Is this organization altered during transfer of cells from tumor to culture? How does the activity of *(de novo)* DNA methyltransferases change in these cells, and how are activity and sequence specificity of these enzymes regulated?

Probably, as in most reviews, this summary has clearly raised many more questions than it has been able to answer. The discussion of results gleaned from work on the adenovirus system, as the discussions of many other systems presented in this volume, have hopefully convinced the reader that studies on the functional implications of specific DNA methylations are well under way. Work in this expanding field may help to answer questions about the organization and modes of expression of the eukaryotic genome.

Acknowledgments

We would like to express our appreciation for the expert editorial assistance rendered by Petra Böhm. Experimental work performed in this laboratory was supported by grants from the Deutsche Forschungsgemeinschaft through SFB74-C1 and from the Ministry of Science and Research of the State of Nordrhein-Westfalen, Düsseldorf, Federal Republic of Germany.

References

Alwine JC, Kemp DJ, Stark GR: Method for detection of specific RNAs in agarose gels by transfer to diazobenzyloxymethyl-paper and hybridization with DNA probes. *Proc Natl Acad Sci (USA)* 1977;74:5350–5354.

Baker CC, Ziff EB: Promoters and heterogeneous 5′ termini of the messenger RNAs of adenovirus serotype 2. *J Mol Biol* 1981;149:189–221.

Behe M, Felsenfeld G: Effects of methylation on a synthetic polynucleotide: The B-Z transition in poly(dG-m^{5}dC)·poly(dG-m^{5}dC). *Proc Natl Acad Sci (USA)* 1981;78:1619–1623.

Behe M, Zimmerman S, Felsenfeld G: Changes in the helical repeat of poly (dG-m^{5}dC)·poly(dG-m^{5}dC) and poly(dG-dC)·poly(dG-dC) associated with the B-Z transition. *Nature* 1981;293:233–235.

Bird AP, Taggart MH, Smith BA: Methylated and unmethylated DNA compartments in the sea urchin genome. *Cell* 1979;17:889–901.

Bos JL, Polder LJ, Bernards R, Schrier, PI, van den Elsen PJ, van der Eb AJ, van Ormondt H: The 2.2 kb E1b mRNA of human Ad12 and Ad5 codes for two tumor antigens starting at different AUG triplets. *Cell* 1981;27:121–131.

Busslinger M, Hurst J, Flavell RA: DNA methylation and the regulation of globin gene expression. *Cell* 1983;34:197–206.

Cate RL, Chick W, Gilbert W: Comparison of the methylation patterns of the two rat insulin genes. *J Biol Chem* 1983;258:6645–6652.

Chow LT, Broker TR, Lewis JB: Complex splicing patterns of RNAs from the early regions of adenovirus-2. *J Mol Biol* 1979;134:265–303.

Christman JK, Price P, Pedrinan L, Acs G: Correlation between hypomethylation of DNA and expression of globin genes in Friend erythroleukemia cells. *Eur J Biochem* 1977;81:53–61.

Christman JK, Weich N, Schoenbrun B, Schneiderman N, Acs G: Hypomethylation of DNA during differentiation of Friend erythroleukemia cells. *J Cell Biol* 1980;86:366–370.

Church, GM, Gilbert W: Genomic sequencing. Proc Natl Acad Sci (USA) 1984;81:1991–1995.

Cohen JC: Methylation of milk-borne and genetically transmitted mouse mammary tumor virus proviral DNA. *Cell* 1980;19:653–662.

Colby WW, Chen EY, Smith DH, Levinson AD: Identification and nucleotide sequence of a human locus homologous to the v-myc oncogene of avian myelocytomatosis virus MC29. *Nature* 1983;301:722–725.

Conklin KF, Coffin JM, Robinson HL, Groudine M, Eisenman R: Role of methylation in the induced and spontaneous expression of the avian endogenous virus ev-1: DNA structure and gene products. *Mol Cell Biol* 1982;2:638–652.

Constantinides PG, Jones PA, Gevers W: Functional striated muscle cells from non-myoblast precursors following 5-azacytidine treatment. *Nature* 1977;267:364–366.

Cook JL, Lewis AM Jr: Host response to adenovirus 2-transformed hamster embryo cells. *Cancer Res* 1979;39:1455–1461.

Cordingley MG, Campbell MEM, Preston CM: Functional analysis of a herpes simplex virus type 1 promoter: identification of far-upstream regulatory sequences. *Nucleic Acids Res* 1983;11:2347–2365.

Creusot F, Acs G, Christman JK: Inhibition of DNA methyltransferase and induction of Friend erythroleukemia cell differentiation by 5-azacytidine and 5-aza-2′-deoxycytidine. *J Biol Chem* 1982;257:2041–2048.

DeSimone J, Heller P, Hall L, Zwiers D: 5-Azacytidine stimulates fetal hemoglobin synthesis in anemic baboons. *Proc Natl Acad Sci (USA)* 1982;79:4428–4431.

Desrosiers RC, Mulder C, Fleckenstein B: Methylation of herpes virus saimiri DNA in lymphoid tumor cell lines. *Proc Natl Acad Sci (USA)* 1979;76:3839–3843.

Deuring R, Klotz G, Doerfler W: An unusual symmetric recombinant between adenovirus type 12 DNA and human cell DNA. *Proc Natl Acad Sci (USA)* 1981;78:3142–3146.

Deuring R, Doerfler W: Proof of recombination between viral and cellular genomes in human KB cells productively infected by adenovirus type 12: Structure of the junction site in a symmetric recombinant (SYREC). *Gene* 1983;26:283–289.

Doerfler W: DNA methylation—A regulatory signal in eukaryotic gene expression. *J Gen Virol* 1981;57:1–20.

Doerfler W, Kruczek I, Eick D, Vardimon L, Kron B: DNA methylation and gene activity: The adenovirus system as a model. *Cold Spring Harbor Symp Quant Biol* 1982;47:593–603.

Doerfler W: DNA methylation and gene activity. *Ann Rev Biochem* 1983;52:93–124.

Doerfler W: DNA methylation and its functional significance: Studies on the adenovirus system. *Curr Top Microbiol Immunol* 1984a;108:79–98.

Doerfler W: DNA methylation: Role in viral transformation and persistence, in Klein G (ed): *Advances in Viral Oncology*. New York, Raven Press, 1984b, vol 4. pp. 217–247.

Drahovsky D, Böhm TLJ: Enzymatic DNA methylation in higher eukaryotes. *Int J Biochem* 1980;12:523–528.

Ehrlich M, Wang RY-H: 5-methylcytosine in eukaryotic DNA. *Science* 1981;212:1350–1357.

Eick D, Stabel S, Doerfler W: Revertants of adenovirus type 12-transformed hamster cell line T637 as tools in the analysis of integration patterns. *J Virol* 1980;36:41–49.

Eick D, Doerfler W: Integrated adenovirus type 12 DNA in the transformed hamster cell line T637: Sequence arrangements at the termini of viral DNA and mode of amplification. *J Virol* 1982;42:317–321.

Eick D, Kemper B, Doerfler W: Excision of amplified viral DNA at palindromic sequences from the adenovirus type 12-transformed hamster cell line T637. *EMBO J* 1983a;2:1981–1986.

Eick D, Fritz H-J, Doerfler W: Quantitative determination of 5-methylcytosine in DNA by reverse-phase high-performance liquid chromatography. *Analyt Biochem* 1983b;135:165–171.

Esche H: Viral gene products in adenovirus type 2-transformed hamster cells. *J Virol* 1982;41:1076–1082.

Esche H, Siegmann B: Expression of early viral gene products in adenovirus type 12-infected and -transformed cells. *J Gen Virol* 1982;60:99–113.

Everett RD, Baty D, Chambon P: The repeated GC-rich motifs upstream from the TATA box are important elements of the SV40 early promoter. *Nucleic Acids Res* 1983;11:2447–2464.

Feinberg AP, Vogelstein B: Hypomethylation distinguishes genes of some human cancers from their normal counterparts. *Nature* 1983;301:89–92.

Flavell RA, Grosveld F, Busslinger M, de Boer E, Kioussis D, Mellor AL, Golden L, Weiss E, Hurst J, Bud H, Bullman H, Simpson E, James R, Townsend ARM, Taylor PM, Schmidt W, Ferluga J, Leben L, Santamaria M, Atfield G, Festenstein H: Structure and expression of the human globin genes and murine histocompatibility antigen genes. *Cold Spring Harbor Symp Quant Biol* 1982;47:1067–1078.

Fradin A, Manley JL, Prives CL: Methylation of simian virus 40 Hpa II site affects late, but not early, viral gene expression. *Proc Natl Acad Sci (USA)* 1982;79:5142–5146.

Gerber-Huber S, May FEB, Westley BR, Felber BK, Hosbach HA, Andres A-C, Ryffel GU: In contrast to other Xenopus genes the estrogen-inducible vitellogenin genes are expressed when totally methylated. *Cell* 1983;33:43–51.

Gjerset RA, Martin DW Jr: Presence of a DNA demethylating activity in the nucleus of murine erythroleukemic cells. *J Biol Chem* 1982;257:8581–8583.

Gorman CM, Moffat LF, Howard BH: Recombinant genomes which express chloramphenicol acetyltransferase in mammalian cells. *Mol Cell Biol* 1982;2:1044–1051.

Groneberg J, Sutter D, Soboll H, Doerfler W: Morphological revertants of adenovirus type 12-transformed hamster cells. *J Gen Virol* 1978;40:635–645.

Groneberg J, Doerfler W: Revertants of adenovirus type 12-transformed hamster cells have lost part of the viral genomes. *Int J Cancer* 1979;24:67–74.

Grosschedl R, Birnstiel ML: Identification of regulatory sequences in the prelude sequences of an H2A histone gene by the study of specific deletion mutants in vivo. *Proc Natl Acad Sci (USA)* 1980;77:1432–1436.

Groudine M, Eisenman R, Weintraub H: Chromatin structure of endogenous retroviral genes and activation by an inhibitor of DNA methylation. *Nature* 1981;292:311–317.

Guntaka RV, Rao PY, Mitsialis SA, Katz R: Modification of avian sarcoma proviral DNA sequences in nonpermissive XC cells but not in permissive chicken cells. *J Virol* 1980;34:569–572.

Günthert U, Schweiger M, Stupp M, Doerfler W: DNA methylation in adenovirus, adenovirus-transformed cells, and host cells. *Proc Natl Acad Sci (USA)* 1976;73:3923–3927.

Günthert U, Jentsch S, Freund M: Restriction and modification in Bacillus subtilis: Two DNA methyltransferases with BsuRI specificity. II. Catalytic properties, substrate specificity, and mode of action. *J Biol Chem* 1981;256:9346–9351.

Harris M: Induction of thymidine kinase in enzyme-deficient Chinese hamster cells. *Cell* 1982;29:483–492.

Hattman S: DNA methylation, in Boyer PD (ed): *The Enzymes*. New York, London, Academic Press, 1981, pp 517–547.

Hearing P, Shenk T: The adenovirus type 5 E1A transcriptional control region contains a duplicated enhancer element. *Cell* 1983;33:695–703.

Holliday R, Pugh JE: DNA modification mechanisms and gene activity during development. Developmental clocks may depend on the enzymic modification of specific bases in repeated DNA sequences. *Science* 1975;187:226–232.

Hynes NE, Rahmsdorf U, Kennedy N, Fabiani L, Michalides R, Nusse R, Groner B: Structure, stability, methylation, expression and glucocorticoid induction of endogenous and transfected proviral genes of mouse mammary tumor virus in mouse fibroblasts. *Gene* 1981;15:307–317.

Jähner D, Stuhlmann H, Stewart CL, Harbers K, Löhler J, Simon I, Jaenisch R: *De novo* methylation and expression of retroviral genomes during mouse embryogenesis. *Nature* 1982;298:623–628.

Johansson K, Persson H, Lewis AM, Pettersson U, Tibbetts C, Philipson L: Viral DNA sequences and gene products in hamster cells transformed by adenovirus type 2. *J Virol* 1978;27:628–639.

Jones N, Shenk T: An adenovirus type 5 early gene function regulates expression of other early viral genes. *Proc Natl Acad Sci (USA)* 1979;76:3665–3669.

Jones PA, Taylor SM: Hemimethylated duplex DNAs prepared from 5-azacytidine-treated cells. *Nucleic Acids Res* 1981;9:2933–2947.

Jones PA, Taylor SM, Mohandas T, Shapiro LJ: Cell cycle-specific reactivation of an inactive X-chromosome locus by 5-azadeoxycytidine. *Proc Natl Acad Sci (USA)* 1982;79:1215–1219.

Kruczek I, Doerfler W: The unmethylated state of the promoter/leader and 5′-regions of integrated adenovirus genes correlates with gene expression. *EMBO J* 1982;1:409–414.

Kruczek I, Doerfler W: Expression of the chloramphenicol acetyltransferase gene in mammalian cells under the control of adenovirus type 12 promoters: Effect of promoter methylation on gene expression. *Proc Natl Acad Sci (USA)* 1983;80:7586–7590.

Kuhlmann I, Doerfler W: Shifts in the extent and patterns of DNA methylation upon explantation and subcultivation of adenovirus type 12-induced hamster tumor cells. *Virology* 1982;118:169–180.

Kuhlmann I, Doerfler W: Loss of viral genomes from hamster tumor cells and nonrandom alterations in patterns of methylation of integrated adenovirus type 12 DNA. *J Virol* 1983;47:631–636.

Langner K-D, Vardimon L, Renz, D, Doerfler W: DNA methylations of three 5′–CCGG–3′ sites in the promoter and 5′ region inactivate the E2a gene of adenovirus type 2. *Proc Natl Acad Sci* (USA) 1984;81:2950–2954.

La Volpe A, Taggart M, Macleod D, Bird A: Coupled demethylation of sites in a conserved sequence of Xenopus ribosomal DNA. *Cold Spring Harbor Symp Quant Biol* 1982;47:585–592.

Ley TJ, DeSimone J, Anagnou NP, Keller GH, Humphries RK, Turner, PH, Young NS, Heller P, Nienhuis, AW: 5-azacytidine selectively increases γ-globin synthesis in a patient with β^+ thalassemia. *N Engl J Med* 1982;307:1469–1475.

Mandel JL, Chambon P: DNA methylation: organ specific variations in the methylation pattern within and around ovalbumin and other chicken genes. *Nucleic Acids Res* 1979;7:2081–2103.

Mann MB, Smith HO: Specificity of HpaII and HaeIII DNA methylases. *Nucleic Acids Res* 1977;4:4211–4221.

McClelland M, Ivarie R: Assymetrical distribution of CpG in an "average" mammalian gene. *Nucleic Acids Res* 1982;10:7865–7877.

McGeady ML, Jhappan C, Ascione R, vande Woude GF: In vitro methylation of specific regions of the cloned Moloney sarcoma virus genome inhibits its transforming activity. *Mol Cell Biol* 1983;3:305–314.

McGhee JD, Ginder GD: Specific DNA methylation sites in the vicinity of the chicken β-globin genes. *Nature* 1979;280:419–420.

McKeon C, Ohkubo H, Pastan I, de Crombrugghe B: Unusual methylation pattern of the $\alpha 2$ (I) collagen gene. *Cell* 1982;29:203–210.

McKnight SL: Functional relationship between transcriptional control signals of the thymidine kinase gene of Herpes simplex virus. *Cell* 1982;31:355–365.

Ortin J, Scheidtmann K-H, Greenberg R, Westphal M, Doerfler W: Transcription of the genome of adenovirus type 12. III. Maps of stable RNA from productively infected human cells and abortively infected and transformed hamster cells. *J Virol* 1976;20:355–372.

Ott M-O, Sperling L, Cassio D, Levilliers J, Sala-Trepat J, Weiss MC: Undermethylation at the 5′ end of the albumin gene is necessary but not sufficient for albumin production by rat hepatoma cells in culture. *Cell* 1982;30:825–833.

Razin A, Riggs AD: DNA methylation and gene function. *Science* 1980;210:604–610.

Razin A, Friedman J: DNA methylation and its possible biological roles. *Progr Nucl Acids Res Mol Biol* 1981;25:33–52.

Rigby PWJ, Dieckmann M, Rhodes C, Berg P: Labeling deoxyribonucleic acid to high specific activity in vitro by nick translation with DNA polymerase I. *J Mol Biol* 1977;113:237–251.

Riggs AD: X inactivation, differentiation, and DNA methylation. *Cytogenet Cell Genet* 1975;14:9–25.

Riggs AD, Jones PA: 5-Methylcytosine, gene regulation and cancer. *Advances in Cancer Research*. 1983;40:1–30, Academic Press, New York and London.

Sager R, Kitchin R: Selective silencing of eukaryotic DNA. A molecular basis is proposed for programmed inactivation or loss of eukaryotic DNA. *Science* 1975;189:426–433.

Sanders Haigh L, Blanchard Owens B, Hellewell S, Ingram VM: DNA methylation in chicken α-globin gene expression. *Proc Natl Acad Sci (USA)* 1982;79:5332–5336.

Schirm S, Doerfler W: Expression of viral DNA in adenovirus type 12-transformed cells, in tumor cells, and in revertants. *J Virol* 1981;39:694–702.

Smith SS, Yu JC, Chen CW: Different levels of DNA modification at 5′CCGG in murine erythroleukemia cells and the tissues of normal mouse spleen. *Nucleic Acids Res* 1982;10:4305–4320.

Southern EM: Detection of specific sequences among DNA fragments separated by gel electrophoresis. *J Mol Biol* 1975;98:503–517.

Stabel S, Doerfler W, Friis RR: Integration sites of adenovirus type 12 DNA in transformed hamster cells and hamster tumor cells. *J Virol* 1980;36:22–40.

Stabel S, Doerfler W: Nucleotide sequence at the site of junction between adenovirus type 12 DNA and repetitive hamster cell DNA in transformed cell line CLAC1. *Nucleic Acids Res* 1982;10:8007–8023.

Stein R, Razin A, Cedar H: In vitro methylation of the hamster adenine phosphoribosyltransferase gene inhibits its expression in mouse L cells. *Proc Natl Acad Sci (USA)* 1982;79:3418–3422.

Stein R, Sciaky-Gallili N, Razin A, Cedar H: Pattern of methylation of two genes coding for housekeeping functions. *Proc Natl Acad Sci (USA)* 1983;80:2422–2426.

Sugisaki H, Sugimoto K, Takanami M, Shiroki K, Saito I, Shimojo H, Sawada Y, Uemizu Y, Uesugi S, Fujinaga K: Structure and gene organization in the transforming Hind III-G fragment of Ad12. *Cell* 1980;20:777–786.

Sutter D, Westphal M, Doerfler W: Patterns of integration of viral DNA sequences in the genomes of adenovirus type 12-transformed hamster cells. *Cell* 1978;14:569–585.

Sutter D, Doerfler W: Methylation of integrated viral DNA sequences in hamster cells transformed by adenovirus 12. *Cold Spring Harbor Symp Quant Biol* 1979;44:565–568.

Sutter D, Doerfler W: Methylation of integrated adenovirus type 12 DNA sequences in transformed cells is inversely correlated with viral gene expression. *Proc Natl Acad Sci (USA)* 1980;77:253–256.

Taylor SM, Jones PA: Multiple new phenotypes induced in 10T1/2 and 3T3 cells treated with 5-azacytidine. *Cell* 1979;17:771–779.

Telford JL, Kressman A, Koski RA, Grosschedl R, Müller F, Clarkson SG, Birnstiel, ML: Delimitation of a promoter for RNA polymerase III by means of a functional test. *Proc Natl Acad Sci (USA)* 1979;76:2590–2594.

van der Ploeg LHT, Flavell RA: DNA methylation in the human $\gamma\delta\beta$-globin locus in erythroid and nonerythroid tissues. *Cell* 1980;19:947–958.

van der Vliet PC, Levine AJ: DNA-binding proteins specific for cells infected by adenovirus. *Nature New Biol* 1973;246:170–174.

Vardimon L, Neumann R, Kuhlmann I, Sutter D, Doerfler W: DNA methylation and viral gene expression in adenovirus-transformed and -infected cells. *Nucleic Acids Res* 1980;8:2461–2473.

Vardimon L, Doerfler W: Patterns of integration of viral DNA in adenovirus type 2-transformed hamster cells. *J Mol Biol* 1981;147:227–246.

Vardimon L, Kuhlmann I, Cedar H, Doerfler W: Methylation of adenovirus genes in transformed cells and in vitro: influence on the regulation of gene expression. *Eur J Cell Biol* 1981;25:13–15.

Vardimon L, Kressmann A, Cedar H, Maechler M, Doerfler W: Expression of a cloned adenovirus gene is inhibited by in vitro methylation. *Proc Natl Acad Sci (USA)* 1982a;79:1073–1077.

Vardimon L, Günthert U, Doerfler W: In vitro methylation of the BsuRI (5′-GGCC-3′) sites in the E2a region of adenovirus type 2 DNA does not affect expression in Xenopus laevis oocytes. *Mol Cell Biol* 1982b;2:1574–1580.

Vardimon L, Renz D, Doerfler W: Can DNA methylation regulate gene expression? *Recent Results Cancer Res* 1983;84:90–102.

Von Acken U, Simon D, Grunert F, Döring H-P, Kröger H: Methylation of viral DNA in vivo and in vitro. *Virology* 1979;99:152–157.

Waalwijk C, Flavell RA: MspI, an isoschizomer of HpaII which cleaves both unmethylated and methylated HpaII sites. *Nucleic Acids Res* 1978;5:3231–3236.

Waechter DE, Baserga R: Effect of methylation on expression of microinjected genes. *Proc Natl Acad Sci (USA)* 1982;79:1106–1110.

Weintraub H, Larsen A, Groudine M: α-globin-gene-switching during the development of chicken embryos: expression and chromosome structure. *Cell* 1981;24:333–344.

Wilks AF, Cozens PJ, Mattaj IW, Jost J-P: Estrogen induces a demethylation at the 5′ end region of the chicken vitellogenin gene. *Proc Natl Acad Sci (USA)* 1982;79:4252–4255.

Wilson VL, Jones PA: Inhibition of DNA methylation by chemical carcinogens in vitro. *Cell* 1983;32:239–246.

12

DNA Methylation and Developmental Regulation of Eukaryotic Globin Gene Transcription

Che-Kun James Shen*

In the last decade, knowledge about the sequence organization of eukaryotic genes and their transcriptional behavior has been greatly increased due to the development of molecular cloning techniques and other biochemical methodology. It is now a relatively easy task to isolate, from the eukaryotic genome, particular genes or gene families along with many kilobases (kb) of flanking DNA—provided that an appropriate DNA library and probes are available. The cloned DNA then can be used, either as substrate or probe, to study problems such as the molecular evolution of the genes and flanking sequences and the transcriptional regulation of the genes.

To many molecular geneticists, perhaps the most challenging question is "What molecular mechanisms control the coordinate and differential expression of specific genes during development and cell differentiation?" A number of gene systems have been used to study the molecular biology of eukaryotic gene regulation; among them are the globin gene families. In all vertebrates studied, it has been found that different members of the globin gene families are turned on and off (switched) during development, and that they are expressed only in certain tissues (erythropoietic tissues). This interesting pattern of expression—coupled with extensive genetic and biochemical data obtained from both protein analysis and the recent availability of clones containing the complete globin gene families from many eukaryotes—makes the globin gene families particularly attractive and tractable for the study of transcriptional regulation in eukaryotes.

The transcriptional regulation of globin genes has been studied in various ways. Transcription of different globin genes in cell-free extract has been used to map the promoters *in vitro* (Proudfoot *et al.* 1980). Transcription of globin

*Department of Genetics, University of California, Davis, California 95616

genes after transfer of the DNA clones into cell cultures (Banerji *et al.* 1981; Mellon *et al.* 1981) has been used to study both the domains and functions of promoters and the effects of enhancer sequences *in vivo*. Analysis of DNA sequences and abnormally spliced globin transcripts isolated from β-thalassemia patients have elucidated the mechanism of RNA splicing (Treisman *et al.* 1982). Most of the globin gene families are arranged in clusters. Therefore, studies also have been initiated on the structure and transcription of the intergenic DNA in the hope that specific structural features (e.g., DNA methylation and chromatin structure (see this chapter) or the nonglobin transcripts (e.g., the small repetitive RNA [Shen and Maniatis, 1982a]) play a role in the *in vivo* regulation of globin gene transcription.

As with many of the gene systems studied, the active transcription of globin genes is associated with the undermethylation of DNA containing the gene and its flanking sequences (see Chapter 16). In this chapter, I will first briefly review and discuss the evidence accumulated within the last few years. Some recent experimental attempts to determine whether methylation is a causative factor in globin gene inactivation will then be presented, along with a discussion of the possible molecular mechanism(s) of this regulation.

In Vivo DNA Methylation Patterns of Globin Gene Clusters During Development

Rabbit β-like Globin Gene Cluster

An mentioned in previous chapters, the study of the relationship between *in vivo* methylation patterns of eukaryotic genes and their activity has been greatly facilitated by the discovery that HpaII and MspI are isoschizomers. The first globin gene investigated in this way was the rabbit β1-globin gene (Waalwijk and Flavell, 1978). The single CCGG site within the large intron of the adult β1-globin gene was found to be methylated to an extent of 50% in spleen, bone marrow, liver, and blood DNA, 80% in brain DNA, and 100% in sperm DNA. However, because only one gene was studied and the DNA was isolated only from adult tissues, the relationship between globin gene activity and DNA methylation was not clear.

By 1979, a number of λ phage-genomic clones have been isolated whose DNA inserts together cover approximately 50 kb of rabbit DNA containing the four closely linked β-like globin genes: β4, β3 (both are embryonic globin genes), β2 (a pseudogene), and β1 (the adult globin gene) (Hardison *et al.* 1979; Lacy *et al.* 1979). These genes, as with most of the other globin gene families (see below), are arranged in the order of their expression during development; they also have the same direction of transcription. Each of the three functional genes has one large intron containing approximately 800 base pairs of unique DNA sequences that could be easily cloned in the plasmid pBR322. Digestion of different rabbit tissue DNA with HpaII or MspI, or a combination

with other restriction enzymes, and subsequent blot hybridization using each one of the three intron-containing subclones as the probe would be expected to give the methylation extent at every CCGG site contained within the cloned gene cluster.

As shown in Figure 12.1, differences in the methylation extents of the four CCGG sites (labeled as 4, 5, 6, and 7) surrounding the $\beta 3$ gene were detected easily. This difference is most dramatic for site number 6, which is located in the 5′ portion of the gene. This site is totally methylated (i.e., it is methylated in all of the cells) in all of the tissues examined; except in blood island—the embryonic erythropoietic tissue—where it is methylated only to the extent of 70%. The extent of methylation at each of the sites was analyzed semiquantitatively as exemplified in Figure 12.1b. Specific sites flanking the $\beta 4$ and $\beta 1$ genes that were totally methylated in nonexpressing and nonerythropoietic tissues were subsequently detected when the intron sequences of these genes were used as the probe (Shen and Maniatis, 1980). Figure 12.2 shows a composite map of the CCGG sites mapped and the methylation extent at each one of these sites. From this map, it is obvious that many sites surrounding the embryonic and adult globin genes are relatively undermethylated in DNA from embryonic and/or adult erythroid tissues. This pattern is most pronounced for three sites (denoted by *) that are undermethylated in erythroid tissues, but are totally methylated in nonerythroid cells. We concluded, at that time, that the degree of CpG methylation in the rabbit β-like globin gene cluster was correlated with gene activity, but that the effect was confined to relatively small regions of DNA.

Human ${}^{G}\gamma$-${}^{A}\gamma$-δ-β Globin Gene Region

The methylation pattern in the human ${}^{G}\gamma$-${}^{A}\gamma$-δ-β globin gene region was studied by van der Ploeg and Flavell in 1979. As with the rabbit β-like globin gene cluster, the human β-like globin gene cluster also contains five functional genes arranged in the order of their expression during development: 5′-ϵ(embryonic)-${}^{G}\gamma$(fetal)-${}^{A}\gamma$(fetal)-δ(adult)-β(adult)-3′ (Maniatis *et al.* 1980). Using both cloned DNA as hybridization probes and various restriction enzymes whose recognition sites contained the dinucleotide sequence CG, van der Ploeg and Flavell (1980) examined the methylation pattern within a 36-kb region that contained the ${}^{G}\gamma$, ${}^{A}\gamma$, δ, and β-globin genes in different tissues and cell cultures. Again, methylation of –CCGG– sequences was studied most extensively. Figure 12.3 includes this information and the methylation pattern of two HhaI sites (GCGC) for cells of three representative tissues: 1) fetal liver (the fetal erythropoietic tissue that expresses the ${}^{G}\gamma$ and ${}^{A}\gamma$ globin genes); 2) bone marrow (the adult erythropoietic tissue expressing mainly the δ- and β-globin genes); and 3) adult liver (a nonerythroid tissue).

Interestingly, the general conclusions are similar to those for the study of rabbit β-like globin genes; that is, in nonexpressing tissues, the CCGG sites

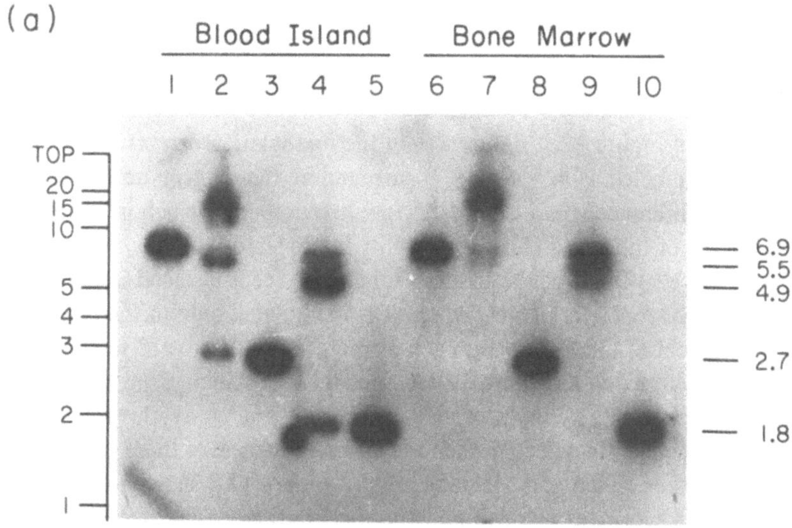

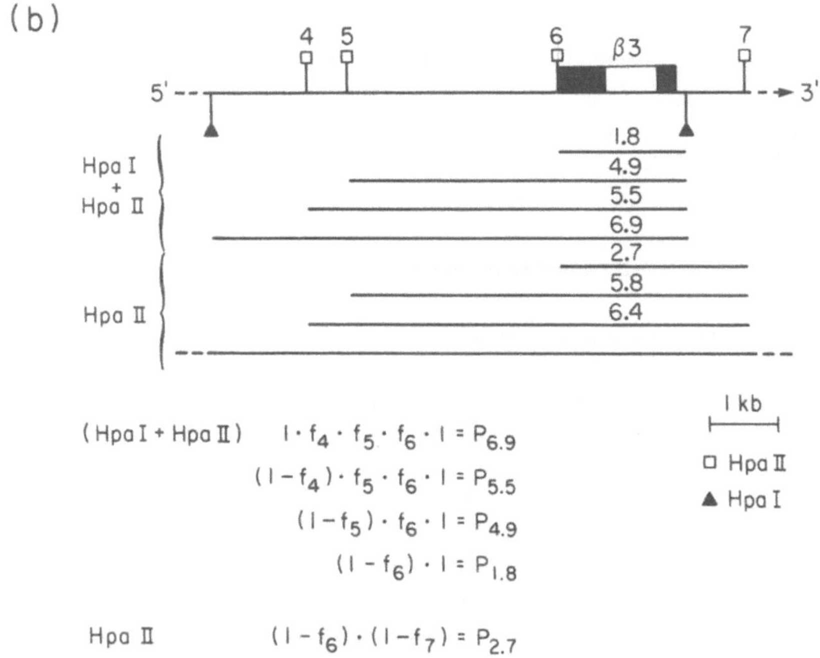

Figure 12.1 Methylation of –CCGG– sequences surrounding the rabbit embryonic β3 globin gene. Blood island and bone marrow DNA were digested with HpaI, HpaII, or MspI and were hybridized to a ^{32}P-labeled restriction fragment containing mainly the β3 large intron. (a) Autoradiograph showing the hybridization pattern of blood

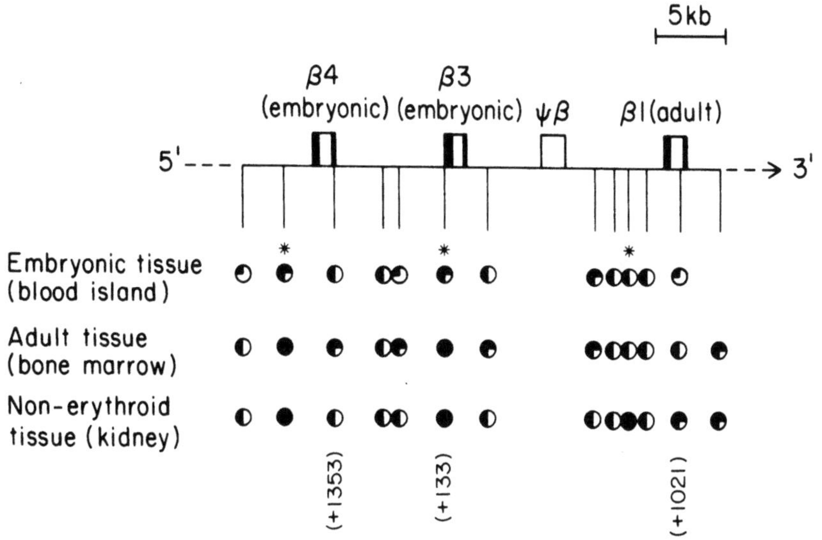

Figure 12.2 Methylation pattern of the rabbit β-like globin gene cluster. The extent of methylation of 13 CCGG sites has been determined as described in detail elsewhere (Shen and Maniatis, 1980). For simplicity, they are shown here as white circles (no methylation), quarter-blackened circles (25% methylation), half-blackened circles (50% methylation), three-quarter blackened circles (75% methylation), and fully blackened circles (100% methylation), respectively. Only the data for DNA extracted from blood island, bone marrow, and kidney are shown here. The sites with "*" are those that show the most dramatic relationship of hypermethylation-gene inactivity. The precise locations of the three sites within the β4, β3, and β1 genes are known from the nucleotide sequence data (β4, Hardison, 1983; β3, Hardison, 1981; β1, Hardison *et al.* 1979). They are indicated in parentheses as their distances in base pairs downstream from the mRNA cap site. (Data adapted with permission from Shen and Maniatis, 1980.)

←

island (lanes 1–5) and bone marrow (lanes 6–10) DNA. Lanes 1 and 6, HpaI alone; lanes 2 and 7, HpaII alone; lanes 3 and 8, MspI alone; lanes 7 and 9, HpaI + HpaII; lanes 5 and 10, HpaI + MspI. Scales on sides are in kilobases. (b) Map of CCGG sites surrounding the β3 globin gene and an expression for estimating methylation frequencies. The CCGG sites are indicated by small white boxes that are numbered 4–7. The lines below the map represent the fragments detected by β3 intron probe. The fragment lengths are indicated in kilobases. The methylation frequencies f_4 through f_7 could be calculated from the five equations in which the weight proportion (P) values are known from the autoradiographs. (Reprinted with permission from Shen and Maniatis, 1980.)

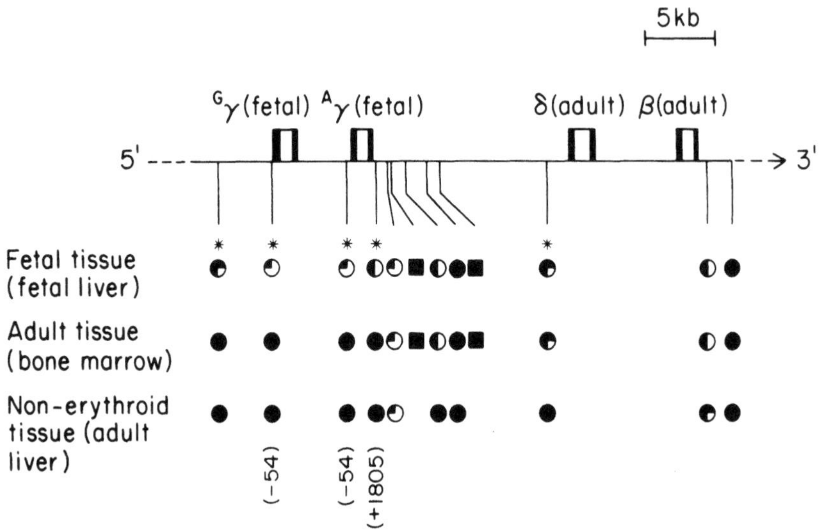

Figure 12.3 Methylation pattern of the human $^G\gamma$-$^A\gamma$-δ-β globin gene region. The extent of methylation at 10 CCGG sites (the circles) and two GCGC sites (the squares) in the human $^G\gamma$-$^A\gamma$-δ-β globin gene region are given here for a fetal erythroid tissue (fetal liver), an adult erythroid tissue (bone marrow), and a nonerythroid tissue (adult liver). Similar to those in Figure 12.2, the methylation levels are indicated by the extent of blackness of the circles and squares. The "*" indicates the five CCGG sites that show an obvious relationship of hypermethylation-gene inactivity among different tissues. (Data derived from van der Ploeg and Flavell, 1980.) The precise locations of the two CCGG sites upstream from the $^G\gamma$ and $^A\gamma$ genes have been determined by DNA sequencing (Shen *et al.* 1981). (Data adapted and revised with permission from van der Ploeg and Flavell, 1980.)

surrounding the inactive genes usually are hypermethylated. On the contrary, the erythroid tissues exhibit a low level of modification in the expressed gene region. Of particular interest in the human $^G\gamma$-$^A\gamma$-δ-β region are two –CCGG– sequences located near the 5′-flanking part of the $^G\gamma$ and $^A\gamma$ genes, respectively. They are resistant to both MspI and HpaII cleavage in nonerythroid or adult erythropoietic tissues (see Figure 12.3). Flavell *et al.* originally proposed that MspI could not cut these two –CCGG– sequences, because their external C's are methylated. This would mean that the two –CCGG– sequences flanking the 5′ side of the two γ-globin genes must be methylated at both C residues. However, more recently, they have discovered that MspI cannot cleave at a –CCGG– sequence if the sequence is preceded on the 5′ side by two G residues and it's internal C is methylated (Busslinger *et al.* 1983a). In fact, both of the –CCGG– sequences in the γ-globin gene regions that are resistant *in vivo* to MspI and HpaII cleavage are flanked on their 5′ side by two guanines (Shen *et al.* 1981). Thus, it is most likely that the MspI resistance at these two sites is due to the *in vivo* methylation at the internal C residue, which also results in their resistance to HpaII cleavage.

Chicken β-like Globin Gene Cluster

As with those in mammals, the chicken genome contains one cluster each of α-like and β-like globin genes. The entire α-cluster and most of the β-cluster have been cloned (Dodgson *et al.* 1981; Dolan *et al.* 1981). One advantage of using the chicken globin gene system to study gene expression is that relatively pure populations of nucleated red blood cells can be easily obtained during various developmental stages. Within the nuclei of these cells, the transcription of many genes, including the globin genes, are still active. In the embryonic red blood cells (e.g., those of a 5-day embryo), the two β-like globin genes (ρ and ϵ) and all of the α-like globin genes (π', α^A, and α^D) are being transcribed. However, in the definitive cells of adult blood, the ρ, ϵ, and π' genes are turned off and the α^A, α^D, and two adult β-like globin genes (β^H and β) are turned on.

In fact, the chicken adult β-globin gene was the first globin gene for which a definite correlation between gene activity and DNA hypomethylation was established (McGhee and Ginder, 1979). Because relatively pure populations of nucleated red blood cells were used in their study, McGhee and Ginder were able to show that certain –CCGG– sequences near the ends of the adult β-globin gene are totally unmethylated in adult red blood cells. In contrast, these sites—whose exact locations were not mapped at that time—were partially methylated in 5-day embryonic cells and nonerythroid tissues such as the brain. These studies were later extended (Ginder and McGhee, 1981; Ginder *et al.* 1983) to the entire gene cluster, and an all-or-none relationship between gene activity and hypomethylation was found for several specific CG-containing sites (those with * in Figure 12.4) that surrounded all four globin genes.

Chicken α-Globin Gene Cluster

The chicken α-like globin gene region consists of three genes (the embryonic gene π', and two adult genes, α^D and α^A) clustered within 8 kb. As in other globin gene clusters, all genes are arranged in order of expression during development, and they are transcribed in the same direction (Dodgson *et al.* 1981). The nucleotide sequences of the three genes and their flanking DNA have been recently determined (Engel *et al.* 1983; Dodgson and Engel, 1983).

Weintraub *et al.* (1981) first studied the CCGG methylation pattern of this gene cluster in adult definitive red blood cells in which the α^D and α^A genes, but not the π' gene, are transcribed. A more extensive study was later performed by Haigh *et al.* (1982), who examined the methylation pattern of –CCGG– and –GCGC– sequences in both embryonic and adult red blood

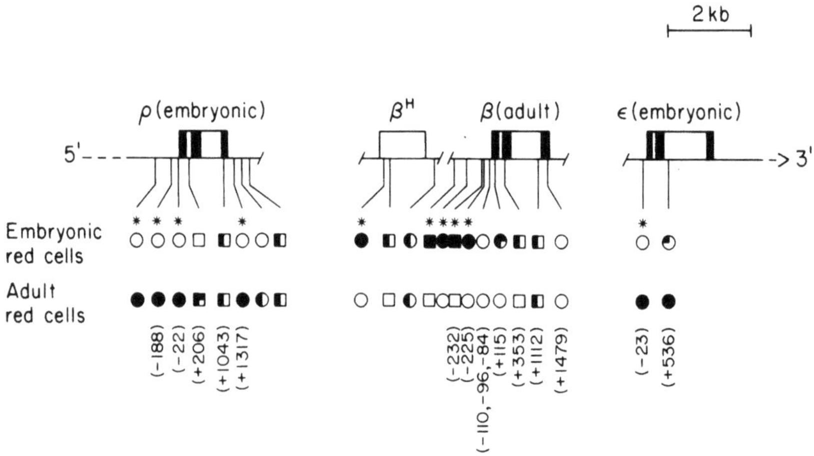

Figure 12.4 Methylation pattern of the chicken β-like globin gene cluster. The four genes are transcribed in the same direction (from left to right) that is similar to the other eukaryotic globin gene clusters. However, the two embryonic genes (ρ and ϵ) are located on the 5′- and 3′-ends of the gene cluster, respectively. The positions of the CCGG sites (circles) and GCGC sites (squares), and their methylation extents in the embryonic and adult red blood cells are derived mainly from the studies by Ginder and McGhee (1981) and Ginder *et al.* (1983). However, because the nucleotide sequences of both the coding regions and their flanking regions are now available for three of the four genes (β, Day *et al.* 1981; McGhee *et al.* 1981; Dolan *et al* 1983; ρ and ϵ, D Engel, personal communication), I have changed the locations of many of these sites that were documented in Ginder *et al.* (1983) according to the sequencing data. In particular, the four CCGG sites (−84, −96, −110, and −225) and one GCGC site (−232) upstream from the adult β-globin gene are derived from McGhee *et al.* (1981). Those sites located within this gene are derived from Dolan *et al.* (1983). Again, the symbol "*" denotes the sites that exhibit an obvious relationship between hypermethylation and transcriptional inactivity. (Data adapted and revised with permission from McGhee *et al.* 1981 and Ginder *et al.* 1983.)

cells, as well as in those of the brain (a nonerythroid tissue) and sperm. In Figure 12.5, I have combined the results obtained by these researchers with appropriate changes of the positions of some of the CCGG and GCGC sites according to the recent nucleotide sequence data. Except for the two -CCGG- sequences on the 5′ side of the α^D gene, the conclusion from the studies of the two groups are basically the same. The nucleotide sequence of the α^A gene (Dodgson and Engel, 1983) indicates that there are four -CCGG- sequences and two -GCGC- sequences located within the gene. This is not consistent with the maps of Weintraub *et al.* (1981) and Haigh *et al.* (1982); thus, it could result from sequence polymorphism. For this reason, the methylation pattern of the α^A gene and its 3′ flanking region is not included in Figure 12.5. In any case, gene activity is again associated with undermethylation of several specific CG-containing sequences (those sites with a * in Figure 12.5).

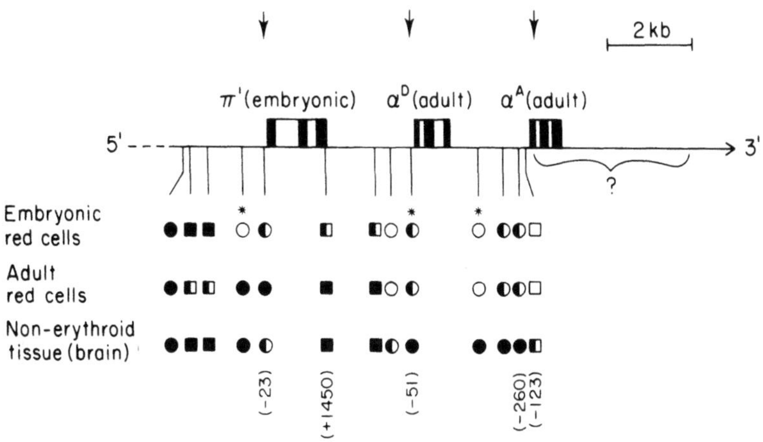

Figure 12.5 Methylation pattern of chicken α-like globin gene cluster. The map shown here combines the DNA methylation study (Haigh *et al.* 1982; Weintraub *et al.* 1981) and the physical map and nucleotide sequences (Engel *et al.* 1983; Dodgson and Engel, 1983). Note that many of the CCGG or GCGC sites mapped by Haigh *et al.* (1982) and Weintraub *et al.* (1981), which are located a long distance (>3 kb) upstream from the gene cluster and do not have a tissue-specific pattern of methylation, are not shown in this map. Also, because there is an apparent discrepancy in the α^A-gene region between CG-containing sites mapped by Haigh *et al.* (1982) and Weintraub *et al.* (1981), and between those determined from the sequencing data (Dodgson and Engel, 1983), this region is marked with a question mark. Again, the "*" denotes the thre CCGG sites that showed the hypermethylation-transcriptional inactivity relationship. The three arrows indicate the three DNAse I-hypersensitive sites mapped by Weintraub *et al.* (1981) that correlate with both the gene activity and undermethylation of the 5′-flanking regions.

Consensus Rules Derived from the Methylation Patterns of Eukaryotic Globin Gene Clusters

Although only a few restriction enzymes have been used in most of the studies described above (i.e., the HpaII/MspI that recognize -CCGG- sequence and HhaI that recognizes GCGC), several interesting conclusions can be drawn from Figures 12.2–12.5.

(1) For all of the species examined, sperm DNA is completely methylated throughout the globin gene clusters.

(2) In general, -CCGG- sequences (and, in some cases, -GCGC- sequences) flanking embryonic globin genes are undermethylated in embryonic erythroid tissue (or cells) relative to both adult erythroid tissues (or cells) and nonerythroid tissues. This is most obvious in the cases of the chicken globin

genes; probably due to the fact that pure populations of red blood cells can be easily fractionated.

(3) The –CCGG– (or –GCGC–) sequences surrounding adult globin genes are always hypermethylated in nonerythroid tissues.

(4) When the extent of methylation of CG-containing sequences flanking adult globin gene regions are compared between adult and embryonic erythropoietic tissues, the pattern is relatively more complex. In the chicken adult β-globin gene region, certain sites are totally methylated in the embryonic red blood cells, but are not methylated at all in the adult red blood cells. However, some of the CG-containing sequences in the rabbit β-, human β-, and chicken α-globin gene clusters do not show significant differences in methylation extent between adult and embryonic (or fetal) erythroid tissues. This is reasonable in the cases of chicken α-globin genes and human β-globin genes. The chicken α^{A}- and α^{D}-globin genes are expressed in both embryonic and adult red blood cells. Similarly, the human β-globin gene is also expressed in the human fetal liver. It is not obvious why the rabbit $\beta 1$-globin gene is undermethylated in both the blood island and the bone marrow, since Rorhbaugh and Hardison (1983) have shown that the $\beta 1$ gene is not transcribed during the early development of the rabbit embryo. This exception to the relationship of hypermethylation-gene inactivity will be discussed in a later section of this chapter.

(5) In each gene cluster, only a few sites show dramatic differences in methylation extent, and most of these "specific" sites are located near the 5′ end of the globin genes. In Figures 12.2–12.5, I have labeled these sites with an *. For example, for each of the three rabbit β-globin genes, there is one –CCGG– sequence located near the 5′ end that showed differences in methylation extent between the active tissue and inactive or nonerythroid tissues (Figure 12.2). The human $^{G}\gamma$-$^{A}\gamma$-δ-β region and the chicken β- and α-globin gene clusters all contain a set of such sites (Figures 12.3–12.5). Because only the enzymes HpaII/MspI and HhaI have been used in these studies, there must exist other CG-containing sequences flanking these "specific" –CCGG– or –GCGC– sequences that also may have a dramatic difference in the methylation extent among different tissues. However, present methodologies do not allow us to detect these differences *in vivo*.

A Causative Effect or a Reflection of Gene Activity?

The studies summarized above and in Figures 12.2–12.5 clearly show that, unlike some other eukaryotic genes (e.g., the two rat insulin genes [Cate *et al.* 1983]), the inactive state of the globin gene(s) is (are) closely associated with hypermethylation of surrounding regions. The critical question is whether hypermethylation is a passive concomitant of gene inactivity or a factor that causes inactivation.

This question was partially answered in other gene systems that also show a hypermethylation-gene inactivity relationship. Vardimon *et al.* (1982) have methylated the E2a region and Ad2 viral DNA with bacterial HpaII methylase, and they injected it into *Xenopus* oocytes. They found that these modified DNA templates remained untranscribed, while the untreated genes were actively transcribed in the oocytes at a normal level. Similar experiments have been performed on the adenine ribosyltransferase *(aprt)* gene. Stein *et al.* (1982a) have co-transformed a cloned hamster *aprt* gene with a *Herpes* thymidine kinase gene *(tk)* into a mouse Ltk^- $aprt^-$ cell line. After propagating cells in selective medium, they found that most of the tk^+ cells are also $aprt^+$. However, if the *aprt* gene was first methylated with bacterial HpaII methylase *in vitro,* then most of the tk^+ cell clones that were selected do not express the *aprt* phenotype. These provide direct evidence that, at least, some of the CCGG sites after methylation inhibit *in vivo* transcription of the genes described above.

Recently, a different approach has been used by Busslinger *et al.* (1983b) to test the effect of methylation at cytosine residues on human globin gene transcription *in vivo.* They have cloned a 3.3-kb DNA fragment containing the entire human $^A\gamma$-globin gene and 1.34 kb of its 5′ flanking sequences in the vector M13 mp8. The single-stranded, recombinant DNA then was converted into a double-stranded form *in vitro,* with the Klenow fragment of *Escherichia coli* DNA polymerase I, dATP, dGTP, dTTP, and 5-methyl dCTP, as first devised by Stein *et al.* (1982b). The hemimethylated DNA was introduced into mouse Ltk^- cells by using the *Herpes tk* gene as a selective marker. The selected Ltk^+ cells were analyzed by blotting-hybridization for the inheritance of the C methylation pattern surrounding the γ-globin gene, and also by nuclease S1 mapping for the transcription of the globin gene. Because different DNA fragments can be used as primers for *in vitro* DNA synthesis, specific segmental methylation patterns can be generated, which allows one to examine the effect of the C methylation in different regions surrounding the γ^A-globin gene on the *in vivo* regulation of transcription.

The results are interesting. It has been found that DNA methylation at C residues in the 5′-region of the $^A\gamma$-globin gene, from bases -760 to $+100$, prevent *in vivo* transcription. This is consistent with the observation that most of the specific -CCGG- sequences hypermethylated in inactive or nonerythroid tissues are located near the 5′ end of the globin genes. It will be important to see whether methylation at only one or a few of the C residues in this region, especially those in the dinucleotide sequence CG, inhibits *in vivo* transcription. It would also be useful to apply this approach to other globin gene systems.

Another line of indirect evidence for the causative role of C methylation in gene expression comes from the use of 5-azacytidine (5-aza-C). This drug was first demonstrated by Johns and Taylor (1980) as inhibiting C methylation when applied to cell cultures; thus, it induces the expressions of some otherwise inactive genes. DeSimone *et al.* (1980) performed the first *in vivo* experiment by administering 5-aza-C to baboons. It was found that the level of hemoglobin F (a fetal hemoglobin) per F cell of injected animals started to increase after approximately 5 days and peaked 5–7 days after the last dose of the drug. Ley *et al.* applied this drug to patients with β-thalassemia (Ley *et al.* 1982) or sickle cell anemia (Ley *et al.* 1983). A transient increase in the transcription of the γ-globin genes also was observed. The increase in the γ-globin mRNA was accompanied by undermethylation around the $^{G}\gamma$-$^{A}\gamma$-globin region. However, the region containing the δ, β, ε, and ζ-globin genes also became undermethylated during this treatment, but the level of transcription of these genes was not increased. In fact, the β-globin gene transcription was decreased. Similar experiments also have been performed on the chicken β-like genes (Ginder *et al.* 1983). In that study, transcription of the embryonic β-globin genes in adults could not be turned on by undermethylation that resulted from the injection of 5-aza-C. These observations could be due to side effects of the 5-aza-C, as well as to the fact that mechanisms other than DNA methylation also control the transcription of the globin genes.

As discussed above, the methylation at C residues surrounding globin genes may very well be one of the factors that causes transcription inactivation *in vivo.* How could this happen? Two mechanisms have been proposed (Stein *et al.* 1982a). First, it is possible that certain methyl groups in the DNA inhibit the formation of an active chromosomal domain (Weisbrod, 1982) that contains the gene(s) to be transcribed. A second possibility is that methylation inhibits transcription through direct interaction with the RNA polymerase and/or other protein factors required for transcription.

The second possibility recently has been tested in an *in vitro* assay system. In that study, recombinant DNA clones containing different mammalian genes and their flanking transcriptional control regions were completely methylated by using highly purified bacterial HpaII methylase or EcoRI methylase. These modified DNA templates were then transcribed *in vitro* in Hela cell extracts. Of all the gene systems tested, which include three different mammalian globin genes (rabbit β1-, β3-, and human α1-globin gene, Shen and Agarwal, unpublished results) and two human Alu family members, (Shen and Maniatis 1982b), the sizes and intensities of the transcripts from the methylated DNA templates were not appreciably different from those of the unmodified DNA. An example, using the human α-globin gene as the substrate, is illustrated in Figure 12.6. These results seem to favor the first mechanism, but do not completely rule out the second possibility. Because these experiments used naked DNA as substrates, while eukaryotic genes are organized into chromatin *in vivo,* it is still possible that RNA polymerase could interact with methylated DNA in the nucleosomal structure in a way that inhibits transcriptional initiation or that causes premature termination. The small differences in efficiency of the *in vitro* transcription between methylated and unmodified DNA templates may be magnified when DNA is organized into an appropriate chromatin state. It also has to be noted that, in this study, only the effect of m^5Cyt within –CCGG– sequences was monitored. The methylation of cytosine in other CG-containing sequences may be equally or more important in regulating the transcription of certain genes, even when they are not associated with histones.

DNA Methylation, Chromatin Structure, and Transcription

Chromatin structure may play an important role in the regulation of globin gene transcription by methylation of the -CG- sequences. This is supported by the congruence of domains of both nuclease sensitivity *in vivo* and DNA regions that are undermethylated.

As shown by Weintraub and Groudine (1976) and others, chromosomal DNA regions containing genes that are transcriptionally active are preferentially digested by DNAse I. The coding regions are even more susceptible to DNAse I digestion than their immediate flanking sequences. These studies have led Weintraub *et al.* to propose a model of "active chromosomal domains" (Stadler *et al.* 1980; Weisbrod, 1982). They have also found that the coding regions and flanking sequences of genes that will be transcriptionally active are also preferentially digested by DNAse I *in vivo*. They suggested that this is due to the existence of a "preactivation" chromatin structure. Interestingly, the DNAse I-sensitive region surrounding the chicken ovalbumin gene contains several CG-containing sites that are undermethylated in expressing tissues (Mandel and Chambon, 1979; Kuo *et al.*, 1979).

Another level of *in vivo* nuclease sensitivity of actively transcribed chromatin was first evidenced by the study of Wu *et al.* (1979), and later by other studies of several gene systems. They have demonstrated that DNAse I generates *in vivo* double-stranded DNA breakage in the vicinity of actively transcribing genes. (These sites have been termed DNAse I-hypersensitive sites). The chicken adult β-globin gene probably is the most extensively studied system with respect to the correlation between undermethylation and DNAse I hypersensitivity. As shown in Figure 12.4, there are four -CCGG- sequences (-84, -96, -110, and -225) and one -GCGC- sequence (-232) upstream from this gene. Although one (or all) of the three closely spaced -CCGG- sequences (-84, -96, -110) is not methylated both in 5-day embryonic red blood cells and in 12-day definitive red blood cells, the methylation patterns of the CG-containing sequences at -225 and -232 and of two more located further upstream are perfectly correlated with gene activity (Ginder and McGhee, 1981); that is, they are completely methylated in 5-day embryonic red blood cells in which the β gene is not transcribed, but are not methylated at all in 12-day red blood cells. McGhee *et al.* (1981) showed that this pattern is correlated with nuclease sensitivity of the same region. Using various nucleases, including DNAse I, micrococcal nuclease, and restriction endonucleases, they showed that the upstream sequence from -60 -260 is hypersensitive to *in vivo* digestion by these enzymes. These results suggest that the DNA in this region, when organized into active chromatin, may not consist of normal nucleosomes. A similar study by Weintraub *et al.* (1981) has been performed on the chicken adult α-like globin gene cluster in 12-day red blood cells. Two interesting points came from that study: 1) there is one DNAse I-hypersensitive site near the 5′ end of each of the three α-like globin genes (see the three arrows in Figure 12.5); and 2) the presence of these three DNAse I-hypersen-

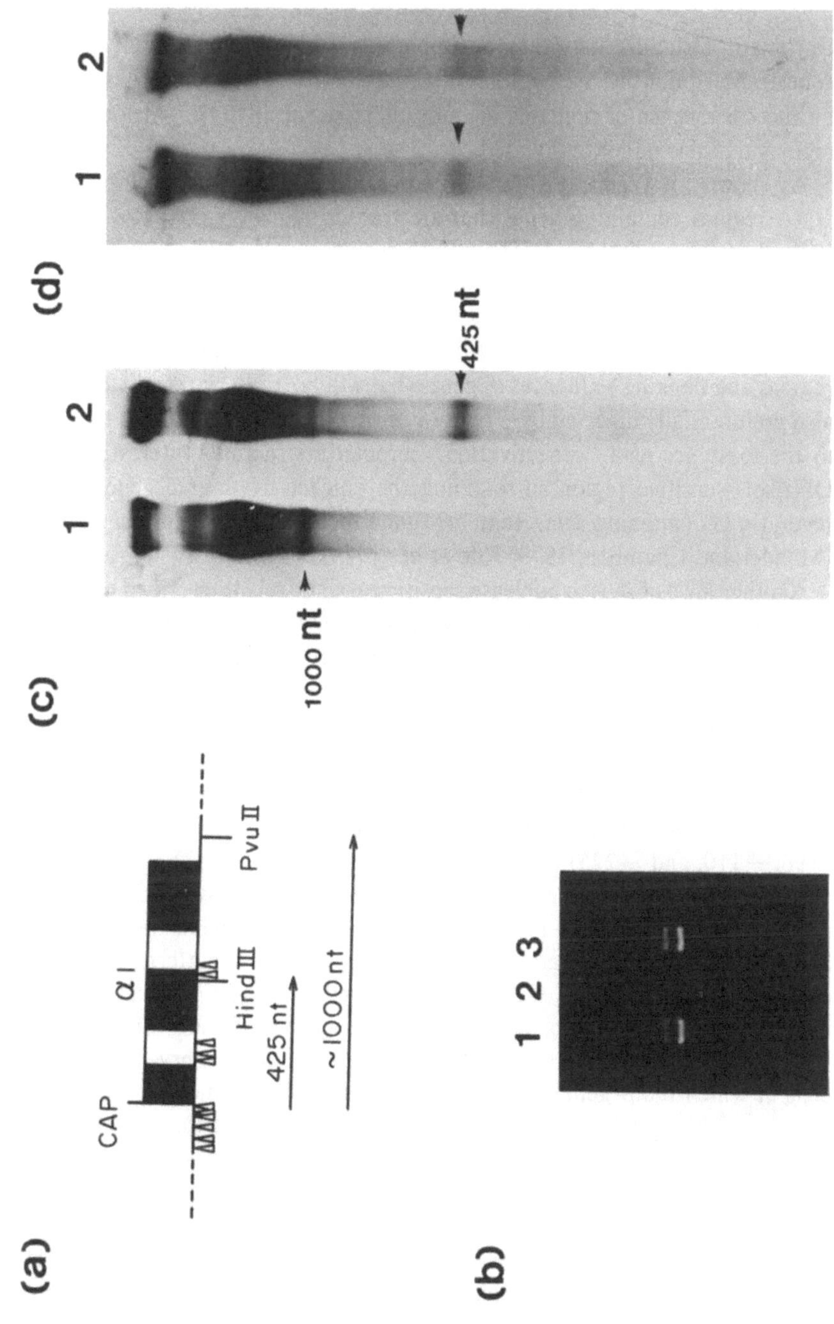
(a)
CAP
α I
Hind III
Pvu II
425 nt
~1000 nt
(b)
1 2 3
(c)
1 2
1000 nt
425 nt
(d)
1 2

sitive sites is inversely correlated with the methylation level in these regions. For example, the DNAse I-hypersensitive site at the 5′ end of the embryonic π' gene is present only in 5-day red blood cells, but not in 12-day red blood cells. Flanking this DNAse I-hypersensitive site are two -CCGG- sequences that are completely methylated in the 12-day red blood cells, but are undermethylated (one of them is totally unmethylated) in embryonic red blood cells. Similarly, the two DNAse I-hypersensitive sites flanking the 5′ ends of the adult α^D- and α^A-globin genes are present in both 5-day embryonic and 12-day definitive red blood cells, but not in the brain. This pattern is correlated again with the existence of several -CCGG- sequences in these two regions that are undermethylated in both embryonic and adult red blood cells, but are completely methylated in the brain (see Figure 12.5).

It seems interesting to mention again the previous observation that DNA regions flanking the rabbit adult β1-globin gene are undermethylated in both the blood island (the embryonic erythropoietic tissue) and the bone marrow (the adult erythropoietic tissue), even though this gene is not transcribed in the blood island. As discussed above, there is a close correlation between DNA undermethylation and "active" chromatin structure. It is possible that chromosomal regions in the "preactivation" state are also undermethylated. This would explain the undermethylation of rabbit β1-globin gene in the blood island, if the DNA region containing this gene is organized into a "preactivation" chromatin structure in the embryonic erythropoietic cells.

Figure 12.6. *In vitro* transcription of HpaII methylase-treated human α1-globin gene. Plasmid subclone pRBα1 (Lauer *et al.* 1980) containing the human α1 globin gene was methylated with HpaII methylase, truncated by HindIII or PvuII restriction enzyme, and transcribed in Hela whole-cell extract (Proudfoot *et al.* 1980). The RNA transcripts were analyzed on 7M urea-5% polyacrylamide gels by using denatured, ^{32}P-end-labeled HaeIII-pBR322 fragments as length markers. (a) Restriction map of α1-globin gene and its flanking DNA sequences. The symbol "Δ" indicates the 11 -CCGG- sequences located both near the cap site and within the gene at positions -137, -116, -86, -56, -4, $+159$, $+176$, $+210$, $+441$, and $+468$, respectively. The arrows show the two RNA transcripts, 425-nt and 1,000-nt long, which are expected to be synthesized from pRBα1 plasmid truncated at HindIII and PvuII restriction sites, respectively. (b) Test of completion of HpaII methylation of pRBα1. Lane 1, pRBα1 methylated by HpaII methylase; lane 2, methylated pRBα1 digested with MspI restriction enzyme; lane 3, methylated pRBα1 digested with HpaII restriction enzyme. (c) *In vitro* transcription pattern of unmethylated pRBα1 DNA truncated at PvuII recognition site (lane 1) or HindIII recognition site (lane 2). (d) Comparison of *in vitro* transcription patterns of HindIII-truncated DNA template of either unmethylated (lane 1) or methylated pRBα1 (lane 2).

What mechanisms could account for the correlation between the methylation pattern of globin genes and their chromatin structure? There are several possibilities:

1. Nucleosomes containing methylated DNA have different stability and structure than those of normal nucleosomes.
2. Nucleosomes containing methylated and unmethylated DNA have different affinities for some nonhistone proteins that are involved in gene regulation.
3. Methylated DNA is organized into a different higher-order chromatin structure than that of unmethylated DNA.

Recently, the *in vitro* biochemical studies by Felsenfeld *et al.* have provided some interesting information with regard to the possible effects of DNA methylation on chromatin structure. First, they have shown that methylation of C residues has very little effect on the conformation of DNA, in the sense that it does not unwind the DNA helix significantly (Felsenfeld *et al.* 1982). This has been confirmed by Cheng *et al.* (1983). They have also performed a series of experiments using the alternating (CpG) polymer as the substrate. They showed that: 1) the methylated polymer undergoes transition from B-DNA to Z-DNA conformation at much lower ionic strength than the ummethylated one (Behe and Felsenfeld, 1981); 2) in 0.2 M NaCl, B-form DNA (nucleosomal core DNA obtained from chicken erythrocyte nucleosomes) has a greater affinity for histones than does the Z-form DNA (methylated core-size CpG polymer); 3) if maintained at B form, the methylated core-size CpG polymer has a twofold greater affinity for histones than the unmethylated CpG core DNA; 4) the methylated (CpG) polymer in the B conformation assembles with histones into nucleosomal core particles that are indistinguishable both from core particles derived from chicken erythrocyte nuclei and those made with unmethylated (CpG) polymer; 5) the normal nucleosome structure is disrupted by the presence of Z conformation of the (CpG) polymer; and 6) core particles containing methylated and unmethylated polymer bind the high-mobility group proteins HMG14 and HMG17 in an identical fashion. These nonhistone proteins may be associated preferentially with transcriptionally active chromatin (Weisbrod *et al.* 1980).

As noted by Felsenfeld *et al.* (1982), all of these conclusions are derived from experiments using (CpG) polymer as the substrate. The nucleotide sequence data of the eukaryotic globin genes and their flanking DNA regions indicate that most, if not all, of the dinucleotide –CpG– sequences are not immediately flanked by other CpG residues; hence, they may not behave as the alternating CpG polymer does *in vitro*. With respect to this point, Felsenfeld *et al.* (1982) have methylated a cloned fragment containing the adult chicken β-globin gene and its flanking sequences with HpaII methylase, and they have incorporated it into nucleosomal core particles. The results showed that there is no difference in the binding of these core particles to HMG14 and HMG17 proteins from those composed of unmethylated DNA.

These results strongly suggest that if DNA methylation regulates globin gene transcription, it probably does so by affecting the interaction of other regulatory nonhistone proteins with the DNA chromatin. The RNA polymerase II and other transcription factor(s) may be such candidates, even though it has been shown that methylation of –CCGG– sequences has no obvious effects on *in vitro* transcription of naked DNA templates (see previous sections of this chapter).

Conclusions

Transcriptional inactivity of many eukaryotic globin genes during development and differentiation is definitely correlated with DNA hypermethylation near the 5′ end of the gene. Limited experimental data suggest that methylation probably is one of the causative factors responsible for repressing transcription. The mechanisms of this transcriptional inhibition *in vivo* may be complex and could involve interaction of chromatin with specific nonhistone proteins. One of the possibilities is that DNA methylation of specific CpG-containing sequences prevents the formation of an active chromatin domain. Extensive isolation and characterization of eukaryotic methylase(s) (see Chapter 3), and the nonhistone proteins involved in the globin gene transcriptions, may be necessary before we understand the molecular basis of the relationship between DNA methylation of specific sequences and the differential transcription of globin genes.

Acknowledgments

I wish to thank Drs. Doug Engel, Dick Flavell, Gordon Ginder, Timothy Ley, and James McGhee for their prompt response to my request for preprints and reprints. In particular, I would like to thank Dick Flavell, Gordon Ginder, and Doug Engel for communicating their results before publication. Dr. Michael Turelli has provided helpful and stimulating discussions. This project is supported, in part, by a grant from the National Institute of Arthritis, Diabetes, and Digestive and Kidney Diseases (AM 29800).

References

Banerji J, Rusconi S, Schaffner W: Expression of a β-globin gene is enhanced by remote SV40 DNA sequences. *Cell* (1981); 27:299–308.

Behe M, Felsenfeld G: Effects of methylation on a synthetic polynucleotide: The B-Z transition in poly(dG-m^{5}dC)·poly(dG-m^{5}dC). *Proc Natl Acad Sci USA* 1981; 78:1619–1623.

Busslinger M, de Boer E, Wright S, Grosveld F G and Flavell R. A. The sequence GGCCGG is resistant to MspI cleavage *Nucleic Acids Res* 1983a; 11:3559–3569.

Busslinger M, Hurst J, Flavell RA: DNA methylation and regulation of globin gene expression. *Cell* 1983b; 34:197–206.

Cate RL, Chick W, Gilbert W: Comparison of the methylation pattern of the two rat insulin genes. *J Biol Chem* 1983; 258:6645–6652.

Cheng S-C, Herman G, Modrich P: Extent of equilibrium helix perturbation upon enzymatic methylation of DNA. (*J Biol Chem*, submitted, 1983).

Day LE, Hirst AJ, Lai EC, Mace M Jr., Woo SLC: 5′-Domain and nucleotide sequence of an adult chicken chromosomal β-globin gene. *Biochemistry* 1981; 20:2091–2098.

DeSimone J, Heller P, Hall L, Zwiers D: 5-Azacytidine stimulates fetal hemoglobin synthesis in anemic baboons. *Proc Natl Acad Sci* 1982; 79:4428–4431.

Dodgson JB, McCune KC, Rusling DJ, Krust A, Engel JD: Adult chicken α-globin genes, α^{A} and α^{D}: no anemic shock α-globin exists in domestic chickens. *Proc Natl Acad Sci USA* 1981; 78:5998–6002.

Dodgson JB, Engel JD: The nucleotide sequence of the adult chicken α-globin genes. *J Biol Chem* 1983; 258:4623–4629.

Dolan M, Sugarman BJ, Dodgson JB, Engel JD: 1981; Chromosomal arrangement of the chicken β-type globin genes. *Cell* 24:669–677.

Dolan M, Dodgsen JB, Engel JD: Analysis of the adult chicken β-globin gene. *J Biol Chem* 1983; 258:3983–3990.

Engel JD, Rusling DJ, McCune KC, Dodgson JB: Unusual structure of the chicken embryonic α-globin gene, π′. *Proc Natl Acad Sci USA* 1983;80:1392–1396.

Felsenfeld G, Nickol J, Behe M, MeGhee J, Jackson D: *Methylation and Chromatin Structure.* Cold Spring Harbor Symposium on *Quantitative Biology,* Cold Spring Harbor Laboratory, 1983, vol 48, in press.

Ginder GD, McGhee JD: DNA methylation in the chicken adult β-globin gene: a relationship with gene expression, Stamatoyannopoulos G, Nienhuis AW (eds): in *Organization and expression of Globin Genes,* New York, Alan R Liss, Inc, 1981; pp 191–201.

Ginder GD, Whitters M, Kelley K, Chase RA: *In vivo* demethylation of chicken embryonic β-type globin genes with 5-azacytidine., Stamatoyannopoulos G, Nienhuis AW (eds): in *Hemoglobin Switching.* New York, Allan R. Liss, Inc, 1983, pp 501–510.

Haigh LS, Owens BB, Hellewell S, Ingram VM: DNA methylation in chicken α-globin gene expression. *Proc Natl Acad Sci USA* 1982; 79:5332–5336.

Hardison RC, Butler ET III, Lacy E, Maniatis T: The structure and transcription of four linked rabbit β-like globin genes. *Cell* 1979; 18:1285–1297.

Hardison RC: The nucleotide sequence of rabbit embryonic globin gene β3. *J Biol Chem* 1981; 256:11780–11786.

Hardison RC: The nucleotide sequence of the rabbit embryonic globin gene β4. *J Biol Chem* 1983; 258:8739–8744.

Jones PA, Taylor SM: Cellular differentiation, cytidine analogs and DNA methylation. *Cell* 1980; 20:85–93.

Kuo T, Mandel J, Chambon P: DNA methylation: correlation with DNase I sensitivity of chicken ovalbumin and conalbumin chromatin. *Nucleic Acids Res* 1979; 7:2105–2113.

Lacy E, Hardison RC, Quon D, Maniatis T: The linkage arrangement of four rabbit β-like globin genes. *Cell* 1979; 18:1273–1283.

Lauer J, Shen C-KJ, Maniatis T: Chromosomal arrangement of human α-like globin genes: sequence homology and α-globin gene duplication. *Cell* 1980; 20:119–130.

Ley TJ, DeSimone J, Anagnou NP, Keller G, Humphries RK, Turner PH, Young NS, Heller P, Nienhuis AW: 5-Azacytidine selectively increases γ-globin synthesis in a patient with β^+-thalassemia. *N Engl J Med* 1982; 307:1469–1475.

Ley TJ, DeSimone J, Noguchi CT, Turner PH, Schechter AN, Heller P, Nienhuis AW: 5-Azacytidine increases γ-globin synthesis and reduces the proportion of dense cells in patients with sickle cell anemia. *Blood* 1983; 62:370–380.

Mandel J, Chambon P: DNA methylation: organ specific variations in the methylation pattern within and around ovalbumin and other chicken genes. *Nucleic Acids Res* 1979; 7:2081–2103.

Maniatis T, Fritsch EF, Lauer J, Lawn RM.: The molecular genetics of human hemoglobins. *Ann Rev Genet* 1980; 14:145–178.

McGhee JD, Ginder GD: Specific DNA methylation sites in the vicinity of the chicken β-globin genes. *Nature* 1979; 280:419–420.

McGhee JD, Wood WI, Dolan M, Engel JD, Felsenfeld G: A 200 base pair region at the 5′ end of the chicken adult β-globin gene is accessible to nuclease digestion. *Cell* 1981; 27:45–55.

Mellon P, Parker V, Gluzman Y, Maniatis T: Identification of DNA sequences required for transcription of the human α-globin gene in a new SV40 host-vestor system. *Cell* 1981; 27:279–288.

Proudfoot NJ, Shander MHM, Manley JL, Gefter ML, Maniatis T: Structure and *in vitro* transcription of human globin genes. *Science* 1980; 209:1329–1335.

Rohrbaugh ML, Hardison RC: Analysis of rabbit β-like globin gene transcripts during development. *J Mol Biol* 1983; 164:395–417.

Shen C-KJ, Maniatis T: Tissue specific DNA methylation in a cluster of rabbit β-like globin genes. *Proc Natl Acad Sci USA* 1980; 77:6634–6638.

Shen C-KJ, Maniatis T: The organization, structure, and *in vitro* transcription of Alu family RNA polymerase III transcription units in the human α-like globin gene cluster: precipitation of *in vitro* transcripts by lupus anti-La antibodies. *J Mol Appl Genetics* 1982a; 1:343–360.

Shen C-KJ, Maniatis T: The nucleotide sequence, DNA modification, and *in vitro* transcription of Alu family repeats in the human-like globin gene cluster, in Huang PC, Kuo TT, Wu R (eds) *Genetic Engineering Techniques: Recent Developments.* New York: Academic Press 1982b; 129–158.

Shen S-H, Slightom JL, Smithies O: A history of human fetal globin gene duplication. *Cell* 1981;26:191–203.

Stalder J, Groudine M, Dodgson JB, Engel JD, Weintraub H: Hb switching in chickens. *Cell* 1980; 19:973–980.

Stein R, Razin A, Cedar H: *In vitro* methylation of the hamster adenine phosphoribosyltransferase gene inhibits its expression in mouse L Cells. *Proc Natl Acad Sci USA* 1982a; 79:3418–3422.

Stein R, Gruenbaum Y, Pollack Y, Razin A, Cedar H: Clonal inheritance of the pattern of DNA methylation in mouse cells. *Proc Natl Acad Sci USA* 1982b; 79:61–65.

Treisman R, Proudfoot NJ, Shander M, Maniatis T: A single-base change at a splice site in a β^{o}-thalassemic gene causes abnormal RNA splicing. *Cell* 1982; 29:903–911.

van der Ploeg LHT, Flavell RA: DNA methylation in the human γ-δ-β globin locus in erythroid and non-erythroid tissues. *Cell* 1980; 19:947–958.

Vardimon L, Kressman A, Cedar H, Maechler M, Doerfler W: Expression of a cloned adenovirus gene is inhibited by *in vitro* methylation. *Proc Natl Acad Sci USA* 1982; 79:1073–1077.

Waalwijk C, Flavell RA: DNA methylation at a CCGG sequence in the large intron of the rabbit β-globin gene: tissue specific variations. *Nucleic Acids Res* 1978; 5:4631–4641.

Weintraub H, Groudine M: Chromosomal subunits in active genes have an altered conformation. *Science* 1976; 93:848–858.

Weintraub H, Larsen A, Groudine M: α-Globin gene switching during the development of chicken embryos: expression and chromosome structure. *Cell* 1981; 24:333–344.

Weisbrod S, Groudine M, Weintraub H: Interaction of HMG14 and 17 with actively transcribed genes. *Cell* 1980; 19:289–301.

Weisbrod S: Active chromatin. *Nature* 1982; 297:289–295.

Wu C, Bingham PM, Livak KJ, Holmgren R, Elgin SCR: The chromatin structure of specific genes: I. Evidence for higher order domains of defined DNA sequences. *Cell* 1979; 16:797–806.

13

X Inactivation, DNA Methylation, and Differentiation Revisited

Arthur D. Riggs*

In 1975, two papers suggested a role for DNA methylation in X chromosome inactivation. In one paper (Riggs, 1975), I argued that: 1) DNA methylation should affect protein-DNA interactions; 2) methylation patterns and a maintenance methylase should exist; and 3) DNA methylation should be involved in mammalian cellular differentiative processes. Holliday and Pugh (1975) argued similarly, although less weight was given to X inactivation and more weight was given to the possibility that 5-methylcytosine (5-meCyt) might be deaminated to thymidine; thus a specific mutational change would be generated, as suggested by Scarano (1971). Recently, several studies of X chromosome inactivation have contributed to the emerging body of evidence supporting a role for DNA methylation in mammalian gene regulation; it is these studies that will be reviewed in this chapter. More comprehensive reviews of X chromosome inactivation have been published recently (Gartler and Riggs, 1983; Graves, 1983).

After fertilization, female mammalian cells have two X chromosomes, and recent evidence suggests that both are active. During embryonic development, one of the X chromosomes in each cell becomes condensed, late replicating, and genetically inactive. In the embryo proper (Figure 13.1), inactivation takes place at about the time of implantation; both the maternal (X^m) and the paternal (X^p) X chromosomes have equal probabilities of becoming inactive. Thus, all somatic cells of a placental mammalian female display what is called random X inactivation. Once the initial determination is made within a given cell, it is permanently fixed; all progeny cells (and resulting cell lineage) will have the same active X chromosome (either X^m or X^p, but not both). This is a clear

*Division of Biology, Beckman Research Institute of the City of Hope, Duarte, California 91010

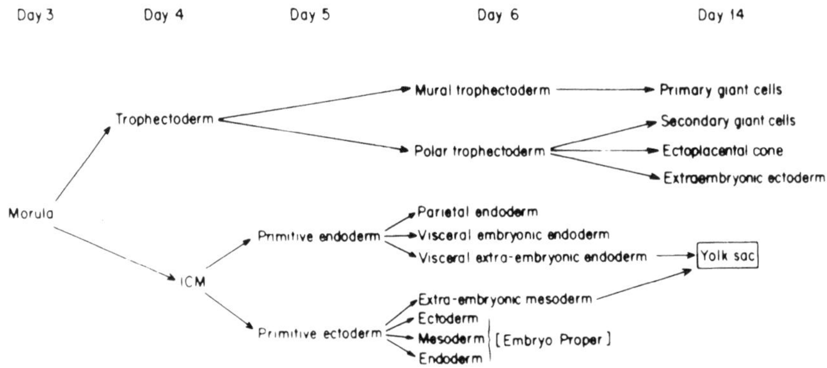

Figure 13.1 Embryonic and extraembryonic cell lineages in the mouse. (From Kratzer *et al.* 1983).

case of somatic heredity, and females are true cellular mosaics. The X chromosomes remain differentiated through repeated mitoses, even though the inactive chromosome is retained in the same nucleoplasm as the active chromosome. How is this possible? At the molecular level, how can two identical DNA sequences (inbred mice) be maintained differentiated from each other during innumerable DNA replication events?

An Inactivation-Reactivation Cycle

A very important question is whether or not an inactivation-reactivation cycle occurs in the germ line. Present evidence strongly indicates that X inactivation occurs in primordial germ cells, and that reactivation regularly occurs in early meiosis. This was first suggested by cytologic studies (Ohno, 1962; Gartler *et al.* 1980) in which a condensed X chromosome is seen in oogonia, but not in oocytes. Additional evidence is provided by XX oocytes having a twofold higher level of X-linked enzymes than XO oocytes (Epstein *et al.* 1972). A recent study using electrophoretic variants of the X-linked enzyme G6PD has provided clear evidence that an inactivation-reactivation cycle occurs in the mouse (Kratzer and Chapman, 1981). G6PD is a dimer; a heterodimer produced by the expression of two alleles in the same cell can be distinguished from both homodimers. In an oogonia cell, either of the two alleles are produced, but not both. However, in oocytes that have entered meiotic prophase, both alleles are expressed. Thus, whatever changes occur in the inactive X chromosome, they must be of a type that can be efficiently erased during meiosis. Thus, base sequence changes (true mutations) and nonreversible chromosome deletions and rearrangements are ruled out for the germ line.

Stability in Somatic Cells

In contrast to the efficient reversibility during meiosis, inactivation in somatic cells is extraordinarily stable. In fact, reactivation of HPRT in normal diploid cells has never been obtained, even when strong selection has been applied. The reversion rate must be less than 10^{-8} in normal cells. Reversion of an *Hpt* gene on the inactive chromosome to the active state has been observed by Kahn and DeMars (1975) in mouse-human hybrid cells containing a functional *Hpt* gene only on the inactive human X chromosome. They were able to select for spontaneous revertants at a rate of about 10^{-6}. The $HPRT^{+}$ reactivants were found to have not reactivated other X-linked enzymes, such as PGK or G6PD. Also, the reactivants were stable and still carried the *Hpt* gene on the human X chromosome. These experiments were the first to establish that the inactive X chromosome carries potentially active genes; no irreversible change (e.g., mutation or deletion) was made in the *Hpt* gene in somatic cells. These experiments also established that X inactivation can be regulated "piecemeal." The active section was still x linked and, therefore, was surrounded by inactive X chromatin.

An *Hpt* gene in DNA from an inactive X chromosome functions poorly in DNA-mediated gene transfer experiments. This was first shown by Liskay and Evans (1980), and it has now been confirmed in several independent experiments with mouse DNA or human chromosomes in mouse cells (see Table 13.1). No productive gene transfer is usually seen when the functional *Hpt* gene is on the inactive X chromosome. Together, these experiments clearly indicate that purified DNA from the inactive X chromosome is at least 20 times and more likely 100 times less efficient in gene transfer for the *Hpt* gene.

Table 13.1 Methylation Maintains X Inactivation

Experimental Result	Reference
DNA-mediated gene transfer of an *Hpt* gene on an inactive X is inefficient relative to that on an active X	Liskay and Evans (1980) Chapman *et al.* (1982) Kratzer *et al.* (1983) Venolia *et al.* (1982, 1983)
5-azacytidine causes efficient reactivation of X-linked genes; up to 1% of surviving cells express a previously silent *Hpt* gene	Mohandas *et al.* (1981) Lester, Korn, and DeMars (1982) Graves (1982)
5-Aza-2′-deoxycytidine is 10 times more effective than 5-azacytidine in X reactivation	Jones *et al.* (1982)
5-Aza-2′-deoxycytidine is most effective in late S for *Hpt* reactivation	Jones *et al.* (1982)
After reactivation by 5-azacytidine, the *Hpt* gene functions normally in DNA-mediated gene transfer	Lester, Korn, and DeMars (1982) Venolia *et al.* (1982)

One study by de Jong *et al.* (1982), with cells from a human Lesch-Nyhan (*Hpt*$^-$) heterozygote transformed with SV40 DNA, failed to show a difference in gene transfer efficiency between the active and inactive chromosomes. However, Venolia and Gartler (1983) have done similar experiments and find, for their cells derived from a Lesch-Nyhan heterozygote, that the normal *Hpt* allele on the inactive X chromosome does not function in gene transfer. It is possible that the cells used by DeJonge *et al.* carried a complete gene on the Lesch-Nyhan chromosome that was silenced by a mechanism other than a mutation within the structural gene for *Hpt*. On balance, these gene transfer experiments using purified DNA provide strong evidence that a change has been made in the inactive X chromosome at the DNA level.

DNA Methylation Maintains the Inactive State

The gene transfer experiments described thus far indicate that the DNA is modified; however, they do not provide evidence about the type of modification or the reversibility of the modification. This type of information has come from experiments using 5-azacytidine (5-aza-C) and related compounds that prevent the formation of 5-meCyt by inactivating DNA methylase after incorporation into DNA (see Chapter 9).

The reactivation of genes on the inactive X chromosome by 5-aza-C was first reported by Mohandas *et al.* (1981). They used a specially constructed mouse-human hybrid cell line that only had an inactive human X chromosome and that was phenotypically HPRT$^-$ because of mutations in the mouse genes. The reversion rate to HPRT$^+$ in the absence of treatment was about 10^{-6}; however, after treatment with 2-μmol of 5-aza-C, the reactivation rate increased more than 1,000-fold. In subsequent experiments with the same cell line (Jones *et al.* 1982), up to 8% of the surviving cells displayed human HPRT activity. Human HPRT is electrophoretically distinguishable from the mouse enzyme. Reactivants for human PGK and G6PD also were observed.

Lester, Korn, and DeMars (1982) did similar experiments with a mouse-human hybrid cell line. They also found that 5-aza-C treatment increased the reactivation rate more than 1,000-fold, from 10^{-6} to 8×10^{-3}. Reactivation of G6PD and PGK was also seen. The reactivation rate of PGK was even higher than that of HPRT. PGK reactivants could be found even by screening unselected cells. Since the reactivation rate for PGK was about the same for the unselected cell population as it was for those that had been selected for HPRT activity, it was concluded that the reactivation on the two genes were independent events.

Graves (1982) tested the effect of 5-aza-C on cells obtained from a hybrid mouse—a cross between *Mus caroli* and *M. musculus*. This is a useful cross, because the two species have electrophoretically distinguishable HPRT, PGK, and G6PD enzymes. The spontaneous reactivation rate for HPRT in these caroli-musculus hybrid cells is much less than that for the mouse-human cells studied previously; no spontaneous reactivants have been seen in more than 10^8 cells. However, after treatment with 5-aza-C, HPRT reactivants are more than 10,000 times more frequent and constitute up to 0.3% of surviving cells. The spectacular sensitivity of hybrid cells to 5-aza-C reactivation is without question.

The reactivants obtained by treatment with 5-aza-C are stable for several months and for up to 200 generations in the absence of selection (Mohandas, 1981; Lester, Korn, and DeMars, 1982; Graves, 1982). When selection against the reactivated *Hpt* allele is applied, HPRT variants appear with about a 10^{-4} frequency, but all of these have lost the human chromosome (Lester, Korn, and DeMars, 1982). Thus, reactivated genes remain on the X chromosome; in somatic cells, the active state is stable. This fact further underscores the extreme stability of the inactive state in normal somatic cells, because—at least in tissue culture—reactivants are viable and stable.

Jones *et al.* (1982) provided evidence that 5-aza-C causes reactivation by a DNA-mediated mechanism. They found that the deoxy derivative of 5-aza-C 5-aza-2^1-deoxycytidine, is 10 times more potent at inducing reactivation. The deoxy analog cannot enter RNA because of the irreversibility of ribonucleotide reductase. 5-azacytidine can be deaminated to 5-azadeoxyuridine, but this agent is not biologically active (see Chapter 9). In keeping with the DNA-mediated mode of action, it was found that deoxyazacytidine is most effective in late S, when the inactive X chromosome is replicating. The maximum reactivation efficiency obtained with deoxyazacytidine was 8%.

When viewed together, these experiments with the X inactivation system probably provide the strongest evidence that DNA methylation is involved in a gene-silencing mechanism used in normal, early embryonic development. Methylation apparently is sufficient to stably maintain X inactivation, because when methylation is blocked, at least three genes efficiently reactivate (probably independently). For emphasis, the various experiments and experimental results are summarized in Table 13.1.

At the Detailed Level, the Methylation Story is Complex

From the experiments discussed above, there seems little doubt that 5-me-Cyt is involved in the maintenance of X inactivation in somatic cells. However, at a more detailed level, we now know that the simple idea that the active X chromosome is more methylated in most DNA regions can be ruled out. Wolf and Migeon (1982) first probed for methylation differences between male (active X only) and female cells (active plus inactive X) by using X-specific hybridization probes and the HpaII/MspI restriction enzyme assay. As probes, they used randomly cloned fragments selected only to be nonrepetitious, specific to the X chromosome, and nonhybridizing to the mouse genome. One of the cloned sequences apparently was transcribed. In total, 18 kb of X-chromosome DNA was probed for methylation changes; no differences between male and female were seen. They also reported that 5-aza-C did not cause detectable reactivation of *Hpt* on the inactive X chromosome in human diploid fibroblasts; in the same paper, hybrid cells between mouse and human cells were shown to have lower methylation levels for the regions probed. This latter result could be an explanation of why reactivation of genes on an inactive chromosome in normal diploid cells has yet to be reported.

Both the above results and the observations that reactivated X-linked genes can exist in at least two states of activity (Migeon *et al.* 1982; Lester, Korn, and DeMars, 1982) suggest that X inactivation may be a multistep process that possibly needs several cooperatively interacting methylated sites for maximal genetic inactivity and stability. We recently have used a cDNA-derived clone for human X-linked PGK (Singer-Sam *et al.* 1983) to probe for methylation in the coding region and introns of the genomic human *PGK* gene. We have not yet seen any simple, reproducible methylation differences between males and females (Keith, Singer-Sam, and Riggs, unpublished). However, we do not as yet have a cloned probe that unambiguously identifies the 5′-flanking region of the X-linked *PGK* gene. Thus, it seems that for the X chromosome, as for several autosomal genes, most methylation may be relatively unimportant with respect to the maintenance of genetic inactivity. In contrast, the idea that methylation of critical sites is a significant factor controlling gene activity is strengthened by the X inactivation data.

Two Forms of X Inactivation

The inactivation picture is further complicated by two different forms of inactivation. Random inactivation of either maternal or paternal X chromosomes takes place in the cells of the embryo proper. This is the type of inactivation discussed thus far. However, nonrandom inactivation takes place in all extraembryonic tissues, with the paternal X chromosome being preferentially inactivated (Takagi and Sasaki, 1975). In marsupials, this type of nonrandom paternal inactivation is also seen in somatic cells. Recent work by Kratzer *et al.* (1983) indicates that the *Hpt* gene on the paternally inactivated X chromosome in the yolk sac (an extraembryonic tissue) does function in DNA-mediated transfer. Although the primary result is that yolk-sac DNA is definitely less efficient in gene transfer than somatic cell DNA, their data also indicate that the paternally inactivated X-chromosome DNA, nonetheless, is three times less efficient in gene transfer than DNA from the active X chromosome. The multistep methylation model suggested above could be used to interpret these data. The prediction would be that the paternally inactivated X chromosome should be less methylated in the 5′-flanking region or other critical regions. This idea is rendered more probable by observations that the extraembryonic tissues are very significantly undermethylated in relation to the embryo proper (Manes and Menzel, 1981; Razin *et al.* 1984).

Chromosome Imprinting

Another very interesting phenomenon, which may be relevant to methylation, exists in the X inactivation system. Since the paternal X chromosome is distinguished form the maternal X chromosome, with respect to inactivation in marsupial embryos and extraembryonic tissues of placental mammals, it must be "imprinted" in some way via passage through the male line. At present, there is no evidence directly connecting chromosome imprinting to methylation, but a connection has been suggested (Gartler and Riggs, 1983). Some intriguing data that are possibly relevant to methylation and imprinting are available in the literature and will be briefly summarized here.

All single-copy genes that have been examined either are as highly or more highly methylated in sperm than they are in nonexpressing somatic cells. In contrast, at least one satellite DNA is very hypomethylated in bull sperm. This same repetitive sequence shows a tissue-specific pattern of methylation in somatic cells (Sano and Sager, 1982; Pages and Roizes, 1982). These important experiments establish that: 1) some sequences in sperm enter the egg undermethylated; and 2) *de novo* methylation occurs during development. One should recall that experiments with retrovirus infection have established that *de novo* methylation may be more efficient in the preimplantation embryo and undifferentiated embryo cell lines (see Chapter 10). Sequences other than bovine satellite are likely to be undermethylated in sperm, because several mammalian species show a reduced level of total DNA methylation in sperm; most noticeably in repetitive sequences (Gama-Sosa *et al.* 1983). The methylation model for imprinting is described below.

Some sequences in sperm are undermethylated in relation to the egg. *De novo* methylation begins in the fertilized egg; however, in those tissues that differentiate early (the extraembryonic tissues; see Figure 13.1), the hypomethylation "imprint" on the paternal X chromosome has not yet been erased. This notion is supported by the evidence that methylation in the early differentiating tissues is lower than in the embryo proper, which probably remains totipotent and does not undergo X inactivation until about the time of implantation (Manes and Menzel, 1981; Razin *et al.* 1984). It should be noted that mechanisms other than methylation may be primarily responsible for keeping genes silent in extraembryonic tissues, which are so generally hypomethylated. One of the first events accompanying early tissue differentiation could be a large decrease in DNA methylation activity. This is in keeping with multiple independent mechanisms for gene control (Weintraub, Larsen, and Groudine, 1981)—an idea that, in my view, is a virtual certainty for complex mammalian cells.

X Inactivation and Cellular Differentiation

Because X chromosome inactivation is a recently evolved mechanism that occurs only in mammals, it is not likely to be completely new; rather, it probably resulted from a minor variation on preexisting mechanisms. It is virtually certain that methylation is involved in the maintenance of X chromosome inactivation. Conversely, no evidence exists that 5-meCyt is involved in the establishment of the differentiated state; this could be done by a mechanism independent of DNA methylation. However, since drastic changes in methylation levels and the methylation system take place in the early differentiation steps of the developing embryo, it seems likely that mammalian cells are making use of DNA methylation as one way of influencing gene expression during early development. If DNA methylation is used primarily as a somatically heritable mechanism for locking genes or DNA regions in a transcriptionally inactive state (Razin and Riggs, 1980), it may be most beneficial to study methylation changes in the early embryo proper shortly after implantation. Unfortunately, this is technically difficult; therefore, little unambiguous information is now available.

References

Chapman VM, Kratzer PG, Siracusa LO, Quarantillo BA, Evans R, Liskay RM: Evidence for DNA modification in the maintenance of X-chromosome inactivation of adult mouse tissues. *Proc Natl Acad Sci USA* 1983;79:5357–5371.

de Jonge AJR, Abrahams PJ, Westerveld A, Bootsma D: Expression of human HPRT gene on the inactive X chromosome after DNA-mediated gene transfer. *Nature* 1982; 295:624–626.

Epstein CJ: Expression of the mammalian X chromosome before and after fertilization. *Science* 1972;175:1467–1468.

Gama-Sosa MA, Wang YH, Kuo KC, Gehrke CW, Ehrlich M: The 5-methylcytosine content of high-repeated sequences in human DNA. *Nucl Acids Res* 1983;11:3087–3095.

Gartler SM, Rivest M, Cole RE: Cytological evidence for an inactive X chromosome in murine oogonia. *Cytogenet Cell Genet* 1980;28:203–207.

Gartler SM, Riggs AD: Mammalian X-chromosome inactivation. *Annu Rev Genet* 1983;17:153–190.

Graves JAM: 5-azacytidine-induced reexpression of alleles on the inactive X chromosome in a hybrid mouse cell line. *Exptl Cell Res* 1982;141:87–94.

Graves JAM: Inactivation and reactivation of the mammalian X chromosome, in Johnson MH (ed): *Development in Mammals*. Amsterdam, Elsevier Science Publishers, 1983, vol 5.

Holliday R, Pugh JE: DNA modification mechanisms and gene activity during development. *Science* 1975;187:226–232.

Jones PA, Taylor SM, Mohandas T, Shapiro LJ: Cell cycle-specific reactivation of X-chromosome locus by 5-azadeoxycytidine. *Proc Natl Acad Sci USA* 1982;79:1215–1219.

Kahan, B, DeMars R: Localized derepression on the human inactive X chromosome in mouse-human cell hybrids. *Proc. Natl Acad Sci USA* 1975;72:1510–1514.

Kaput J, Sneider TW: Methylation of somatic vs germ-cell DNAs analyzed by restriction endonuclease digestions. *Nucl Acids Res* 1979;7:2303–2322.

Kratzer PG, Chapman VM X-chromosome reactivation in oocytes of *Mus caroli*. *Proc Natl Acad Sci USA* 1981;78:3093–3097.

Kratzer PG, Chapman VM, Lambert H, Evans R, Liskay RM: Differences in the DNA of the inactive X chromosome of fetal and extraembryonic tissues of mice. *Cell* 1983;33:37–42.

Lester SC, Korn NJ, DeMars R: Derepression of genes on the human inactive X chromosome: evidence for differences in locus-specific rates of derepression and rates of transfer of active and inactive genes after DNA-mediated transformation. *Somatic Cell Genet* 1982;8:265–284.

Liskay RM, Evans RJ: Inactive X-chromosome DNA does not function in DNA-mediated cell transformation for the hypoxanthine phosphoribosyltransferase gene. *Proc Natl Acad Sci USA* 1980;177:4895–4898.

Manes C, Menzel P: Demethylation of CpG sites in DNA of early rabbit trophoblast. *Nature (London)* 1981;293:589–590.

Migeon BR, Wolf SF, Mareni C, Axelman J: Derepression with decreased expression of the G6PD locus on the inactive X chromosome in normal human cells. *Cell* 1982;29:595–600.

Mohandas T, Sparkes R, Shapiro LJ: Reactivation of an inactive X chromosome: evidence for X inactivation by DNA methylation. *Science* 1981;211:393–396.

Ohno S, Klinger HP, Atkin NB: Human oogenesis. *Cytogenetics* 1962;1:42–51.

Pages M, Roizes G: Tissue specificity and organization of CpG methylation in calf satellite DNA. *Nucl Acids Res* 1982;10:565–576.

Razin A, Webb C, Szyf M, Yisraeli J, Rosenthal A, Naveh-Many T: Variations in DNA methylation during mouse cell differentiation in vivo and in vitro. *Proc Natl Acad Sci USA* 1984;81:2275–2279.

Riggs, AD: X inactivation, differentiation, and DNA methylation. *Cytogenet Cell Genet* 1975;14:9–25.

Sano H, Sager R: Tissue specificity and clustering of methylated cytosines in bovine satellite I DNA *Proc Natl Acad Sci USA* 1982;79:3584–3588.

Scarano E: The control of gene function in cell differentiation and in embryogenesis. *Adv Cytopharmacol* 1971;1:13–23.

Singer-Sam J, Simmer RL, Keith DH, Shively L, Teplitz M, Itakura K, Gartler SM, Riggs AD: Isolation of a cDNA clone for human X-linked 3-phosphoglerate kinase by use of a mixture of synthetic oligodeoxyribonucleotides as a detection probe. *Proc Natl Acad Sci USA* 1983;80:802–806.

Sturm KS, Taylor JH: Distribution of 5-methylcytosine in the DNA of somatic and germline cells from bovine tissues. *Nucl Acids Res* 1981;9:4537–4546.

Takagi N, Sasaki M: Preferential inactivation of the paternally-derived X chromosome in the extraembryonic membranes of the mouse. *Nature (London)* 1975;256:640–642.

Venolia L, Gartler SM, Wassman ER, Yen P, Mahandas T, Shapiro LJ: Transformation with DNA from 5-azacytidine-reactivated X chromosomes. *Proc Natl Acad Sci USA* 1982;79:2352–2354.

Venolia L, Gartler SM: Comparison of the transformation frequency of human active and inactive X-chromosome DNA. *Nature (London)* 1983;302:82–83.

Weintraub H, Larsen A, Groudine M: Alpha-globin gene switching during the development of chicken embryos: expression and chromosome structure. *Cell* 1981;24:333–344.

Wolf SF Migeon BR: Studies of X-chromosome DNA methylation in normal human cells. *Nature (London)* 1982;295:667–671.

14

Left-handed Z-DNA and Methylation of d(CpG) Sequences

Alexander Rich*

DNA is a macromolecule that can adopt several different conformations. Until recently, it was believed that all of these represented variations of right-handed, double helical conformations with some differences associated with the amount of helical twist per base pair. It was believed that all of these variations preserved basic features of the molecule—including the fact that all of the nucleotides were similar. We now know that this is not true. DNA can exist in several conformations, both right- and left-handed; there can be different forms of the nucleotides existing simultaneously within the same molecule.

In this chapter, we consider the possible relevance of conformational changes in DNA and the methylation of d(CpG) sequences that are known to influence gene expression, especially in higher eukaryotes (Razin and Riggs, 1980). Methylation of these sequences has a great influence on the stability of the left-handed Z-DNA conformation.

Molecular Structure of Left-handed Z-DNA

Z-DNA is a left-handed form of the double helix in which the two antiparallel sugar-phosphate chains are held together by Watson-Crick base pairs (Rich et al, 1984). It was discovered in single crystal x-ray diffraction analyses of oligonucleotides that contained alternations of cytosine and guanine residues (Wang et al, 1979). Left-handed Z-DNA has 12 base pairs per turn with a helical repeat at 44.6 Å, which is comparable with right-handed B-DNA,

*Department of Biology, Massachusetts Institute of Technology Cambridge, Massachusetts 02139

which has 10 base pairs per turn with a helical repeat of 34Å. Z-DNA is somewhat thinner than B-DNA and is elongated. Figure 14.1 shows van der Waals diagrams of both B-DNA and Z-DNA. The heavy line passing from *phosphate* to phosphate allows one to trace the polynucleotide chain. Its zig-zag arrangement in Z-DNA contributed to the naming of the molecule.

One does not go from right-handed B-DNA to left-handed Z-DNA simply by twisting the molecule in the other direction. Instead, a number of internal conformational changes have to occur; the most significant one involves flipping

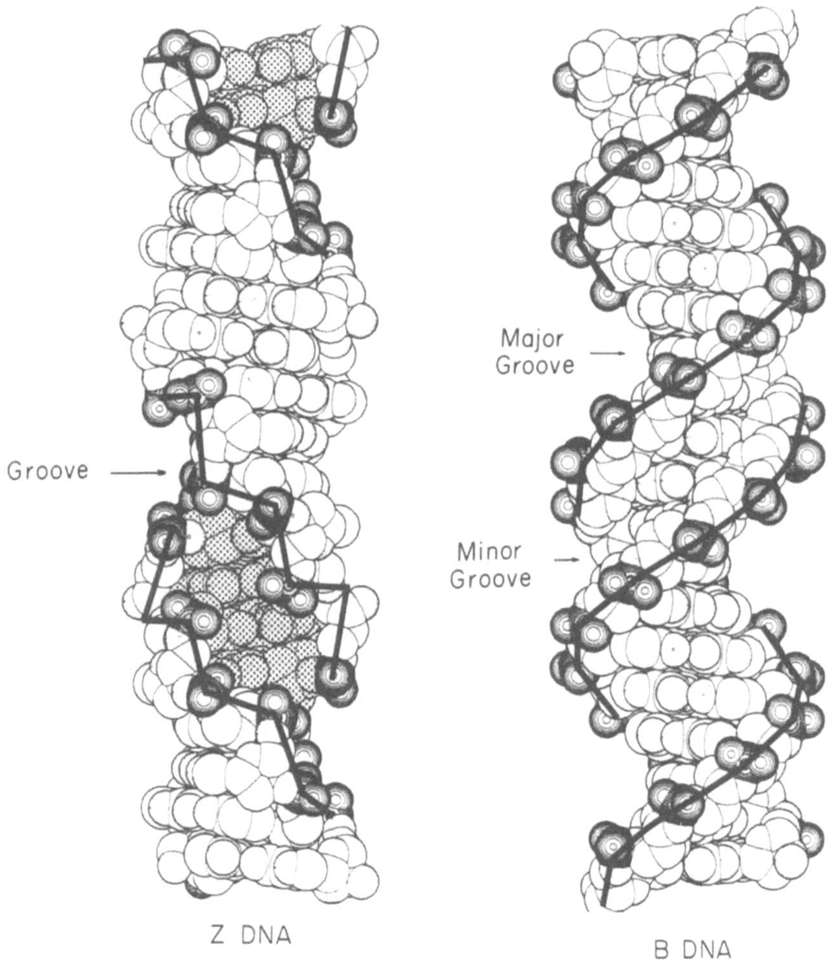

Figure 14.1 van der Waals models of Z-DNA and B-DNA. The irregularity of the Z-DNA backbone is illustrated by the heavy lines that go from phosphate-to-phosphate residue along the chain. The Z-DNA diagram shows the molecules as they appear in the hexamer crystal (Wang *et al.* 1979). The groove in Z-DNA is quite deep, extending almost to the axis of the double helix. In contrast, B-DNA has a smooth line connecting the phosphate groups, and two grooves are present; neither one of which extends to the helix axis of the molecule.

of the planar purine-pyrimidine base pairs, so that they have an upside down orientation relative to what is found in B-DNA.

An internal conformational change occurs when the base pairs flip over in changing from B-DNA to Z-DNA. In the Z-DNA crystal structure that contains alternating guanine and cytosine bases, the guanine residues were found to have adopted the syn conformation through a rotation about the glycosyl bond, as shown in Figure 14.2. The cytosine residues remain in the anti conformation, which is found in all of the nucleotides of B-DNA. However, the

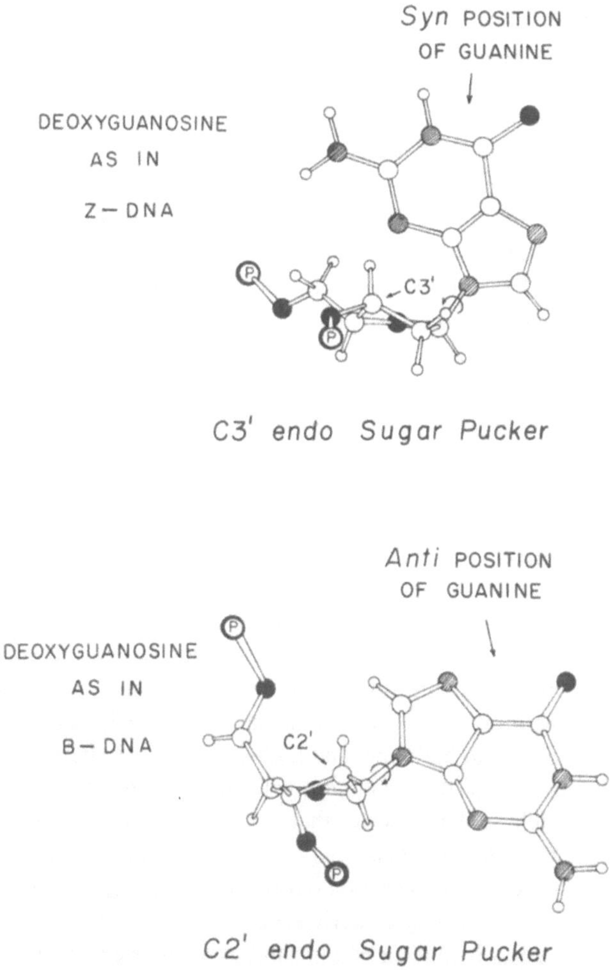

Figure 14.2 Conformation of deoxyguanosine in B-DNA and Z-DNA. The sugar is oriented so that the plane defined by C1′-01′-C4′ is horizontal. Atoms lying above this plane are in the endo conformation. The C3′ is endo in Z-DNA, while the C2′ is endo in B-DNA. In addition, Z-DNA has guanine in the syn position, in contrast to the anti position in B-DNA. The curved arrow around the glycosyl carbon-nitrogen linkage indicates the site of rotation.

cytidine residues also flip over in the process. It is the rotation of both this base and its sugar that gives rise to the irregular zig-zag character of the backbone. Purine residues can form the syn conformation without any energy loss; pyrimidine residues also can form the syn conformation, but there is some energy penalty in doing this because of crowding between the pyrimidine oxygen (02) atom and the sugar ring. Because of the presence of alternating syn conformations, Z-DNA is favored in sequences that have alternations of purines and pyrimidines.

In Z-DNA, there is one helical groove seen in Figure 14.3, which shows a cross-section of the molecule. It can be seen that the groove extends almost to the axis of the molecule, and the guanine-cytosine base pairs are positioned away from the axis. In Z-DNA, they form part of the outer convex surface of the molecule. In Figure 14.3, it can be seen that the C8 atom in the imidazole ring of guanine is in the outer part of the molecule, while the hydrogen attached to the C8 residue in B-DNA is in van der Waal's contact with the sugar-phosphate backbone.

The relative positions of the base pairs in Z-DNA and B-DNA are shown in the end views of Figure 14.4. These diagrams represent idealized and regular versions of both Z-DNA and B-DNA. It can be seen that the axis of B-DNA goes through the center of the base pair, while the axis of Z-DNA is removed from the base pair. For this reason, B-DNA has two grooves (one slightly broader than the other), while Z-DNA has one deep groove and an outer convex surface that is analogous to the major groove of DNA; but, in Z-DNA, it is no longer seen as a groove.

Z-DNA in Solution

In a low-salt solution, poly(dG-dC) has a circular dichroism that is characteristic of B-DNA. However, when the salt concentration is raised, there is a near inversion of the circular dichroism. This shows that a conformational change has occurred in the molecule (Pohl and Jovin, 1972). Analysis of the Raman spectra of poly(dG-dC) in both low- and high-salt solutions (Pohl *et al.* 1973), as well as a comparison of these solutions with the crystals that form Z-DNA, have made it possible to demonstrate that the high-salt form of poly(dG-dC) is Z-DNA (Thamann *et al.* 1981).

Considerable work has been carried out on the changes in conformation of polynucleotides under a variety of environments (Rich *et al.* 1984). The conclusion of these studies is that an equilibrium exists between B-DNA and Z-DNA, such that B-DNA is at the lower energy level under physiologic conditions. The equilibrium is influenced by a number of factors. The first is the nucleotide sequence. Alternating purines and pyrimidines favor Z-DNA, and the CG base pairs are the most favorable for forming Z-DNA. However, sequences of DNA that do not have strict alternations of purines and pyrimidines still are able to form Z-DNA, although it takes more energy (Nordheim

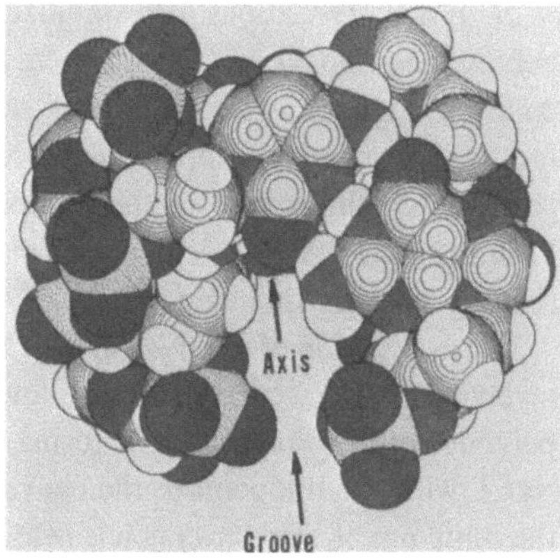

Figure 14.3 End view of Z-DNA in which 3 base pairs are shown in a van der Waals diagram. The groove extends almost down to the helical axis, which passes near the cytosine 02 atom. As the base pairs come toward the reader, they rotate in a clockwise direction, thereby exposing three phosphate groups on the left. Oxygen atoms are shaded with circles, phosphorous atoms have spiked circles, and nitrogen atoms are indicated by dashed circles.

et al. 1982). Chemical modifications are known to change the equilibrium. For example, bromination of poly(dG-dC), during which bromine atoms are attached to the C8 atom of guanine, stabilizes Z-DNA. It does this by stabilizing the guanine syn conformation relative to the anti. Polymers of brominated poly(dG-dC) exist as Z-DNA under physiologic salt solutions. It has been shown to be highly immunogenic, thus producing antibodies that are specific for Z-DNA (Lafer *et al.* 1981. These antibodies have been used in a variety of ways to demonstrate the existence of Z-DNA not only in the genome of cells, but also in plasmids.

Several factors are known that can stabilize Z-DNA *in vivo:* 1) Z-DNA is stabilized by Z-DNA-binding proteins that have been isolated from a variety of cell types—these bind to Z-DNA specifically, but not to B-DNA (Nordheim

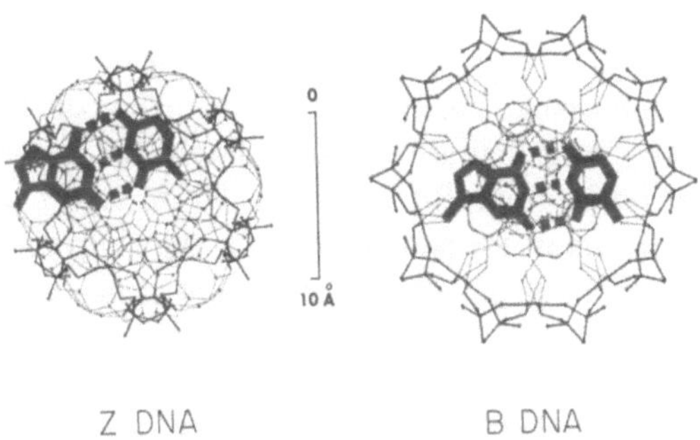

Figure 14.4 End views of regular idealized helical skeletal diagrams of Z-DNA and B-DNA. Heavier lines are used for the phosphate-sugar backbone. A guanine-cytosine base pair is shown by shading, and the difference in the positions of the base pairs is quite striking; they are near the center of B-DNA, but at the periphery of Z-DNA.

et al. 1982); 2) Cations also stabilize Z-DNA, especially spermine and other polyamines; and 3) A strong element for stabilizing Z-DNA is negative supercoiling (Singleton *et al.* 1982; Peck *et al.* 1982). This stabilization is due to the fact that a negatively supercoiled DNA has a supercoiling energy that is proportional to the square of the number of negative supercoils (Wang, 1980). Forming Z-DNA in a segment of plasmid reduces the number of supercoils, thereby strongly reducing the negative supercoiling energy.

The conversion of B-DNA to Z-DNA is most easily monitored by the near inversion of the circular dichroism found in the ultraviolet (uv) spectrum of the polynucleotide. Poly(dG-dC) forms Z-DNA at elevated concentrations of NaCl, with the midpoint of the conversion near 2.2 mol of NaCl. For $MgCl_2$, the midpoint of the conversion is near 0.7 mol (Pohl and Jovin, 1972). A major difference between Z-DNA and B-DNA (see Figure 14.1) is the closest approach distance between phosphate groups on the two polynucleotide strands. In B-DNA, the distance of closest approach is 11.7 Å; in Z-DNA, the closest approach between phosphates is 7.8 Å. In a solution at higher salt concentration, the cations cluster around the negatively charged phosphate groups and reduce the phosphate-phosphate repulsion energy. This leads to Z-DNA stabilization.

Methylation of CpG Sequences Stabilizes Z-DNA

Chemical modifications have been known to alter the B-Z equilibrium. As mentioned above, bromination of poly(dG-dC) stabilizes the guanine syn conformation and alters the B-Z equilibrium in favor of Z-DNA. Methylation of cytosine on the 5 position of cytosine in poly(dG-dC) produces a polymer that is made up of methylated d(CpG) units. This modification had a dramatic effect on the B-Z equilibrium, and it strongly stabilizes the molecule in the Z conformation (Behe and Felsenfeld, 1981). Indeed, in a physiologic salt solution, Z-DNA is the stable form of poly (dG-dC)—not B-DNA. The effect of methylation can be monitored by changes in the circular dichroism; however, a number of other physical parameters can also be measured that clearly indicate that Z-DNA is forming. The effect of the methyl group on the B-Z equilibrium can be described quantitatively; for example, the midpoint in the conversion of poly(dG-dC) from B-DNA to Z-DNA is 0.7 mol $MgCl_2$ as mentioned above. However, for poly(dG-m^5dC), the midpoint is 0.6 mmol; a change of three orders of magnitude in the amount of magnesium required to stabilize the polymer!

Polamines stabilize Z-DNA and the original crystal from which the structure of Z-DNA emerged was a spermine complex of a hexanucleotide (Wang *et al.* 1979). For the methylated polymer, poly(dG-m^5dC), Z-DNA is stabilized by micromolar amounts of spermine and spermidine (Behe and Felsenfeld, 1981). This is significant in view of the fact that millimolar amounts of polyamines are found in the nuclei of cells.

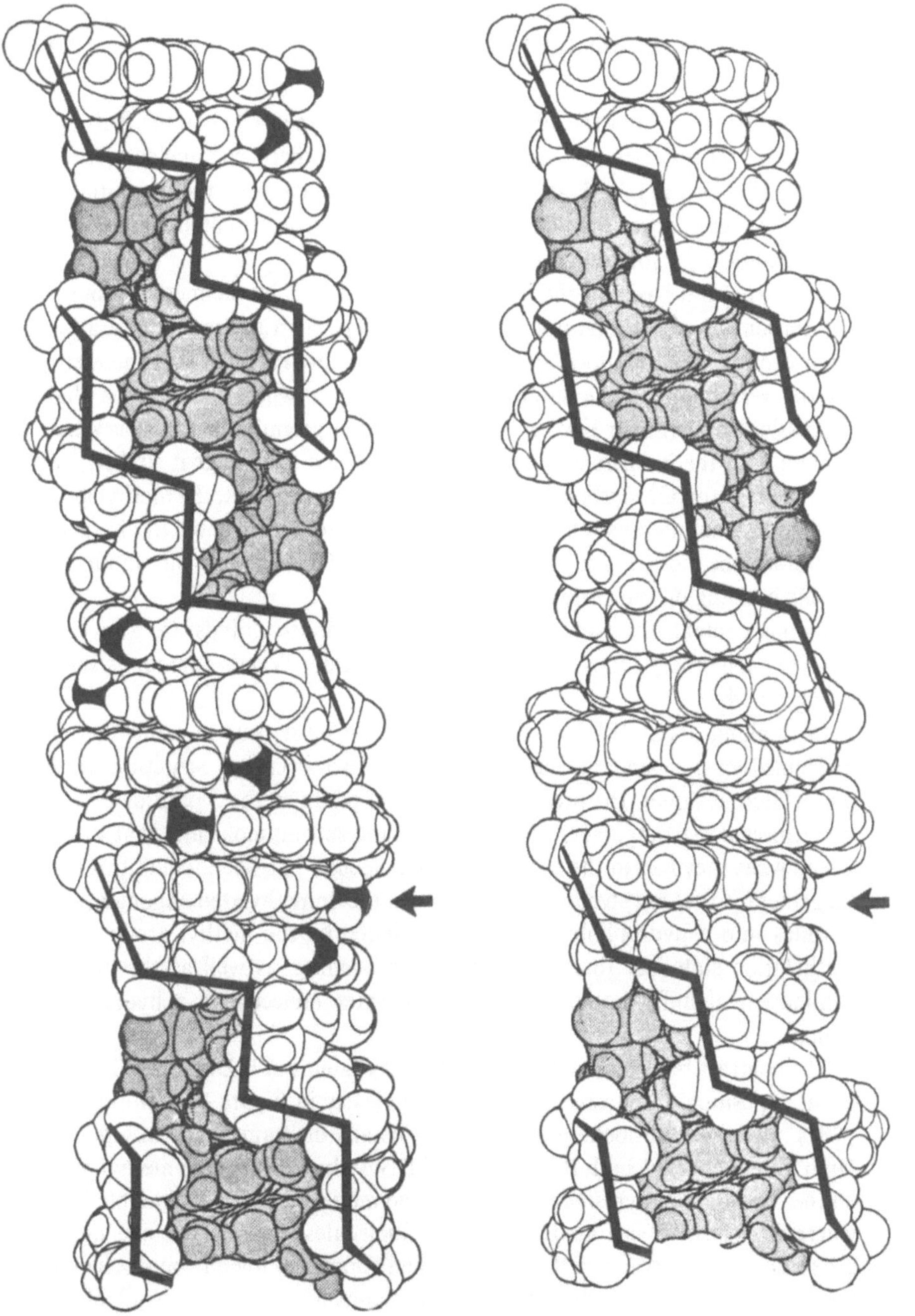

Figure 14.5 Two van der Waals drawings showing the structure of Z-DNA in both its unmethylated and methylated forms, as determined in single-crystal structures of $(dC\text{-}dG)_3$ and $(m^5dC\text{-}dG)_3$, respectively. The groove in the molecule is shown by the shading. The black zigzag lines go from phosphate group to phosphate group to show the arrangement of the sugar-phosphate backbone. The methyl groups on the C5 position of cytosine in the methylated molecule (left) are drawn in black. The arrow shows that a depression on the surface of the unmethylated polymer (right) is filled by methyl groups (left). The methyl group is in close contact with the imidazole ring of guanosine and the C1′ and C2′ atoms of the sugar ring.

The conclusion about the effect of methyl groups in m^5CpG on the B-Z equilibrium is that they strongly stabilize Z-DNA in synthetic polynucleotides. The stabilization occurs to a degree that is somewhat surprising; the magnitude of the effect seems larger than one might have intuitively estimated, based on the simple addition of methyl group.

The structural basis for the stabilization of Z-DNA by the methyl group on the carbon C5 of cytosine was revealed in a crystal structure study of a hexamer d(m^5dC-dG)$_3$ (Fujii *et al.* 1982). The structure revealed that the methyl groups were found on the outer surface of Z-DNA, where they filled a slight depression as shown in Figure 14.5. The methyl group is in van der Waal's contact with the imidazole group of guanine, which overhangs it slightly, and with the C1′ and C2′ carbon atoms of the sugar ring. In short, the methyl group is filling a small depression, the walls of which are hydrophobic in character. Therefore, the methyl group itself is stabilizing Z-DNA largely through hydrophobic bonding on the surface.

In this context, the two methyl groups on opposite strands are also fairly close to each other; almost, but not quite, in van der Waal's contact as shown in Figure 14.5.

In thinking about B-DNA and Z-DNA, one must bear in mind that the molecules exist in an aqueous environment. In the crystal structure of the methylated hexamer, the presence of a methyl group on the 5 position of cytosine causes a rearrangement of the surrounding solvent molecules that is relative to what they had in the unmethylated molecule. The net effect of this rearrangement is that a water molecule, which in the unmethylated oligonucleotide intrudes into a hydrophobic region, is excluded by the methyl group. Thus, methylation stabilizes through hydrophobic bonding, as well as through the elimination of unfavorable solvation by a water molecule in a hydrophobic region.

The contrast between the positions of methyl goups in B-DNA and Z-DNA in the 5 position of cytosine is shown more clearly in Figure 14.6 (Fujii *et al.* 1982). Here, we are looking at a van der Waal's diagram of the m^5CpG and Gpm^5C sequences of both B-DNA and Z-DNA. In B-DNA, the methyl group on the 5 position of cytosine projects into the major groove of the double helix, where it is largely surrounded by water molecules. In Z-DNA, the methyl group does not protrude; rather it is blocked almost completely on one side by the imidazole group of guanine and is consequently much less available for contact with water molecules.

A comparative overview of the Z-DNA structure with and without methyl groups is shown in the two stereodiagrams in Figures 14.7a and 14.7b. These illustrate more clearly the manner in which the methyl group is tucked into the surface of the molecule, where it is less accessible to a solvent.

The Z-DNA-stabilizing effect of d(CpG) methylation is also seen in other sequences. AT base pairs form Z-DNA less readily than GC base pairs (Rich *et al.* 1984); and, the self-complementary sequence d(CGTACG) does not crystallize as Z-DNA in its unmodified form. However, if the cytosines are

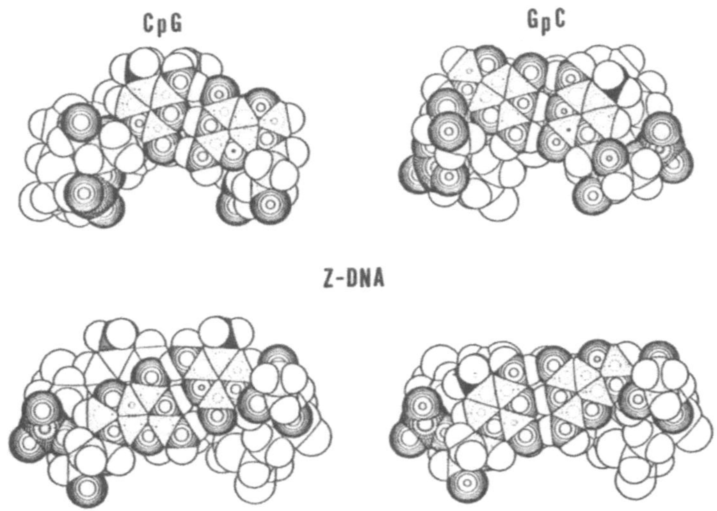

Figure 14.6 Fragments of Z-DNA and B-DNA are shown in van der Waals diagram containing two base pairs with the sequences m^5CpG and Gpm^5C. The base pair closer to the reader has shaded atoms, while the base pair facing away from the reader shows the atoms only in outline. The methyl group on 5-methylcytosine in the upper base pair closer to the reader is solid black, while the methyl group attached to the lower base pair is shaded gray. The two diagrams at the bottom show the methyl groups in B-DNA, which are more exposed to solvent water molecules than the methyl groups in the upper two base pairs in Z-DNA.

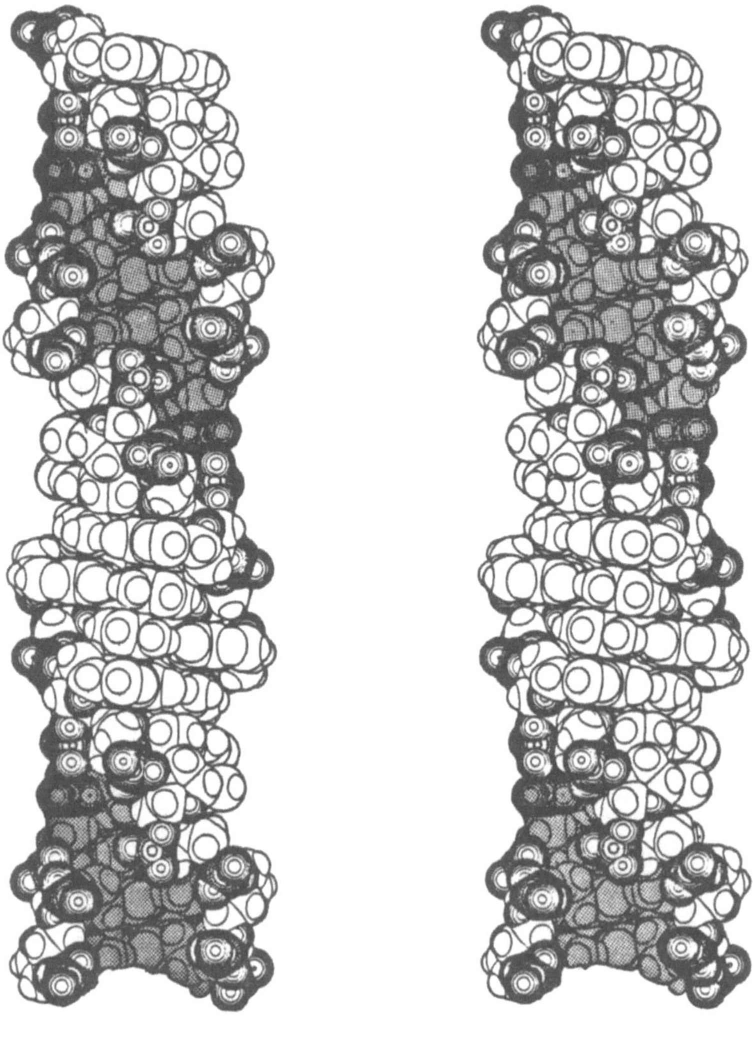

Figure 14.7 Stereodiagrams of Z-DNA. Three dimensionality can be seen either with stereo viewers or simply by looking at the diagram and relaxing the eye muscles until the two images merge. The zig-zag array of phosphate groups shows up clearly.

A. Stereodiagrams of the Z-DNA helix, as found in the unmethylated hexamer crystals. Van der Waal's models are drawn in which the oxygens in the backbone are indicated by circles and the phosphate groups are indicated by circles with crossed lines. Note that a slight depression is seen on the convex outer surface of the molecule due to the fact that the guanine imidazole rings project further away from the axis than the cytosine rings.

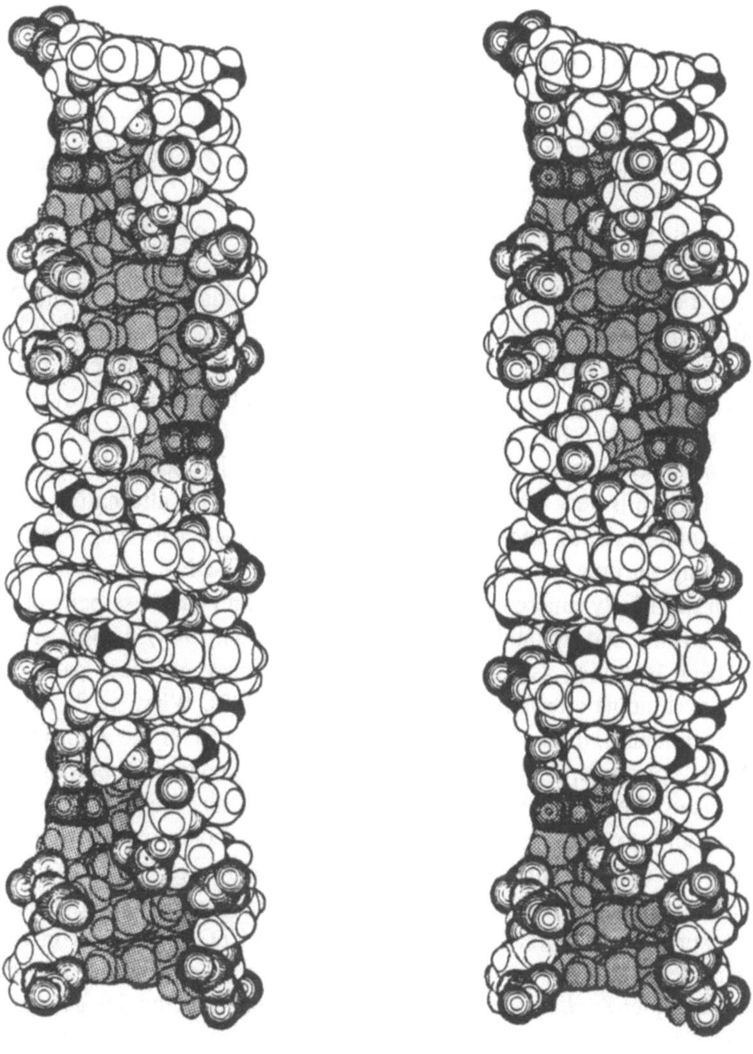

B. Stereodiagram of the methylated hexamer. The methyl groups are shaded solid black. Three hexamer segments are shown in each diagram just as they appear in the crystal. The methyl groups are close together, and they fill part of the depression on the surface that is caused by the overhanging imidazole rings of guanine, which protrude from the center of the molecule.

methylated, the fragment crystallizes as Z-DNA (Wang *et al.* 1984). The stabilization of the Z-DNA is attributed to the presence of the methyl groups in the oligonucleotide fragments.

Does Methylation of d(CpG) *In Vivo* Induce Z-DNA Formation?

At present, we are left with something in the nature of a dilemma. There is reasonable evidence that C5 methylation of cytosine in d(CpG) sequences is used as a signal to inhibit expression of genetic information (Razin and Riggs, 1980). Removal of the methyl group is associated with activation of the gene in many systems, although not in all systems. There is good evidence that in a deoxypolynucleotide with alternating cytosine and guanine, methylation of cytosine is a powerful component for stabilizing the Z conformation. This dilemma arises from the fact that we do not know what occurs *in vivo*. The *in vivo* substrate is very far removed from a polymer with alternating cytosine and guanine residues, and it is even far from sequences with long stretches of alternating purines and pyrimidines. Segments of Z-DNA can be formed *in vitro* in negatively supercoiled plasmids that contain eight base pairs of alternating purines and pyrimidines (Nordheim and Rich, 1983). Longer stretches have been identified as forming Z-DNA when there are base pairs out of purine-pyrimidine alternation (Nordheim *et al.* 1982). What we do not know is what influence, if any, methylation may have *in vivo* on the ability to form Z-DNA. For example, it is known that negatively supercoiled plasmids, with an insert of alternating methylated cytosine and guanine, require less negative superhelicity to stabilize the insert in the Z-DNA conformation than if the insert were unmethylated (Klysik *et al.* 1983). However, the biological evidence strongly indicates that very few methyl groups are required to inactivate gene expression. For example, methylation of d(CpG) sequences in the 5′-flanking sequences of hemoglobin genes (Busslinger *et al.* 1983) or adenovirus genes (Langner *et al.* in press, 1984) are adequate for inactivating gene expression in transfection experiments. Furthermore, the number of methyl groups introduced in the region may be as small as three or even one in the promotor region. Could this possibly influence Z-DNA formation? An attempt to address this question has been carried out by measuring the extent to which methylated 5′-flanking sequences of hemoglobin genes influence Z-DNA formation (Nickol and Felsenfeld, 1983). These experiments were carried out with plasmids to see whether methylation would make it easy for the segment of DNA to form Z-DNA, as indicated by changes in superhelicity on negative supercoiling. It was found that methylation did not induce the ability to form Z-DNA. Thus, we are left with a body of data that generally does not support the concept that Z-DNA formation may be occurring *in vivo* at methylated d(CpG) sites. The only reason that it seems like a paradox is because the physical evidence of its effect on polymers is quite strong; often, such *in vitro* experiments provide important clues regarding things that happen *in vivo*.

From a physical point of view, it might seem a little difficult for methylation of cytosine to be readily detected in d(CpG) sequences in B-DNA. The reason for this is that the methyl group lies in the major groove of B-DNA, where it may be difficult to distinguish from the other methyl groups in the major groove that arise from thymine residues in AT base pairs. As viewed from the major groove, the major structural difference between a T-A and an m^5C-G base pair is the orientation of the NH ... 0 hydrogen bonds in the major groove. In the CG base pairs, the amino groups are on cytosine; in the AT base pairs, they are on adenine. However, even though this seems like a small change in structure, it may be large enough to be detected by enzymes or other proteins.

Conversely, if methylation of d(CpG) sequences were to appreciably shift the equilibrium from B-DNA to Z-DNA, it would then represent a type of structural amplification mechanism in which a rather minor change in the molecule (an addition of the methyl group) nucleates or helps to promote a larger transformation that involves a reorganization of the molecule. From a physical point of view, it seems that this could only be possible if there were proteins found in the organism that specifically bind to Z-DNA containing methylated d(CpG) segments. The existence of such proteins, which bind strongly to a methylated Z-DNA segment, would make it possible to stabilize a small perturbation in the B-Z equilibrium that is induced by one or a small number of methyl groups. If the binding were strong enough, it could stabilize Z-DNA, even in the absence of large stretches with alternations of purines and pyrimidines. The existence of such proteins is presently unknown. Z-DNA-binding proteins have been isolated; as yet, however, none have been reported that have a stronger affinity for methylated Z-DNA molecule than for a nonmethylated form. If such proteins were found, it could represent a bridge connecting the physical studies on the stabilization of Z-DNA due to methylation with the biological studies demonstrating the role of methylation in inactivating gene expression. However, in the absence of such proteins, the relationships between Z-DNA and methylation-induced gene inactivation must remain only an interesting hypothesis at best.

Acknowledgments

This research was supported by grants from the National Institutes of Health, the American Cancer Society, and the National Aeronautics and Space Administration.

References

Behe M, Felsenfeld G: Effects of methylation on a synthetic polynucleotide: The B-Z transition in Poly(dG-m^5dC)·poly(dG-m^5dC). *Proc Natl Acad Sci USA* 1981; 78:1619–1623.

Busslinger M, Hurst J, Flavell RA: DNA methylation and the regulation of globin gene expression. *Cell* 1983;34:197–206.

Fujii S, Wang AH-J, van der Marel G, van Boom JH, Rich A: Molecular structure of (m^5dC-dG)$_3$: The role of methyl group on 5-methyl cytosine in stabilizing Z-DNA. *Nucl Acids Res* 1982;10:7879–7892.

Klysik J, Stirdivant SM, Singleton CK, Zacharias W, Wells RD: Effects of 5 cytosine methylation on the B-Z transition in DNA restriction fragments and recombinant plasmids. *J Mol Biol* 1983;168:51–71.

Lafer EM, Moller A, Nordheim A, Stollar BD, Rich A: Antibodies specific for left-handed DNA. *Proc Natl Acad Sci USA* 1981;78:3546–3550.

Langner KD, Vardimon L, Reuz D, Doerfler W: DNA methylations of Three 5′-CCGG-3′ sites in the promoter and 5′ regions inactivate the E2a gene of adenovirus type 2. *Proc. Natl Acad Sci USA* 1984;81:2950–2954.

Nickol JM Felsenfeld G: DNA conformation at the 5^1 end of the chicken adult β-globin gene. *Cell* 1983;35:467–477.

Nordheim A, Lafer EM, Peck LJ, Wang JC, Stollar BD, Rich A: Negatively supercoiled plasmids contain left-handed Z-DNA segments as detected by specific antibody binding. *Cell* 1982;31:309–318.

Nordheim A, Tesser P, Azorin F, Kwon YH, Moller A, Rich A: Isolation of *Drosophila* proteins that bind selectively to left-handed Z-DNA. *Proc Natl Acad Sci USA* 1982;79:7729–7733.

Nordheim A, Rich A: Negatively supercoiled simian virus 40 DNA contains Z-DNA segments within transcriptional enhancer sequences. *Nature* 1983;303:674–679.

Peck LJ, Nordheim A, Rich A, Wang JC: Flipping of cloned d(pGpG)$_n$·d(pCpG)$_n$ DNA sequences from right to left-handed helical structure by salt, Co(III), or negative supercoiling. *Proc. Natl Acad Sci USA* 1982;79:4560–4564.

Pohl FM, Jovin TM: Salt-induced co-operative conformational change of a synthetic DNA: Equilibrium and kinetic studies with Poly(dG-dC). *J Mol Biol* 1972;67:375–396.

Pohl FM, Ranade A, Stockburger M: Laser raman scattering of two double-helical forms of Poly(dG-dC). *Biochim Biophys Acta* 1973;335:85–92.

Razin A, Riggs AD: DNA methylation and gene function. *Science* 1980;10:604–610.

Rich A, Nordheim A, Wang AH-J: The chemistry and biology of left-handed Z-DNA. *Ann Rev Biochem* 1984;53:791–846.

Singleton CK, Klysik J, Stirdivant SM: Left-handed Z-DNA is induced by supercoiling in physiological ionic conditions. *Nature* 1982;299:312–316.

Thamann TJ, Lord RC, Wang AH-J, Rich A: The high salt form of Poly(dG-dC)·Poly(dG-dC) is left-handed Z-DNA: Raman spectra of crystals and solutions. *Nucl Acids Res* 1981;9:5443–5457.

Wang, AH-J, Hakoshima T, van der Merel G, van Boom JH, Rich A: AT base pairs are less stable than GC base pairs in Z-DNA: The crystal structure of d (m^5CGTAm^5CG). *Cell* 1984;37:321–331.

Wang JC: Superhelical DNA. *Trends Biochem Sci* 1980;5:219–221.

Wang, AH-J, Hakoshima T, van der Merel G, van Boom JH, Rich A: AT base pairs are less stable than GC base pairs in Z-DNA: The crystal structure of d(m^5CGTAm^5CG). *Cell* 1984;37:321–331.

15

Chromatin Structure and Gene Expression

Kathleen F. Conklin*
Mark Groudine*†

Approximately 10 years ago, it was established that the bulk of the eukaryotic genome is organized in a repeating unit that consists of histones and DNA in a structure called the nucleosome. Since that time, a vast effort has been made to define more completely the structure, composition, and organization of DNA and proteins within the nucleus. This effort has resulted in a detailed understanding of histone-histone and histone-DNA interactions within the nucleosome. It also has revealed a marked variability in the content and higher-order organization of chromatin fractions engaged in various cellular processes. By microscopy, for example, the structure and organization of eukaryotic chromatin is dynamic. Chromosomes condense, extend, and migrate during replication and cell division. In addition, dramatic differences in chromatin ultrastructure accompany transcription; active regions extend and loop out from the chromosome axis. Biochemical analyses of chromatin also have revealed changes in the composition of histone and nonhistone proteins that correlate not only with the transcriptional activity of a region of chromatin, but also with its chromatin structure. Similarly, the distribution of 5-methylcytosine (5-meCyt) residues among different chromatin fractions is selective in that transcriptionally active regions generally are depleted in methylated sequences. There is also evidence that changes in DNA topology and DNA conformation are associated with the transcriptional activation of eukaryotic DNA, as well as with changes in chromatin structure.

The observations that these alterations in chromatin structure are associated with transcriptional activation have led to the propostion that they may provide a regulatory function in differential gene expression. This notion is supported by several independent observations. First—in contrast to the major mode of

*Hutchinson Cancer Research Center, 1124 Columbia Street, Seattle, Washington 98104

†Department of Radiation Onocology, University of Washington Medical School, Seattle, Washington 98105.

regulation of prokaryotic gene expression, in which an RNA polymerase and other regulatory proteins bind to specific DNA sequences and initiate transcription—eukaryotic genes appear to require additional elements for their proper regulation. For example, while the expression of certain eukaryotic genes normally is restricted to specific lineages during development, cloned copies of several tissue-specific genes fail to show any such restriction either in *in vitro* systems using purified transcription factors and extracts from different cell types (Luse and Roeder, 1980; Wasylyk *et al.* 1980; Tsujimoto *et al.* 1981) or after the DNA-mediated transformation of these genes into various cell types (Mantei *et al.* 1979; Wold *et al.* 1979; Robins *et al.* 1982; Kioussis *et al.* 1983). The hypothesis that one component responsible for this abnormal expression is either regulated by or, at least, communicated by structure, is derived from the findings that correct expression apparently is dependent on regions of DNA located at distant sites along the DNA fiber. For example, deletions occuring in cis, several kilobases 5′ to the adult human β-globin gene, can result in a failure to activate this gene developmentally—even though the β-globin gene promoter is transcriptionally competent as assayed *in vitro* (Kioussis *et al.* 1983). Similarly, it has been reported that the transcriptional competence of retroviruses may be dependent on the chromatin structure of the region into which they integrate (Feinstein *et al.* 1982; Ucker *et al.* 1983). Additional support for this notion is derived from the effects that specific elements of DNA exert on the transcriptional activity of genes located within several kilobases of the sites of insertion of these elements (e.g., Roeder *et al.* 1980; Payne *et al.* 1982; Levinson *et al.* 1982; Rechavi *et al.* 1982; Luciw *et al.* 1983; McGinnis *et al.* 1983; Williamson *et al.* 1983). Thus, distant sequences appear to be able to exert both positive and negative effects on the transcriptional activity of nearby genes without interfering with the sequences required for activity per se.

Within the context of the observations cited above, any element that is involved in such long-range effects either must be able to act over a distance or must initiate events that can be propagated over a distance. This type of communication could take many forms. For example, such a signal could be mainly structural (e.g., altering the topology of DNA within a defined region) or could be compositional (e.g., dictating a change in the protein composition either locally or over a region of chromatin). Either of these events could have profound effects on the substrate suitability of regions of the genome for RNA polymerase, DNA methylases, or other molecules involved in transcriptional activation.

Any attempt to understand the potential functional properties of the different conformations of chromatin that, thus far, have been identified requires a fundamental knowledge of the interactions and composition of the basic chromatin components. In this chapter, therefore, we will first attempt to provide this baseline information; then we will consider the possible structure-function relationships.

Basic Chromatin Organization

The Nucleosome

Nucleosome Constituents

The basic repeating chromatin subunit is the nucleosome—an approximately 200 bp length of DNA associated with the five major histones (H1, H2A, H2B, H3, and H4; reviewed in Kornberg, 1974, 1977; Finch *et al.* 1977; Felsenfeld, 1978; McGhee and Felsenfeld, 1980; Sperling and Wachtel, 1981; Igo-Kemenes *et al.* 1982). Mild digestion of chromatin with micrococcal nuclease releases individual nucleosomes, as well as dimers, trimers, and larger oligomers. The deproteinized DNA products can be separated by agarose gel electrophoresis, and they are seen as a "ladder" after staining with ethidium bromide. Each step represents one additional nucleosome length. The monomeric nucleosome contains H1, as well as a histone octomer composed of two each of the four core histones, H2A, H2B, H3, and H4 (Kornberg 1974, 1977; Finch *et al.* 1977). 146 bp of DNA wraps externally around the octomer, making almost two complete turns (for a review, see Wang 1982); the remainder of DNA, called the linker, varies in length from approximately 0–100 bp, depending on the cell type or organism from which it is isolated. The significance and basis for the length of linker DNA is unknown, but it apparently is governed by one or several nonhistone nuclear components (Laskey *et al.* 1978; Laskey and Earnshaw, 1980; McGhee and Felsenfeld, 1980).

Under mild digestion conditions, micrococcal nuclease preferentially cleaves chromatin within the linker DNA. More extensive treatment of nucleosomes degrades the linker DNA (Noll *et al.* 1975) and results in the release of histone H1 to generate the nucleosome core particle (Noll and Kornberg, 1977). This has been interpreted as an indication that the linker DNA is involved, at least, in stabilizing the association of H1 to the nucleosome fiber. In addition to the association between H1 and linker DNA, *in situ* cross-linking experiments have demonstrated specific contact points between H1 and histones H2A (Bonner and Stedman, 1979) and H3 (Ring and Cole, 1979). Since H1 is located at the region where DNA enters and exits the nucleosome core (Simpson 1978; Thoma *et al.* 1979; Moyne *et al.* 1981), and since elution of H1 from nucleosomal chromatin can lead to movement of core histone octomer components (Tatchell and Van Holde, 1978; Weischet and Van Holde, 1980), H1 is thought to be intimately involved in maintaining the integrity of the core/linker spacing of nucleosomes. As described in later sections, it also appears to play a major role in higher levels of organization of eukaryotic chromatin.

The basic nucleosomal organization described above is virtually ubiquitous in bulk cellular chromatin. Therefore, it would not be expected to find a high degree of sequence specificity in histone-DNA interactions. However, early investigations of nucleosome formation *in vitro* revealed that assembly can occur at preferred sites along defined segments of DNA (Chao *et al.* 1979; Wasylyk *et al.* 1979). Since that time, nonrandom positioning of nucleosomes has been observed in chromatin assembled *in vitro* on both natural and synthetic DNA templates, as well as in cellular chromatin. These findings have led to proposals that inclusion (or exclusion) of nucleosomes from specific segments of DNA provides a regulatory element in determining chromatin function. The evidence that suggests a relationship between transcription regulation and nucleosome formation, organization, and/or content will be discussed in the section in this chapter on "Protein Composition and Organization."

Higher-order Chromatin Structure

Under mild isolation conditions, chromatin can be visualized microscopically as a 30nm in diameter fiber. This structure, which is referred to as the thick fiber, has a compaction ratio (i.e., the length of extended B form DNA divided by the length of DNA assembled into chromatin) of apporoximatley 40. The organization of nucleosomal chains within the 30nm fiber may be as a helix, which forms a solenoid structure (Finch and Klug, 1976; McGhee *et al.* 1983, and references therein), or as aggregates of multiple nucleosomes, which form superbeads (Renz *et al.* 1977; Kiryanov *et al.* 1982). It should be noted that whichever arrangement is assumed within the thick fiber (as well as in nucleosomes), DNA sequences that are quite distant along the extended DNA fiber can be brought into close apposition when assembled into chromatin. The possible significance of this arrangement has received considerable attention in light of the effects that such an organization could have on protein-DNA interactions (see the section on "Protein Composition and Organization").

In low ionic strength solutions, the 30nm fiber reversibly extends to form the 10nm (thin) fiber (Thoma *et al.* 1979) in which nucleosome cores and linker DNA are distinct and appear as "beads on a string" (Olins and Olins, 1974). This more extended fiber has a packing ratio of approximately 6–7. Elution of H1 from chromatin also can induce a transition from the 30nm to the 10nm fiber. Under these conditions, however, the nucleosome core particles shift and appear to slide into close proximity (Tatchell and Van Holde, 1978); the resultant fiber can no longer properly recondense into the 30nm fiber (Moyne *et al.* 1981). These data suggest that H1 is involved in maintaining both the integrity of nucleosomal chains and the organization of these chains within the thick fiber. The latter function could be based on H1-H1 interactions that have been inferred from cross-linking data (Ring and Cole, 1979; see also McGhee and Felsenfeld, 1980, and references therein). This affinity, combined with the association of H1 with core histone components, could provide (at least) a portion of the driving force required to bring nucleosome cores into proximity, thereby facilitating (or initiating) chromatin condensation.

Several orders of further compaction are required to generate the mitotic chromosomes of metaphase from the rather amorphous-looking chromatin visualized during interphase. Although it is beyond the scope of this chapter to review the literature that is relevant to chromosome ultrastructure, one important observation should be noted. Visualization of histone-depleted interphase (Benyajati and Worcel, 1976) and metaphase (Paulson and Laemmli, 1977; Marsden and Laemmli, 1979) chromatin revealed the presence of large (30–90 kb) loops or domains of DNA that appeared to radiate from a central nonhistone core or matrix. Similar loops also have been seen in lampbrush chromosomes and in Balbiani rings of polytene chromsomes (Lamb and Daneholt, 1979). The notion that these structural domains may define functional units is the focus of extensive investigation and will be discussed in later sections of this chapter.

Altered Chromatin Conformations

To define the potential relationship between chromatin structure and gene activity, it is first important to describe the structural differences that distinguish active from bulk chromatin. The determination of such differences has been accomplished by several types of experimental approaches. Variations in ultrastructure have been observed via microscopic visualization of intact chromosomes and mildly dispersed chromatin; both local and global perturbations in structure have been detected by the use of a variety of nucleases. In this section, we will discuss conformational alterations associated with transcriptionally active regions of chromatin; and we will present data that relates to the temporal and hierarchical relationships between structure and transcriptional activity.

Microscopy

Transcriptionally active segments of chromatin can be readily identified microscopically by the spreading technique pioneered by Miller and co-workers (e.g. Miller and Beatty, 1976). Extended regions of chromatin often are seen that extrude from the longitudinal axis of the condensed chromosome, with identifiable polymerase molecules and nascent RNA chains being clearly evident. The lampbrush chromosomes of amphibians (e.g., Franke *et al.* 1977; Scheer, 1978), chromatin prepared from *Drosophila* embryos (McKnight and Miller, 1976; McKnight *et al.* 1977), amplified nucleolar ribosomal DNAs of several organisms (e.g., Foe, 1977; Franke *et al.* 1977; Scheer *et al.* 1977; Scheer, 1978), and chromatin containing the Balbiani ring 75S RNA unit (Lamb and Daneholt, 1979) all include examples of the types of chromatin that have been used to examine the ultrastructure of active chromatin. Via microscopic techniques, there is general agreement that transcriptionally active regions are structurally distinct from bulk chromatin, as they have a more relaxed or extended conformation (reviewed in Mathis *et al.* 1980).

In the types of analyses described above, it also has been observed that chromatin extension associated with active units does not need to be limited to those sequences that are actually being transcribed; it can include significant regions of chromatin flanking the expressed gene. This observation has led to suggestions that these extended loops might represent functional chromatin domains, and that activation (or suppression) of a gene within such a domain could result in passive activation (or suppression) of adjacent genes (see the section in this chapter on "Chromatin Structure and Gene Expression"). The validity of this type of organization and control is currently being evaluated by attempting to elucidate both the events and components involved in initiating chromatin extension as well as the composition of DNA and protein within, and at the base of, chromatin loops. In addition, it is of interest to determine the relationship, if any, between the extended loops associated with transcription and those observed in dispersed chromatin (see the previous section on "Higher-Order Chromatin Structure").

Nuclease Sensitivity

Micrococcal Nuclease

Digestion of bulk chromatin with micrococcal nuclease generates a characteristic population of cleavage products that are typical of nucleosome-associated DNA (see the previous section on "Nucleosome Constituents"). Analysis of individual genes or specific chromatin fractions, however, reveals that micrococcal nuclease is able to distinguish between at least some transcriptionally active and inactive regions of chromatin. The altered specificity is evident not only in cleavage rates, but also (in some cases) in cleavage products.

Bloom and Anderson (1978, 1979) demonstrated that under mild digestion conditions (1–3% acid solubilization of chromatin-associated DNA), subunits released from oviduct-derived chromatin were enriched five-to-sixfold for ovalbumin DNA sequences; the nuclease-resistant fraction was correspondingly depleted for these sequences. Since this preferential release was not observed in chromatin from liver, where ovalbumin is not expressed, the increased sensitivity of this region is coincident with its expression and is not a property of ovalbumin gene chromatin per se. This finding has been confirmed for ovalbumin (Bellard *et al.* 1980, 1982) and has been extended for other active sequences. These include the *Drosophila melanogaster* heat-shock protein locus 70, (Wu, 1980; Levy and Noll, 1981), the sea urchin α-histone genes (Spinelli *et al.* 1982), murine leukemia proviral DNA (Panet and Cedar, 1977), chick globin in immature red blood cells (Bloom and Anderson, 1979), nucleolar chromatin in *Physarum polycephalum* (Butler *et al.* 1978), and for DNA sequences complimentary to polyA RNA in trout testes (Levy-Wilson and Dixon, 1979).

Some reports, including several for ovalbumin-containing chromatin in oviduct (e.g., Garel and Axel, 1976), for globin-containing chromatin in erythrocytes (Weintraub and Groudine, 1976), and for ribosomal DNA-containing chromatin in *Tertrahymena pyriformis* (Mathis and Gorovsky, 1976) indicated that micrococcal nuclease did not show a preference for active regions. The discrepancy, at least for ovalbumin, apparently is based on the extent of nuclease digestion; specificity is evident only at low levels of digestion and is abolished at higher levels. However, this explanation may not account for the negative results obtained with other active units. It is possible that the variable susceptibilities of active chromatin to digestion by micrococcal nuclease reflect structural heterogeneities among different transcriptionally active regions of chromatin. Although there are several other explanations, one possibility is that the rate of transcription determines the relative susceptibility of chromatin to digestion by micrococcal nuclease. This suggestion is supported by the cleavage rates of several developmentally regulated or inducible genes: 1) the chicken globin genes are preferentially sensitive to micrococcal nuclease digestion only in immature red blood cells, where they are expressed—but not in mature red blood cells (Bloom and Anderson, 1979), which transcriptionally are virtually inert; 2) the *D. melanogaster* heat-shock protein locus chromatin is micrococcal nuclease-sensitive only during heat shock (Wu, 1980; Levy and Noll, 1981); and 3) the micrococcal nuclease sensitivity of the chicken ovalbumin gene chromatin progressively decreases in oviduct after hormone withdrawal (Bloom and Anderson, 1979). These data are compatible with the hypothesis that the increased sensitivity to digestion by micrococcal nuclease reflects chromatin alterations that are related to the transcription process per se.

DNAseI

In contrast to the regularly sized DNA fragments generated after micrococcal nuclease digestion of chromatin, DNA purified from DNAseI-digestion chromatin exhibits a smeared pattern after agarose gel electrophoresis. However, detailed investigations of the initial cleavage products generated by DNAse I have revealed that transcriptionally active regions of chromatin are preferentially sensitive to digestion by the enzyme (Weintraub and Groudine, 1976). Preferential sensitivity to digestion by DNAseI appears to be a general property of transcriptionally active regions of chromatin. It has been documented for many regions including 1) the globin genes in avian red blood cells, but not in fibroblasts (Weintraub and Groudine, 1976); 2) the ovalbumin gene in oviduct (Garel and Axel, 1976; Palmiter *et al.* 1978; Bellard *et al.* 1980), but not in liver (Garel and Axel, 1976); 3) active retrovirus proviral DNAs (Panet and Cedar, 1977; Groudine *et al.* 1981a; Chiswell *et al.* 1982); 4) globin genes in sheep erythroblasts (Young *et al.* 1978); 5) *Xenopus laevis* (Williams and Tata, 1983) and chicken (Burch and Weintraub, 1983) vitellogenin genes in liver; 6) *Xenopus laevis* ribosomal DNA (Macleod and Bird, 1982); 7) immunoglobulin genes in lymphocytes (Storb *et al.* 1981b; Parslow and Granner,

1982); and 8) yeast alcohol dehydrogenase genes (Sledziewski and Young, 1982). In addition, sequences complimentary to poly A mRNA in trout testes are found in chromatin that is preferentially sensitive to digestion by DNAseI (Levy and Dixon, 1977), as are sequences complimentary to low-abundance mRNAs in rat liver chromatin (Garel *et al.* 1977).

The increased nuclease sensitivity of transcriptionally active segments of chromatin has been interpreted as a reflection of the increased accessability of DNA within these regions due to a relaxed or open conformation of active versus bulk chromatin. It should be noted that this more open conformation does not necessarily correspond to the extended conformation that was defined microscopically.

Nuclease-Hypersensitive Sites

In addition to the general extension or relaxation of active chromatin, local perturbations in chromatin structure have also been identified. These sites, which can be up to 500 bp in length, are hypersensitive to digestion by one or several of a variety of endonucleases, including DNAseI (e.g., Wu *et al.* 1979a; for a review, see Elgin, 1981), restriction endonucleases, micrococcal nuclease and DNAseII (e.g. McGhee *et al.* 1981), nuclease S1 (e.g., Larsen and Weintraub, 1982), as well as endogenous cellular nucleases (e.g., Waldeck *et al.* 1978; Burch and Weintraub, 1983; Senear and Palmiter, 1983). Some regions are also recognized by chemical reagents that bind preferentially to single-stranded DNA (Kohwi-Shigematsu *et al.* 1983).

The majority of nuclease-hypersensitive sites have been identified with DNAseI (Elgin, 1981). The sequence specificity of this enzyme is low (Bernardi *et al.* 1975). Although rare DNAseI-hypersensitive sites can be detected in naked cellular DNA (e.g., Storb *et al.* 1981a), virtually all DNAseI-hypersensitive sites are chromatin-specific. The unambiguous designation of a hypersensitive site with other nucleases often is less straightforward; the sequence specificity of micrococcal nuclease has been established (Bernardi *et al.* 1975; Dingwall *et al.* 1981; Horz and Altenberger, 1981), and its usefulness in these and other types of analyses has been questioned (Keene and Elgin, 1981). Restriction enzymes, by definition, are site-specific; the designation of a site as hypersensitive to restriction enzyme cleavage, therefore, is determined by comparison of the relative cleavage rates of the same or other enzymes in different regions of chromatin. The single-strand-specific probes, nuclease S1 and bromoacetaldehyde, also have been used in a small number of chromatin analyses (Larsen and Weintraub, 1982; Kohwi-Shigematsu *et al.* 1983); the implications and significance of these apparently single-stranded regions of DNA within eukaryotic chromatin is discussed in later sections.

The vast majority of hypersensitive regions associated with developmentally controlled genes are tissue-specific: 1) globin chromatin-hypersensitive sites are evident in red blood cell lineage cells, but are absent in chromatin from brain and fibroblast tissues (Stalder *et al.* 1980a,b); 2) chicken vitellogenin chromatin contains several 5′-associated hypersensitive sites in liver chromatin from laying hens, but not from male liver chromatin, embryonic liver chromatin, or brain or lymphocyte chromatin (Burch and Weintraub, 1983); and 3) the rat preproinsulin II gene contains a nuclease-hypersensitive region upstream of the gene in a pancreatic B cell tumor that secretes insulin, but not in chromatin from liver, spleen, kidney, or brain (Wu and Gilbert, 1981).

The identification of chromatin structural alterations that accompany transcriptional activation of eukaryotic genes raises two immediate questions. First, do any or all of these conformations constitute regulatory elements? Second, what is the molecular basis for this apparent chromatin reorganization?

Chromatin Structure and Gene Expression

The changes in chromatin conformation discussed above could be a consequence of gene activation, could reflect events required for gene activation, and/or could (themselves) contribute directly to the regulation of gene expression. Attempts to determine the hierarchic relationship(s) between chromatin structural alterations and chromatin function have focused on: 1) the documentation of correlations between changes in chromatin conformation and transcriptional activation of genes present in their normal cellular location; and 2) the analysis of the effects exerted by different segments of DNA on both the structure and activity of associated genes.

In Vivo Correlations

One approach taken to determine the relationship between chromatin structural alterations and gene activity has been to correlate the nuclease sensitivity and ultrastructure of chromatin that contains inducible or developmentally regulated genes before, during, and after transcriptional activity. If chromatin structural alterations are required for gene expression, a prediction from these types of analyses would be that chromatin reorganization would be established before the onset of detectable transcription. Via microscopic techniques, such a requirement for chromatin reorganization would be supported by the identification of transcription units that were included within extended chromatin before detection of bound polymerase molecules within that transcription unit. Analyses of ribosomal DNA-containing chromatin have revealed the presence of maximally extended units that are devoid of identifiable ribonucleoprotein fibrils (Foe *et al.* 1976; Foe, 1977; Franke *et al.* 1977; Scheer, 1978), which suggests that chromatin reorganization can occur in the absence of transcription. However, since it has not been established whether the chromosomal regions examined were completely devoid of polymerase molecules, these data indicate only that the extended conformation of rDNA is not dependent on continued high levels of transcription; they do not resolve the question of whether chromatin extension is initiated before polymerase binding (for discussion, see Foe, 1977). Of the single-copy genes that have been examined, pre-extended regions have not been convincingly demonstrated. In fact, it appears that the extension of non-rDNA-containing chromatin is roughly proportional to the number of bound, active polymerase molecules (see McKnight *et al.* 1977; Lamb and Daneholt, 1979; see also the section in this chapter on "Core Histones and Nucleosomes"). Therefore, it remains unresolved whether some degree of relaxation or extension exists before transcription initiation, and also whether chromatin extension of ribosomal versus nonribosomal units are subject to different regulatory mechanisms.

Another approach to investigating the temporal relationship between chromatin structural alterations and gene activation has been to correlate the appearance of nuclease-sensitive and nuclease-hypersensitive regions with transcriptional activity. Studies of this type on the chicken adult α and β-embryonic globin genes have been conducted in detail; therefore, they will be reviewed here (see also Groudine *et al.* 1981c; Weintraub *et al.* 1981b; Groudine and Weintraub, 1981).

In analyzing a population of cells estimated to contain at least 50% precursor erythroblasts from 22-hour chicken embryos, it was found that no globin gene sequences were transcriptionally active, and that the α-and β-globin gene clusters were neither sensitive to DNAseI nor contained red blood cell-specific hypersensitive sites. Within two generations (at approximately 35 hours of development), progeny of these precursor cells were found to produce large amounts of globin mRNA and to contain an active globin chromatin structure, as assayed by the presence of specific hypersensitive sites and preferential sensitivity to DNAseI. By day 5 of development, it is possible to isolate relatively pure populations of red blood cells that express the embryonic, but not the adult, globin genes. Analysis of the chromatin conformation of these genes revealed that both the embryonic and adult globin genes are in a DNAseI-sensitive conformation, although the expressed embryonic genes are quantitatively more sensitive than the adult genes. Adult (14-day) red blood cells, which do not express detectable levels of the embryonic globin genes, also contain both classes of globin genes in a sensitive chromatin conformation; however, in these adult cells, the sensitivity of the embryonic gene is less than the adult. Each cell type also contains characteristic hypersensitive sites in the α- and β-chromosomal regions, most of which are unique to—and coincident with—the expression of the embryonic or adult globin genes (Stalder *et al.* 1980a,b).

The first conclusion that can be drawn from these data is that DNAseI sensitivity and hypersensitivity are not dependent on transcription—at least as assayed in this system (see below). The second point to consider is that the active conformation is not confined to the transcribed region. In the case of the avian globin genes, a relatively large (approximately 30-kb) domain of chromatin that contains a cluster of adult and embryonic globin genes is in a more nuclease-sensitive conformation in all erythroid lineage cells present after 22 hours of development. Other genes, such as the ovalbumin gene in oviduct tissue, also appear to be embedded in relaxed chromatin domains that are significantly longer than the coding regions (Lawson *et al.* 1980, 1982). In the case of the chicken α-globin (Weintraub *et al.* 1981a) and β-globin genes (Stalder *et al.* 1980a) the regions flanking the transcription unit(s) are less nuclease-sensitive than the coding regions, but are more sensitive than bulk DNA. These findings are reminiscent of those described by microscopic techniques, in which case extended chromatin domains included apparently inactive segments of chromatin that surrounded the actively transcribed gene (see the previous section on "Microscopy"). Although compatible with the notion of chromatin domains, it should be mentioned again that there is no direct evidence that microscopically defined chromatin extension and the more open chromatin conformation defined by nuclease sensitivity both share a common basis (see Lamb and Daneholt, 1979, for a brief discussion).

The results cited above do not provide direct evidence that chromatin structural alterations are required for gene activity. In particular, although the intermediate sensitivity of the adult globin gene in embryonic red blood cells has been interpreted as a preactivation state of the adult globin gene chromatin—and, therefore as a potential prerequisite for transcription—such an interpretation is not strictly valid, since 5-day red blood cells are not the direct precursors to 14-day red blood cells (discussed in Groudine *et al.* 1981c, and in references therein).

The requirement for a population of cells that contains a purportedly activated, yet transcriptionally silent, gene is a major difficulty in analyzing the temporal relationship between chromatin structure and expression of developmentally regulated genes. In an attempt to circumvent the problem of obtaining pure populations of precursors to avian red blood cells, the activation of globin gene expression was studied in clones of cultured erythroblasts derived from chicken bone marrow cells that were transformed *in vitro* by a temperature-sensitive avian erythroblastosis retrovirus (tsAEV) (Weintraub *et al.* 1982). These cells fail to produce hemoglobin at 36°C; however, when the product or products of the viral-transforming gene *(erb)* are inactivated by a temperature shift to 42°C, several different cloned lines of these cells begin to make hemoglobin. This change in phenotype is associated with an increase in the levels of α-and β-globin mRNAs. Since one such tsAEV-transformed erythroblast clone displayed erythroid-specific, β-globin-hypersensitive sites at 36°C before overt transcription of the β-globin gene at 42°C, these experiments were interpreted to support the assertion that certain chromatin changes precede transcriptional activation and are not a consequence of transcription per se. While these experiments clearly support the notion that specific aspects of chromatin structure can be independent of concomitant transcription, the extrapolation of this observation to normal events in erythroid differentiation is difficult. One possibility is that the appearance of globin-hypersensitive sites in specific tsAEV erythroblast clones could be a consequence of viral infection or transformation, rather than a reflection of a developmentally significant event (see the section in this chapter on "Chromatin Structure and Development"; see also Groudine and Weintraub, 1982).

An interesting observation that may relate to the concept of cell lineage-specific preactivation of certain genes was made regarding the nuclease hypersensitivity of the chicken vitellogenin *(VTGII)* gene chromatin (Burch and-Weintraub, 1983). Vitellogenin, which is the major chicken yolk protein gene, is an estrogen-inducible gene that normally is expressed in the livers of egg-laying hens. A major hypersensitive region is located within, and several are located 3′ to, the apparent coding region of *VTGII* in chromatin isolated from liver of 9-day-old chickens. At this stage of development, the estrogen receptor system is not yet established (Lazier, 1978) and no detectable steady-state *VTGII*-related mRNA is present in liver cytoplasm (Jost *et al.* 1978). In addition, although none of these sites are present in chromatin from brain, fibroblast, or red blood cell, they are present in another estrogen-responsive organ (oviduct). However, this potential activation in oviduct is not completed. As assayed by quantitation of steady-state RNA (Wilks *et al.* 1982) and by nuclear run-off transcription (J. Burch and C. Casimir, personal communication), the *VTGII* gene in oviduct is never detectably expressed. In addition, chromatin that includes the *VTGII* gene does not form additional hypersensitive sites at the 5′-end of the gene in oviduct; sites that characteristically form in liver on *VTGII* expression after exposure to hormone. This raises the possibility that the gene has been marked for activation in two estrogen-responsive organs (liver and oviduct), only one of which (liver) can complete activation. This hypothesis is a matter of conjecture and is currently under study.

It is significant to note that in the studies on the chicken α- and β-globin genes, and on the *VTGII* gene in oviduct, transcription was monitored by the technique of nuclear run-off transcription (McKnight and Palmiter, 1979; Groudine *et al.* 1981b). This technique, which allows only polymerase molecules that initiated transcription *in vivo* to complete elongation of nascent RNA transcripts *in vitro,* is the most direct measure of transcription because it obviates the difficulties in attempting to distinguish between rates of transcription and subsequent RNA metabolism. Other studies that attempt to correlate nuclease sensitivity with the developmental activation of individual genes are severely limited by the lack of reliable transcription data. Assays that rely on quantitation of steady-state levels of mRNA, which unfortunately constitute the bulk of information on most genes, are subject to significant error: 1) cytoplasmic RNAs have diverse half-lives that can be drastically influenced by cellular conditions and the cellular environment; and 2) rates and pathways of RNA processing can give misleading information on the actual rate of transcription. This is a severe problem when analyzing transcripts of hormone-responsive genes, since it has been shown in several cases that hormone treatment can significantly increase RNA stability (e.g., Wiskocil *et al.* 1980; Brock and Shapiro, 1983). Evaluation of gene activity based on quantitation of protein products is so far removed from transcription as to be virtually useless in determining transcriptional regulatory phenomena. Determining the onset of transcription is crucial, since it has been reported that sequences that code for low-abundancy mRNAs and therefore, may represent infrequently transcribed genes, can be as sensitive to digestion with DNAseI as are highly transcribed genes (Garel *et al.* 1977).

Thus, the significance of many of the correlations between chromatin structural alterations and gene activity are difficult to interpret. For example, several readily inducible genes (e.g., the *D. melanogaster* heat-shock protein loci [Wu, 1980; Keene *et al.* 1981] and the mouse metallothionein gene [Senear and Palmiter, 1983]) contain hypersensitive sites before induction. A similar finding has been reported for the developmentally regulated chicken *VTGII* gene, which contains specific chromatin alterations in liver before full activation by hormone (Burch and Weintraub, 1983). In addition, several reports indicate that nuclease sensitivity (or hypersensitivity) persists after cessation of transcription; the *Xenopus laevis* (Williams and Tata, 1983) and chicken (Burch and Weintraub, 1983) vitellogenin genes remain DNAseI-sensitive or hypersensitive after hormone withdrawal, as does the chicken ovalbumin gene after the addition of a potent hormone inhibitor (Palmiter *et al.* 1978). However, the interpretation that these altered conformations are independent of transcription or represent preactivation of a region of chromatin must be qualified, since a low level of expression of these genes in the "uninduced" state either has not been convincingly ruled out (*VTGII* in liver, heat-shock proteins) or has (in fact) been observed (metallothionein).

While the preceding discussion has included examples of chromatin structural alterations detected as nuclease-hypersensitive sites that apparently are not dependent on overt high level transcription, other sites have been described whose appearance directly correlates with induction of expression of the associated gene. Examples include one site that is evident near the 5′ end of the *VTGII* gene (Burch and Weintraub, 1983) and another near the 5′ end of the metallothionein gene (Senear and Palmiter, 1983). In both cases, detection is coincident with presentation of a hormone that apparently is required for high-level expression. This raises the possibility that these sites might relate to hormone binding itself. It is possible that binding of regulatory molecules, at least in some cases, will generate what is detected as a hypersensitive site, constituting a virtual footprint of chromatin; and also that such an event could be inseparable from detectable transcription.

Although these types of analyses reveal a correlation between changes in chromatin structure and the transcriptional activation of several eukaryotic genes, they do not establish a regulatory role for structure in gene activation. However, they have provided a basis for the design and interpretation of related experiments that allow a more direct evaluation of the effect that structural features of chromatin might have on the regulation of transcription (see below).

Sequence Alterations and Rearrangements

The notion that one level of control of gene activity resides in structural features of chromatin, reflected by induction of nuclease sensitivity or hypersensitivity or by the opening or relaxation of domains of chromatin, is supported by the so-called position effects that flanking regions exert on genes in cis—often over considerable distances. In a narrow sense, an examination of the effect of position on expression could be limited to those cases in which a gene is moved to an entirely new location. For this discussion, we will consider not only these examples, but also more general cases in which a subset of sequences that distantly flank a gene—but that are distinct from the sites minimally required for expression—have undergone deletions, rearrangements, translocations, or insertions. The classic example of apparently nonspecific cis effects of neighboring DNA is provided by the inactive X chromosome; analyses of chromosomes generated by X-autosome translocations show that somatic genes located near recombination breakpoints can become inactivated (reviewed in Eicher, 1970). Similar effects have been described for genes translocated to positions adjacent to heterochromatic regions of chromatin (e.g., Kioussis *et al.* 1983; see also below). Conversely, increased expression of individual genes has been described either after their insertion into an active X-chromosome (dosage compensation; Spradling and Rubin, 1983) or when placed in cis—either in plasmid or chromosomal DNA—with one of the so-called enhancer elements (see below). The effect of position also has been proposed as one parameter involved in the variable levels of expression of activity of DNA integrated into the cellular genome after microinjection (Luciw *et al.* 1983), of retrovirus proviral DNA integrated at distinct sites in chromosomal DNA (Feinstein *et al.* 1982; Ucker *et al.* 1983), and of cellular genes near sites of somatic translocations and recombination. In this section, we describe a few examples of those position effects for which there is evidence that chromatin structural alterations may be involved in altering and/or regulating gene expression.

One of the best-studied examples of position-mediated regulation is that of the mating-type genes of the yeast *Saccharomyces cerevisiae* (Nasmyth, 1982; Klar *et al.* 1981). In this case, the activity of the a- or α-mating-type genes is dependent on their location. Of the three possible locations that these genes occupy (*MAT, HML,* or *HMR*), only those at *MAT* are active; the *HML* and *HMR* loci are repressed through the action of the *SIR* (or *MAR*) gene products. Of particular interest are the findings that the *SIR* gene products act at a distance and that their binding induces chromatin structural alterations not only at their apparent initial site of interaction with DNA, but also over a region including *HML* (or *HMR*). This example is unusual, because both the DNA sequences and the proteins that appear to initiate such long range effects have been identified. The relevant components responsible for position effects in the additional examples cited below are not as well characterized; yet, they provide supportive evidence for the notion that some sequences that are capable of forming altered chromatin conformations also induce chromatin structural alterations and influence the transcriptional activity of adjacent genes.

One gene whose expression appears to be influenced by the identity of nearby flanking sequences in mice (Sheng-Ong *et al.* 1982), humans (Della-Favera *et al.* 1982), and chickens (Hayward *et al.* 1981; see also below) is the cellular oncogene *myc*. In the case of the chicken *myc* gene, integration of retrovirus proviral DNA in at least some nearby positions is associated with the development of B-cell lymphomas. This integration event also correlates with elevated levels of *myc*-encoded RNA relative to a variety of untransformed lymphocytic and nonlymphocytic cell populations (Hayward *et al.* 1981; Payne *et al.* 1982; Gonda *et al.* 1982). Since the level of *myc* gene expression is unknown in the as yet unidentified lymphocytic precursor cells that give rise to bursal lymphomas, it is impossible to determine whether the integration of proviral DNA results in increased expression of the *myc* gene or whether it prevents the potential inactivation of these sequences, which might occur during the maturation of precursor cells into mature lymphocytes. In some cases, this "activation" apparently is achieved by downstream promotion, whereby *myc*-containing transcripts are initiated in the LTR region of the provirus and read through into the *myc* gene (Hayward *et al.* 1981). In other cases, such a mechanism is not likely, since LTRs are found in the opposite transcriptional orientation of *myc* or, in one case, downstream of the *myc* gene (Payne *et al.* 1982). In several cases for which downstream promotion of *myc* gene sequences is not occurring, the evidence that increased expression of the *myc* gene may be influenced by chromatin structural alterations induced by retrovirus DNA is suggested by the pattern of nuclease-hypersensitive sites flanking the *myc* gene (Schubach and Groudine, 1984). In normal cells, distinct DNAseI-hypersensitive sites are detectable within a region 2.5-kbp upstream from the first coding *myc* gene exon. In contrast, none of these endogenous hypersensitive sites are evident in several bursal lymphoma cell lines that contain proviral sequences near the *myc* gene. Instead, only one major DNAseI-hypersensitive site is present, and it is located within the proviral LTR. Examination of *myc*-containing transcripts from these cell lines showed that they did not contain any viral information, indicating that they were not initiated in the proviral LTR. Therefore, these data indicate that the presence of proviral DNA in this region is associated both with chromatin reorganization and with apparently altered expression of adjacent cellular sequences. In addition, they also suggest that dominant elements in the viral LTR, which can act at a distance, may be responsible for inducing these changes in structure and activity. Given the enhancing effect that proviral LTRs exert on the transcriptional activity of other genes located in cis (Luciw *et al.* 1983; Levinson *et al.* 1982), these findings are compatible with the notion that such activation is induced or communicated via a mechanism that is recognized by different promoter element; rather than by a mechanism that is gene-specific (see below).

The *Sgs4* glue gene of *D. melanogaster* provides a striking example of the apparent role of chromatin structural alterations in gene expression. The *Sgs4* glue protein is expressed in the salivary gland during the third instar larval stage. As with other developmentally regulated genes, the *Sgs4* glue gene contains several DNAseI-hypersensitive sites in a region of approximately 500 bp surrounding the transcription start site (−480, −405, −330, −70 and +30; Shermoen and Beckendorf, 1982); these sites are present only in cells in which *Sgs4* glue *mRNA* is detectable. Several mutant strains of *Drosophila* are available that contain undetectable or low (approximately 2%) levels of glue mRNA relative to the"wild-type" Oregon strain (Muskavitch and Hogness, 1982). Three low-producing strains (Oriental or *ORL*) were found to contain a deletion of 50 bp between −305 to −356; also, not surprisingly, they did not contain a hypersensitive site at the −330 position, presumably due to the absence of sequences in this region. Since the low levels of glue mRNA found in these cells are detectable only at the appropriate time in development, it was concluded that sequences required for efficient expression—but not for the control of the onset of expression—of glue mRNA were deleted (Muskavitch and Hogness , 1982). The Berkeley strain *(BER)*, in which no detectable *Sgs4*-encoded RNA is ever found, contains a deletion extending from −392 through −486; interestingly, no DNAseI-hypersensitive sites are found at the 5′ end of the *Sgs4* gene in this strain. These results suggest a hierarchic ordering of the formation of structures in this region, with an event approximately 500-bp upstream from a transcription unit that is required for structural alterations between about 200–500-bp away. It is a matter of conjecture whether this reflects (for example) the binding of a protein at −500 (which either proceeds along the chromatin to −300 or to the start site), whether a structural alteration of −500 induces localized conformational alterations in the region, or whether some other type of long-range communication is involved. What is clear is that deletion of two regions—both of which contain hypersensitive sites and both of which lie a significant distance from the *Sgs4* "promoter"—have profound, although variable, effects on the expression of the associated gene. The expression of several other eukaryotic genes, including both the sea urchin histone H2A gene (Grosschedl and Birnstiel, 1980a,b) and the *herpes* thymidine kinase gene (McKnight *et al.* 1981), also are dependent on the identity of sequences a distance from the transcription initiation start site; although the involvement of structure has not been specifically addressed in these cases.

A final and potentially more direct approach to probing the regulatory significance of chromatin structure on gene expression, is to artifically introduce sequences that form alternate chromatin conformations into recombinant DNA molecules, and then to monitor their effect on the transcriptional competence of genes in cis. One such region, which has been analyzed in detail, is within the heavy chain constant region of the mouse immunoglobulin gene. In germ-line DNA, the variable and constant portions of immunoglobulin genes reside at distant sites within the genome; production of a functional immunoglobulin heavy chain protein is dependent on a chromosomal rearrangement in somatic cells that brings the variable and constant regions of this gene into proximity (Bernard *et al.* 1978; Early *et al.* 1980). The variable region, which provides the promoter for the heavy chain gene after rearrangement, apparently is inactive in its germ-line configuration and is expressed only after rearrangement. The immunoglobulin heavy chain constant region, however, is sensitive to digestion with DNAseI before rearrangement (Storb *et al.* 1981b; Parslow and Granner, 1982); it also exhibits a low level of transcriptional activity in immature lymphocytes before rearrangement, although transcripts are rapidly degraded before transport to the cytoplasm (Kemp *et al.* 1980; Van Ness *et al.* 1981). In addition, both the rearranged immunoglobulin genes in B lymphocytes (Parslow and Granner, 1982) and the unrearranged genes in T lymphocytes (Storb *et al.* 1981a) contain DNAseI-hypersensitive sites in a region 5′ to the heavy chain constant region. These findings led to the suggestion that sequences within the heavy chain constant region hypersensitive site might serve to activate the variable region promoter after rearrangement (Queen and Baltimore, 1983; see references therein). Support for this hypothesis is suggested by the fact that sequences included within this hypersensitive site are able to enhance the expression of neighboring genes when present in recombinant plasmid molecules (Gillies *et al.* 1983; Banerji *et al.* 1983; Queen and Baltimore, 1983). This immunoglobulin region is one of several elements that not only form hypersensitive structures in chromatin, but also enhance expression of genes in cis. Others include the 72 bp repeats of SV40 and polyoma and sequences in retrovirus LTRs. In the context of developmentally controlled regulation, it is particularly interesting that these elements can display at least partial cell-type (Gillies *et al.* 1983; Banerji *et al.* 1983) or species (deVilliers *et al* 1982; Laimins *et al.* 1982) specificity. This suggests that similar elements may be responsible, at least in part, for the lineage-specific expression of some developmentally controlled genes. In addition, while recognition of these elements can be subject to cell-type restriction, the fact that these enhancers show little specificity with regard to the gene on which they act suggests that their activity is independent of gene-specific regulatory factors that act only on unique promoters; instead their activity is based on a more generally recognizable, long-range-inducing event. In this regard, it is important to stress the fact that the specificity of the immunoglobulin enhancer element is maintained by the DNA sequences themselves; that is, this region continues to show preferential activity in lymphocyte cell populations even after removal from its endogenous location, insertion into plasmid molecules, replication in bacterial cells, and reintroduction into eukaryotic cells. This point is crucial, since it suggests that the element supplies a dominant regulatory signal to adjacent promoters in the absence of signals that might operate during development to mark this gene for activation (see also the section in this chapter on "DNA-hypersensitive Sites").

Although a limited number of other cloned genes are subject to at least partial regulation after reintroduction into cells via transfection or microinjection or into animals via the germ line (Corces *et al.* 1981; Goldberg *et al.* 1983;

Mayo *et al.* 1982; Spradling and Rubin, 1983), cloned copies of several other developmentally regulated genes can override the regulatory factors that act on their endogenous, chromosome-associated homologs. For example, the rabbit β-globin (Wold *et al.* 1979) and human growth hormone (Robins *et al.* 1982) genes are transcriptionally competent after introduction into mouse fibroblast cell lines, while neither of these genes are active in the equivalent cell type *in vivo,* nor are the mouse homologs of these genes active in the cells used for analysis. A cloned copy of the rabbit β-globin gene is also expressed in nonerythroid tissues after introduction into mice via germ-line transfection (Constantini and Lacey, 1981). In addition, the α_{2u} globulin gene—normally active in male livers—is transcriptionally competent in cultured mouse cells after transfection of the cloned gene (Kurtz, 1981).

One further example is provided by a cloned copy of the human globin gene derived from a patient deficient in β-globin synthesis. This particular gene, which fails to be activated developmentally in its inherited location, is present on a chromosome that contains a deletion that is approximately 3 kbp upstream of the β-globin gene. Interestingly, the deletion results in the juxtaposition of the β-globin gene 5′-flanking sequences to a region of chromatin that is highly resistant to digestion by DNAseI, and that apparently is transcriptionally inert (Kioussis *et al.* 1983). This would either suggest that positive regulatory elements required for globin gene activation had been removed by the deletion and/or that the inactive translocated chromatin exerted an inhibitory effect on globin gene activation that is similar to the position effects described above. In an attempt to identify the elements responsible for this abnormal expression, genomic clones of normal and affected genes were introduced into HeLa cells *in vitro.* In this case, it was found that clones containing the β-globin gene and approximately 3 kbp of normal 5′-flanking DNA and a clone containing these sequences, as well as an additional 2.5 kbp of the potentially inhibitory translocated DNA, were expressed with equal efficiency. In this example then, the element(s) responsible for the differential expression of the normal and affected genes seen *in vivo* was not evident after transfection of cloned copies of these genes. However, this result might at least be partially explained by the fact that HeLa cells were used to monitor the transcriptional competence of the cloned genes. That is, if an erythrocyte-specific factor is responsible for the difference in expression of the normal and affected genes, it is possible that HeLa cells, which presumably would lack such a factor, would not discriminate between the two genes. Resolution of this question awaits further experimentation.

The results cited above raise another very central question: why are cloned copies of tissue specific genes transcriptionally active in cells that repress the expression of their endogenous homologs? One possibility is that genes that display a tissue-specific pattern of expression *in vivo* might be associated with distant sequences that play a role in repressing their activity in nonexpressing tissues. In this case, such sequences may frequently be eliminated from the relatively small clones of genomic DNA used in these experiments. If such sequences normally are responsible for repressing the expression of genes in specific cell types, then it might be predicted that all genes have the potential to be expressed in all cell types when disassociated from their inhibitory elements. However, such a negative model cannot explain the additional data that the activity of cloned copies of tissue specific genes appears to be significantly decreased relative to that of their developmentally activated endogenous counterparts. In this context therefore, it is also reasonable to assume that positive cell or species-specific elements, including DNA sequences and regulatory proteins, contribute to gene regulation *in vivo.* The data cited above would suggest that if such positive elements are required for regulated expression during normal development, then some feature of cloned, plasmid-associated DNAs can (at least in some cases) either partially eliminate the need for such positive control or provide a functional substitute. This latter possibility may relate to one or several of the potentially significant differences between cloned and endogenous copies of a gene. For example, the "position" of cloned genes is defined not only in terms of the amount of endogenous flanking DNA included in the clone, but also in terms of the vector and other DNA sequences present; cloned DNAs that have integrated into the recipient genome are subject to the additional effects contributed by chromosome location. Furthermore, unintegrated, small circular DNA molecules can be topologically distinct from genes present in large chromosomal domains; growth in bacteria leads to an abolition of the DNA methylation pattern generated *in vivo.* Finally, naked DNA, when introduced into cells, is subject to *de novo* chromatin assembly in the absence of the normal developmental signals directed either by location (i.e., flanking sequences) or by the lineage history of the cell in which the gene is normally activated or silenced. Each of these parameters has been implicated not only in contributing to the local chromatin structure, but also as elements that could be involved in directing the long-range effects evident in the regulated expression of eukaryotic genes. In the following section, we will review some of the elements thought to be involved in establishing an active chromatin conformation. We also will discuss how these parameters might be involved in the abnormal expression of DNAs reintroduced into cells, as well as ways in which each might contribute to the positive and/or negative long-range effects observed *in vivo.*

Formation and Maintenance of an Active Chromatin Conformation

Elucidation of the topology, organization, and composition of transcriptionally active regions of chromatin would certainly contribute to an understanding of the steps required to effect the transition from an inactive to an active chromatin structure; it also would allow more direct evaluation of the role that chromatin structural components might play in the regulation of chromatin function. Several potentially signficant parameters have been identified that may influence chromatin structure and gene expression, although the evidence for their direct participation in regulation is limited. These include: 1) the nucleosome content and organization of chromatin, 2) the composition of non-histone chromatin-associated proteins; 3) levels of DNA methylation; and 4) DNA conformation and topology. The following sections will provide a brief review of the evidence that implicates each of these parameters as regulatory components, with an emphasis on the possible ways in which they might interact to influence chromatin structure and function.

DNA-Hypersensitive Sites

In a preceding discussion (see the previous section on "Nuclease-hypersensitive Sites"), nuclease-hypersensitive regions were designated as chromatin-specific based on the fact that the same sites were not preferentially cleaved in naked cellular DNA. Recently, however, analyses of several segments of naked DNA have revealed the presence of sites that are specifically recognized by nuclease S1 (Beard *et al.* 1973; Lilley, 1980; Panayotatos and Wells, 1981; Larsen and Weintraub, 1982) and by the single-strand-specific chemical reagent bromoacetaldehyde (Lilley, 1983; Kohwi-Shigematsu *et al.* 1983); this suggests that these sites are at least partially composed of regions with single-stranded character. Of particular interest to this discussion is the finding that some of these sites in naked eukaryotic DNA correspond to the ncuclease-hypersensitive sites that are detectable in the same region of DNA when it is present in active chromatin (Larsen and Weintraub, 1982). This suggests that DNA, under certain conditions, can play a dominant role in the formation of nuclease-hypersensitive sites in chromatin (Weintraub, 1983; Elgin, 1982). Detailed comparisons of the plasmid and chromatin cleavage sites, however, show that they are not identical, which indicates that chromatin-specific factors can induce, suppress, or displace DNA-based nuclease-sensitive conformations. Analyses of the DNA conformations recognized by S1, and also of the factors that influence the formation of S1 nuclease-sensitive sites, currently are rudimentary. However, a short review of what has been established will serve to describe the avenues being followed to elucidate the potential significance of DNA conformation and topology in regulating gene expression.

Nuclease S1 preferentially degrades single-stranded, rather than double-stranded, DNA (Vogt, 1973). However, it also has been shown to introduce specific internal cleavages in both linear (Chan and Wells, 1974; Dodgson and Wells, 1977) and circular (see below) double-stranded DNA molecules. Its activity on linear DNA is largely determined by the local base composition; discrete cutting sites are localized to regions of high A-T content (Moreau *et al.* 1982). Interestingly, similar sites also are preferred binding sites for RNA polymerase (Jones *et al.* 1977). In an investigation of S1 cleavage of linear DNAs that contain defined mismatches, Dodgson and Wells (1977) reported that approximately three to five contiguous base pairs must be unpaired for efficient cleavage by S1.

S1 cleavage of circular double-stranded DNA molecules was initially described several years ago (Beard *et al.* 1973; Mechali *et al.* 1973; Germond *et al.* 1974); only recently, however, have the requirements for such cleavage or the structure(s) of the cleavage sites been examined in any detail. S1 cleavage of circular molecules is dependent on the overall topology of the DNA substrate in that site-specific cleavage is limited to molecules that are supercoiled (Beard *et al.* 1973; Shishido, 1980; Lilley, 1980; Panayotatos and Wells, 1981). The degree of supercoiling required for efficient S1 cleavage of circular molecules falls within a range of superhelical densities that are compatible with the physiologic range reported for prokaryotic cells (Singleton and Wells, 1982; see also Courey and Wang, 1983). One effect that supercoiling is known to have on covalently closed circular DNA molecules is to favor the formation of non-B DNA conformations such as hairpins, cruciforms (Hsieh and Wang, 1975; Singleton and Wells, 1982; Mizuuchi *et al.* 1982), or Z-DNA (Wang, *et al.* 1979; Klysik *et al.* 1981; Peck *et al.* 1982; Zacharias *et al.* 1982; Singleton *et al.* 1982; Stirdivant *et al.* 1982). This correlation raises the possibility that such conformations constitute substrates for nuclease S1. Detailed investigations of S1 cleavage sites have provided evidence that some non-B DNA conformations are substrates for S1. S1 cleavage sites in prokaryotic plasmid DNAs appear to be localized to regions that are bounded by inverted repeat sequences and that form relatively stable cruciform structures (Lilley, 1980, 1981; Panayotatos and Wells, 1981); the S1 cleavage site in supercoiled SV40 DNA is within a region that contains sequences able to form Z DNA under certain conditions (Nordheim and Rich, 1983). The hypothesis that Z-form DNA is a determinative factor in conferring S1 sensitivity to supercoiled SV40 DNA is also suggested by independent experiments indicating that S1 is able to cleave DNA at B-Z DNA junction regions (Singleton *et al.* 1983).

S1-sensitive sites have been mapped for several segments of eukaryotic DNA present in supercoiled molecules, including SV40 (Beard *et al.* 1973), several yeast genes (Carnevali *et al.* 1983), sea urchin histone genes (Hentschel, 1982), the chicken β-globin gene (Larsen and Weintraub, 1982; Nickol and Felsenfeld, 1983; Schon *et al.* 1984), chicken endogenous virus (Conklin, Skalka and Groudine, unpublished), and *Drosophila* heat shock protein genes (Mace *et al.* 1983). Analyses of the cleavage sites present in these and other molecules imply that other conformations can also serve as substrates for cleavage by S1; these include regions of high AT content, out-of-register base-paired structures formed in homopolymer or polypurine/polypyrimidine sequences, and noncruciform single-stranded loops.

The significance of these conformations in contributing to regulation *in vivo* currently is circumstantial. The first suggestion that they are functionally significant is (as mentioned above) their relationship to chromatin hypersensitive sites. Second, non-B DNA conformations, such as those described above, frequently have been proposed as candidate regulatory sites (Wells *et al.* 1980; see also Wartell, 1977; Cantor, 1981; McKnight *et al.* 1981; Hall *et al.* 1982; Nordheim *et al.* 1982), which might act through several different mechanisms. For example, formation or release of non-B DNA conformations could be modulated *in vivo* to influence the superhelical density within a restrained segment of DNA. In fact, it has been observed that the S1 cleavage sites detected in a circular molecule can be diminished or eliminated by the introduction of additional segments of DNA that contain sites cleaved by S1 (Lilley, 1981; Carnevali *et al.* 1983). Thus, if the degree of strain on DNA induced by supercoiling influences its ability to serve as a substrate for transcription (or replication) in eukaryotes as it does in prokaryotes (Helinski and Clewell, 1971; Smith, 1981; see also the following section in this chapter), then formation or release of non-B DNA conformations could serve to modulate the torsional strain and (therefore) the activity of promoter elements included within a restrained segment of DNA. Further speculation leads to the notion that such a restrained segment in chromatin could be the loops or domains in which cellular DNA apparently is organized (see the previous section on "Higher-order Chromatin Structure"). This type of regulation is consistent with the observed long-range effects that enhancer elements exert on promoters in cis (Yaniv, 1982), on the long-range position effects described for expression of several cellular genes, and on the transcriptional activity of coresident genes in plasmid DNA (Gregory and Butterworth, 1983; see also the previous section on "Sequence Alterations and Rearrangements"). The finding that some enhancers contain sequences and/or conformations that are sensitive to cleavage by S1, together with the finding that the activity of some enhancers appears to be (at least) partially species- or cell-type-specific (de Villiers *et al.* 1982; Laimins *et al.* 1982; Gillies *et al.* 1983; Banjeri *et al.* 1983), raises the possibility that specific factors may promote, stabilize, or induce alternate conformations that, in turn, may affect gene activity. In this context, it is important to note that in addition to topologic considerations, the conformation of a segment of DNA could certainly influence its capacity to interact with proteins. In fact, there are indications that some S1-sensitive regions preferentially lose or exclude nucelosomes (Waskylyk *et al.* 1979; Weintraub, 1983; and see the section in this chapter on "Protein Composition and Organization"). In addition, the identification of Z DNA binding proteins (Nordheim *et al.* 1982) suggests not only that this DNA form may exist *in vivo,* but also that there are some situations in which the Z form may be stabilized and/or provide a conformational substrate for the binding of such proteins (Nordheim *et al.* 1981; Hamada *et al.* 1982; Queen *et al.* 1981; Saffer and Lerman, 1983; but see Hill and Stoller, 1983, for a discussion). Identification of additional factors that might modulate or recognize DNA conformation *in vivo* currently is a major point of interest. A difficulty in this regard is the question of which conformations identified *in vitro* might reflect those that actually form *in vivo.* For example, a problematic—but potentially significant—aspect of S1 cleavage of supercoiled DNAs is that cleavage specificity varies depending on the salt concentration, ionic conditions, temperature, solvent conditions, and pH (Larsen and Weintraub, 1982; Kohwi-Shigematsu *et al.* 1983; Courey and Wang, 1983). Although problematic to the investigator, this type of flexibility would be expected, and probably would be advantageous, for any factor that must respond to developmental signals for differential gene expression (see also the

the section in this chapter on "Speculations About Chromatin Structure and Development").

DNA Topology

The fact that several proposed regulatory regions of eukaryotic DNA have the potential to form alternate DNA conformations, when placed under torsional strain induced by supercoiling *in vitro* (see the previous section), suggests that supercoiling might play a regulatory role in eukaryotes as it does in prokaryotes. In bacteria, a minimum level of DNA supercoiling apparently is required for efficient transcription, replication, and recombination (Helinski and Clewell, 1971; Smith, 1981; see also Wang, 1974). The basis of this requirement has been interpreted as being a mechanism that primes a DNA molecule for any event that would relieve the torsional strain induced by supercoiling. These include the opening of DNA strands by melting, by interstrand binding of molecules such as polymerase, by formation of structures such as cruciforms or hairpins, or by conformations such as Z DNA. However, as reviewed below, the question of whether torsional strain induced by supercoiling, or even of supercoiling per se, exists and/or is used in regulation of chromatin function in eukaryotes remains experimentally unresolved.

In prokaryotes, the activities of DNA gyrase, which introduces negative supercoils into DNA (reviewed in Cozzarelli, 1980a) and topoisomerases (which release them) (Cozzarelli, 1980b; Gellert, 1981), combine to maintain an overall level of strain. In eukaryotes, however, a gyrase-like activity has not been identified, although topoisomerase activities appear to be abundant. Therefore, if eukaryotic chromatin does contain supercoiled DNA, a mechanism distinct from that employed by prokaryotes might be required for its induction and maintenance. One proposed mechanism (Larsen and Weintraub, 1982; Weintraub, 1983) for formation of supercoiled DNA in eukaryotic chromatin takes into account the fact that the loss or removal of nucleosomes from a restrained segment of DNA results in the introduction of negative superhelical turns (Germond *et al.* 1975). Whatever mechanism is used, however, the presence of high levels of topoisomerase activity in eukaryotic cells suggests that torsional strain induced by supercoiling would have to be used immediately. Alternatively, such strain would need to be maintained, perhaps by sequestration of the supercoiled DNA from topoisomerases or by the effective storage of supercoil-induced strain via formation of alternate DNA conformations not recognized by, or protected from, topoisomerases.

Several lines of evidence originally suggested that eukaryotic DNA was relaxed *in vivo,* and that the level of supercoiling of either circular molecules isolated from cells or of intact cellular DNA, could be accounted for by the removal of nucleosomes from restrained DNA (Germond *et al.* 1975). In addition, attempts to directly measure DNA supercoiling *in situ* with the photoreactive chemical cross-linking agent psorilen failed to detect significant levels of underwound DNA (Sinden *et al.* 1980). However, as pointed out by the authors, the absolute significance of the latter finding is questionable. The difficulty in interpretation rests on the notion that supercoiling might be confined to active segments of DNA. Since only approximately 10% of the eukaryotic genome is estimated to be transcriptionally active at any time, and since the psorilen-binding experiments required a value in excess of 10% for detection, the negative findings could simply reflect the sensitivity of the technique applied. Partially for this reason, attempts to directly demonstrate a requirement for torsional strain in the function of eukaryotic chromatin have been limited primarily to investigations on the structure and transcriptional activity of relatively short DNA molecules introduced into eukaryotic cells.

Analysis of the topology of SV40 minichromosomes isolated from infected cells indicates that torsional strain is associated with active chromatin. Transcriptionally active SV40 minichromosomes can be separated from inactive molecules by sedimentation (Green and Brooks, 1976). After such a fractionation, it was found that the active—but not inactive—molecules were efficient substrates for topoisomerase I (Luchnik *et al.* 1982), which is an enzyme that relaxes supercoiled DNA. Thus, in a case where the contribution of inactive chromatin is minimized, torsionally strained active chromatin is detectable. These data also imply that a mechanism does exist *in vivo* to selectively inhibit topoisomerase-mediated relaxation of at least some torsionally strained molecules. It was also reported that treatment of active, but not inactive, SV40 minichromosomes with topoisomerase I resulted in a release of virtually all histones from the DNA (Luchnik *et al.* 1982). This indicates that torsional strain is required at least to maintain chromatin protein-DNA interaction on a transcriptionally active molecule; this implies that such strain would be required for transcription as well. These findings are consistent with analyses on the conformation and/or transcriptional activity of molecules injected into *Xenopus* oocyte nuclei described below.

Several reports have demonstrated that linear DNA is virtually inactive as a transcription template after injection into *Xenopus* oocytes, while circular templates of identical DNA content are efficiently transcribed (Laskey *et al.* 1978; Gurdon and Melton, 1981; Mertz, 1982; Harland *et al.* 1983). Recent investigations have further demonstrated that the inactivity of linear DNA is not completely accounted for by trivial factors such as template degradation (Mertz, 1982; Harland *et al.* 1983). In addition, analyses on the fate of both linear and circular molecules of SV40 DNA after introduction into *Xenopus* oocytes or incubation in HeLa cell extracts demonstrated that circular, but not linear, DNAs assembled into ordered chromatin with a regular 200 bp nucleosome organization (Mertz, 1982). Since the differences in nucleosome organization and transcriptional efficiency of these molecules is independent of sequence, yet dependent on topology, these data have been interpreted in light of the obvious topologic distinction between circular and linear molecules; linear DNAs have unrestrained ends. This difference would be critical if torsional strain is required at any stage in the generation or maintenance of a transcriptionally competent template, since linear DNA, which is free to rotate about its ends, could immediately dissipate strain.

In an attempt to investigate the role of chromatin topology in transcription, experiments have been conducted to find out if circular transcriptionally active DNA (the *herpes* thymidine kinase gene [Harland *et al.* 1983] or *Xenopus* rDNA [Pruitt and Reeder, 1984] cloned into pBR322) retained transcriptional activity when linearized *in vivo.* By injecting restriction enzymes into oocytes, they were able to demonstrate that transcription of tk or rDNA was inhibited on linearization. In light of the results of Luchnik *et al.* (1982) described above, this parameter could be proper protein-DNA interaction (either RNA polymerase or chromatin protein), torsional strain per se, or both.

Additional evidence that torsional strain might be required to generate a transcriptionally competent molecule comes from the fate of nicked circles after injection into *Xenopus* oocyte nuclei. Nicked circles normally are efficiently ligated, assembled into chromatin, and able to serve as templates for transcription after introduction into occyte nuclei (Laskey *et al.* 1978; reviewed in Gurdon and Melton, 1981). Typical of nucleosome associated DNA, they are recovered as supercoiled DNA after deproteinization. Pruitt and Reeder (1984), however, demonstrated that introduction of nicked circular molecules into oocytes, under conditions that delayed (but did not prevent) repair of nicked DNA, resulted in the generation of molecules that apparently were transcriptionally inert. That these molecules were associated with chromatin components is suggested by the fact that they were recoverable as supercoiled DNAs after extraction. In the view of the authors, chromatin was presumed to assemble on the nicked molecules, and ligation occurred after this assembly due to the delay in repair. In these experiments, the organization of chromatin assembled on nicked molecules was not analyzed. Therefore, it is not possible to distinguish between several possible explanations for the inactivity of these molecules. First, chromatin assembled on nicked (and therefore relaxed) molecules might be irregular, similar to the findings described with linear DNAs (Mertz, 1982), and might have remained irregular even after the subsequent ligation of chromatin-associated DNA. If proper protein-DNA organization is a prerequisite for template use, then this abnormal assembly could result in the observed transcriptional incompetence. Another possibility is that chromatin assembly on nicked molecules might proceed normally, but might generate chromatin whose conformation and/or composition is incompatible with transcription. This could be viewed within the context of the local or global topology of the molecules, or of chromatin constituents that assembled on relaxed molecules (see the following section in this chapter). In either case, these results indicte that chromatin assembled on relaxed molecules is inefficient as a template for transcription, thus reinforcing the idea that torsional strain may be required to generate a transcriptionally competent molecule.

The studies cited above all employed relatively short molecules to investigate the role of DNA conformation on chromatin assembly and transcriptional competence. However, it is reasonable to ask if such relatively small molecules provide a relevant system with which to evaluate the role of chromatin topology in gene regulation. This question arises in light of the potential organizatin of eukaryotic chromosomes into large loops or domains 30–100 kb in length. This issue recently was addressed (Pruit and Reeder, 1984) by asking if the endogenous ribosomal DNA genes of *Xenopus* oocytes were transcriptionally inactivated as a consequence of cleavage by restriction enzyme digestion, as were the small circular copies of this gene. The use of several restriction enzymes failed to inhibit either elongation or initiation of transcription of the endogenous genes. However, since a restriction enzyme that should have inhibited transcription (i.e., one that cleaved within an essential promoter sequence) was not included in these experiments, it cannot be unequivocally stated that all rDNA units were cleaved. If verified, however, these results, when taken together with those cited above (see this section; see also the previous section on "Sequence Alterations and Rearrangements"), reinforce the notion that endogenous genes are subject to additional levels of organization influencing their expression that are not available to small, circular, plasmid molecules. This may be due to size, circularity, topology, or (as described below) to the protein content, organization, and/or DNA methylation patterns established on endogenous genes during development.

Protein Composition and Organization

A central focus of any discussion about the regulation of gene expression is the role of protein-DNA interactions. Investigations of the basic organizational requirements for expression (e.g., those described both in the previous section and below) suggest that the arrangement and composition of histones, of variant and modified histones, and of other abundant chromatin-associated proteins differs between active and inactive chromatin; it also implies that these different protein constituents may be involved in regulation. Although nonabundant proteins undoubtedly play a major role in controlling gene activity, few candidate regulatory proteins have been identified in eukaryotes. Therefore, our focus will be on reviewing some of the variations in chromatin composition that have been described between active and inactive chromatin, with an emphasis on how other factors implicated in regulation (e.g., DNA topology, conformation, sequence, and modification) might relate to the nonrandom distribution of chromatin proteins. In addition, emphasis will be placed on how the protein variations described might act to initiate the formation of, or to maintain, an active chromatin conformation.

Core Histones and Nucleosomes

The basic repeating organization of nucleosome-associated DNA described in the previous section on "The Nucleosome" is typical of bulk eukaryotic chromatin. However, as evidenced both by *in vitro* reconstitution experiments and by examination of chromatin assembled *in vivo,* it is now apparent that this very ordered structure is not uniform and that both local and long-range disruptions are evident at many sites. For example, two segments of eukaryotic chromatin appear to be nucleosome-free *in vivo.* One of these includes the region proximal to the origin of replication of SV40 (Saragosti *et al.* 1980; Jakobovitz *et al.* 1980); the other is located 5′ to the adult chicken globin gene (McGhee *et al.* 1981). In addition, an apparently nucleosome-free region near the 5′ end of the 75S RNA transcription unit has also been identified in a Balbiani ring of chromosome IV in *Chironomus* tentants (Lamb and Daneholt, 1979). Interestingly, the nucleosome-free regions of both SV40 and the chicken β-globin gene not only are hypersensitive to nuclease digestion *in vivo* (Scott and Wigmore, 1978; Varshavsky *et al.* 1979; Saragosti *et al.* 1980; McGhee *et al.* 1981; Larsen and Weintraub, 1982), but also are associated with S1 nuclease-hypersensitive sites when analyzed as supercoiled DNAs (Beard *et al.* 1973; Larsen and Weintraub, 1982). Analyses of additional supercoiled molecules suggests that at least some other S1 sensitive regions appear to preferentially lose or exclude nucleosomes in *in vitro* reconstitution experiments (Weintraub, 1983). It recently has been demonstrated that the nucleosome-free region of SV40, which is detectable in approximately 20% of the molecules isolated from infected cells (Scott and Wigmore 1978; Varshavsky *et al.* 1979; Saragosti *et al.* 1980; Jakobovitz *et al.* 1980), contains three segments of alternating purine-pyrimidine tracts 8 bp in length, which is a composition consistent with the ability to flip into Z DNA (Wang *et al.* 1979); these sequences also specifically bind anti-Z-DNA antibodies under certain conditions (Nordheim and Rich, 1983). The finding that this region preferentially excludes nucleosomes in *in vitro* reconstitution experiments (Wasylyk *et al.* 1979) suggests that the absence of nucleosomes *in vivo* could be at least partially accounted for by DNA sequence and/or conformation. However, this region is also the site for T-antigen binding, thus raising the possibility that (at least *in vivo*) the significant event for nucleosome exclusion is associated with T antigen and not with formation of Z DNA per se. In an attempt to further investigate the role that Z form DNA might play in affecting nucleosome assembly, *in vitro* reconstitution experiments have been conducted by using synthetic alternating purine/pyrimidine polymers; sequences that can readily make the B- to Z-form transition (Wang *et al.* 1979; Klysik *et al.* 1981; Haniford and Pulleyblank, 1983). The alternating poly(dA-dT) and poly(dG-dC) polymers, when in the B conformation, are efficient substrates for nucleosome assembly (Simpson and Kunzler, 1979). However, Nickol *et al.* (1982) were unable to demonstrate assembly on poly(dG-dC) if the polymer was first converted to Z form DNA. In addition, it was reported that nucleosome-associated B form DNA could no longer make the transition into the Z form conformation. Although one report (Miller *et al.* 1982) suggested that Z-DNA could assemble into nucleosomes, these data are controversial. In particular, it was reported that nucleosomes could form on poly(dG-dC) when this polymer was in either the B or Z conformation. Moreover, once assembled into nucleosomes, the DNA could flip into the Z form; inclusion in a nucleosome structure actually facilitated this transition. The controversial aspect of this report is based on the finding that although the nucleosomal DNA appeared to be in the Z form by several criteria, electron-microscopic analysis showed that nucleosome-like structures were evident only on the ends of the DNA molecules (Miller *et al.* 1982). This raises the possibility (as discussed by the authors) that conformational flexibility at the ends of the polymer permitted nucleosome assembly in these regions, while the internal region, which apparently maintained the more typical Z-DNA form, was not efficiently assembled into nucleosomes. Therefore, these data tend to support the hypothesis that Z-form DNA is an inefficient substrate for nucleosome assembly.

Poly(dA)-poly(dT) and poly(dG)-poly(dC) also fail to form regular nucleosomes (Simpson and Kunzler, 1979; Kunkel and Martinson, 1981), which implies that homopolymeric stretches in eukaryotic DNA might tend to exclude nucleosomes *in vivo.* Interestingly, the nucleosome-free region near the 5′ end of the adult chicken β-globin gene contains a stretch of 16 guanosine residues; this suggests that this feature may constitute a partial basis for the lack of nucleosomes in this region. However, since this region is nucleosome-free only in cells in which this gene is activated, sequence alone cannot explain the tissue-specific chromatin disruption; other parameters, such as changes in DNA topology and/or conformation, therefore might also be involved in nucleosome exclusion. Finally, since the S1-sensitive site in this region of the chicken β-globin gene is displaced in cellular chromatin relative to supercoiled DNA (Weintraub, 1983), the binding of other cellular proteins(s) must also contribute to the observed chromatin disruption (see Emerson and Felsenfeld, 1984, and below).

The latter point raises the more general question regarding the nucleosome organization of transcriptionally active versus inactive chromatin. On a purely mechanical level, it could be imagined that the persistence of the highly organized repeating structure of nucleosome associated DNA would be incompatible with the presence (and movement) of the transcriptional apparatus. It has been suggested that the typical nucleosome structure would impede the path of RNA polymerase; models have been proposed suggesting that the core histone octomer is permanently absent from transcriptionally active regions, slides, or disassociates from one strand, transiently disassociates from DNA, or "reorganizes" to allow polymerase to pass (Weintraub *et al.* 1976; Gould *et al.* 1980). In addition to the question regarding physical exclusion of nucleosomes by other proteins such as polymerase, a (perhaps) more important point in the context of regulation is whether the typical nucleosome structure is compatible with the extended more-open conformation associated with active chromatin, or, whether changes in nucleosome organization, spacing, or content might be required for, or associated with, transcriptional activation.

Attempts to determine the organization of histones within transcriptionally active regions have been made both by examining active chromatin segments microscopically—to determine if extended and/or polymerase-associated regions have the beaded appearance typical of nucleosomal DNA—and by determining if active gene sequences are included in subunit monomers generated by micrococcal nuclease analogous to those characterized for bulk DNA.

Microscopic techniques clearly distinguish active from inactive regions of chromatin based on the presence or absence of polymerase-associated ribonucleoprotein fibrils along segments of extended chromatin. Examination of several gene regions (both ribosomal and nonribosomal units) has provided a consensus view that highly active units are devoid of nucleosome-like structures (Mathis *et al.* 1980; and references therein); these units, in fact, are so tightly packed with polymerase molecules that there is not room for nucleosomes to form. However, at least a subset of the core histones appear to be present in such segments, as evidenced by the fact that some highly active gene regions efficiently bind antibodies directed against histones H2A and H2B (McKnight *et al.* 1977). These data indicate first that nucleosome components do not need to completely disassociate from highly active units; second, that the persistence of nucleosome structures is not required for high-level transcription. If, as discussed in previous sections, one role of chromatin assembly is to generate an organized—perhaps torsionally strained—molecule suitable as a transcription template, then these findings suggest that tightly packed polymersase molecules might effectively be able to replace nucleosomes in this function. Attempts have been made to investigate this possibility by examining several moderately or infrequently transcribed units to determine if nucleosome-like particles can be discerned along a transcription unit that also contains polymerase molecules. Of the nonribosomal DNA units that have been examined, all appear to contain nucleosome-like particles positioned between active polymerase molecules. The density of nucleosomes, however, is decreased relative to bulk chromatin (Foe *et al.* 1976; McKnight and Miller, 1976; Laird *et al.* 1976; Foe, 1977; McKnight *et al.* 1976, 1977; Lamb and Daneholt, 1979), perhaps to a degree inversely proportional to the number of active polymerase molecules within that unit (McKnight *et al.* 1977; Lamb and Daneholt, 1979). In addition, the nucleosomes that are present often appear to be irregularly spaced relative to each other. In contrast, nucleolar ribosomal DNA genes from several species have been seen as being nucleosome-free both in the absence of transcription and under moderate levels of transcription (Foe, 1977; Franke *et al.* 1977; Scheer, 1978). Therefore, although the data obtained with nonribosomal DNA units could be interpreted as an indication that the presence of nucleosomes is not only compatible with transcription, but may be required to provide structural integrity to polymerase-free segments of moderately transcribed genes, the data from ribosomal units is at variance with this conclusion. Several explanations have been offered to resolve this apparent discrepancy. For example, it has been suggested that the differences in nucleosome organization of ribosomal and nonribosomal units may reflect a true difference in their requirements for, or their response to, transcription. Alternatively, these differences also might reflect the fact that some nucleosomes are more labile than others and, therefore, are selectively lost during preparation of chromatin for spreading. These difficulties remain unresolved, and they have been discussed in detail (Mathis *et al.* 1980; Igo-Kemenes *et al.* 1982; see also Franke *et al.* 1977; Scheer *et al.* 1977; Labhart *et al.* 1982).

Since micrococcal nuclease preferentially cleaves chromatin in the internucleosomal linker DNA under moderate digestion conditions, another approach to probing the organization of chromatin at the structural level is to determine if active gene sequences are located in nucleosome-sized subunits, if these sequences are included within the nucleosome ladder typically generated from bulk chromatin, and if the content of active nucleosomes resembles that of nucleosomes generated from bulk chromatin.

Chromatin subunits analogous to nucleosome monomers have been shown to contain DNA sequences of genes transcribed in the cell types from which the chromatin was isolated (Lacy and Axel, 1975; Garel and Axel, 1976; Weisbrod *et al.* 1980; Weisbrod and Weintraub, 1981; Reeves, 1976; Kuo *et al.* 1976). In addition, such monomer-sized subunits generated from active regions appear to contain the four core histones (e.g. Levy-Wilson and Dixon, 1979; Weisbrod, 1980), although their stoichiometry reportedly can vary from that of bulk chromatin monomers (Weisbrod *et al.* 1980). The fact that sequences included in subunit monomers from two active genes were sensitive to digestion with DNAseI (Weisbrod *et al.* 1980; Senear and Palmiter, 1981) indicates that these monomers most likely originated from activated chromatin and not from an inactive fraction that might have been present in the population. However, since the active conformation defined by sensitivity to digestion with DNAseI is not always confined to units that are being actively expressed (see the previous section on "DNAseI"), these data must be interpreted in light of the distinction between chromatin organized in an active conformation and chromatin being actively transcribed. Given this qualification, these data are consistent with those obtained for nonribosomal units by microscopy in that apparently typical nucleosome-like particlies can be included within activated regions of chromatin.

One technique has been described that allows the isolation of a chromatin fraction enriched in actively transcribed sequences (Gottesfeld *et al.* 1974). This fractionation is dependent on the increased solubility at low Mg^{++} ion concentrations of chromatin associated with nascent RNA transcripts. Analyses of the protein composition of this transcriptionally active fraction have provided data consistent with that described above for activated fractions of chromatin (Gottesfeld *et al.* 1974; Gottesfeld and Butler, 1977; Gottesfeld and Partington, 1977). However, analysis of such active fractions has been limited to a small number of cases, and the generality of these results awaits further investigation (Mathis *et al.* 1980, for a complete discussion).

Micrococcal nuclease digestion of intact chromatin also has been used in attempts to probe the relative spacing and absolute positioning of nucleosomes in active regions. As described below, although several potentially interesting differences in the nucleosome organization of individual genes has been described, a generally applicable active nucleosome conformation has not been identified. For example, examination of the relative spacing of nucleosomes along the repeated α-histone genes of sea urchin (Spinelli *et al.* 1982) provided evidence that could support the microscopically defined organization of non-rDNA units cited above. In correspondance to the timing of expression of these genes, it was noted that a complete loss of an identifiable micrococcal nuclease-banding pattern was evident in 32–64 cell-stage embryos (which express these genes), while a distinct pattern that is typical of nucleosome-associated bulk chromatin was seen in nonexpressing sperm and mesenchyme blastulas. Therefore, in this case, there is evidence that the basic nucleosome organization of these genes is severely compromised during transcriptional activity. This disruption might reflect the decreased density and irregular spacing of nucleosome structures on active transcription units, as seen microscopically. In addition, the intermediate size of digestion products from active-stage embryos was not consistent with the histone DNA sequences being naked DNA; under the conditions employed in these experiments, protein-free DNA would have been completely degraded to acid-insoluble material. This conclusion is reminiscent of the findings of Foe (Foe *et al.* 1976), who noted the presence of extended smooth transcription units that also were not protein-free, but that apparently were complexed with basic proteins. Although this might reflect an association of histone proteins in an altered conformation, the identity of these proteins has not been established in either case.

A similar conclusion can be drawn from examination of chromatin containing the *Drosophila melanogaster* heat-shock protein locus 70 gene. In this case, nucleosomes appeared to be uniquely and regularly positioned along the gene; however, during heat shock, there was a dramatic blurring of the banding pattern (Wu *et al.* 1979b), suggesting that nucleosomes either may be displaced or may slide during periods of transcriptional activity.

Apparent disruption of the basic chromatin organization of active genes, however, is not universal. Examination of chromatin containing two active single-copy genes (the ovalbumin gene in oviduct (Bellard *et al.* 1982) and single-copy *Herpes* thymidine kinase gene transfectants (Camerini-Otero and Zasloff, 1980) revealed that the coding sequence regions were included in a 200 bp micrococcal nuclease-generated ladder indistinguishable from that identified for bulk chromatin. These results clearly are in contrast to those generally obtained by microscopy and, in some instances, by micrococcal nuclease cited above.

Together, these data suggest several possibilities. First, different genes may be organized in distinct nucleosome conformations, all of which are compatible with an active chromatin conformation and/or with transcription. Second, the distinctions noted above could reflect differences either in the transcription rate of genes being analyzed or in the fraction of active units in the popoulation. Finally, it is possible that the presence or absence of a regular 200bp repeat ladder is an inadequate parameter by which to evaluate nucleosome organization. For example, regularly spaced molecules of RNA polymerase (or other proteins associated with active regions) might generate a similar pattern. Alternatively, the 200 bp repeat organization of nucleosomes might be maintained in all of these units, but its detection might be obscured if other factors block cleavage of DNA by micrococcal nuclease.

A final point regarding nucleosome organization and its potential relationship to transcription involves the absolute placement of nucleosomes along a segment of DNA. Early investigations into nucleosomes formation in *in vitro* reconstitution experiments revealed that nucleosomes could assemble on some segments of DNA at unique positions or at a limited number of sites (Chao *et al.* 1979; Wasylyk *et al.* 1979). In addition, nucleosomes also can be nonrandomly located along relatively long segments of chromatin *in vivo* (Wittig and Wittig, 1979; Levy and Noll, 1980; Louis *et al.* 1980; Samal *et al.* 1981). This alignment of nucleosomes in a limited number of positions is referred to as phasing (reviewed in Kornberg, 1981; Zachau and Igo-Kemenes, 1981; Igo-Kemenes *et al.* 1982). The reports that nucleosomes may be nonrandomly positioned have led to speculation that nucleosome phasing might constitute a regulatory element in chromatin function (Wittig and Wittig, 1982; Gargiulo *et al.* 1982). For example, the positioning of nucleosomes could result in the exposure or sequestration of specific DNA sequences or conformations, thereby facilitating or hampering their interaction with regulatory molecules (but see Chao *et al.* 1980a,b). In addition, this type of organization could also bring DNA sequences (or conformations) into the close proximity required in concert to interact with regulatory molecules. The hypothesis that nucleosome phasing might be involved in regulation has received support by some investigators based on the findings that the absolute positioning of nucleosomes may shift on transcriptional activation (Bryan *et al.* 1981; Samal *et al.* 1981; see Gargiulo *et al.* 1982, for a discussion). Currently, however, the significance of phasing is questionable. The basis for this uncertainty is twofold. First, although phasing clearly is evident in some cases, several reports that describe phasing remain controversial (Fittler and Zachau, 1979; Singer, 1979; Musiich *et al.* 1982). The basic disagreement surrounds the fact that micrococcal nuclease, which has been used to map the location of nucleosomes along a segment of DNA, introduces site-specific cleavages in naked DNA—a phenomenon not always adequately taken into account when analyzing nucleosome placement (Keene and Elgin, 1981). In addition, the identification of specific DNA sequences and/or conformations that preferentially form or exclude nucleosomes suggests that phasing might be a passive consequence of the presence of such regions in chromatin. In this view, sites that have high or low affinities for nucleosome formation could constitute nucleating sites that could result in alignment of the nucleosomes surrounding it, since the relative spacing of nucleosomes can be relatively constant over discrete segments of chromatin. A nucleating site also could be formed by the binding of a protein to a specific DNA sequence and/or conformation, with concomitant exclusion of nucleosomes from this binding site (see Bloom and Carbon, 1982, for a potential example). Therefore, although nucleosome phasing has been proposed as a candidate mechanism to communicate long-range effects along the DNA fiber, it is possible that the important regulatory event resides at the nucleating site, not at the positions of nucleosomes surrounding the nucleating site.

Other Chromatin-associated Proteins

Extensive analyses of chromatin-associated proteins have led to the identification of several constituents that are reportedly generally enriched in transcriptionally active regions of chromatin or whose presence appears to be directly involved in the regulated expression of a particular gene. The current literature on this subject has been thoroughly discussed in several recent reviews (Mathis *et al.* 1980; Weisbrod, 1982; Igo-Kemenes *et al.* 1982; Cartwright *et al.* 1982). Therefore, our discussion will be limited to a small number of examples of chromatin-associated proteins that may contribute to transcriptional regulation via structure.

A particularly interesting variation in protein composition between some transcriptionally active and inactive chromatin fractions is the apparent depletion of histone H1 in active chromatin (Lau and Ruddon, 1977; see also Levy-Wilson and Dixon, 1979). Chromatin fractions that are depleted in histone H1 also are reportedly depleted in methylated cytosine residues (Ball *et al.* 1983)—another feature associated with transcriptionally active chromatin (see the following section). As described in the previous section on "Basic Chromatin Organization," H1 appears to play a major role *in vitro* in maintaining the integrity of chromatin and in effecting some degree of chromatin compaction. Therefore, although the effects of H1 depletion *in vivo* are not well understood (Weischet and VanHolde, 1980), it could be envisioned that removal of H1 is one component responsible for relaxing chromatin to facilitate transcription. However, H1 depletion also has been reported for chromatin containing transcriptionally inactive satellite DNA (Weber and Cole, 1982). In addition, recent data indicates that H1 is easily lost during chromatin isolation (Schlaeger, 1982), therefore suggesting that the stoichiometry of H1 in purified chromatin may not reflect its true representation *in vivo.* More controlled methods of chromatin isolation may help to resolve this confusion in the future.

Two of the chromatin-associated high-mobility group (HMG) proteins, HMG 14 and 17 (Goodwin *et al.* 1978), or their counterpart in trout (H6), are enriched in certain active chromatin fractions (Levy *et al.* 1977; Weisbrod and Weintraub, 1979, 1981; Weisbrod *et al.* 1980). What is particularly interesting about these proteins is the correlation between their presence in chromatin and the sensitivity of that chromatin to digestion by DNAseI. In particular, it has been demonstrated that after elution of HMGs from chromatin, active sequences lose their preferential sensitivity to digestion by DNAseI; readdition of the eluate results in restoration of sensitivity (Weisbrod *et al.* 19080; Gazit *et al.* 1980). In addition to their association with active regions *in vivo,* HMGs apparently bind preferentially to active chromatin *in vitro* (Sandeen *et al.* 1980; Weisbrod and Weintraub, 1981). This was demonstrated by showing that HMG proteins preferentially bound to (Sandeen *et al.* 1980), and reconstituted, the DNAseI-sensitive conformation to (Weisbrod and Weintraub, 1979) globin gene-containing micrococcal nuclease monomers generated from erythrocyte chromatin, even in the presence of a 20-fold excess of bulk monomers. The fact that HMG proteins can distinguish between active and inactive monomers supports the notion that active monomers are somehow distinct from bulk monomers (Albanese and Weintraub, 1980; see also Varshavsky *et al.* 1982; Swerdlow and Varshavsky, 1983).

Other chromatin-associated proteins have been identified that apparently are nonrandomly distributed among active and bulk chromatin: 1) the D1 protein of *Drosophila* is localized to heterochromatic regions, suggesting that it may serve to condense regions of chromatin (Rodriguez-Alfageme *et al.* 1981; Levinger and Varshavsky, 1982; reviewed in Brutlag, 1980); 2) proteins that apparently are localized to the puffs of *Drosophila* polytene chromosomes (Mayfield *et al.* 1978) provide a potential set of chromatin constituents that may relate to structure and/or activity; and 3) the existence of proteins that bind specifically to Z-DNA (Nordheim *et al.* 1982) suggests that maintenance of this DNA conformation may be important intracellularly (see also Hill and Stollar, 1983). One striking example of chromatin proteins that influence not only chromatin structure, but also gene expression, are the *SIR* gene products of yeast, which control the activity of mating-type loci. In this case, the gene products of *SIR* maintain an inactive chromatin conformation on the silent copy of the yeast-mating type cassette by binding at sites removed from those minimally required for expression (Klar *et al.* 1981; Nasmyth, 1982; see also the previous section on "Sequence Alterations and Rearrangements").

Recently, Emerson and Felsenfeld (1984) have identified a protein fraction which contains components that bind specifically to the 5′ flanking region of the chicken adult β-globin gene, and which induce a nuclease sensitive conformation to this region. The significance of this result is strongly suggested by the fact that the component responsible for inducing the alternate conformation is detected only in cells that normally express the chicken adult β-globin gene. Clearly, analyses of this type will be instrumental in the understanding of protein DNA interactions and their role in chromatin structure and gene activity.

As stated earlier, the examples provided here represent only a small number of the variations in chromatin composition that have been described. A direct role for most of these alternate or additional constituents remains uncertain; clearly, the focus of current and future investigations will be to identify proteins of regulatory importance and to elucidate not only their direct contribution to regulation, but also their relationship to the other parameters implicated in gene activation.

DNA Methylation

One of the most striking variations in eukaryotic chromatin is the distribution of 5-methylcytosine (5-meCyt) residues among different chromatin fractions.

Cytosine methylation, which occurs primarily at CpG dinucleotides (Doscocil and Sorm, 1962; Grioppo *et al.* 1968), is the only known DNA modification in higher eukaryotes; it is one parameter that may be involved in regulating gene expression (Razin and Riggs, 1980; Ehrlich and Wang, 1981; Felsenfeld and McGhee, 1982; Doerfler, 1983; and this volume). The suggestion that DNA methylation may have a regulatory function is based on several findings. These include the nonrandom distribution of 5-meCyt among active and inactive regions of chromatin, the effects that compounds interfering with DNA methylation have on gene expression, and the ability to alter expression of cloned DNA by introduction of methylated residues *in vitro*. In addition, it also has been found that cytosine methylation may influence DNA conformation and topology (Behe and Felsenfeld, 1981; Behe *et al.* 1981; and see the previous sections on "DNA-hypersensitive Sites" and "DNA Topology" for a discussion of the significance of DNA conformation and topology on gene expression). Although the correlation between DNA hypomethylation and transcription has been established for some time, a convincing cause-and-effect relationship has not been demonstrated. A brief review of the difficulties encountered in establishing such a regulatory role should illustrate the problems in determining causality.

On a global level, the degree of DNA methylation parallels the relative transcriptional potential of different chromatin compartments. For example, regions of constitutive heterochromatin selectively bind significantly greater amounts of anti-5-me-Cyt antibodies than do regions of euchromatin (Miller *et al.* 1974). Analyses of chromatin fractions (Razin and Cedar, 1977) and many individual genes also support the general correlation between transcriptional activity and hypomethylation (see reviews listed above and in this volume). Furthermore, using a novel procedure of "nick translating"-intact chromatin, it was demonstrated that the DNAseI-sensitive fraction of chromatin generally is depleted in 5-me-Cyt-relative to the resistant fraction (Naveh-Many and Cedar, 1981). Analogous results have been reported for the DNAseI-sensitive fraction of both chromatin-containing ribosomal DNA genes in mouse liver nuclei (Bird *et al.* 1981a) and the ovalbumin gene in oviduct (Kuo *et al.* 1979).

However, transcriptionally active DNA is not typically completely unmethylated. Instead, it contains specific sites that are unmethylated or hypomethylated (Van der Ploeg and Flavell, 1980; Bird and Taggart, 1980; Bird *et al.* 1981b; Jones *et al.* 1981; Grainger *et al.* 1983). This suggests that it may be the pattern of methylated (or unmethylated) residues, which is important in modulating the activity of associated genes. Unfortunately, descriptions of sites of methylation have been largely limited to those that lie within recognition sequences of methyl-sensitive restriction endonucleases; techniques to monitor the methylation status of each cytosine within a defined region of DNA as it exists *in vivo* are not practically useful in most cases. This limitation, together with the nonuniform level of methylation at many sites that are detectable, has made it difficult to unequivocally assign a regulatory role to DNA methylation in transcription. An additional difficulty has been introduced by inadequate determinations of transcriptional activity of the gene of interest (see the previous section on "*In Vivo* Correlations" for a discussion). Therefore, although there are reports of active genes that contain 5-meCyt in coding (Jones *et al.* 1981) or potential control (Burch and Weintraub, 1983; Grainger *et al.* 1983) regions—as well as "inactive" genes that are hypomethylated (Kuo *et al.* 1979; McKeon *et al.* 1982)—neither serve to invalidate the correlation between activity and hypomethylation nor to support a regulatory role for methylation.

Another approach that has been taken to ascribe functional significance to methylation has been to metabolically alter total levels of 5-methyldeoxycytosine in cellular DNA, and then to determine any subsequent changes in gene activity. Two reagents that lead to an overall reduction in levels of DNA methylation are 5-azacytidine (5-aza-C) (or the more potent doexy analog; Jones and Taylor, 1980) and L-ethionine (Christman *et al.* 1977; Boehm and Drahovsky, 1981). These reagents do lead to changes in gene activity, as documented in a variety of cell types; growth in 5-aza-C results in induction of differentiated cell types from cultured mouse 10T1/2 cells (Jones and Taylor, 1980), in activation of several loci on the inactive human X chromosome (Mohandas *et al.* 1981), and in induction of silent proviruses in both mouse (Niwa and Sugahara, 1981) and chicken (Groudine *et al.* 1981a) cells. Although these results are very striking, the phenomenon has been difficult to interpret due to the pleiotropic effects engendered by introducing such metabolites into critical cellular pathways (Cihak, 1974).

One potential approach to determining the contribution of any nonspecific effects that these reagents may have on gene expression would be to test their effect on expression in *Drosophila.* Flies do not contain detectable levels of 5-meCyt in their DNA (Urieli-Shoval *et al.* 1982); therefore, if such agents are acting through changes in DNA modification, then gene expression in *Drosophila* should not be affected by these reagents. If these reagents do lead to changes in gene expression in *Drosophila,* this would indicate that the phenomenon may not relate to DNA methylation but that it is therefore more complex than originally suggested.

A somewhat more controlled approach to determining the possible effects DNA methylation might have on transcription regulation is provided by the introduction of methylated cytosines into purified DNA molecules, followed by determination of the transcriptional potential of such molecules after their introduction into various cells. Studies of this type have indicated that methylation of discrete segments of DNA (often in promoter regions) can severely impair the capacity of the molecule to function as an efficient substrate for transcription. This type of analysis has been successfully carried out on segments of the SV40 geneome (Fradin *et al.* 1982), murine retrovirus proviral DNA (McGeady *et al.* 1983), and the human γ-globin gene (Busslinger *et al.* 1983). These findings directly demonstrate that DNA methylation can affect the transcriptional competence of a gene. However, they do not reveal whether the cell normally uses DNA methylation to control gene activity.

The difficulty in interpreting these results is that methylation of cytosine residues can be viewed in the context of mutation. In this viewpoint, methylation of certain cytosine residues could serve to identify bases that are crucial for expression, rather than to reveal physiologically significant effects of methylation. It could be envisioned that cytosine residues that lie in crucial regulatory sites are never methylated *in situ* (see the following section), but are protected in some way. Therefore, while methylation of such sites (either *in vivo* or *in vitro*) will destroy promoter function, this may not reflect a mechanism used by the cell to modulate the activity of these sequences.

The notion that cytosine methylation could be recognized as a mutation is exemplified by base substitution experiments in the *Escherichia coli lac* gene operator DNA (Fisher and Caruthers, 1979). In these analyses, it was reported that substitution of a specific thymine residue with a cytosine or uracil residue in the *E. coil lac* operator gene effectively abolished binding of repressor. However, introduction of a methylated cytosine residue restored binding function. In this case, therefore, it is the methyl group that is critical to competent binding; this indicates that the identity of the base itself is, in some cases, less important than the presence or absence of a methyl group. Given such a specificity, *in vitro* introduction of methylated cytosine residues—either of selected CpGs (with HpaII methylase) or of all cytosine residues—may be of no greater physiologic significance than substitution of cytosines with thymine.

The *lac* repressor-binding experiments also have been cited as evidence that methylated residues can affect protein-DNA interactions. An effect of methylation on protein recognition also is clearly evident in the substrate specificity of methyl-sensitive restriction enzymes. Similarly, introduction of methyl groups into eukaryotic cellular DNA *in vivo* undoubtedly could alter the character of a DNA recognition site; therefore, it could be envisioned as a potential DNA-dependent regulatory element. However, since virtually no gene-specific regulatory proteins have been identified in eukaryotes, nothing is known about the effects that eukaryotic DNA methylation might have on most potentially significant protein-DNA interactions. Limited investigations on histone-DNA interactions suggest that DNA methylation per se does not grossly affect nucleosome formation (Felsenfeld *et al.* 1982). However, the finding that DNA methylation can enhance the ability of some DNAs to form Z-DNA, which apparently is not an efficient substrate for nucleosome assembly (see the previous section on "Core Histones and Nucleosomes"), suggests at least one mechanism by which DNA methylation might alter basic protein-DNA interactions in eukaryotes.

Therefore, although it has been clearly established that DNA methylation can affect the structure of some DNA molecules, the specificity of certain protein-DNA interactions, and the transcriptional competence of several segments of DNA, what remains to be elucidated is how and if the cell may use these effects. In this context, it will be important to extend our limited knowledge on the interrelationships among DNA methylation, DNA topology, and DNA conformation; and, also on how these parameters might combine to influence chromatin structure, composition, and function.

Speculations about Chromatin Structure and Development

In the preceding sections, we have attempted to present a critical review of past and fairly recent experimental observations that have led to models regarding the relationships among DNA conformation and modification, chromatin composition and structure, and gene expression. In this section, we will discuss, in somewhat greater detail, several observations from our own laboratory that may provide a basis for contemplating the mechanisms by which chromatin structure could serve as a template for the transmission of information regarding gene activity from cell generation to cell generation, as well as from germ line to fertilized egg.

Propagation of Chromatin Structure and Gene Expression

In attempting to understand the developmental significance of the appearance of changes in chromatin structure during embryogenesis, we (Groudine and Weintraub, 1982) attempted to investigate a specific feature of the differentiation process common to many multicellular organisms. Certain determinative events appear to be induced in precursor cells at one time in development and, independent of the concurrent action of the original inducing influence, the effects of these events are expressed in progeny cells that begin overt differentiation some time later (Alberts *et al.* 1983). Thus, it would seem that a specific event occurs at one time in development, and, in some way, the consequences of that event can be propagated to progeny cells in the absence of the original stimulus. One possible mechanism for the transmission of early developmental signals to progeny cells could be the setting up of certain structural features of chromatin that have the ability to self-propagate. In this viewpoint, these structural features, once established in progenitor cells, might then provide a substrate for the products of genes expressed later in a specific lineage; or, they might respond to other embryonic cues present at a later developmental time. As discussed previously, the formation of local perturbations in DNA structure assayed as nuclease hypersensitive sites has been reported to be influenced by salt, pH, and other ionic variables (see the previous section on "DNA-hypersensitive Sites"). In addition, it has been reported that the induction of specific puffs in polytene chromosomes isolated from unfixed salivary glands of *Chironomous* is ion- and pH-dependent (Lezzi and Robert, 1976). Since gradients and/or transient temporal changes in such variables have been described in many developmental systems (Alberts *et al.* 1983), we wondered if hypersensitive sites might have the property of self-propagation, and hence, a property that could serve as a structural analog for the transmission of early developmental signals to progeny cells. In an attempt to test this hypothesis, we investigated the possibility that nuclease hypersensitive sites 5′ to chicken globin genes might have the ability to form in cells that are removed from the stimuli that led to the formation of such hypersensitive sites in progenitor cells.

Since the transitions from progenitor stem cell to precursor red blood cell to erythroblast occur so rapidly and asynchronously in the developing chicken embryo, we were unable to address this question during the course of hematopoesis in normal development. Instead, we used a cell culture system that we felt might harbor the potential to address this question. It was reported by Garry *et al.* (1981) that growth of normal chick embryo fibroblasts (CEF) in medium containing high NaCl (0.2–0.25 mol) resulted in the acquisition by these cells of many of the phenotypic properties of CEF that are transformed by various oncogenic retroviruses, including Rous sarcoma virus (RSV). It was also shown that the actual level of NaCl within these cells increased commensurately as a result of this treatment, mimicking the increase in internal NaCl observed after oncogenic transformation. On returning these cells to normal medium, the intracellular NaCl concentration reverted to the levels present before exposing the cells to NaCl shock, and these cells no longer displayed any of the phenotypic characteristics of transformed CEF. Thus, given the possible influence of ionic conditions on hypersensitive site formation, the acquisition of a transformed phenotype by CEF exposed to high NaCl, and the reported activation of globin genes by CEF transformed by RSV (Groudine and Weintraub, 1975; 1980), we wondered if high salt could induce DNAseI hypersensitive sites within the globin containing chromatin in CEF.

These experiments revealed that after treatment of CEF with high NaCl for 12 hours, the normal DNAseI hypersensitive sites usually observed in red blood cells appeared at the 5′ and 3′ side of the β^A-globin gene. In addition these sites persisted after CEF were treated with high concentrations of NaCl for 12 hours, were returned to medium containing physiologic concentrations of NaCl (0.15 mol), and then were allowed to grow logarithmically for about 20 generations in the normal medium. Thus, hypersensitive sites in globin chromatin induced by NaCl were propagated to progeny cells over 20 generations in the absence of continued exposure to high NaCl; that is, in the absence of the activity that was initially responsible for generating the altered chromosomal structure assayed by DNAse hypersensitivity.

Analysis of run-off transcription products and of steady-state RNA from these salt-shocked CEF (either during incubation in high NaCl or after return to normal medium) did not reveal the presence of globin transcripts. Thus, while changes in the ionic environment may be sufficient for the initiation (and stabilization) of a hypersensitive site, this inheritable change in chromatin structure does not seem to be sufficient for the initiation of RNA synthesis. This uncoupling between the hypersensitive structure and transcription is a characteristic that would be compatible with a molecular analog of a determinative event in development.

Clearly, not all hypersensitive sites are stably propagated from cell generation to cell generation (e.g., some of the vitellogenin and metallothionein sites described in the section on "Chromatin Structure and Gene Expression"). As already discussed, one possibility is that since these unstable sites are associated with inducible genes, these sites may represent footprints of the receptor complexed to DNA. In this viewpoint, these sites would be dependent on the continued presence or absence of the inductive event (e.g., hormone introduction or withdrawal).

A Simple Mechanism for Propagation

A reasonable mechanism for the apparent ability of hypersensitive sites to propagate stems directly from one specific aspect of their structure. As discussed in previous sections, some chromatin-hypersensitive sites are sensitive to S1 nuclease, which suggests that they contain regions on non-B-form DNA. Similarly, the globin-hypersensitive sites in NaCl-CEF also are detected by S1 nuclease (Groudine and Weintraub, 1982). A simple mechanism for propagation based on this type of structure is depicted in Figure 15.1. For ease of presentation of the model, we have depicted this structure as a hairpin, although any one of a number of non-B-DNA conformations could be substituted. In this figure, we presume that once induced, the structure is stabilized by the binding of a protein (triangle) that specifically binds to this altered conformation. The fact that these conformations can be induced in fibroblasts by NaCl suggests that the stabilizing factors are, in fact, very general and probably are not tissue-specific or DNA sequence-specific. The factor also may be in great excess, since induction of hypersensitive sites occurs so rapidly (within 30 minutes) after exposure to high NaCl and is independent of RNA or protein synthesis (Groudine and Weintraub, 1982). If this protein is analogous to the T4 gene 32 protein (Alberts and Frey, 1976), then it is reasonable that as the replication fork passes through the structure (Figure 15.1), the protein remains bound to the parental DNA strand; presumably to the sugar-phosphate backbone. This is an absolute requirement of the model. After the fork passes through this region, the binding of the protein would stabilize the original structure and promote hairpin formation (Figure 15.1) in the daughter duplex. As a result, the newly made strand would become non-B form in this region, another protein that is specific for this altered conformation would bind to the new strand (Figure 15.1 d and e), and the basic elements of the hypersensitive structure would be restored. The structure would be propagated to both daugh-

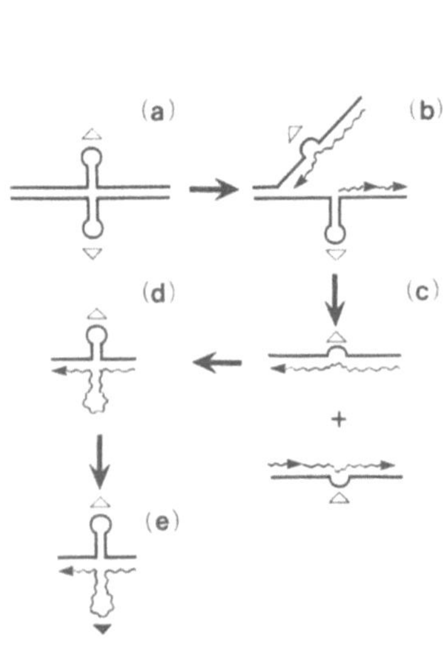

Figure 15.1. A model for propagation of hypersensitive sites based on both the single-stranded character of these sites and a postulated class of single-strand-specific DNA-binding proteins. Details of the model are discussed in the text. Triangles: single-strand-specific DNA-binding proteins with the capacity to remain bound to the DNA as the replication fork copies the bases. Wavy lines: newly made DNA strands.

ter chromosomes in a way that would depend only on a class of very general non-B-form DNA-binding proteins. The mechanism also would be independent of the original factors needed to induce it. This type of mechanism would seem to have many of the properties expected for a molecular analog of a determination event in development.

Propagation of Structural Information Through the Germ Line

In thinking about the relationships among gene expression, DNA conformation in chromatin, and DNA methylation, it is interesting to note that significant changes in all of these parameters occur during vertebrate spermatogenesis. For example, before the condensation of the nucleoprotein complex in mature sperm of various species, protamines replace some classes of histones and transcriptional activity ceases (Utakoji, 1966; Candido and Dixon, 1972; Platz *et al.* 1975; Kierszenbaum and Tres, 1975; Grimes *et al.* 1975; Mezquita and Teng, 1977). In addition, an increased level of methylation of sperm DNA, as compared to somatic tissue DNA, has been found in most vertebrates examined thus far (reviewed in Razin and Riggs, 1980). Thus, an interesting problem is presented by these phenomena. If structural signals, such as undermethylation and hypersensitive sites, are important in propagating information for gene expression, how is this information maintained in the hypermethylated, hypercondensed sperm chromatin in a form that might be important in permitting expression of specific genes on the paternal chromosomes during early development? To approach this question, we reasoned that if hypersensitive sites were maintained and propagated in a stable fashion by a tightly bound protein (see the model presented above), then such a protein might sterically hinder the procession of a methylase in this region. Alternatively, if most hypersensitive sites are conformationally distinct frm bulk DNA, the *de novo* methylase that operates during spermatogenesis might lack specificity for such conformations and might leave these specific regions of active or potentially active genes unmethylated. In this view, the methylation of DNA during spermatogenesis could play a role in protein-DNA (or protamine) interactions that are important in condensation of the sperm nucleus; the resultant hypomethylation of specific regions of DNA would be a passive consequence of this process. This type of model would then predict that by virtue of the maintenance of these hypersensitive sites in an unmethylated state, this could "mark" such sites as signals for gene activation early in development.

Our first approach in testing this hypothesis was to analyze the methylated state in sperm DNA of a gene expressed in all tissues in the chicken, including testes. This specific gene, *ev-3,* is an active endogenous virus whose chromatin structure and methylated state in somatic tissue have been reported previously (Groudine *et al.* 1981a). Essentially, the gene contains five hypersensitive sites; one in each of the long-terminal repeats (*LTRs*) at it's 5′ and 3′ ends, two in the *gag* gene, and one in the *env* gene. In addition, the provirus contains over

15 HpaII sites, all of which are unmethylated in somatic tissue. While no CpGs are present in the LTRs in restriction enzyme recognition sequences—thereby not permitting the analysis of the relationship of methylation and hypersensitive sites in these elements—each of the sites in *gag* and *env* are located in regions containing HpaII sites. By comparing the locations of sub-bands generated by various restriction enzyme digests of S1- or DNAseI-treated DNA from nuclei containing *ev-3* with the unmethylated sites in this gene in sperm, we (M. Groudine and R. Eisenman, unpublished observations) have found that only those HpaII sites that map to the same locations as hypersensitive sites in somatic tissue remain unmethylated in sperm DNA. This type of correlation between the location of hypersensitive sites in somatic cell chromatin and unmethylated point sites in sperm DNA has been found for all ubiquitously expressed genes that we have examined thus far (M. Groudine, unpublished observations).

Reports from other laboratories investigating the methylated state of specific genes in several species (e.g. Stein *et al.* 1983) also suggest that 5′ regions of genes that would be expected to be expressed in most cells also are undermethylated in sperm DNA. While the relationships of these other unmethylated sites in sperm DNA and hypersensitive sites in somatic tissues have not been addressed directly, the location of these unmethylated sites in sperm is consistent with the region often associated with nuclease hypersensitivity.

Our finding, that regions of DNA corresponding to hypersensitive sites of ubiquitously expressed genes remain unmethylated in sperm DNA, also raises the question of the possible temporal relationship between undermethylation and the generation of hypersensitive sites in development. One possibility is that on fertilization and decondensation of paternal DNA, the unmethylated regions assume a different conformation than bulk DNA, are recognized by a specific class of proteins, and thereby are marked as different—as well as being propagated during division, as postulated in the preceding section. Clearly, however, such hypersensitive site-stabilizing proteins could remain bound to the parental DNA on fertilization and, throughout development, could remain a steric hinderance to methylase activity. In either case, these specific regions of DNA would be maintained in an undermethylated state by the continued presence of such proteins and/or altered DNA conformations; also, other regions of a transcription unit could become progressively unmethylated coincident with overt transcription of the gene.

Concluding Remarks

A basic concept that has emerged from investigations on the requirements for expression of eukaryotic genes is that the promoter, as defined in prokaryotic systems, is inadequate to account for the differential expression of most eukaryotic genes during normal development. In particular, while control elements for prokaryotic genes generally are localized to relatively small regions surrounding the transcription initiation start site, the controlled expression of eukaryotic genes appears dependent on positive and/or negative regulatory sequences at distant sites along the DNA fiber. In addition, the observations that some such sequences apparently can alter the expression of different genes suggests that they lead to the induction and transmission of regulatory signals that can be recognized by several distinct promoter elememts. In this discussion, we have attempted to review the evidence that the conformation of chromatin is important in influencing its substrate suitability for transcription, and also that modulation of chromatin structure is used as a mechanism to regulate gene expression in eukaryotes. As described throughout this chapter, recent observations suggest that this type of regulation must certainly involve the conformation and topology of DNA within chromatin; hopefully, ongoing investigations will clarify not only the contribution of DNA to this structural-based mechanism of regulation, but also will provide insight into the role of specific proteins in effecting changes in chromatin structure and gene expression.

Acknowledgments

We would like to thank our colleagues, particularly John Burch, Steve Pruitt, Ron Reeder, Ginger Zakian, and Hal Weintraub, for their criticisms, comments, and helpful discussions. A special thanks to Helen Devitt for invaluable assistance in preparation of this manuscript. Work in the authors' laboratory has been supported by grants from the National Science Foundation (PCM 82-04696), National Cancer Institute (RO1 CA 28151-04) and National Institute of Health (1 PO1 AM 31232-02 AMSC) to Mark Groudine and National Institute of Health (1-T32-CA 09229) to Kathleen F. Conklin. Mark Groudine is a Scholar of the Leukemia Society of America.

References

Albanese I, Weintraub H: Electrophoretic separation of a class of nucleosomes enriched in HMG 14 and 17 and actively transcribed globin genes. *Nucl Acids Res* 1980;8:2787–2805.

Alberts BM, Frey L: T4 bacteriophage gene 32: A structural protein in the replication and recombination of DNA. *Nature* 1976;227:1313–1318.

Alberts BM, Bray D, Lewis J, Raff M, Roberts K, Watson J: *Molecular Biology of the Cell*. New York, London, Garland Publishing, Inc, 1983, pp 689–690, 834–890.

Ball DJ, Gross DS, Garrard WT: 5′-methylcytosine is localized in nucleosomes that contain histone H1. *Proc Natl Acad Sci USA* 1983;80:5490–5494.

Banerji J, Olson L, Schaffner W: A lymphocyte-specific cellular enhancer is located downstream of the joining region in immunoglobulin heavy chain genes. *Cell* 1983;33:729–740.

Beard P, Morrow JF, Berg P: Cleavage of circular superhelical simian virus 40 DNA to a linear duplex by S1 nuclease. *J Virol* 1973;12:1303–1313.

Behe M, Felsenfeld G: Effects of methylation on a synthetic polynucleotide: The B-Z transition in poly(dG m^{5}dC)-poly(dG m^{5}dC). *Proc Natl Acad Sci* 1981;78:1619–1623.

Behe M, Zimmerman S, Felsenfeld G: Changes in the helical repeat of poly(dG m^{5}dC)-poly(dG m^{5}dC) and poly(dGdC)-polydGdc associated with the B-Z transition. *Nature* 1981;293:233–235.

Bellard M, Kuo MT, Dretzen G, Chambon P: Differential nuclease sensitivity of the ovalbumin and β-globin chromatin regions in erythrocytes and oviduct cells of laying hen. *Nucl Acids Res* 1980;8:2737–2750.

Bellard M, Dretzen G, Bellard F, Oudet P, Chambon P: Disruption of the typical chromatin structure in a 2500 base pair region at the 5′ end of the actively transcribed ovalbumin gene. *EMBO J* 1982;1:223–230.

Benyajati C, Worcel A: Isolation, characterization, and structure of the folded interphase genome of *Drosophila melanogaster*. *Cell* 1976;9:393–407.

Bernard O, Hozumi N, Tonegawa S: Sequences of mouse immunoglobulin light chain genes before and after somatic changes. *Cell* 1978;15:1133–1144.

Bernardi A, Gaillard C, Bernardi G: The specificity of five DNAases as studied by the analysis of 5′-terminal doublets. *Eur J Biochem* 1975;52:451–457.

Bird AP, Taggart MH: Variable patterns of total DNA and rDNA methylation in animals. *Nucl Acids Res* 1980;8:1485–1497.

Bird AP, Taggart MH, Gehring CA: Methylated and unmethylated ribosomal RNA genes in the mouse. *J Mol Biol* 1981a;152:1–17.

Bird AP, Taggart MH, Macleod D: Loss of rDNA methylation accompanies the onset of ribosomal gene activity in early development of *X. laevis*. *Cell* 1981b;26:381–390.

Bloom KS, Anderson JN: Fractionation of hen oviduct chromatin into transcriptionally active and inactive regions after selective micrococcal nuclease digestion. *Cell* 1978;15:141–150.

Bloom KS, Anderson JN: Conformation of ovalbumin and globin genes in chromatin during differential gene expression. *J Biol Chem* 1979;254:10532–10539.

Bloom KS, Carbon J: Yeast centromere DNA is in a unique and highly ordered structure in chromosomes and small circular minichromosomes. *Cell* 1982;29:305–317.

Boehm TL, Drahovsky D: Elevated transcriptional complexity and decrease in enzymatic DNA-methylation in cells treated with L-ethionine. *Cancer Res* 1981;41:4101–4106.

Bonner WM, Stedman JD: Histone H1 is proximal to histone 2A and to A_{24}. *Proc Natl Acad Sci USA* 1979;76:2190–2194.

Brock ML, Shapiro DJ: Estrogen stabilizes vitellogenin mRNA against cytoplasmic degradation. *Cell* 1983;34:207–214.

Brutlag DL: Molecular arrangement and evolution of heterochromatic DNA. *Ann Rev Genet* 1980;14:121–144.

Bryan PN, Hofstetter H, Birnstiel ML: Nucleosome arrangement on tRNA genes of *Xenopus laevis*. *Cell* 1981;27:459–466.

Burch JBE, Weintraub H: Temporal order of chromatin structural changes associated with activation of the major chicken vitellogenin gene. *Cell* 1983;33:65–76.

Busslinger M, Hurst J, Flavell RA: DNA methylation and the regulation of globin gene expression. *Cell* 1983;34:197–206.

Butler MJ, Davies KE, Walker IO: The structure of nucleolar chromatin in *physarum polycephalum*. *Nucl Acids Res* 1978;5:667–678.

Camerini-Otero RD, Zasloff MA: Nucleosomal packaging of the thymidine kinase gene of herpes simplex virus transferred into mouse cells: An actively expressed single copy gene. *Proc Natl Acad Sci USA* 1980;77:5079–5083.

Candido EPM, Dixon GH: Trout testis cells 3. Acetylation of histones in different cell types from developing trout testis. *J Biol Chem* 1972;247:5506–5510.

Cantor C: DNA choreography. *Cell* 1981;25:293–295.

Carnevali F, Caserta M, DiMauro E: Topological modifications and template activation are induced in chimaeric plasmids by inserted sequences. *J Mol Biol* 1983;165:59–77.

Cartwright IL, Abmayr SM, Fleischmann G, Lowenhaupt K, Elgin SCR, Keene MA, Howard GC: Chromatin structure and gene activity: The role of non-histone chromosomal proteins. *CRC Crit Rev Biochem* 1982;13:1–86.

Chan HW, Wells RD: Structural uniqueness of lactose operator. *Nature* 1974;252:205–209.

Chao MV, Gralla JD, Martinson HG: DNA sequence directs placement of histone cores on restriction fragments during nucleosome formation. *Biochemistry* 1979;18:1068–1074.

Chao MV, Gralla JD, Martinson HG: *Lac* operator nucleosomes. I. Repressor binds specifically to operator within the nucleosome core. *Biochemistry* 1980a;19:3254–3260.

Chao MV, Martinson HG, Gralla JD: *Lac* operator nucleosomes. II. *Lac* nucleosomes can change conformation to strengthen binding. *Biochemistry* 1980b;19:3260–3269.

Chiswell DJ, Gillespie DA, Wyke JA: The changes in proviral chromatin that accompany morphological variation in avian sarcoma virus-infected rat cells. *Nucl Acids Res* 1982;10:3967–3980.

Christman J, Price P, Pedriman L, Acs G: Correlation between hypomethylation of DNA and expression of globin genes in Friend erythroleukemia cells. *Eur J Biochem* 1977;81:53–61.

Cihak A: Biological effects of 5-azacytidine in eukaryotes. *Oncology* 1974;30:405–422.

Corces V, Pellicer A, Axel R, Meselson M: Integration, transcription, and control of a *drosophila* heat shock gene in mouse cells. *Proc Nat Acad Sci USA* 1981;78:7038–7042.

Costantini F, Lacey E: Introduction of a rabbit β-globin gene into the mouse germ line. *Nature* 1981;294:92–94.

Courey A, Wang JC: Cruciform formation in a negatively supercoiled DNA may be kinetically forbidden under physiological conditions. *Cell* 1983;33:817–829.

Cozzarelli NR: DNA gyrase and the supercoiling of DNA. *Science* 1980;207:953–960.

Cozzarelli NR: DNA topoisomerases. *Cell* 1980b;22:327–328.

Della-Favera R, Bregni M, Erikson J, Patternson D, Gallo RC, Croce CM: Human c-*myc onc* gene is located on the regions of chromosome eight that is translocated in Burkitt lymphoma cells. *Proc Natl Acad Sci USA* 1982;80:519–523.

deVilliers J, Schaffner W: A small segment of polyoma virus DNA enhances expression of a cloned β-globin gene over a distance of 1400 base pairs. *Nucl Acids Res* 1981;9:6251–6264.

deVilliers J, Olson L, Tyndall C, Schaffner W: Transcriptional "enhancers" from SV40 and polyoma virus show cell type preference. *Nucl Acids Res* 1982;10:7965–7976.

Dingwall C, Lomonossoff GP, Laskey RA: High sequence specificity of micrococcal nuclease. *Nucl Acids Res* 1981;9:2659–2673.

Dodgson JB, Wells RD: Action of single-strand specific nucleases on model DNA heteroduplex of defined size and sequence. *Biochemistry* 1977;16:2374–2379.

Doerfler W: DNA methylation and gene activity. *Ann Rev Biochem* 1983;52:93–124.

Doscocil J, Sorm F: Distribution of 5-methylcytosine in pyrimidine sequences of deoxyribonucleic acids. *Biochem Biophys Acta* 1962;55:953–959.

Early P, Huang H, Davis M, Calame K, Hood L: An immunoglobulin heavy chain variable region gene is generated from three segments of DNA: VH, D and JH. *Cell* 1980;19:981–999.

Ehrlich M, Wang RY: 5-methylcytosine in eukaryotic DNA. *Science* 1981;212:1350–1357.

Eicher EM: X-autosome translocations in the mouse: Total inactivation versus partial inactivation of the X-chromosome. *Adv Genet* 1970;15:175–259.

Elgin SCR: DNase I-hypersensitive sites of chromatin. *Cell* 1981;27:413–415.

Elgin SCR: Chromatin structure, DNA structure. *Nature* 1982;300:402–403.

Emerson BM, Felsenfeld G: Specific factor conferring nuclease hypersensitivity at the 5′ end of the chicken adult β-globin gene. *Proc Natl Acad Sci USA* 1984;81:95–99.

Feinstein SC, Ross SR, Yamamoto KR: Chromosomal position effects determine transcriptional potential of integrated mammary tumor virus DNA. *J Mol Biol* 1982;156:549–565.

Felsenfeld G: Chromatin. *Nature* 1978;271:115–122.

Felsenfeld G, McGhee J: Methylation and gene control. *Nature* 1982;296:602–603.

Felsenfeld G, Nickol J, Behe M, McGhee J, Jackson D: Methylation and chromatin structure. *Cold Spring Harbor Symp Quant Biol* 1982;47:577–584.

Finch JT, Klug A: Solenoidal model for superstructure in chromatin. *Proc Natl Acad Sci USA* 1976;73:1897–1901.

Finch JT, Lutter LC, Rhodes D, Braun RS, Rushton B, Levitt M, Klug A: Structure of nucleosome core particles of chromatin. *Nature* 1977;269:29–36.

Fisher EF, Caruthers MH: Studies on gene control regions. XII: The functional significance of a *lac* operator constitutive mutation. *Nucl Acids Res* 1979;7:401–416.

Fittler F, Zachau G: Subunit structure of α-satellite DNA containing chromatin from African green monkey cells. *Nucl Acids Res* 1979;7:1–13.

Foe VE, Wilkinson LE, Laird CD: Comparative organization of active transcription units in *oncopeltus fasciatus*. *Cell* 1976;9:131–146.

Foe VE: Modulation of ribosomal RNA synthesis in *oncopeltus fasciatus:* An electron microscopic study of the relationship between changes in chromatin structure and transcriptional activity. *Cold Spring Harbor Symp Quant Biol* 1977;42:723–740.

Fradin A, Manley JL, Prives CL: Methylation of simian virus 40 Hpa II site affects late, but not early, viral gene expression. *Proc Natl Acad Sci USA* 1982;79:5142–5146.

Franke WW, Scheer U, Trendelenburg M, Zentgraf H, Spring H: Morphology of transcriptionally active chromatin. *Cold Spring Harbor Symp Quant Biol* 1977;42:755–772.

Garel A, Axel R: Selective digestion of transcriptionally active ovalbumin genes from oviduct nuclei. *Proc Natl Acad Sci USA* 1976;73:3966–3970.

Garel A, Zolan M, Axel R: Genes transcribed at diverse rates have a similar conformation in chromatin. *Proc Natl Acad Sci USA* 1977;74:4867–4871.

Gargiulo G, Wassermann W, Worcel A: Properties of the chromatin assembled on DNA injected into *Xenopus* oocytes and eggs. *Cold Spring Harbor Quant Biol* 1982;47:549–556.

Garry R, Moyer M, Bishop J, Moyer R, Waite M: Transformation parameters induced in chick cells by incubation in media of altered NaCl concentration. *Virology* 1981;111:427–439.

Gazit B, Panet A, Cedar H: Reconstitution of a deoxyribonuclease I-sensitive structure on active genes. *Proc Natl Acad Sci USA* 1980;77:1787–1790.

Gellert M: DNA topoisomerases. *Ann Rev Biochem* 1981;50:879–910.

Germond JE, Vogt VM, Hirt B: Characterization of the single-strand-specific nuclease S1 activity on double-stranded supercoiled polyoma. *Eur J Biochem* 1974;43:591–600.

Germond JE, Hirt B, Oudet P, Gross Bellard M, Chambon P: Folding of the DNA double helix in chromatin-like structures from simian virus 40. *Proc Natl Acad Sci USA* 1975;72:1843–1847.

Gillies SD, Morrison SL, Odi VT, Tonegawa S: A tissue specific transcription enhancer element is located in the major intron of a rearranged immunoglobin heavy chain gene. *Cell* 1983;33:717–728.

Goldberg DA, Posakony JW, Maniatis T: Correct developmental expression of a cloned alcohol dehydrogenase gene transduced into the drosophila germ line *Cell* 1983;34:59–73.

Gonda TJ, Sheiness DK, Bishop JM: Transcripts from the cellular homologs of retroviral oncogenes: distribution among chicken tissues. *Mol Cell Biol* 1982;2:617–624.

Goodwin GH, Walker JM, Johns EW: in Busch H (ed): *The Cell Nucleus*. New York, Academic Press, 1976, vol 6, pp 181–219.

Gottesfeld JM, Garrard WT, Bagi G, Wilson RF, Bonner J: Partial purification of the template-active fraction of chromatin: a preliminary report. *Proc Natl Acad Sci USA* 1974;71:2193–2197.

Gottesfeld JM, Butler PJG: Structure of transcriptionally-active chromatin subunits. *Nucl Acids Res* 1977;4:3155–3173.

Gottesfeld JM, Partington GA: Distribution of messenger RNA coding sequences in fractionated chromatin. *Cell* 1977;12:953–962.

Gould HJ, Cowling GJ, Harborne NR, Allan J: An examination of models for chromatin transcription. *Nucl Acids Res* 1980;8:5255–5266.

Grainger RM, Hazard-Leonards RM, Samaha F, Hougan LM, Lesk MR, Thomsen

GH: Is hypomethylation linked to activation of δ-crystallin genes during lens development? *Nature* 1983;306:88–91.

Green MH, Brooks TL: Isolation of two forms of SV40 nucleoprotein containing RNA polymerase from infected monkey cells. *Virology* 1976;72:110–120.

Gregory SP, Butterworth PHW: A comparison of the promoter strengths of two eukaryotic genes *in vitro* reveals a region of DNA that can influence the rate of transcription in cis over long distances. *Nucl Acids Res* 1983;11:5317–5326.

Grimes SR, Chae C-B, Irvin JL: Acetylation of histones of rat testis. *Arch Biochem Biophys* 1975;168:425–435.

Grippo P, Iaccarino M, Parisi E, Scarano E: Methylation of DNA in developing sea urchin embryos. *J Mol Biol* 1968;36:195–208.

Grosschedl R, Birnstiel ML: Identification of regulatory sequences in the prelude sequences of an H2A histone gene by the study of specific deletion mutants *in vivo*. *Proc Natl Acad Sci USA* 1980a;77:1432–1436.

Grosschedl R, Birnstiel ML: Spacer DNA sequences upstream of the T-A-T-A-A-A-T-A sequence are essential for promotion of H2A histone gene transcription *in vivo*. *Proc Natl Acad Sci USA* 1980b;77:7102–7106.

Groudine M, Weintraub H; Rous sarcoma virus activates embryonic globin genes in chicken fibroblasts. *Proc Natl Acad Sci USA* 1975;72:4464–4468.

Groudine M, Weintraub H. Activation of cellular genes by avian RNA tumor viruses. *Proc Natl Acad Sci USA* 1980;77:5351–5354.

Groudine M, Weintraub H: Activation of globin genes during chicken development. *Cell* 1981;24:393–401.

Groudine M, Eisenman R, Weintraub H: Chromatin structure of endogenous retroviral genes and activation by an inhibitor of DNA methylation. *Nature* 1981a;292:311–317.

Groudine M, Peretz M, Weintraub H: Transcriptional regulation of hemoglobin switching in chicken embryos. *Mol Cell Biol* 1981b;1:281–288.

Groudine M, Peretz M, Weintraub H: The structure and expression of globin chromatin during hematopoiesis in the chicken embryo. in Stamatoyannopoulos G, Nienhuis AW (eds): *Organization and Expression of Globin Genes*. New York, Alan R. Liss, 1981c, pp 163–173.

Groudine M, Weintraub H: Propagation of globin DNase I-hypersensitive sites in absence of factors required for induction: A possible mechanism for determination. *Cell* 1982;30:131–139.

Gurdon JB, Melton DA: Gene transfer in amphibian eggs and oocytes. *Ann Rev Genet* 1981;15:189–218.

Hall BD, Clarkson SG, Tocchini-Valenti G: Transcription initiation of eukaryotic transfer RNA genes. *Cell* 1982;29:3–5.

Hamada H, Petrino MG, Kakunago T: A novel repeated element with Z-DNA forming potential is widely found in evolutionarily diverse eukaryotic genomes. *Proc Natl Acad Sci USA* 1982;79:6465–6469.

Haniford DB, Pulleyblank DE: Facile transition of poly [d(TG)-d(CA)] into a left-handed helix in physiological conditions. *Nature* 1983;302:632–634.

Harland RM, Weintraub H, McKnight SL: Transcription of DNA injected into *Xenopus* oocytes is influenced by template topology. *Nature* 1983;301:38–43.

Hayward WS, Neel BG, Astrin SM: Activation of a cellular *onc* gene by promoter insertion in ALV-induced lymphoid leukosis. *Nature* 1981;290:475–480.

Helinski DR, Clewell DB: Circular DNA. *Ann Rev Biochem* 1971;40:899–942.

Hentschel CC: Homocopolymer sequences in the spacer of a sea urchin histone gene repeat are sensitive to S1 nuclease. *Nature* 1982;295:714–716.

Hill RJ, Stollar BD: Dependence of Z-DNA antibody binding to polytene chromosomes on acid fixation and DNA torsional strain. *Nature* 1983;305:338–340.

Horz W, Altenberger W: Sequence specific cleavage of DNA by micrococcal nuclease. *Nucl Acids Res* 1981;9:2643–2658.

Hsieh T, Wang JC: Thermodynamic properties of supercoiled DNAs. *Biochemistry* 1975;14:527–535.

Igo-Kemenes T, Horz W, Zachau HG: Chromatin. *Ann Rev Biochem* 1982;51:89–121.

Jakobovitz EB, Bratosan S, Aloni Y: A nucleosome-free region in SV40 minichromosomes. *Nature* 1980;285:263–265.

Jones BB, Chan H, Rothstein S, Wells, RD, Reznikoff WS: RNA polymerase binding sites in λplac5 DNA. *Proc Natl Acad Sci USA* 1977;74:4914–4918.

Jones PA, Taylor SM: Cellular differentiation, cytidine analogues and DNA methylation. *Cell* 1980;20:85–93.

Jones RE, DeFeo D, Piatigorsky J: Transcription and site-specific hypomethylation of the delta-crystallin genes in the embryonic chicken lens. *J Biol Chem* 1981;256:8172–8176.

Jost JP, Ohno T, Panyim S, Scheurch AR: Appearance of vitellogenin mRNA sequences and rate of vitellogenin synthesis in chicken liver following primary and secondary stimulation by 17 β-estradiol. *Eur J Biolchem* 1978;84:355–361.

Keene MA, Elgin SCR: Micrococcal nuclease as a probe of DNA sequence organization and chromatin structure. *Cell* 1981;27:57–64.

Keene MA, Corces V, Lowenhaupt K, Elgin SCR: DNase I hypersensitive sites in *Drosophila* chromatin occur at the 5′ ends of regions of transcription. *Proc Natl Acad Sci USA* 1981;78:143–146.

Kemp D, Harris A, Cory S, Adams J: Expression of the immunoglobulin C gene in mouse T and B lymphoid and myeloid cell lines. *Proc Natl Acad Sci USA* 1980;77:2876–2880.

Kierszenbaum AL, Tres LL: Structural and transcriptional features of the mouse spermatid genome. *J Cell Biol* 1975;65:258–270.

Kioussis D, Vanin E, deLange T, Flavell RA, Grosveld FG: Globin gene inactivation by DNA translocation in γβ-thalassaemia. *Nature* (in press, 1984).

Kiryanov GJ, Smirnova TA, Polyakov VYu: Nucleomeric organization of chromatin. *Eur J Biochem* 1982;124:331–338.

Klar AJS, Strathern JN, Hicks JB: A position-effect control for gene transposition: state of expression of yeast mating-type genes affects their ability to switch. *Cell* 1981;25:517–524.

Klysik J, Stirdivant SM, Larson JE, Hart PA, Wells, R: Left handed DNA in restriction fragments and a recombinant plasmid. *Nature* 1981;290:672–677.

Kohwi-Shigematsu T, Gelinas R, Weintraub H: Detection of an altered DNA conformation at specific sites in chromatin and supercoiled DNA. *Proc Natl Acad Sci USA* 1983;80:4389–4393.

Kornberg RD: Chromatin structure: A repeating unit of histones and DNA. *Science* 1974;184:868–871.

Kornberg RD: Structure of chromatin. *Ann Rev Biochem* 1977;46:931–954.

Kornberg RD: The location of nucleosomes in chromatin: specific or statistical? *Nature* 1981;292:579–580.

Kunkel G, Martinson HG: Nucleosomes will not form on double-stranded RNA or over poly(dA)-poly(dT) tracts in DNA. *Nucl Acids Res* 1981;9:6869–6888.

Kuo MT, Sahasrabuddhe CG, Saunders GF: Presence of messenger specifying sequences in the DNA of chromatin subunits. *Proc Natl Acad Sci USA* 1976;73:1572–1575.

Kuo MT, Mandel JL, Chambon P: DNA methylation: correlation with DNase I sensitivity of chicken ovalbumin and conalbumin chromatin. *Nucl Acids Res* 1979;7:2105–2114.

Kurtz DT: Hormonal inducibility of α-$_{2u}$ globulin genes in transfected mouse cells. *Nature* 1981;291:629–631.

Labhart P, Ness P, Banz E, Parish R, Koller T: Model for the structure of the active nucleolar chromatin. *Cold Spring Harbor Symp Quant Biol* 1982;42:557–564.

Lacy E, Axel R: Analysis of DNA of isolated chromatin subunits. *Proc Natl Acad Sci USA* 1975;72:3978–3982.

Laimins LA, Khaury G, Gorman C, Hawara B, Gruss P: Host-specific activation of transcription by tandem repeats from simian virus 40 and moloney murine sarcoma virus. *Proc Natl Acad Sci USA* 1982;79:6453–6457.

Laird CD, Wilkinson LE, Foe VE, Chooi WY, Flanagan JR: Analysis of chromatin-associated fiber arrays. *Chromosoma* 1976;58:169–192.

Lamb MM, Daneholt B: Characterization of active transcription units in Balbiani rings of chironomus. *Cell* 1979;17:835–848.

Larsen A, Weintraub H: An altered DNA conformation detected by S1 nuclease occurs at specific regions in active chick globin chromatin. *Cell* 1982;29:609–622.

Laskey RA, Honda BM, Mills AD, Morns NR, Wyllie AH, Mertz JE, DeRobertis EM, Gurdon JB: Chromatin assembly and transcription in eggs and oocytes of *Xenopus laevis*. *Cold Spring Harbor* Sumposium of Quantitative Biology 1978;42:171–178.

Laskey RA, Earnshaw WC: Nucleosome assembly. *Nature* 1980;286:763–767.

Lau AF, Ruddon RW: Proteins of transcriptionally active and inactive chromatin from Friend erythroleukemia cells. *Exp Cell Res* 1977;107:35–46.

Lawson, GM, Tsai M, O'Malley BW: Deoxyribonuclease I sensitivity of the nontranscribed sequences flanking of the 5′ and 3′ ends of the ovomucoid gene and the ovalbumin and its related X and Y genes in hen oviduct nuclei. *Biochemistry* 1980;19:4403–4411.

Lawson GM, Knoll BJ, March CJ, Woo SLC, Tsai M-J, O'Malley BW: Definition of 5′ and 3′ structural boundaries of the chromatin domain containing the ovalbumin multigene family. *J Biol Chem* 1982;257:1501–1507.

Lazier CB: Ontogeny of the vitellogenic response to oestradiol and of the soluble nuclear oestrogen receptor in embryonic-chic liver. *Biochem J* 1978;174:143–152.

Levinger L, Varshavsky A: Selective arrangement of ubiquinated and D1 protein-containing nucleosomes within the *Drosophila* genome. *Cell* 1982;28:375–385.

Levinson B, Klhoury G, Vande Woude G, Gruss P: Activation of SV40 genome by 72 base pair tandem repeats of Moloney sarcoma virus. *Nature* 1982;295:568–572.

Levy A, Noll M: Multiple phases of nucleosomes in the hsp70 genes of Drosophila melanogaster. *Nucl Acids Res* 1980;8:6059–6068.

Levy A, Noll M: Chromatin fine structure of active and repressed genes. *Nature* 1981;289:198–203.

Levy B, Dixon GH: Renaturation kinetics of cDNA complementary to cytoplasmic polyadenylated RNA from rainbow trout testis. Accessibility of transcribed genes to pancreatic DNase. *Nucl Acids Res* 1977;4:833–898.

Levy B, Wong NCW, Dixon GH: Selective association of the trout-specific H6 protein with chromatin regions susceptible to DNase I and DNase II. Possible location of HMG-T in the spacer region between core nucleosomes. *Proc Natl Acad Sci USA* 1977;74:2810–2814.

Levy-Wilson B, Dixon GH: Limited action of micrococcal nuclease on trout testis nuclei generates two mononucleosome subsets enriched in transcribed DNA sequences. *Proc Natl Acad Sci USA* 1979;76:1682–1686.

Lezzi M, Robert M: Chromosomes isolated from unfixed salivary glands of *Chironomous;* in Beerman W. (ed) *Developmental Studies on Giant Chromosomes: Results and Problems in Cell Differentiation.* New York, Springer-Verlag, 1976; vol 4, pp 35–57.

Lilley DMJ: The inverted repeat as a recognizable structural feature in supercoiled DNA molecules. *Proc Natl Acad Sci USA* 1980;77:6648–6472.

Lilley DMJ: Hairpin-loop formation by inverted repeats in supercoiled DNA is a local and transmissible property. *Nucl Acids Res* 1981;9:1271–1288.

Lilley DMJ: Structural perturbation in supercoiled DNA: Hypersensitivity to modification by a single-strand selective chemical reagent conferred to inverted repeat sequences. *Nucl Acids Res* 1983;11:3097–3112.

Louis C, Schedl P, Samal B, Worcel A: Chromatin structure of the 5S RNA genes of D. melanogaster. *Cell* 1980;22:387–392.

Luchnik AN, Bakayev VV, Zbarsky IB, Georgiev GP: Elastic torsional strain in DNA within a fraction of SV40 minichromosomes: Relation to transcriptionally active chromatin. *EMBO J* 1982;1:1353–1358.

Luciw PA, Bishop JM, Varmus HE, Capecchi M: Location and function of retroviral and SV40 sequences that ehnahce biochemical transformation after microinjection of DNA. *Cell*1983; 33:705–716.

Luse DS, Roeder RG: Accurate transcription initiation on a purified mouse β-globin DNA fragment in a cell-free system. *Cell* 1980;20:691–699.

Mace HAF, Pelham HRB, Travers AA: An S1 nuclease sensitive structure associated with short direct repeats in the 5′ flanking regions of *Drosophila* heat shock genes. *Nature* 1983;304:555–557.

Macleod D, Bird A: DNase I sensitivity and methylation of active versus inactive rRNA genes in *Xenopus* species hybrids. *Cell* 1982;29:211–218.

Mantei N, Boll W, Weissmann C: Rabbit β-globin mRNA production in mouse L cells transformed with cloned rabbit β-globin chromosomal DNA. *Nature* 1979;281:40–46.

Marsden MPF, Laemmli UK: Metaphase chromosome structure: Evidence for a radial loop model. *Cell* 1979;17:849–858.

Mathis DJ, Gorovsky MA: Subunit structure of rDNA-containing chronmatin. *Biochemistry* 1976;15:750–755

Mathis DJ, Oudet P, Chambon P: Structure of transcribing chromatin. *Prog Nucl Acid Res Mol Biol* 1980;24:1–55.

Mayfield JE, Serunian LA, Silver LM, Elgin SCR: A protein released by DNase I digestion of *Drosophila* nuclei is preferentially associated with puffs.*Cell* 1978;14:539–544.

Mayo KE, Warren R, Palmiter RD: The mouse metallothionein-I gene is transcriptionally regulated by cadmium following transfection into human or mouse cells. *Cell* 1982;29:99–108.

McGeady ML, Jhappan C, Ascione R, Vande Woude GF: *In vitro* methylation of specific regions of the cloned moloney sarcoma virus genome inhibits its transforming activity. *Mol Cell Biol* 1983;3:305–314.

McGhee JD, Felsenfeld G: Nucleosome structure. *Ann Rev Biochem* 1980;49:1115–1156.

McGhee JD, Wood WI, Dolan M, Engel JD, Felsenfeld G: A 200 base pair region at the 5′ end of the chicken adult β-globin gene is accessible to nuclease digestion. *Cell* 1981;27:45–55.

McGhee JD, Nickol JM, Felsenfeld G, Rau DC: Higher order structure of chromatin: Orientation of nucleosomes within the 30nm chromatin solenoid is independent of species and spacer length. *Cell* 1983;33:831–841.

McGinnis W, Shermoen AW, Beckendorf SK: A transposable element inserted just 5′ to a drosophila glue protein gene alters gene expression and chromatin structure. *Cell* 1983;34:75–84.

McKeon C, Ohkubo H, Pastan I, de Crombrugghe B: Unusual methylation pattern of the alpha 2 (1) collagen gene. *Cell* 1982;29:203–210.

McKnight GS, Palmitter RA: Transcriptional regulation of the ovalbumin and conalbumin genes by steroid hormones in chick oviduct. *J Biol Chem* 1979;254:9050–9058.

McKnight SL, Miller OL Jr: Ultrastructural patterns of RNA synthesis during early embrygenesis of *Drosphila melanogaster*. *Cell* 1976;8:305–319.

McKnight SL, Sullivan NL, Miller OL, Jr: Visualization of the silk fibroin transcription unit and nascent silk fibroin molecules on polyribosomes of bombyx mori. *Prog Nucl Acids Res Mol Biol* 1976;19:313–318.

McKinght SL, Bustin M, Miller OL Jr: Electron microscopic analysis of chromosome metabolism in the *Drosophila melanogaster* embryo. *Cold Spring Harbor Symp Quant Biol* 1977;42:741–754.

McKnight SL, Gavis ER, Kingsbury R, Axel R: Analysis of transcriptional regulatory signals of the HSV thymidine kinase gene: Identification of an upstream control region. *Cell* 1981;25:385–398.

Mechali M, de Recondo A-M, Girard M: Action of the S1 endonuclease from aspergillus oryzae on simian virus 40 supercoiled component I DNA *BBRC* 1973;54:1306–1320.

Mertz J: Linear DNA does not form chromatin containing regularly spaced nucleosomes. *Mol Cell Biol* 1982;2:1608–1618.

Mezquita C, Teng CS: Changes in nuclear and chromatin composition and genomic activity during spermatogenesis in the maturing rooster testis. *Biochem J* 1977;164:99–111.

Miller FD, Rattner JB, Van de Sande: Nucleosome-core assembly on B and Z forms of poly[d(G-m^{5}C)]. *Cold Spring Harbor Symp Quant Biol* 1982;47:571–575.

Miller OJ, Schnedl W, Allen J, Erlanger BF: 5-methylcytosine localized in mammalian constitutive heterochromatin. *Nature* 1974;251:636–637.

Miller OL Jr, Beatty BR: Visualization of nucleolar genes. *Science* 1969;164:955–957.

Mizuuchi K, Mizuuchi M, Gellert M: Cruciform structures in palindromic DNA are favored by DNA supercoiling. *J Mol Biol* 1982;156:229–243.

Mohandas T, Sparkes RS, Shapiro LJ: Reactivation of an inactive human X chromosome: Evidence for X inactivation by DNA methylation. *Science* 1981;211:393–396.

Moreau J, Marcaud L, Maschat F, Kejzlarova-Lepesant J, Lepesant J-A, Scherrer K: A + T rich linkers define functional domains in eukaryotic DNA. *Nature* 1982;295:260–262.

Moreau P, Hen R, Wasylylk B, Everett R, Gaub MP, Chambon P: The SV40 72 base pair repeat has a striking effect on gene expression both in SV40 and other chimeric recombinants. *Nucl Acids Res* 1981;9:60-47–6068.

Moyne G, Freeman R, Saragosti S, Yaniv M: A high resolution electron microscopy study of nucleosomes from simian virus 40 chromatin. *J. Mol Biol* 1981;149:735–744.

Musiich PR, Brown FL, Maio JJ: Nucleosome phasing and micrococcal nuclease cleavage of African green monkey component α-DNA. *Proc Natl Acad Sci USA* 1982;79:118–122.

Muskovitch MAT, Hogness DS: An expandable gene that encodes a *Drosophila* glue protein is not expressed in variants lacking remote upstream sequences. *Cell* 1982;29:1041–1051.

Nasmyth KA: The regulation of yeast mating-type chromatin structure by SIR: An action at a distance affecting both transcription and transposition. *Cell* 1982;30:567–578.

Naveh-Many T, Cedar H: Active gene sequences are undermethylated. *Proc Natl Acad Sci USA* 1981;78:4246–4250.

Nickol H, Behe M, Felsenfeld G: Effect of the B-Z transition in poly(dG^{m5}dC)-poly(dG^{m5}C) on nucleosome formation. *Proc Natl Acad Sci USA* 1982;79:1771–1775.

Nickol JM, Felsenfeld G: DNA conformation at the 5′ end of the chicken adult β-globin gene. *Cell* (*35*:467–477, 1983).

Niwa O, Sugaharo T: 5′-azacytidine induction of mouse endogenous type C virus and suppression of DNA methylation. *Proc Natl Acad Sci USA* 1981;78:6290–6294.

Noll M, Thomas JO, Kornberg RD: Preparation of native chromatin and damage caused by shearing. *Science* 1975;187:1203–1206.

Noll M, Kornberg RD: Action of micrococcal nuclease on chromatin and the location of histone H1. *J Mol Biol* 1977;109:393–404.

Nordheim A, Pardue ML, Lafer EM, Moller A, Stollar BD, Rich A: Antibodies to left-handed Z-DNA bind to interband regions of *Drosophila* polytene chromosomes. *Nature* 1981;294:417–422.

Nordheim A, Tesser P, Azorin F, Kwon YH, Moller A, Rich A: Isolation of *Drosophila* proteins that bind selectively to left-handed Z-DNA. *Proc Natl Acad Sci USA* 1982;79:7729–7733.

Nordheim A, Rich A: Negatively supercoiled SV40 viral DNA contains Z-DNA segments within the transcriptional enhancer sequences. *Nature* 1983;303:674–679.

Olins OL, Olins DE: Spheroid chromatin units (ν bodies). *Science* 1974;183:330–332.

Palmiter R, Mulvihill E, McKnight S, Senear A: Regulation of gene expression in the chick oviduct by steroid hormones. *Cold Spring Harbor Symp Quant Biol* 1978;42:639–647.

Panayotatos N, Wells RD: Cruciform structures in supercoiled DNA. *Nature* 1981;289:466–470.

Panet A, Cedar H: Selective degradation of integrated murine leukemia proviral DNA by deoxyribonucleases. *Cell* 1977;11:933–940.

Parslow TG, Granner DK: Chromatin changes accompany immunoglobulin kappa gene activation: a potential control region within the gene. *Nature* 1982;299:449–451.

Paulson JR, Laemmli UK: The structure of histone-depleted metaphase chromosomes. *Cell* 1977;12:817–828.

Payne GS, Bishop JM, Varmus HE: Multiple arrangements of viral DNA and an activated host oncogene in bursal lymphomas. *Nature* 1982;295:209–214.

Peck LJ, Nordheim A, Rich A, Wang JC: Flipping of cloned $d(pCpG)_n$-$d(pCpG)_n$ DNA sequences from right- to left-handed helical structure by salt, Co(III), or negative supercoiling. *Proc Natl Acad Sci USA* 1982;79:4560–4564.

Platz RD, Grimes SR, Meistrich ML, Hnilica LS: Changes in nuclear proteins of rat testis cells separated by velocity sedimentation. *J Biol Chem* 1975;250:5791–5800.

Pruitt SC, Reeder RH: Effect of topological constraint on transcription of rDNA in *Xenopus* oocytes: Comparison of plasmid and endogenous genes. *J Mol Biol* 1984;174:121–139.

Queen C, Lord ST, McCutchan TF, Singer M: Three segments from the monkey genome that hybridize to simian virus 40 have common structural elements. *Mol Cell Biol* 1981;1061–1068.

Queen C, Baltimore D: Immunoglobulin gene transcription is activated by downstream sequence elements. *Cell* 1983;33:741–748.

Razin A, Cedar H: Distribution of 5-methylcytosine in chromatin. *Proc Natl. Acad Sci USA* 1977;74:2725–2728.

Razin A, Riggs AD: DNA methylation and gene function. *Science* 1980;210:604–610.

Rechavi G, Givol D, Canaani E: Activation of a cellular oncogene by DNA rearrangement: possible involvement of an Is-like element. *Nature* 1982;300:607–611.

Reeves R: Ribosomal genes of *Xenopus laevis:* Evidence of nucleosome in transcriptionally active chromatin. *Science* 1976;194:529–531.

Renz M, Nehls P, Hozier J: Involvement of histone H1 in the organization of the chromosome fiber. *Proc Natl Acad Sci USA* 1977;74:1879–1883.

Ring D, Cole RD: Chemical cross-linking of H1 histone to the nucleosomal histones. *J Biol Chem* 1979;254:11688–11695.

Robins DM, Paek I, Seebury DH, Axel R: Regulated expression of human growth hormone genes in mouse cells. *Cell* 1982;29:623–631.

Rodriguez-Alfageme C, Rudkin GT, Cohen LH: Isolation, properties and cellular distribution of D1, a chromosomal protein of *Drosophila. Chromosoma* 1981;78:1–31.

Roeder GS, Farabaugh P, Chaleff D, Fink G: The origins of gene instability in yeast. *Science* 1980;209:1375–1380.

Saffer JD, Lerman MI: Unusual class of Alu sequences containing a potential Z-DNA segment. *Mol Cell Biol* 1983;3:960–964.

Samal B, Worcel A, Louis C, Schedl P: Chromatin structure of the histone genes of D. melanogaster. *Cell* 1981;23:401–409.

Sandeen G, Wood WI, Felsenfeld G: The interaction of high mobility proteins HMG 14 and 17 with nucleosomes. *Nucl Acids Res* 1980;8:3757–3778.

Saragosti S, Moyne G, Yaniv M: Absence of nucleosomes in a fraction of SV40 chromatin between the origin of replication and the region coding for the late leader RNA. *Cell* 1980;20:65–73.

Scheer U, Trendelenburg MF, Krohne G, Franke W: Lengths and patterns of transcription units in the amplified nucleoli of oocytes of *Xenopus laevis. Chromosoma* 1977;60:147–167.

Scheer U: Changes of nucleosome frequency in nucleolar and nonnucleolar chromatin as a function of transcription: an electron microscopic study. *Cell* 1978;13:535–549.

Schlaeger EJ: Replicative conformation of parental nucleosomes: salt sensitivity of deoxyribonucleic acid-histone interaction and alteration of histone H1 binding. *Biochemistry* 1982;21:3167–3174.

Schon E, Evans T, Welsh J, Efstradiadis A: Conformation of promoter DNA: I. Fine mapping of S1 hypersensitive sites. *Cell* (*35:*837–848, 1984).

Schubach W, Groudine M: Alteration of c-*myc* chromatin structure by avian leukosis virus integration. *Nature* (in press, 1984).

Scott WA, Wigmore DJ: Sites in simian virus 40 chromatin which are preferentially cleaved by endonucleases. *Cell* 1978;15:1511–1518.

Senear AW, Palmiter RD: Multiple structural features are responsible for the nuclease sensitivity of the ovalbumin gene. *J Biol Chem* 1981;256:1191–1198.

Senear AW, Palmiter RD: Expression of the mouse metallothionein-1 gene alters the nuclease hypersensitivity of its 5′ regulatory region. *Cold Spring Harbor Symp Quant Biol* 1983;47:539–547.

Shen-Ong GLC, Keath EJ, Piccoli SP, Cole MD: Novel *myc* oncogene RNA from abortive immunoglobulin-gene recombination in mouse plasmacytomas. *Cell* 1982;31:443–452.

Shermoen AW, Beckendorf SK: A complex of interacting DNase I hypersensitivity sites near the *Drosophila* Glue Protein Gene, Sgs4. *Cell* 1982;29:601–607.

Shishido K: Relationship between S1 endonuclease-sensitivity and number of superhelical turns in a negatively-twisted DNA. *FEBS Lett* 1980;111:333–336.

Simpson RT: Structure of the chromatosome, a chromatin particle containing 160 base pairs of DNA and all the histones. *Biochemistry* 1978;17:5524–5531.

Sinden RR, Carlson JO, Pettijohn DE: Torsional tension in the DNA double helix measured with trimethylpsoralen in living *E. coli* cells: Analogous measurements in insect and human cells. *Cell* 1980;21:773–783.

Singer DS: Arrangement of a highly repeated DNA sequence in the genome and chromatin of the african green monkey. *J Biol Chem* 1979;254:5506–5514.

Singleton CK, Wells RD: Relationship between superhelical density and cruciform formation in plasmid pVH51. *J Biol Chem* 1982;257:6292–6295.

Singleton CK, Klysik J, Stirdivant SM, Wells RD: Left-handed Z-DNA is induced by supercoiling in physiological ionic conditions. *Nature* 1982;299:312–316.

Singleton CK, Klysik J, Wells RD: Conformational flexibility of junctions between contiguous B- and Z-DNAs in supercoiled plasmids. *Proc Natl Acad Sci USA* 1983;80:2447–2451.

Sledziewski A, Young ET: Chromatin conformational changes accompany transcriptional activation of a glucose-repressed gene in *Saccharomyces cerevisiae. Proc Natl Acad Sci USA* 1982;79:253–256.

Smith GR: DNA supercoiling: another level for regulating gene expression. *Cell* 1981;24:599–600.

Sperling R, Wachtel EJ: The histones. *Adv Protein Chem* 1981;34:1–60.

Spinelli G, Albanese I, Anello L, Ciaccio M, DiLiegro I: Chromatin structure of histone genes in sea urchin sperm and embryos. *Nucl Acids Res* 1982;10:7977–7991.

Spradling AC, Rubin GM: The effect of chromosomal position on the expression of the drosophila Xanthine dehydrogenase gene. *Cell* 1983;34:47–57.

Stadler J, Larsen A, Engel JD, Dolan M, Groudine M, Weintraub H: Tissue-specific DNA cleavages in the globin chromatin domain introduced by DNase I. *Cell* 1980a;20:451–460.

Stalder J, Groudine M, Dodgson JB, Engel JD, Weintraub H: Hb switching in chickens. *Cell* 1980b;19:973–980.

Stein R, Sciaky-Gallili N, Razin A, Cedar H: Patterns of methylation of two genes coding for housekeeping functions. *Proc Natl Acad Sci USA* 1983;80:2422–2426.

Stirdivant SM, Klysik J, Wells, RD: Energetic and structural inter-relationship between DNA supercoiling and the right-to-left-handed Z helix transitions in recombinant plasmids. *J Biol Chem* 1982;257:10159–10165.

Storb U, Arp B, Wilson R: The switch region associated with immunoglobin Cμ genes is DNase I hypersensitive in T lymphocytes. *Nature* 1981a;294:90–92.

Storb U, Wilson R, Selsing E, Walfield A: Rearranged and germline immunoglobulin K genes: different states of DNase I sensitivity of constant K genes in immunocompetent and non-immune cells. *Biochemistry* 1981b;20:990–996.

Swerdlow PS, Varshavsky A: Affinity of HMG 17 for a mononucleosome is not influenced by the presence of ubiquitin-H2A semihistone but strongly depends on DNA fragment size. *Nucl Acids Res* 1983;11:387–401.

Tatchell K, Van Holde KE: Compact oligomers and nucleosome phasing. *Proc Natl Acad Sci USA* 1978;75:3583–3587.

Thoma F, Koller T, Klug AL: Involvement of histone H1 in the organization of the nucleosome and of the salt-dependent superstructures of chromatin. *J Cell Biol* 1979;83:403–427.

Tsujimoto Y, Hirose S, Tsuda M, Suzuki Y: Promoter sequence of fibroin gene assigned by *in vitro* transcription system. *Proc Natl Acad Sci USA* 1981;78:4838–4842.

Ucker DS, Firestone GL, Yamamoto KR: Glucocorticoids and chromosomal position modulate murine mammary tumor virus transcription by affecting efficiency of promoter utilization. *Mol Cell Biol* 1983;3:551–561

Urieli-Shoval S, Gruenbaum Y, Sedat J, Razin A: The absence of detectable methylated bases in drosophila melanogaster DNA. *FEBS Lett* 1982;146:148–152.

Utakoji T: Chronology of nucleic acid synthesis in meiosis of the male chinese hamster. *Exp Cell Res* 1966;42:585–596.

Van der Ploeg LHT, Flavell RA: DNA methylation in the human $\gamma\delta\beta$-globin locus in erythroid and non-erythroid tissues. *Cell* 1980;19:947–958.

Van Ness BJ, Weigert M, Coleclough C, Mather EL, Kelley DE: Transcription of the unrearranged mouse C_k locus sequence of the initiation region and comparison of activity with a rearranged V_k-C_k gene. *Cell* 1981;27:593–602.

Varshavsky AJ, Sundin O, Bohn M: A stretch of "late" SV40 viral DNA about 400 bp long which includes the origin of replication is specifically exposed in SV40 minichromosomes. *Cell* 1979;16:543–466.

Varshavsky AJ, Levinger L, Sundin O, Barsoum J, Ozkaynak E, Swerdlow P, Finley D: Cellular and SV40 chromatin replication, segration, ubiquitination, nuclease-hypersensitive sites, HMG-containing nucleosomes and heterochromatin-specific protein. *Cold Spring Harbor Symp Quant Biol* 1982;47:511–528.

Vogt VM: Purification and further properties of single-strand specific nuclease from aspergillus oryzae. *Eur J. Biochem* 1973;33:192–200.

Waldeck W, Fohring B, Chowdhury K, Gruss P, Sauer G: Origin of DNA replication in papovavirus chromatin is recognized by endogenous endonuclease. *Proc Natl Acad Sci USA* 1978;75:5964–5968.

Wang AH-J, Quigley GJ, Kalpak FJ, Crawford JL, van Boom JH, van der Marel G, Rich A: Molecular structure of a left-handed double helical DNA fragment at atomic resolution. *Nature* 1979;282:680–686.

Wang J-C: Interactions between twisted DNAs and enzymes: The effects of superhelical turns. *J Mol Biol* 1974;87:797–816.

Wang JC: The path of DNA in the nucleosome. *Cell* 1982;29:724–726.

Wartell RM: The transmission of stability or instability from site specific protein DNA complexes. *Nucl Acids Res* 1977;4:2779–2797.

Waskylyk B, Oudet P, Chambon P: Preferential *in vitro* assembly of nucleosome cores on some AT-rich regions of SV40 DNA. *Nucl Acids Res* 1979;7:705–719.

Waskylyk B, Kedinger C, Corden J, Brison O, Chambon P: Specific *in vitro* initiation of transcription on conalbumin and ovalbumin genes and comparison with adenovirus-2 early and late genes. *Nature* 1980;285:367–373.

Weber JL, Cole RD: Chromatin fragments containing bovine 1.715 g/ml satellite DNA. Nucleosome structure and protein composition. *J Biol Chem* 1982;257:11784–11790.

Weintraub H, Groudine M: Chromosomal subunits in active genes have an altered conformation. *Science* 1976;193:848–856.

Weintraub H, Worcel A, Alberts B: A model for chromatin based upon two symmetrically paired half-nucleosomes. *Cell* 1976;9:409–417.

Weintraub H, Larsen A, Groudine M: α-Globin gene switching during the development of chicken embryos: Expression and chromatin structure. *Cell* 1981a;24:333–344.

Weintraub H, Weisbrod S, Larsen A, Groudine M: Changes in globin chromatin structure during red cell differentiation in chick embryos, in Stamatoyannopoulos G, Nienhuis AW (eds): *Organization and Expression of Globin Genes*. New York, Alan R Liss, 1981b, pp 175–190.

Weintraub H, Beug H, Groudine M, Graf T: Temperature-sensitive changes in the structure of globin chromatin in lines of red cell precursors transformed by ts-AEV. *Cell* 1982;28:931–940.

Weintraub H: A dominant role for DNA secondary structure in forming hypersensitive structures in chromatin. *Cell* 1983;32:1191–1203.

Weisbrod S, Weintraub H: Isolation of a subclass of nuclear proteins responsible for conferring a DNase I sensitive structure on globin chromatin. *Proc Natl Acad Sci USA* 1979;76:639–634.

Weisbrod S, Groudine M, Weintraub: Interaction of HMG 14 and 17 with actively transcribed genes. *Cell* 1980;19:289–301.

Weisbrod S, Weintraub H: Isolation of actively transcribed nucleosomes using immobilized HMG 14 and 17 and an analysis of α-globin chromatin. *Cell* 1980;19:289–301.

Weisbrod S: Active chromatin. *Nature* 1982;297:289–295.

Weischet WO, Van Holde KE: Nuclease digstion promotes structural rearrangements in H1-depleted chromatin. *Nucl Acids Res* 1980;8:3743–3755.

Wells RD, Goodman TC, Hillen W, Horn GT, Klein RD, Larson JE, Muller UR, Neuendorf SK, Panayotatos N, Stirdivant SM: DNA structure and gene regulation. *Prog Nucl Acid Res Molec Biol* 1980;24:167–267.

Wilks AJ, Cozens PJ, Mattaj IW, Jost JP: Estrogen induces a demethylation at the 5′ end region of the chicken vitellogenin gene. *Proc Natl Acad Sci USA* 1982;79:4252–4255.

Williams JL, Tata JR: Simultaneous analysis of conformation and transcription of A and B groups of vitellogenin genes in male and female *Xenopus* during primary and secondary activation by estrogen. *Nucl Acids Res* 1983;11:1151–1166.

Williamson VM, Cox D, Young ET, Russel DW, Smith M: Characterization of transposable element-asociated mutations that alter yeast alcohol dehydrogenase I expression. *Mol Cell Biol* 1983;3:21–31.

Wiskocil R, Bensky P, Dower W, Goldberger RF, Gordon J, Deeley RG: Coordinate regulation of two estrogen-dependent genes in avian liver. *Proc Natl Acad Sci USA* 1980;77:4474–4478.

Wittig B, Wittig S: A phase relationship asociates tRNA structural gene sequences with nucleosome cores. *Cell* 1979;18:1173–1183.

Wittig S, Wittig B: Function of a tRNA gene promoter depends on nucleosome position. *Nature* 1982;297:31–38.

Wold B, Wigler M, Lacy E, Maniatis T, Silverstein S, Axel R: Introduction and expression of a rabbit β-globin gene in mouse fibroblasts. *Proc Natl Acad Sci USA* 1979;76:5684–5688.

Wu C, Bingham PM, Livak KJ, Holmgren R, Elgin SCR: The chromatin structure of specific genes: I. Evidence for higher order domains of defined DNA sequence. *Cell* 1979a;16:797–806.

Wu C, Wong YC, Elgin SCR: The chromatin structure of specific genes: II. Disruption of chromatin structure during gene activity. *Cell* 1979b;16:807–814.

Wu C: The 5′ ends of *Drosophila* heat shoc, genes in chromatin are hypersensitive to DNase I. *nature* 1980;286:854–860.

Wu C, Gilbert W: Tissue-specific exposure of chromatin structure at the 5′ terminus of the rat prepoinsulin II gene. *Proc Natl Acad Sci USA* 1981;75:1577–1580.

Yaniv M: Enhancing elements for activation of eukaryotic promoters. *Nature* 1982;297:17–18.

Young NS, Benz EJ Jr, Kantor JA, Kretschner P, Nienhuis AW: Hemoglobin switching in sheep: Only the γ gene is in the active conformation in fetal liver but all the β and γ genes are in the active conformation in bone marrow. *Proc Natl Acad Sci USA* 1978;75:5884–5888.

Zacharias W, Larson JE, Klysik J, Stirdivant SM, Wells RD: Conditions which cause the right-handed to left-handed DNA conformational transitions. *J Biol Chem* 1982;257:2775–2782.

Zachau HG, Igo-Kemenes T: Face to phase with nucleosomes. *Cell* 1981;24:597–598.

16

Gene Methylation Patterns and Expression

Joel Yisraeli*
Moshe Szyf*

Three basic approaches have been employed in the study of the effect of DNA methylation of particular genes on their expression. One approach uses chemicals that cause demethylation (e.g., 5-azacytidine [5-aza-C]), thereby changing the methylation pattern of the cellular genome (see Chapter 9). Using the proper selection techniques, the expression of particular genes following hypomethylation can be assayed. This method is limited, however, by the fact that the sequences of interest may not undergo hypomethylation in a fashion that will mimic normal developmental processes. Using the second approach, genes can be methylated *in vitro* and can be introduced into cellular genomes, thus allowing the study of the effect of methylation at specific sites on gene expression (see Chapter 8). Again, it remains unclear how relevant the results of such experiments are to the processes occurring *in vivo*. The third approach involves careful observation of the naturally occurring variations in the methylation pattern of a particular gene in various tissues. A large number of genes have been studied using this approach. In this chapter, the available data is brought together and discussed in an attempt to obtain a general concept of the pattern of methylation of eukaryotic genes and to examine the interrelationship between DNA methylation patterns of particular genes and their expression.

The basic procedure for analyzing methylation patterns in gene sequences is simple. It involves the use of restriction endonucleases, which do not cleave the DNA when their recognition site is methylated, and the examination of the restriction patterns. In eukaryotes, the only methylated nucleotide is cytosine; almost all of the methylated cytosines are found in the dinucleotide CpG. No adequate method is available yet for determining the level of methylation of

Department of Cellular Biochemistry, The Hebrew University-Hadassah Medical School, Jerusalem, Israel all CpGs in a gene sequence. Subsets of the –CpG– sequences may be analyzed, however, by using methylation-sensitive restriction enzymes. Thus, HpaII (whose recognition site is CCGG), HhaI (GCGC), and AvaI (CPyCGPuG) are three enzymes that do not cleave if the internal C is methylated. HpaII has been very useful, because an isoschizomer, MspI, cleaves regardless of the state of methylation of its site. The procedure for assaying methylation patterns is to digest DNA with a restriction enzyme that is sensitive to methylation (possibly in conjunction with an enzyme or enzymes that do not recognize 5-meCyt), to electrophorese this digest on an agarose gel, and to perform a standard Southern blot and filter hybridization. Using the proper probes for this hybridization, one can analyze the percentage of the molecules that were cleaved (i.e., unmethylated) at any particular site of the restriction enzyme that was used.

The frequency of a particular site's methylation can be computed from the strength of the bands on the autoradiogram (Figure 16.1). It is important to remember that this frequency is a statistical average of approximately 10^6–10^7 molecules. Thus, a "25% hypomethylated site" implies that 75% of the molecules are methylated at this site.

The data presented in Figures 16.2–16.30 are a compilation of the available data that came to our knowledge up to the end of 1983. We have presented the methylation pattern maps in a uniform fashion. For every gene, the methylation pattern in at least one expressing and one nonexpressing tissue or cell type is shown. In every case, the level of expression of the gene is indicated (generally based on determination of mRNA levels). A great part of the published data required considerable extrapolation to derive all the data necessary for our mapping technique. In some cases, the maps were extracted by us from the raw data in the Southern blots. The manipulations made for mapping each gene are explained in the corresponding legend.

Methylation Pattern of Eukaryotic Genes

The methylation patterns of 29 eukaryotic genes in various expressing and nonexpressing tissues or cell lines that have been determined by different investigators are presented as methylation maps according to the following technique (see Figures 16.2–16.30). Each map consists of a series of lines. The first line is a physical map of the gene with the exons symbolized by shaded boxes. Each succeeding line represents the methylation pattern in the tissues or cell lines as indicated. The degree of hypomethylation at every restriction site is represented by the length of a line descending from the site. A full-length line indicates 100% hypomethylation (zero methylation). The absence of a line at a specified site indicates that the site is fully methylated. The sites will be referred to as sites a, b, c, and so on, starting at the 5′ end of the map.

The genes analyzed represent a wide array of tissue-specific and housekeeping genes from a number of eukaryotic species that range from amphibians to primates. Therefore, the maps presented in Figures 16.2–16.30 allow us to examine how methylated sites are distributed among the eukaryotic genes in various tissues. Even a cursory view of the data presented reveals that genes generally lose some methyl groups during the differentiation process ("hypomethylation"). For every gene, we define the resulting distribution of methyl groups as its pattern of methylation in that tissue (see Figure 16.31). As can be seen for most genes, the pattern of methylation varies from tissue to tissue. By comparing the degree of methylation at a particular site in various tissues or cell lines, one can classify the sites into two groups. Sites that exhibit "site-specific" methylation are methylated to the same extent in every tissue (see Figure 16.31, II, site c; III, sites a, b; V, site c) Sites that demonstrate "tissue-specific" methylation are methylated to different extents in different tissues (see Figure 16.31, I, sites a-e; II, sites a, d; V, site e).

Site Specificity

Site specificity is demonstrated at many sites in the analyzed genes, and those sites may be subdivided into three distinct groups:

(1) Sites that are fully methylated in all tissues, including sperm (see Figure 16.31, II, sites b, e). This group includes sites c and e in the human growth hormone gene (Figure 16.25), site a in the rat insulin I gene (Figure 16.3), and sites f, h, and i in the human globin gene cluster (Figure 16.6). In a similar vein, Satz and Singer have described a site that specifically seems to undergo *de novo* methylation when transfected into L-cells (Satz and Singer, 1983).

(2) Sites that are unmethylated in all differentiated tissues examined (see Figure 16.31, II, site c). Here one finds sites a-f in the chicken $\alpha 2$ (1) collagen gene (Figure 16.21), sites b, c, and f in the rat α-actin gene (Figure 16.29), and sites a-e in the Chinese hamster *aprt* gene (Figure 16.20).

In a number of cases, unmethylated sites could be found even in sperm DNA; for example, in Chinese hamster *aprt* gene (Figure 16.20), chicken $\alpha 2$ (1) collagen gene (Figure 16.21) and in *Xenopus* rRNA genes (Bird *et al.* 1981). These sites remained unmethylated in all of the differentiated tissues that were analyzed.

(3) Sites that are partially methylated in all differentiated tissues studied. Examples representing this group are sites f, g, and i in the Chinese hamster *aprt* gene (Figure 16.20); sites c, f, and g in the rat α-fetoprotein gene (Figure 16.28), and sites d-g in the *Xenopus* albumin gene (Figure 16.27).

Tissue Specificity

The other class of sites are methylated in a tissue-specific fashion and may be subdivided into three overlapping subgroups:

(1) Fully methylated sites, which are fully methylated in at least one tissue (Figure 16.31, I, sites a-e; II, sites a and d). In the rat pepck gene (Figure

16.5), sites d and e are methylated in spleen and fetal liver, but not in adult liver, while site f is fully methylated exclusively in the spleen. In the chicken δ-crystallin gene (1) (Figure 16.15), site g is methylated in sperm, adult liver, and 16d-embryonic neural retina, but not in 16d-embryonic lens and kidney; while site e is methylated in all tissues, but not in the lens.

(2) Partially methylated sites. Sites that are partially methylated in some, but not all, tissues (Figure 16.31 II, sites a and d). In the chicken δ-crystallin gene (2) (Figure 16.30), sites d and e are partially methylated in kidney and lens, but not in retina or sperm. In the rat pepck gene (Figure 16.5), sites f, g, and s are partially methylated in fetal liver, but are fully methylated in spleen. In the rat α-actin gene (Figure 16.28), site g is partially methylated in kidney, but not in other tissues.

(3) Unmethylated sites, which are unmethylated in at least one tissue (Figure 16.31, I, sites a-e; V, site a). In rat γ-casein gene (Figure 16.2), site c is unmethylated in mammary gland, but is methylated in liver.

Partial Methylation—Tissue and Site Specificity

As mentioned above, partial methylation is a statistical phenomenon in which a particular site is methylated in some, but not all, of the cells in a tissue. In some cases, a number of partially methylated sites exist in a single gene of a particular tissue. Several examples can be demonstrated that represent both tissue-specific undermethylations—as in the rat insulin II gene (Figure 16.18, sites d, e, and f) or the chicken δ-crystallin gene (2) (Figure 16.30, sites d, e)—or site-specific undermethylations—as is the case with the *aprt* gene (Figure 16.20, sites f, g, and i). The obvious question that this phenomenon poses is whether the partially methylated sites derive from a subpopulation of cells that are unmethylated at these sites or whether the undermethylated sites are randomly distributed in all cell types. In at least two cases, the Chinese hamster *aprt* gene (Stein *et al.* 1983) and the rat insulin II gene (Cate *et al.* 1983), it has been shown that coexisting unmethylated sites are randomly distributed among the cells of a tissue. Thus, partial methylation of coexisting sites may represent independent events resulting from a "tendency" of sites to lose their methyl groups.

An interesting observation is that some sites that become unmethylated in expressing tissues tend to be partially unmethylated in nonexpressing tissues. For example, in the chicken ovalbumin gene, site d, which is partially methylated in both erythrocytes and liver (where the gene is not expressed), is fully unmethylated in the oviduct (where it is active) (Figure 16.17). Similar changes are observed in the rat insulin II gene; sites d and f, which are partially unmethylated in brain and kidney, are fully unmethylated in insulinomas and liver (Figure 16.18). Therefore, partially methylated sites demonstrate a "site-specific" tendency to lose methyl groups, which may be strengthened in some tissues—mainly those expressing the gene in which the site is found.

Correlation of Eukaryotic Gene Methylation Patterns and Gene Expression

Site specificity, by definition, cannot correlate with differential gene expression, because the same hypomethylation occurs in both expressing and nonexpressing tissues. However, in the case of housekeeping genes, which are constitutively expressed in all tissues, such a correlation can exist. In the *aprt* gene, for example (Figure 16.20), all 5′-end CpG sites that were examined were hypomethylated in all tissues, including sperm.

Tissue specificity may correlate with differential expression. Indeed, most of the sites that exhibit tissue specificity are hypomethylated in tissues that express the genes in which they are found. For example, in the human globin gene cluster (Figure 16.6), sites b, c, and d are hypomethylated in fetal liver (where these genes are active), but not in bone marrow, adult liver, or sperm (where the genes do not express). In the rat pepck gene (Figure 16.5), all of the analyzed sites are hypomethylated in adult liver (where the gene is active), but not in spleen (where it does not express). However, there also exists methylatable sites that undergo tissue-specific changes in their methylation pattern and that show no correlation with gene expression. One clear example of a tissue-specific hypomethylation of a gene that does not correlate with its expression is the rat α-actin gene (Figure 16.29) in which site e is fully hypomethylated only in the kidney, where this gene is inactive.

The genes presented in this chapter contain various combinations of sites that exhibit these types of tissue specificity, along with others that exhibit site specificity. Despite these variations, we were able to divide the 29 analyzed genes into five paradigmatic groups. In our first class, the active gene is fully unmethylated, while the nonactive gene is fully methylated (Figure 16.31, I). The second group contains genes in which some tissue-specific hypomethylations occur in active tissues exclusively, while other hypomethylations are site-specific and occur in nonexpressing tissues as well (Figure 16.31, II). In some cases, the specific changes are in the 5′ end, while these changes appear in the 3′ end in others. In the third group, the genes are undermethylated, at least in some sites in sperm, as well as in all other tissues. In some cases, hypomethylation of other sites may occur in expressing tissues (Figure 16.31, III). Group IV describes a situation in which a gene remains fully methylated in all tissues. In this group, we included genes that undergo a change in methylation in response to hormone induction. Group V describes all of the cases in which tissue-specific hypomethylations of active and nonactive genes can be detected, while no correlation with the state of activity may be determined (Figure 16.31, V).

While all genes classified in groups I–III exhibit some correlation with gene expression, only the first class describes striking correlations, in which all tissue-specific hypomethylations correlate with gene activity. In groups II and III, some tissue-specific hypomethylations do correlate with gene expression, while others do not. If these changes that correlate with gene expression are to be considered significant, one must say that these other hypomethylations are essentially background noise. We have divided all of the analyzed genes according to our five paradigms and have presented them in this order in Figures 16.2–16.30 and in Table 16.1.

A number of other possible patterns could be imagined, for example, a fully methylated gene in sperm, which undergoes tissue-specific hypomethylation at its 5′ end in the tissue where the gene is expressed. As no examples for such patterns were found, we did not include them in our classification.

While most genes fit into one of the five paradigms, some problems and exceptions should be noted:

Number of Analyzed Sites

The determination of the methylation state of genes is based on the use of CpG methylation-sensitive restriction enzymes, mainly the isoschizomers HpaII and MspI (see the introductory section of this chapter). In some cases, the number of available sites for such an analysis is low. This is the case with *Xenopus* globin genes, where only two HpaII sites in the 3′ end were studied and were found to be undermethylated (Gerber-Huber *et al* 1983). A similar problem arises with the human β-globin gene cluster (Figure 16.6), the rabbit β-like globin genes (Figure 16.19), and the mouse immunoglobin heavy chain genes (Figures 16.9–16.14).

Contamination with Other Cell Populations

In some cases, the active cells are only a subpopulation of the analyzed tissue, and the purification of active cells is technically difficult. In such cases, the methylation maps do not represent the real situation in the active cells. In some studies, measures have been taken to purify the active cells, as in the case of rat α-fetoprotein gene (Vedel *et al.* 1983). It is probable that the problem of cell population may have impeded the methylation analysis of rabbit β-like globin genes, in which hematopoietic cells in fetal liver could not be purified from the fetal liver tissue (Shen and Maniatis, 1980).

The rat pepck gene (Figure 16.5, sites f, g, k, l, p, and r) is partially methylated in fetal liver. It is feasible that contamination with embryonic hematopoietic cells masks a more massive hypomethylation in fetal liver cells or vice versa. The same problem may have complicated the analysis of human growth hormone gene in the anterior pituitary gland (Figure 16.25).

Chicken δ-crystallin Gene (1)

This gene was assigned to group II, although most of the hypomethylation in this gene does not seem to correlate with its active state. It recently has been shown that a specific hypomethylation accompanies the activation of δ-crystallin in chicken lens tissue during embryogenesis. A similar change has not been found in nonexpressing tissue (Grainger *et al.* 1983).

Table 16.1 Methylation Pattern Paradigms

Paradigm	Species	Gene	Figure/Reference
I	Rat	γ-casein	16.2
		Insulin I	16.3
		Albumin	16.4
		Pepck[a]	16.5
	Human	Globin	16.6
	Chicken	α-globin gene cluster	16.7
		β-globin gene cluster (ρ ϵ)	16.8
	Xenopus	Globin	Gerber-Huber *et al.* (1983)
	Mouse	Immunoglobulin heavy chain, C_γ, δ, μ	16.9–16.11
		IgK heavy chain region	16.12–16.14
II	Chicken	δ-crystallin I	16.15
		β-like globin gene cluster (β^H, β)	16.16
		Ovalbumin	16.17
	Rat	Insulin II	16.18
	Rabbit	β-like globin gene cluster	16.19
III	Chinese hamster	aprt[b]	16.20
	Mouse	dhfr[c]	Stein *et al.* (1983)
	Chicken	α-2 (I) collagen	16.21
IV	Chicken	Vitellogenin	16.22
	Xenopus	Vitellogenin A-1	16.23
		Vitellogenin A-2	16.24
V	Human	Growth hormone	16.25
		Chorionic somatomammotropin	16.26
	Xenopus	Albumin	16.27
		rRNA	Bird *et al.* (1981)
	Rat	α-fetoprotein	16.28
		α-actin	16.29
		Myosin	M Shani (personal communication)
	Chicken	δ-crystallin II	16.30

[a]Phosphoenolpyruvate carboxy kinase.

[b]Adenine phosphoribosyltransferase.

[c]Dihydrofolate reductase.

Human Growth Hormone Gene

This gene belongs to group V, although site a is partially undermethylated in the anterior pituitary, where it is active (Figure 16.25). Contamination with other cell populations may explain this slight undermethylation; however, because the same site is fully unmethylated in hydatiform mole and placenta (both are nonexpressing tissues), it is not feasible that this site plays a role in activation of this gene.

Xenopus Albumin Gene

This gene was placed in group V, although one site (Figure 16.27a) in its 5′ end was slightly unmethylated in hepatocytes, where this gene is expressed. It is improbable that this site, which is unmethylated in 10% of *Xenopus* liver cells, plays a role in the activation of this gene, since it is unlikely that only 10% of liver cells express albumin (Gerber-Huber *et al.* 1983). It should be noted that many other sites exhibit partial undermethylation in both active and nonactive tissues (sites d-j).

Hemimethylation

The restriction enzyme analysis of methylation of CpG sites does not enable us to identify hemimethylated sites (see the section on "Discussion"). Rat α-actin and myosin genes do not demonstrate methylation changes after induction of gene activity (Figure 16.29; M Shanni, personal communication). As this activation results in cessation of DNA replication after one round of replication, hemimethylated sites may have been created. Hemimethylation may be sufficient to activate this gene.

Xenopus Ribosomal RNA Genes

These genes were characterized as the group V type, although most of the CpG sites in the nontranscribed sequences undergo a slight (15%) undermethylation by the time they are activated (Bird *et al.* 1981). However, a recent experiment (Macleod and Bird, 1983) has demonstrated that even fully methylated genes may be transcribed in *Xenopus* oocytes.

C_k Region

The C_k region of the mouse immunoglobulin heavy chain genes of two plasmacytomas (pC 8701 and pC 7183) that undergo rearrangement, but do not transcribe, are undermethylated (Figure 16.14). These plasmacytomas are the only case in which undermethylation of C_k genes does not correlate with their expression. It is feasible that chromatin changes associated with the rearrangement process can cause hypomethylation, even when the rearranged structure does not express.

Chicken Vitellogenin Gene

In this gene, one site undergoes partial undermethylation in response to hormone induction (Figure 16.22). This undermethylation does not seem to play a role in the expression of this gene, because hypersensitive sites can be established on genes that have not been demethylated at this MspI/HpaII site (Burch and Weintraub, 1983) (see the section on "Discussion"). Moreover, a similar undermethylation may be observed in the vitellogenin gene of estrogen-treated hen oviducts, which is not active (Wilks *et al.* 1982). It is feasible that this methylation change results from chromatin changes that accompany hormone induction.

Sperm DNA

In most of the examined genes, a methylation analysis of genes in sperm DNA was not performed. In the paradigms presented in Figure 16.31, we assume that the genes in classes I, II, IV, and V are fully methylated in sperm DNA.

Discussion

To understand the significance of the data we present here, it is important that the limitations of the approach are understood. The Southern blot analysis, employed to map methylation patterns suffers from two types of problems. The first type is of a technical nature, while the second type is more general. To obtain interpretable data, appropriate restriction enzymes and probes must be chosen for the analysis. If the DNA sample is cut with only one restriction enzyme, the presence of a large number of sites can make it almost impossible to identify the various bands that result from methylation of a subset of the sites. This proper identification of the bands is necessary to accurately analyze the methylation frequencies. By codigesting the DNA with a second or, possibly, even a third enzyme that is not sensitive to methylation, one often can better define and distinguish the possible bands. If this is done with the appropriate small probe, the analysis becomes much more manageable and accurate. Analysis of the site closest to the probe is always more accurate than analysis of those further away, because it is only possible to look at these more distant sites when the closer sites are methylated.

With some enzymes, the problem of polymorphisms occurring within restriction sites can be significant. While the cloned DNA that was used to construct a restriction map may have a particular site, this site may have been lost in the analyzed DNA and would appear to be methylated. Methylation at HhaI and AvaI sites must always be viewed in this light. Conversely, the isoschizomer pair of HpaII/MspI is a boon to those who do these analyses, because MspI generally prevents this ambiguity. The one exception in which ambiguity can still exist, even when cutting with MspI, is when the recognition site is in a particular context and the internal C is methylated (GGC^{m}CGG). It recently has been shown that this site is refractory to low concentrations of MspI (Keshet and Cedar, 1983; Busslinger *et al.* 1983).

As shown by even a cursory glance over the maps presented in this chapter, a large number of different tissues and cell lines have been used in these studies. It is very important to consider the particular traits of the cell type being analyzed when looking at methylation patterns and correlation of expression. The techniques used to isolate these various tissues or lines differ in their ability to yield pure populations. Of course, contamination of a population of cells with other cells that are not synthesizing the gene being assayed could complicate the analysis considerably. Certain tissues and cell lines present other sorts of considerations. Extraembryonic tissues in mouse, rabbit, and humans (see Chapter 7; Razin *et al.* 1984; Manes and Menzel, 1981; van der Ploeg and Flavell, 1980), as well as in a number of primary tumors (Feinberg and Vogelstein, 1983), have been shown to be significantly undermethylated on a general level. Therefore, in such tissues, one must remember that what we have called the background noise is greater. The significance of this massive demethylation is unclear, but one must proceed with caution when searching for a correlation between methylation patterns and gene expression in these cells. For example, the genes coding for DHFR and β-globin in the mouse are significantly undermethylated in the yolk sac, compared to either sperm or liver, although there is no change in their expression (DHFR is expressed in both, β-globin is not expressed in either) (Razin *et al.* 1984).

As clearly pointed out in the above discussion and from the fact that there are sites whose degree of demethylation is invariant from tissue to tissue (see above), undermethylation is, in some sense, a relative term. To say that a gene is hypomethylated in a particular tissue is ambiguous, since this can mean that this gene either is not fully methylated in this tissue or it exists in a more fully methylated state in another tissue. This more restricted use of the concept of undermethylation is clearly of greater importance to one attempting to correlate it with gene expression. It is for this reason that the methylation pattern in sperm of a particular gene is of interest, because it provides a starting point for examining hypomethylation.

The general impression conveyed by the maps in this chapter is that genes are less methylated in differentiated tissues than in sperm. However examples of *de novo* methylation do exist. It has been shown that satellite DNA in calf thymus is more methylated than in any other tissue, including sperm (Sturm and Taylor, 1981; Pages and Roizes, 1982). The Chinese hamster *aprt* gene presents an apparent case of *de novo* methylation as well (Stein *et al.* 1983). The Xho site in the middle of the gene is about 20% methylated in sperm, and the pattern is similar in other tissues. This implies that 20% of the sperm population is methylated at these sites. Unless the oocyte population is 100% methylated at these sites (and barring some unusual selection of sperm), at least some of the zygotes fertilized by this sperm will be unmethylated at these sites on both alleles. *De novo* methylation must be occurring, therefore, to bring the methylation level back up to 20%. To this end, it is important to note that Stein *et al.* performed their analysis of the *aprt* gene with DNAs isolated from individual Chinese hamsters.

The method of analyzing methylation patterns of genes provides the end-result of a dynamic process. Assuming that *de novo* methylation does not occur on a very large scale during the normal process of differentiation, the methylation pattern in a mature tissue reflects the result of all the hypomethylation events that have occurred over its developmental history. The orchestration of these events, however, is lost in such an analysis. Considering this possibility, Grainger *et al.* (1983) have looked at the pattern of methylation of the chicken δ-crystallin genes in the lens right before, during, and after synthesis of the protein begins. They find that one site appears to undergo hypomethylation concurrent with the appearance of the protein, while a set of several other sites becomes demethylated significantly later. Furthermore, Burch and Weintraub (1983) have shown that there is a specific hypomethylation that occurs in the chicken vitellogenin gene when its synthesis is induced by estrogen treatment of immature hen livers. This hypomethylation, however, is seen 72 hours (two cell divisions) posthormone treatment, while the mRNA is already detected after 16 hours (before the first division). Synthesis of vitellogenin mRNA, therefore, must be occurring while this site is still fully methylated. In the pepck gene, a small undermethylation of some sites in the fetal liver precedes expression of the gene. This undermethylation is greatly increased in the adult liver, where the gene is expressed (Figure 16.5). Clearly, to analyze the order of events, one must take into account the dynamism of the system.

A second type of problem is posed by the dynamic nature of the system. A hypomethylation event that has been involved in the expression of a gene that has since been silenced could quite possibly remain unmethylated. This phenomenon could explain why the chicken collagen α2 (1) gene is unmethylated in nonexpressing tissues (Figure 16.21). Collagen synthesis may occur in a large variety of cells in early stages of development and is then turned off at a later stage (Wartiovaara *et al.* 1980).

It is important to remember that all of the maps of methylation patterns presented in this chapter are, in fact, just sets of restriction sites of a very select group of enzymes. HpaII sites, for example, represent only about 6% of all the CpG dinucleotides present in the genome. If only one or a few sites were important for activating a gene, they could be quite easily missed by this method (i.e., not be within a recognition site for one of these sets of restriction enzymes). In general, the number of HpaII sites (or, for that matter, HhaI or AvaI sites) within the immediate vicinity of the gene varies a great deal from gene to gene, sometimes making a statistical comparison difficult. In addition, when comparing methylation patterns with expression, it is not always clear what is within that immediate vicinity of the gene. The picture is further complicated by the fact that the enzymes that are sensitive to 5-meCyt are sensitive to hemimethylation (i.e., methylation on only one strand) as well (Gruenbaum *et al.* 1982). In the case of rat α-actin and myosin, the cells in the *in vitro* system undergo only one division before they begin to synthesize actin or myosin. In as much as hypomethylation is thought to be an inhibition of methylation after DNA replication, and because only one strand (the new one) is assumed to become methylated out of each daughter molecule, two rounds of replication is the minimal length of time to see the appearance of HpaII sensitivity. In other words, hemimethylation may be sufficient to allow activation of a gene, but another round of replication (without methylation) is required before the demethylation is detected. For example, in the actin/myosin systems, hemimethylation at certain sites may be connected to the onset of transcription, but they are not detectable because of the lack of replication.

Conclusions

From methylation patterns of the 30 genes analyzed here, it became clear that, in general, a process of hypomethylation occurs as a cell develops from a germline to a fully differentiated state. Thus, sperm is always the most highly methylated tissue in all of the genes examined. We find no site that is hypomethylated in sperm as compared to DNA of the somatic tissue.

For many of the genes described here, hypomethylation occurs in both expressing and nonexpressing tissues. When one looks at these hypomethylations as independent events, one finds some sites whose level of methylation depends on the tissue being examined and others that are methylated to the same extent in every tissue. When one attempts to correlate these hypomethylations with expression, one finds that approximately two-thirds of the genes examined here show one of two basic types of correlation. In a large majority of cases, the correlation with expression is exhibited by undermethylation of all or almost all of the sites throughout the gene (Figure 16.31, I). In other genes, however, only a specific site or subset of sites demonstrate a correlation with gene expression (Figure 16.31, II and III). The remaining one-third of the cases shows either no obvious correlation (Figure 16.31, V) or no hypomethylations at all (Figure 16.31, IV). However, it is important to remember that because of the limitations of the analysis mentioned above, we cannot be certain that all of the genes of type V show no correlation between their methylation pattern and expression. When the hypomethylation is over the entire gene, as in type I, the correlation is more certain.

While tissue-specific hypomethylations of individual sites exist that are not correlated with gene expression, we find no case where a gene is totally unmethylated and not expressed. Conversely, several genes are expressed while being totally methylated. Although the interrelationship between DNA methylation and gene expression is not unequivocable, the paradigms presented here may represent different programs of involvement of methylation in gene control.

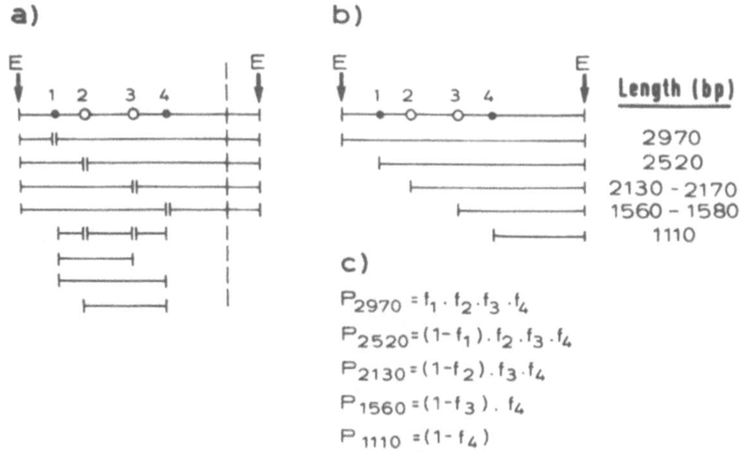

Figure 16.1 Method for analyzing levels of methylation from Southern blots. To describe the method of analysis, we have detailed how we analyzed the human chorionic somatomammotropic gene (see Figure 16.26) (Hjelle *et al.* 1982). (a) shows all the possible fragments from a double digestion with EcoRI (E) and HpaII (sites 1–4). The top line is a map of the HpaII- and EcoRI-relevant restriction sites. If all of the HpaII sites are methylated, a single fragment is produced, which is shown on the top line. If site 1 is unmethylated, while sites 2–4 are methylated, two bands are generated whose sizes are diagrammed on the second line. If site 2 is the only unmethylated site, one sees the two fragments shown on the third line, and so on. Various combinations of demethylations are shown on lines 6–9. In general, in a double digest with n-methyl-sensitive restriction sites, which are flanked on either side by sites that are always cut, the number of possible fragments is $\binom{n+2}{2} = (n + 2)!/(2n!)$. To calculate the frequency of methylation at each site (f), one has to set up a series of equations that give the probabilities (P) of generating a particular fragment in terms of these frequencies. These probabilities are determined by normalizing the densities (as determined by a scanning densitometer) of the bands seen on the autoradiogram of the hybridized Southern blot. However, because of a problem with redundancy, one must select a subset of the bands to normalize. The fragments that "cross" the dotted line in (a) can be normalized together without the problem of redundancy. (While the dotted line could have been placed anywhere along the gene, this position yields the easiest subgroup to analyze.) (b) shows the bands selected for analysis, along with their lengths. (c) shows how the probability of each band is calculated from the frequencies of methylation. Knowing these probabilities (the normalized densities), one can calculate the methylation frequencies.

Figures 16.2–16.30 Methylation patterns of eukaryotic genes in expressing and nonexpressing tissues. The methylation patterns of 29 genes in a number of different tissues were mapped. The published data, on which our mapping is based, is referenced in the legend to each figure. In some cases, the percentage methylation of each methylatable site was determined by the authors of the original paper by using specific small probes, double digestions, and quantitation of the resulting fragments via scanning. In the remaining cases, when only partial analysis of the data was available, we analyzed the published autoradiograms of the Southern blot and computed the level of methylation of every site from the strength of the bands on the autoradiograms (see legend to Figure 16.1).

The method used for the analysis of each gene in each respective figure is listed in the corresponding legend. All of the results were mapped in a uniform fashion. The first line in each map is a physical map of the gene with the exons symbolized by shaded boxes. Noncoding regions in the resulting mRNAs are symbolized by empty boxes. When no information was available on the location of introns, the whole gene is symbolized by ▨. Each succeding line represents the methylation patterns in the tissues or lines that were tested. The various methylation sites are as follows: HpaII, ●; HhaI, □; AvaI, △ ; When a number of adjacent sites were present and the independent analysis of the site was difficult, we regarded them as one site and symbolized them as an open circle, O. Sites that were not analyzed for various reasons are denoted as an X.

The degree of methylation at every restriction site tested is represented by the length of a line descending from the site. A full line indicates 100% hypomethylation.* The relative level of expression is indicated when the level of the mRNA was quantitated in the corresponding papers. In all other cases a + symbolizes expressing tissues.

The methylated sites will be referred to according to alphabetical order from the 5′ end to the 3′ end. The first analyzed site at the 5′ end will be site a; the second, site b, and so on.

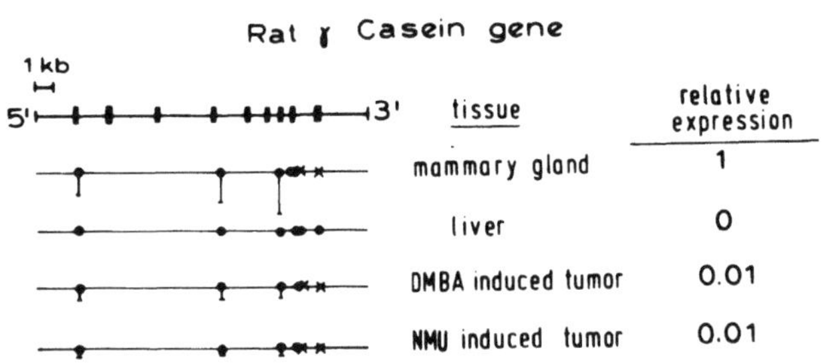

Figure 16.2 Rat γ-casein gene. Determination of the relative level of expression of γ-casein gene was based on Supowit and Rosen (1982). Our analysis was based on visual examination of the autoradiograms. (From Johnson *et al.* 1983.)

* Full line = 100% hypomethylation

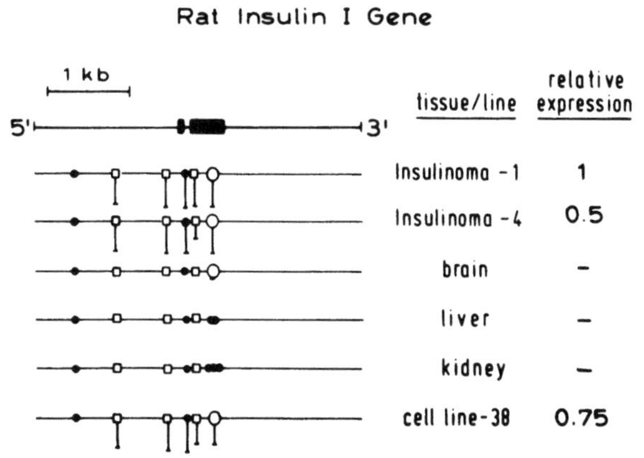

Figure 16.3 Rat insulin I gene. The authors performed a detailed analysis (double digestion, specific probes, and scanning of restriction fragments). The level of methylation of all methylated sites indicated in this map are according to their determinations. (From Cate *et al.* 1983.)

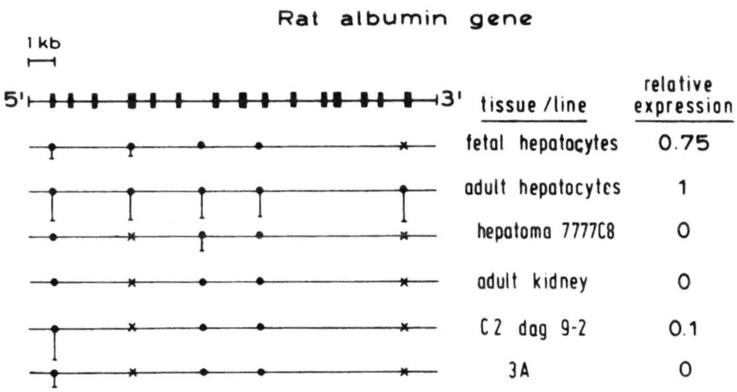

Figure 16.4 Rat albumin gene. C_2 dag 9-2 and 3A are hepatoma cell lines. The relative expression in the first four tissues was calculated on the basis of mRNA levels, while the level of secreted albumin was measured in the other two hepatoma cell lines. It should be noted that only a subset of all the HpaII sites has been examined. (From Vedel *et al.* 1983, first four maps; Marie Odile *et al.* 1982, last two maps.)

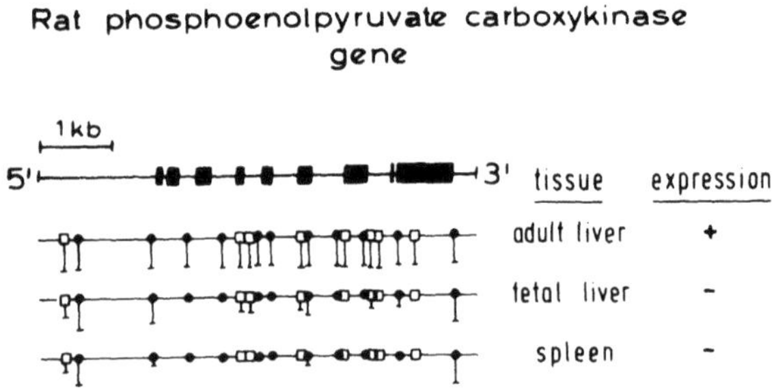

Figure 16.5 Rat phosphoenolpyruvate carboxykinase gene. Methylation levels were determined by visual examination of the autoradiograms. (From Benvenisty *et al.* 1984.)

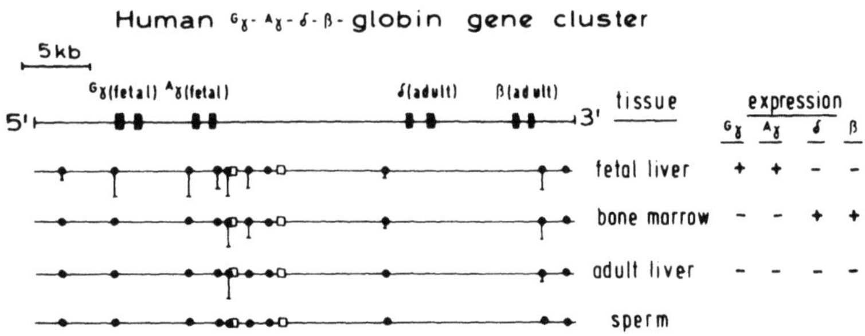

Figure 16.6 Human globin gene cluster. The authors performed double digestions of the DNA and determined the level of methylation of the various sites by a statistical analysis (see Figure 16.1). (From van der Ploeg and Flavell, 1980; Chapter 11.)

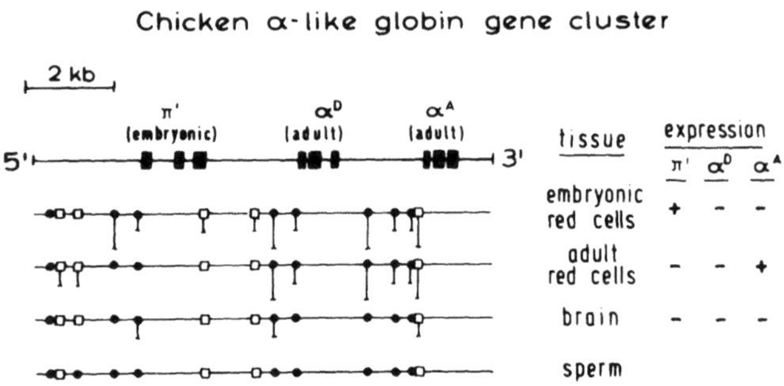

Figure 16.7 Chicken α-like globin gene cluster. (From Haigh *et al.* 1982; Weintraub *et al.* 1981.)

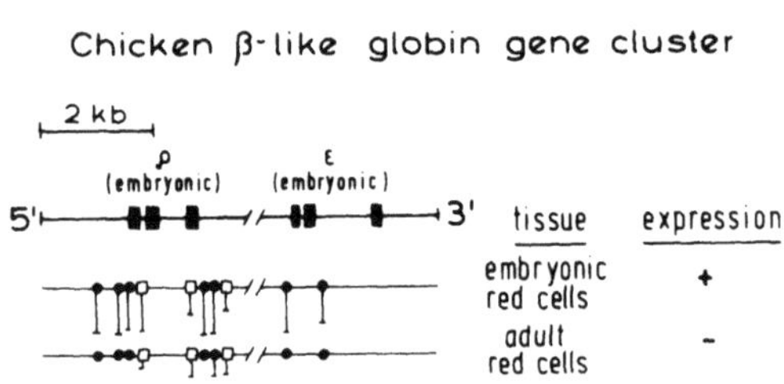

Figure 16.8 Chicken β-like globin gene cluster. (From McGhee and Ginder, 1979; Ginder *et al.* 1983.)

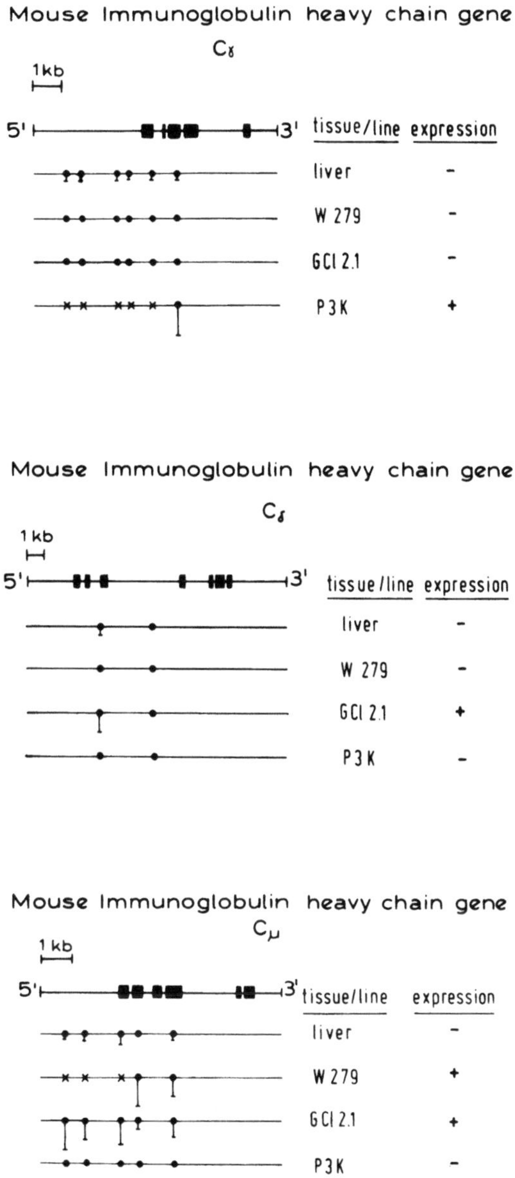

Figure 16.9–16.11 Mouse immunoglobulin heavy chain genes Cγ, C $_{\delta}$, Cμ. The percentage methylation at each site was derived from the relative intensities of the bands in a double digest with HpaII and EcoRI and with hybridization to one EcoRI-MspI fragment. This method does not allow detection of sites upstream to an unmethylated HpaII site; therefore, many critical sites go undetected. W-279 is a cell line that originated in an inbred strain derived from a (Balb/c × N2B)F, mouse. GCl 2.1 is a hybridoma between mouse lipopolysaccharide-stimulated spleen cells and a Syrian hamster B-cell lymphoma. P3K is a myeloma cell line. (From Rogers and Wall, 1981.)

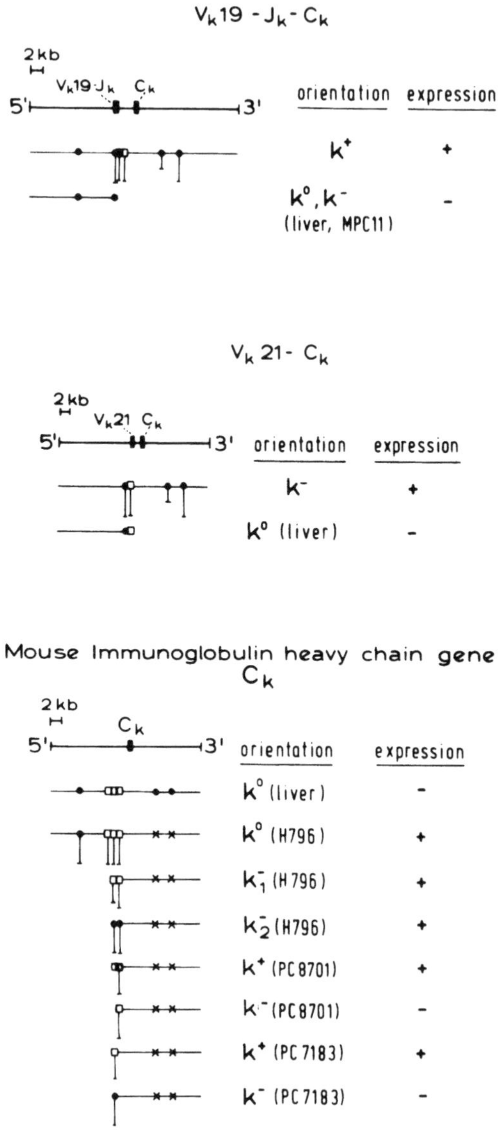

Figures 16.12–16.14 Mouse immunoglobulin heavy chain genes V_k19-J_k-C_k, V_k21-C_k and C_k. All the maps in Figure 16.14 are orientated to the C_k gene, even when rearrangements have occurred (lines 3–8). K^- (PC 8701) and k^- (PC 7183) alleles are not transcribed because of a lack of initiation signals from the V_k gene. However, they have undergone rearrangement. H 796 is a hybridoma of a fetal liver cell and a myeloma line (A_5 8653), which has lost its k^+ allele. PC 8701 and PC 7183 are plasmacytomas. V_k 19-J_k-C_k and V_k 21-C_k: The k^0k^- methylation maps represent either the liver gene or the chromosome where the k^- rearrangement has taken place. In the second case, the V_k19 gene is upstream from the splice; therefore, it remains in a germ-line orientation. (From Mather and Perry, 1983.)

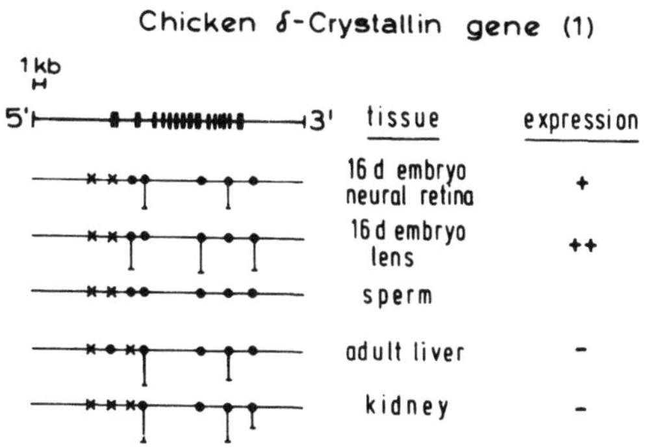

Figure 16.15 Chicken δ-crystallin gene (1). The authors have not performed a detailed analysis of the percentage methylation of specific sites (using specific probes and double digests). Our methylation map is based on a visual examination of the raw data. Because we lack more detailed data, some modifications to the picture we have drawn are possible. (From Jones *et al.* 1981; Bower *et al.* 1983.)

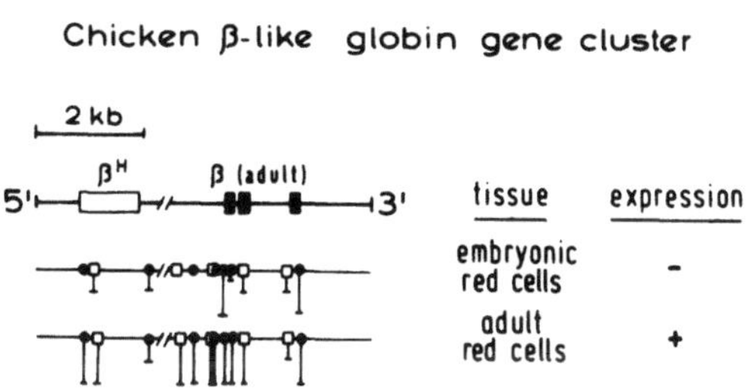

Figure 16.16 Chicken β-like globin gene cluster (β_H, β). (See legend to Figure 16.8.)

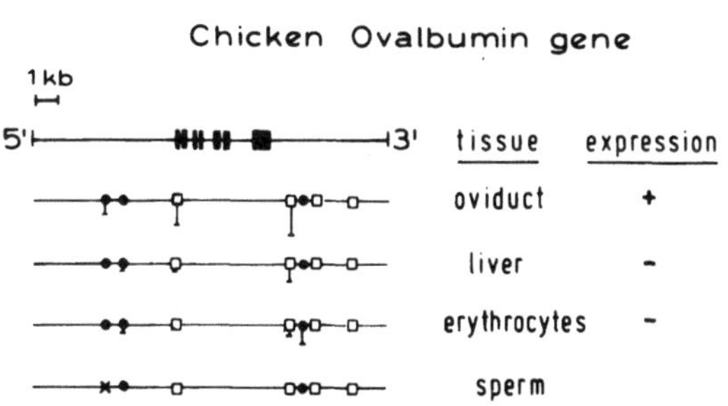

Figure 16.17 Chicken ovalbumin gene. A detailed quantitation of the percentage methylation of the sites was performed (double digests and scanning). Correction was made for heterozygosity resulting from polymorphic sites. (From Mandel and Chambon, 1979.)

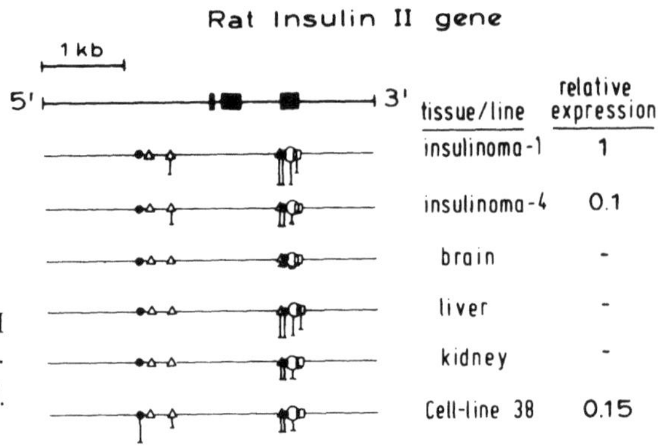

Figure 16.18 Rat insulin II gene. See the legend to Figure 16.3. (From Cate *et al.* 1983.)

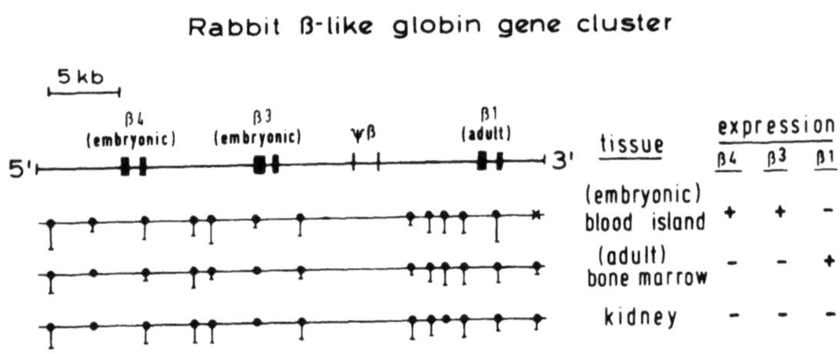

Figure 16.19 Rabbit β-like globin gene cluster. A detailed statistical analysis of the percentage methylation of the various sites has been performed (double digests and scanning). (From Shen and Maniatis, 1980.)

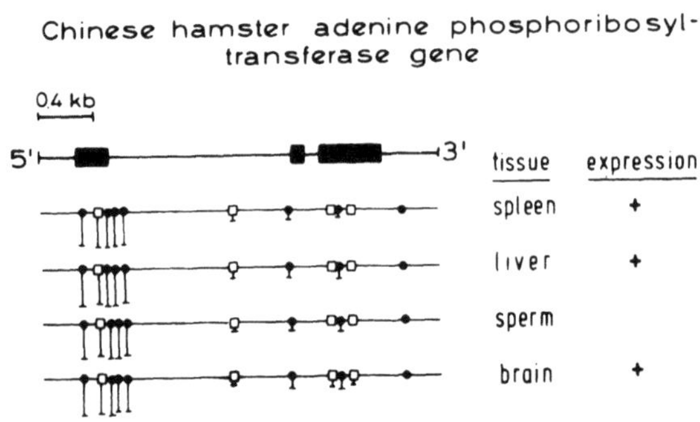

Figure 16.20 Chinese hamster adenine phosphoribosyl transferase gene *(aprt)*. *Aprt* is a key enzyme in the biosynthesis of nucleotides; therefore, it is expected to be synthesized in all cells. A detailed analysis of the percentage of methylation of all sites has been performed (double digest and regional probes). Quantitation of band intensities in the autoradiogram was by visual examination. (From Stein *et al.* 1983.)

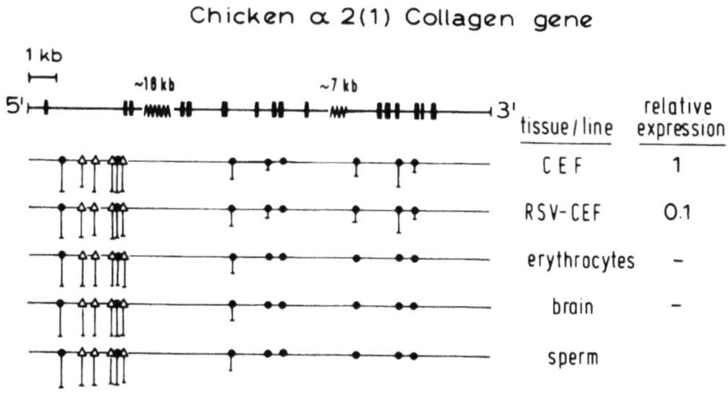

Figure 16.21 Chicken $\alpha 2(1)$ collagen gene. The percentage methylation was estimated by visual examination of the raw data. The authors have performed a detailed analysis by using regional probes and double digests. CEF, cultured chick embryo fibroblasts. RSV-CEF, CEF transformed with Rous sarcoma virus (RSV). In this system, the synthesis of type I collagen is severely inhibited. The relative expression of the various lines was determined by measuring mRNA levels. (From McKeon *et al.* 1982.)

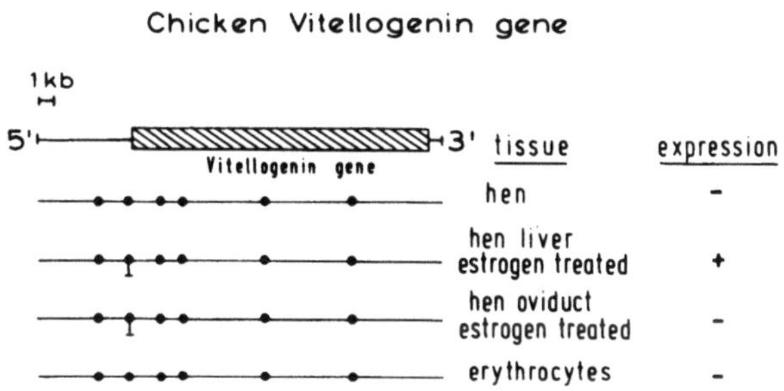

Figure 16.22 Chicken vitellogenin gene. The percentage methylation was established by visual examination. A double digest was performed. Vitellogenin mRNA was detected 16 hours after hormone treatment. The change in methylation appears to commence at 72 hours posttreatment. The lag may indicate a requirement for dividing twice before the loss of methyl groups can be assayed. (From Wilks *et al.* 1982; Burch and Weintraub 1983.)

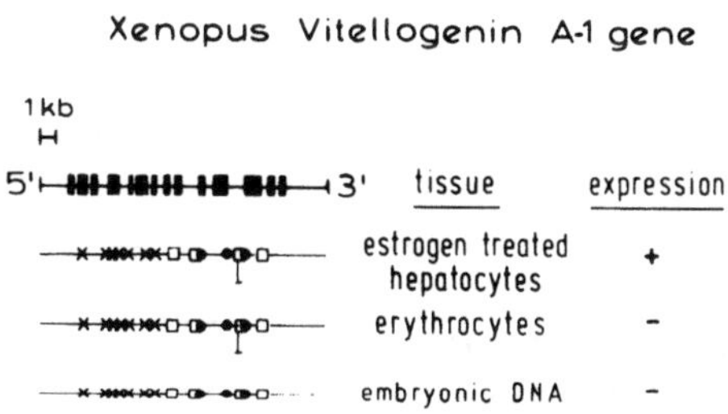

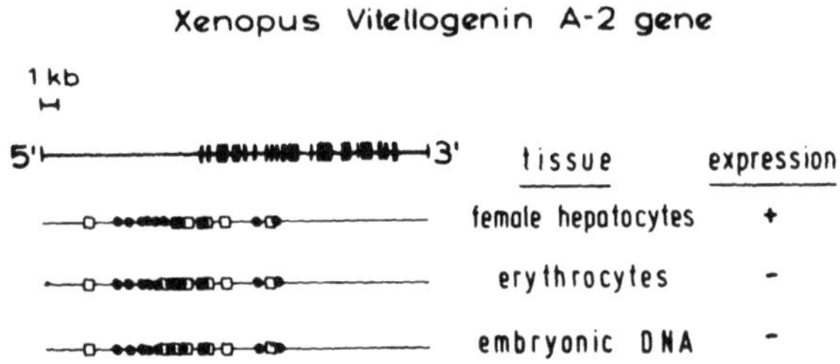

Figures 16.23–16.24 *Xenopus* vitellogenin A-1 and A-2. The percentage methylation was estimated by visual examination of the autoradiograms. The authors used regional probes and performed double digests. In gene A-1, the 5′ region was not analyzed because of a lack of an adequate probe. These sites are symbolized in our map by an X. (From Gerber-Huber *et al.* 1983.)

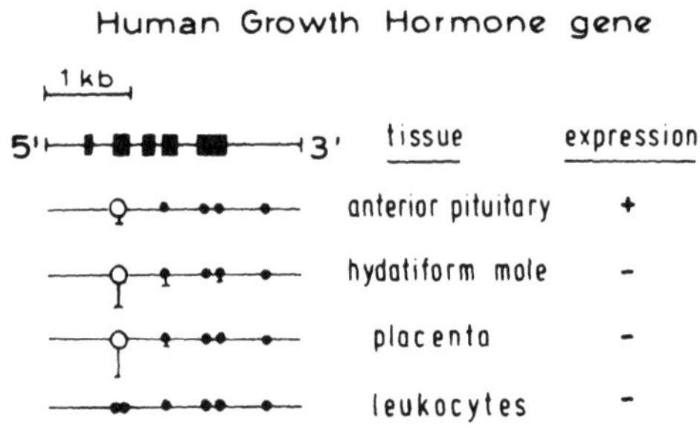

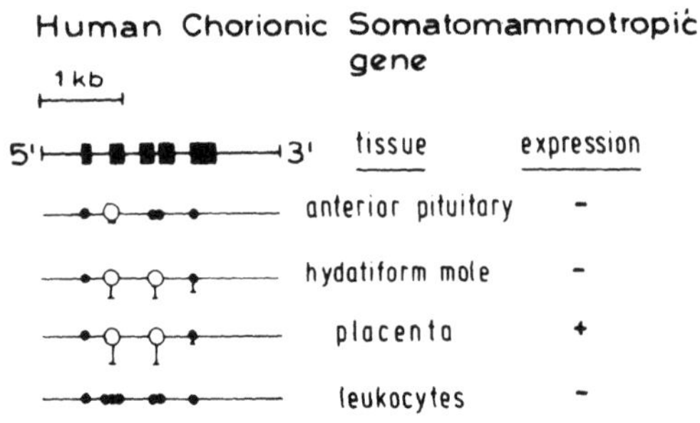

Figures 16.25–16.26 Human growth hormone and human chorionic somatomammotropic genes. Human growth hormone and human chorionic somatomammotropic genes are 85% homologous judged by the amino acid sequence. Therefore, probes cross-hybridize extensively. Thus, quantitation is approximate. The authors performed double digests and estimated methylation level by visual examination. (From Hjelle *et al.* 1982.)

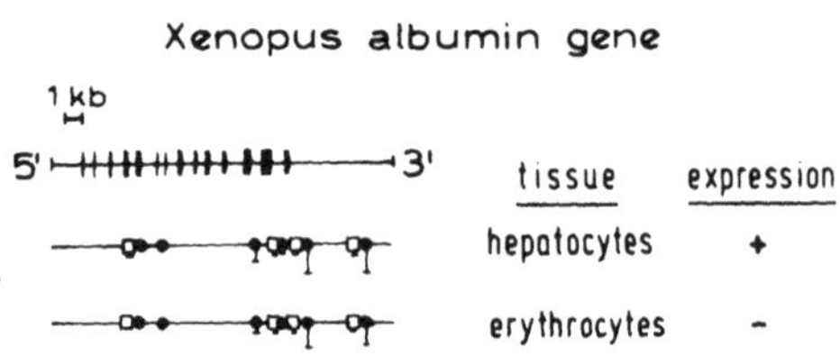

Figure 16.27 *Xenopus* albumin gene. (From Gerber-Huber *et al.* 1983.)

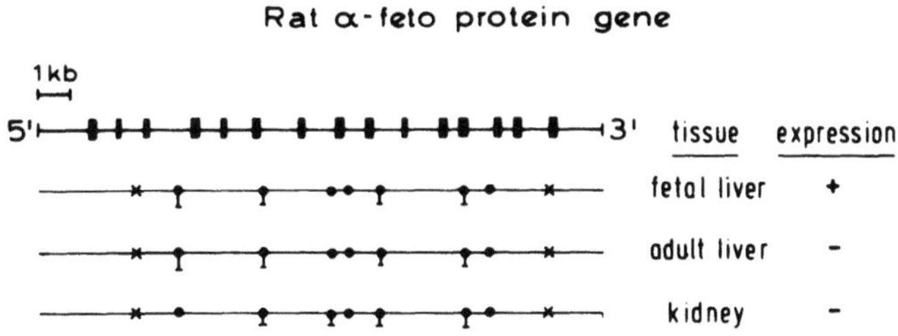

Figure 16.28 Rat α-fetoprotein. Methylation levels were determined by visual examination. (From Vedel *et al.* 1983; Kunnath and Locker, 1983.)

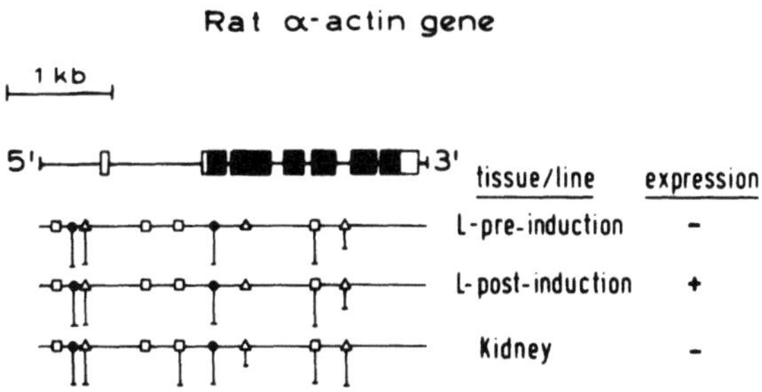

Figure 16.29 Rat α-actin gene. Methylation levels were determined by visual examination of the autoradiograms. (From M Shani, personal communication.)

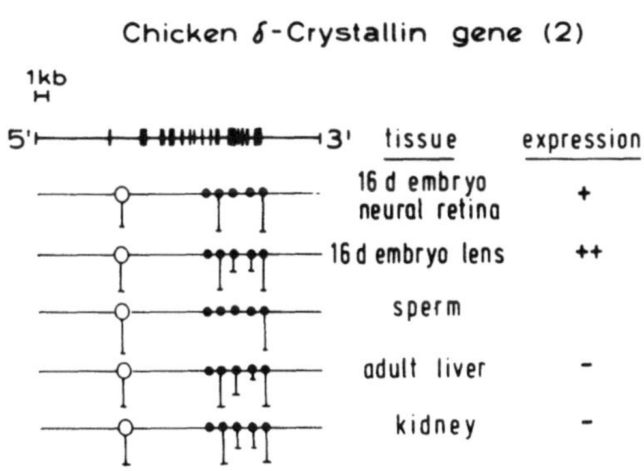

Figure 16.30 Chicken δ-crystallin gene (2). (See legend to Figure 16.15.) (From Jones *et al.* 1981; Bower *et al.* 1983.)

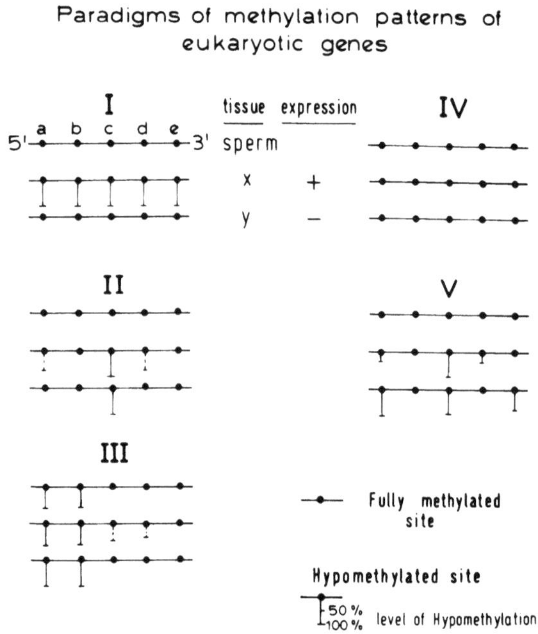

Figure 16.31 Paradigms of methylation patterns of eukaryotic genes. The methylation patterns of eukaryotic genes that were analyzed in this chapter may be divided into the five paradigmatic groups that are described in this figure. Every methylation pattern is symbolized by three lines. The first line symbolizes the methylation patterns of the genes that fall under this paradigm in the sperm. The second horizontal line represents the methylation pattern of these genes in tissues that express them, while the third line represents the state of methylation of these genes in tissues where they are silent. The methylatable sites are symbolized by a closed circle. The degree of hypomethylation at every site is represented by the length of a line descending from the site. A full line indicates 100% hypomethylation. The absence of a line indicates a fully methylated site.

Acknowledgments

This chapter is a result of the inspiration and encouragement of Drs. Howard Cedar and Aharon Razin. We gratefully acknowledge their help in every aspect of its preparation. We also wish to thank Caroline Gopin, whose patience was surpassed only by her perseverance in the preparation of this manuscript. Finally, we also thank Drs. Moshe Shani and Nissim Benvenisty for providing the restriction maps and methylation data of the rat actin and myosin and pepck genes before publication.

References

Benvenisty N, Mencher D, Meyuhas O, Razin A, and Reshef L: Methylation of rat cytosolic phosphoenolpyruvate carboxykinase gene: Patterns associated with tissue specificity and development (submitted for publication, 1984).

Bower DJ, Errington LH, Cooper DN, Morris, S, Clayton RM: Chicken δ-crystallin gene expression and methylation in several non-lens tissues. *Nucl Acids Res* 1983; 11:2513–2527.

Burch JBE, Weintraub H: Temporal order of chromatin structural changes associated with activation of the major chicken vitellogenin gene. *Cell* 1983; 33:65–76.

Busslinger M, de Boer E, Wright S, Grosveld FG, and Flavell RA: The sequence GGC^{m}CGG is resistant to MspI cleavage. *Nucl Acids Res* 1983; 11:3559–3570.

Cate RL, Chick W, Gilbert W: Comparison of the methylation patterns of the two rat insulin genes. *J Biol Chem* 1983; 258:6645–6652.

Feinberg AP, Vogelstein B: Hypomethylation distinguishes genes of some human cancers from their normal counterparts. *Nature* 1983; 301:89–91.

Gerber-Huber S, May FEB, Westley BR, Felber BK, Hosbach HA, Andres A-C, Ryffel GV: In contrast to other Xenopus genes the estrogen-inducible vitellogenin genes are expressed when totally methylated. *Cell* 1983; 33:43–51.

Ginder DD, Whitters M, Kelley K, Chase RA: in Starmatoyannopoulos G, Neinhuis AW (eds): *Hemoglobin Switching.* New York, Alan R Liss, Inc, 1983.

Grainger RM, Hazard-Leonards RM, Samaha F, Hougan LM, Lesk MR, Thomsen GH: Is hypomethylation linked to activation of δ-crystallin genes during lens development. *Nature* 1983; 306:88–91.

Gruenbaum Y, Cedar H, Razin A: Restriction enzyme digestion of hemimethylated DNA. *Nucl Acids Res* 1981; 11:2509–2515.

Haigh LS, Owens BB, Hellwell S, Ingram VM: DNA methylation in chicken α-globin gene expression. *Proc Natl Acad Sci USA* 1982; 79:5332–5336.

Hjelle BL, Phillips JA III, Seeburg PH: Relative levels of methylation in human growth hormone and chorionic somatomammotropin genes in expressing and nonexpressing tissues. *Nucl Acids Res* 1982; 10:3459–3474.

Johnson LM, Levy J, Supowit SC, Yu-Lee LY, Rosen JM: Tissue and cell specific casein gene expression. *J Biol Chem* 1983; 258:10805–10811.

Jones RE, De Feo D, Piatgorsky J: Transcription and site-specific hypomethylation of the δ-crystallin genes in the embryonic chicken lens. *J Biol Chem* 1981; 256:8172–8176.

Keshet I, Cedar H: Effect of CpG methylation on MspI. *Nucl Acids Res* 1983; 11:3571–3580.

Kunnath L, Locker J: Developmental changes in the methylation of the rat albumin and α-fetoprotein genes. *EMBO J* 1983; 2:317–324.

Macleod D, Bird A: Transcription in oocytes of highly methylated rDNA from Xenopus laevis sperm. *Nature* 1983; 306:200–203.

Mandel JL, Chambon P: DNA-methylation: Organ specific variations in the methylation pattern within and around ovalbumin and other chicken genes. *Nucl Acids Res* 1979; 7:2081–2090.

Manes C, Menzel P: Demethylation of CpG sites in DNA of early rabbit trophoblast. *Nature* 1981; 293:589–590.

Mather EL, Perry RP: Methylation status and DNase I sensitivity of immunoglobulin genes: Changes associated with rearrangement. *Proc Natl Acad Sci USA* 1983; 80:4689–4693.

McGhee JD, Ginder GD: Specific DNA methylation sites in the vicinity of the chicken β-globin genes. *Nature* 1979; 280:419–420.

McKeon C, Ohkubo H, Pastan I, de Crombrugge B: Unusual methylation pattern of the $\alpha 2$(I) collagen gene. *Cell* 1982; 29:203–210.

Pages M, Roizes G: Tissue specificity and organization of CpG methylation in calf satellite DNA I. *Nucl Acids Res* 1982; 10:565–576.

Razin A, Webb C, Szyf M, Yisraeli J, Rosenthal A, Naveh-Many T, Sciaky-Gallili N, and Cedar H: Variations in DNA methylation during mouse cell differentiation *in vivo* and *in vitro. Proc Natl Acad Sci USA,* 1984; 81:2275–2279.

Rogers J, Wall R: Immunoglobulin heavy chain genes: Demethylation accompanies class switching. *Proc Natl Acad Sci USA* 1981; 78:7497–7501.

Satz M Leonardo, Singer Dinah S: Differential expression of porcine major histocompatibility DNA sequences introduced into mouse L-cells. *Molecular and Cellular Biology* 1983; 3:2006–2016.

Shen CKJ, Maniatis T: Tissue specific DNA methylation in a cluster of rabbit β-like globin genes. *Proc Natl Acad Sci USA* 1980; 77:6634–6638.

Stein R, Sciaky-Gallili N, Razin A, Cedar H: Pattern of methylation of two genes coding for housekeeping functions. *Proc Natl Acad Sci USA* 1983; 80:2422–2426.

Sturm KS, Taylor JH: Distribution of 5-methylcytosine in the DNA of somatic and germline cells from bovine tissues. *Nucl Acids Res* 1981; 9: 4537–4546.

Supowit SC, Rosen JM: Hormonal induction of casein gene expression limited to a small subpopulation of 7.12 dimethlbenz (a) anthracene induced mammary tumor cells. *Cancer Res* 1982; 42:1355–1360.

van der Ploeg LHT, Flavell RA. DNA methylation in the human β-globin locus in erythroid and nonerythroid tissues. *Cell* 1980; 19:947–958.

Vedel M, Gomes-Garcia M, Sala M, Sala-Trepat JM: Changes in methylation pattern of albumin and of α-fetoprotein genes in developing rat liver and neoplasma. *Nucl Acids Res* 1983; 11:4335–4354.

Wartiovaara J, Leivo L, Vaheri A: Matrix glycoproteins in early mouse development and in differentiation of teratocarcinoma cells, in Subtelny S, Wessells NK (eds): *The Cell Surface: Mediator of Developmental Processes.* New York, Academic Press, 1980, pp 305–324.

Weintraub H, Larsen A, Groudine M: α-globin switching during the development of chicken embryos: expression and chromosome structure. *Cell* 1981; 24:333–344.

Wilks AF, Cozens PJ, Mattaj IW, Jost JP: Estrogen induces a demethylation of the 5′ end region of the chicken vitellogenin gene. *Proc Natl Acad Sci USA* 1982; 79:4252–4255.

Index

A

N-Acetoxy-2-acetylaminofluorene, 120, 121
Actin gene, 159
  methylation of
    gene expression and, 357
    hemimethylation and, 360, 364
    map of, 375*f*
    site specificity for, 355
    tissue specificity of, 356
  in type II restriction-modification system inhibition, 64
  in type I restriction modification system inhibition, 21
Adenine phosphoribosyl-transferase gene
  in developmental regulation of transcription, 259
  methylation of
    map of pattern of, 372*f*
    transcription inactivation with, 232, 235
S-Adenosylmethionine, 132
Adenovirus
  DNA methylation studies in, 228–230
  E2a gene of type 2 DNA, 222
    methylation effects on, 226, 232–235, 233*f*, 235–237
    promoter regions for, 232
  type 12 genes of
    map of, 238*f*
    methylation of promoter regions of, 229, 235, 237–238, 239*f*
AdoMet binding
  in type I restriction-modification systems, 18, 19, 23, 78
  in type II restriction-modification systems, 47, 48, 53, 55
  in type III restriction-modification systems, 28, 78
Albumin gene, 157
  methylation of
    map of, 371*f*, 374*f*
    paradigm for, 359*t*, 360
    site specificity for, 355
    transcription inactivation with, 232
Alcohol dehydrogenase genes, 300
Aminopurine, 98–99, 98*t*
Antibodies
  histone, 321
  Z-DNA specific, 283
ATP
  in cytidine/uridine phosphorylation, 167
  in type I restriction-modification mechanism, 18, 21–22, 22*f*, 78
  in type III restriction-modification mechanism, 29, 78
5-Aza-deoxycytidine, 153, 166, 168

5-Aza-deoxycytidine (*cont.*)
in DNA methyltransferase inhibition, 169, 171
structure of, 172*f*
X chromosome inactivation and, 273
5-Azacytidine
carcinogenicity of, 179–180
chemistry of, 166–167, 166*f*
chromosomal damage by, 181–182
decomposition of, 165, 166*f*
in differentiation
mesenchymal cell, 173–175
systems induced by, 175–179, 175*t*
DNA methylation inhibition from, 225
with cytosine bases, 88, 168
eukaryotic, 169–171
mechanism of, 170*f*, 171–173, 173*f*
prokaryotic, 88, 168
gene activation by, 165–182, 225
via DNA methylation inhibition, 177–179, 178*t*
selectivity of, 153
globin gene transcription with, 259–260
mutagenic activity of, 180–181
phosphorylation of, 167–168
retroviral genome methylation with, 202
in RNA methylation inhibition, 169
structure of, 165, 166*f*
X chromosome inactivation and, 272–273

B

B-cell lymphomas, 307
Bacillus subtilis methylases, 43*t*, 46–47, 52, 103
Bacteria
antirestriction systems of, 79–80
DNA methylation in, *see* Prokaryotic DNA methylation
eukaryotic expression of methylases of, 65–66
in evolution of restriction-modification systems, 76–77
type I restriction-modification systems of, 12, 24
type III restriction-modification systems of, 24
Balbiani rings, 297
Bromodeoxyuridine, 113
Butanedione, 47

C

Cancer
5-azacytidine in, 165, 176, 178*t*, 179–180
DNA methylation in, 7–8, 120–121, 225, 227
genes for, sequence alterations and rearrangements in, 307
Carcinoma cells, embryonal
for study of embryonic development, 193
viral expression in
de novo methylation with, 203–205, 204*t*
with differentiation, 196–198, 198*t*
retroviral types for, 194, 195*t*
Casein gene methylation
map of, 366*f*
paradigm for, 359*t*
tissue specificity for, 356
Caulobacter bacteroides methylases, 103
Chloramphenical acetyltransferase gene, 226, 235, 237–238, 239*f*
Chorionic somatommamotropic gene, 373*f*
Chromatin structure
active *vs* bulk, 297
altered conformations of, 297–301
DNAse I and, 299–300
micrococcal nuclease and, 298–299
microscopy of, 297–298
nuclease sensitivity with, 298–299
nuclease-hypersensitive sites in, 300–301
in development, 329–334
for DNA methylation, 155–156, 157
gene expression and, 206, 293–335
formation and maintenance of, 311–329
histones and nucleosomes in, 318–325

hypersensitive sites, 312–315
in vivo correlations of, 301–305
methylation and, 326–329
in prokaryotes *vs* eukaryotes *vs* viruses, 294
propagation of, 330–334
protein composition and organization with, 318–326
sequence alterations and rearrangements, 306–311
topology, 315–318
in globin gene transcription regulation, 261–265
higher-order, 296–297
at nucleosome level, 295–296
solenoid, 296
in superbeads, 296
thick *vs* thin fiber, 296
Chromosomes
Balbiani rings of, 297
lampbrush, 297
X, *see* X chromosome
Collagen gene methylation
gene expression from, 150, 224
map of, 372*f*
in nonexpressing tissues, 363
site specifity for, 355
Crystallin gene methylation
gene expression with, 160
maps of patterns of, 370*f*, 374*f*
paradigm for, 359*t*
protein appearance and, 363
tissue specificity for, 356
Cytidine deaminase, 168

D

Demethylation
in differentiation, 138, 190
in embryogenesis, 150, 191
fron maintenance methylase inhibition, 206
in methylation pattern formation, 137
prokaryotic, 83
substances causing, 153
Determinator proteins in DNA methylation, 7
Differentiation
5-azacytidine effect on, 173–179, 175*t*, 178*t*
de novo methylation in, 208–209, 209*t*
DNA methylation in, 111–112, 138, 149–150
globin, methylation correlation to, 148, 152
X chromosome inactivation and, 276
Differentiator proteins, 7
Dihydro-5-azacytidine, 167
Dihydrofolate reductase genes, 232
DNA gyrase, 315
DNA methylation
antirestriction systems affecting, 30–32, 31*t*, 79
assays for, 151
in bacteria, *see* Prokaryotic DNA methylation
bromodeoxyuridine effect on, 113
in carcinogenesis, 7–8, 120–121, 225, 227
characteristics of, in various organisms, 128–129, 129*t*
chromatin structure and, 155–156, 157, 264–265, 326–329
chromosomal imprinting and, 275
de novo, 4, 32, 79
in adenovirus systems, 229
biological role for, 209–212
in differentiation, 149, 150, 208–209, 209*t*
in *E. coli* K-12, 82, 83
in embryonic development, 157–158, 202–206, 203*t*, 204*t*, 208–209, 209*t*
in inactivation of foreign DNA, 210
methylases for, 3–4, 79, 132–133
in methylation pattern formation, 137, 189
as post-transcriptional event, 157
repair function of, 210
in repression of genes for oogenesis, 210–212
of satellite DNA sequences, 156
timing of, 128
in differentiation, 111–112, 138, 149–150

DNA methylation (*cont.*)
in DNA replication and repair, 111–124
maintenance or alteration of patterns of, model for, 112–113, 112*f*
nascent strands and, 113
at Okazaki fragments, 116–118
patterns of, 135–136
timing of, 114–116
in embryonic development, 191, 192*t*
in eukaryotes, *see* Eukaryotes, DNA methylation in
in adenovirus system, 228–230, 234–235
5-azacytidine and, 225
in vitro studies, 224, 226
inverse correlations, 223–224, 232–234, 233*f*
organ-specificity for, 233
at promoter and 5′ -ends, 230–232, 231*f*, 235–239, 238*f*, 239*f*
as regulatory signals, 226–228
variations in, 156–158
in gene expression, 111, 147–160, 206, 221
in gene inactivation, 221
with cytosine bases, 153
at promoter regions, 228
regulatory signals and, 227–228
of globin genes, *see* Globin genes, methylation of
heritability of, 112, 121–122
histones and, 264
historical perspectives on, 1–2
inhibition of, 7
azacytidine in, 15, 168–173, 170*f*, 177–179, 178*t*
in *E. coli* K-12, 83
magnesium ions in, 132
N-methyl-*N*-nitrosourea in, 120
nicotinamide in, 117–118
in type I restriction-modification systems, 19, 21
in type II restriction-modification systems, 47
in type III restriction-modification systems, 28–29
interconversions between various states of, 127–128, 128*f*
maintenance, 3–4, 4*f*, 79
biological role for, 210
in clonal inheritance of methylation patterns, 139–140, 189, 190
inhibitors of, 153
methylases for, 3–4, 79, 132–133, 134
timing of, 128
mechanism of action of, 154–156
methyl transferases in, *see* Methyl transferases, DNA
nucleosome stability and structure with, 264
patterns of, 6–8, 127–142
for actin gene, 375*f*
for adenine phosphoribosyl transferase gene, 372*f*
in adenovirus systems, 229
for albumin gene, 360, 367*f*, 374*f*
biological functions of, 127
for casein gene, 366*f*
cell population contamination and, 358
of chicken globin gene cluster, 255–256, 256*f*, 257*f*
clonal inheritance of, 136–137, 137*f*, 139–141
for collagen gene, 373*f*
for crystallin gene, chicken, 358, 371*f*, 375*f*
determinator proteins in, 7
differentiated, 137*f*, 138
for fetoprotein gene, 375*f*
formation of, 136–138, 136*f*, 137*f*, 139*t*
gene expression and, 131, 353–364
for globin gene clusters, 368*f*, 371*f*, 372*f*
with hemimethylation, 360
of housekeeping genes, 131, 131*t*
for human chorionic somatomammotropic gene, 374*f*
of human globin gene region, 251–255, 254*f*
for human growth hormone gene, 360, 374*f*

for immunoglobulin gene, 360, 369*f*
for insulin gene, 367*f*, 371*f*
maps of, for eukaryotic genes, 366–374*f*
methyl group distribution along DNA and, 129–130
number of sites analyzed and, 358
for ovalbumin gene, 371*f*
paradigms of, 358, 359*t*, 374*f*
for phosphoenolpyruvate carboxykinase gene, 367*f*
in plants *vs* vertebrates *vs* bacteria, 128–129, 129*t*
procedures for analysis of, 353–354, 361
of rabbit beta-like globin gene clusters, 250–251, 252*f*, 253*f*
in replication, 135–136
for ribosomal RNA genes, 360
site specificity in, 355, 356
Southern blot analysis of, 361, 365*f*
in sperm, 355, 362, 374*f*
tissue-specific, 138, 139*t*, 148, 149, 159, 355, 356
for vitellogenin gene, 361, 372*f*
of retroviral genomes, 198–202, 201*t*
RNA polymerase interaction with, 260
significance of, biological, 4–8, 111
somatic and germ cell evidence of, 190, 190*t*
species specificity of, 190, 191*t*
structural alterations of DNA from, 156, 221
in type I restriction-modification systems, 18, 18*f*, 19, 20
AdoMet binding in, 18, 19, 23, 78
enzyme mechanism in, 78–79
recognition sequences for, 77–78, 77*t*
in type II restriction-modification systems, 40
enzyme mechanism for, 78–79
recognition sequences for, 77–78, 77*t*
specificity of, 42–44, 43*t*
in type III restriction-modification systems, 29
AdoMet binding in, 28, 78
enzyme mechanism for, 78–79
recognition sequences for, 77–78, 77*t*
in X chromosome inactivation, 271–276
in Z-DNA formation induction, 290–291
Z-DNA stability with, 284–290, 286–289*f*
DNA recombination, *see* Recombinant DNA
DNA repair
damage-inductible (SOS), 99
DNA methylation and, 119
invisibility of dam with, 101
DNA methylation and, 111–124
adenine bases in, 91–93, 96
cytosine bases in, 88
model for maintenance or alteration of patterns of, 112–113, 112*f*
de novo, 210
mismatch
genetic recombination and, 91–92
phage T7 and, 92
phage lambda and, 92, 95
DNA replication
in adenovirus systems, 228–229
DNA methylation and, 111–124
adenine bases in, 94–95
cytosine bases in, 89
model for maintenance or alteration of patterns of, 112–113, 112*f*
for nascent strands, 113
patterns of, 135–136
restriction enzymes and requirements for, 114
timing of, 114–116
of retroviruses, 193–194, 196–198, 198*t*
DNA structure
B-form
base pairing in, 282, 283*f*
conformational changes in, 281–282
conversion of, in solution, 282, 284

DNA structure (*cont.*)
deoxyguanosine conformation in, 281, 281*f*
DNA methylation and, 264
helical grooves of, 282, 283*f*
methylation effects on, 284–290, 287*f*, 291
in nucleosome assembly, 319
supercoiling and, 313
van der Waals models of, 280, 280*f*
bases in, 2–3, 2*f*
chromatin, *see* Chromatin structure
cruciform, 313
hairpin, 313
linker, 295
methylation effects on, 156, 221
nucleosome association with, 319–321
supercoiling
in eukaryotes, 315–316
nuclease sensitivity and, 313, 314
nucleosome association with, 319
regulatory role of, 315
topology, 315–318
Z-form
antibodies specific for, 283
base pairing in, 282, 283*f*
bromination effects on, 283
cations affecting, 284
composition of, 279
conformational changes in, 281–282, 283*f*
deoxyguanosine conformation in, 281, 281*f*
discovery of, 279
DNA methylation and, 264
induction of, 290–291
methylation and, 284–291, 286–289*f*
nuclease sensitivity and, 313
in nucleosome assembly, 319
in solution, 282–284
van der Waals models of, 280, 280*f*
DNAse
chromatin high mobility group proteins and, 325–326
chromatin pattern of digestion with, 299–300
DNA methylation effects on, 155
in evolution of restriction-modification systems, 76–77
in globin gene transcription, 261
hypersensitive sites for, 300
oncogenes and, 307

E

Embryonic development
chicken globin gene clusters in, 255, 258
chromatin structure during, 330–333
demethylation in, 150, 191
DNA methylation in, 5, 7, 189–212
chromatin structure and regulatory sequences with, 206–208
constancy of patterns of, 190–192, 190*t*, 191*t*
de novo, 157–158, 202–206, 203*t*, 204*t*, 208–209, 209*t*
formation of patterns of, 136–137, 138
viral expression and infectivity with, 198–202, 200*t*, 201*t*
experimental tools for study of, 193
genes specific for, 211–212
human globin gene region during, 251, 258
Moloney murine leukemia virus expressed in, 194–196, 197*t*
rabbit globin gene clusters in, 251, 258
repression of genes for oogenesis in, 210–211
retroviruses expressed in, 193
with pre- *vs* postimplantation, 196–198, 198*t*
types of, reflecting stages of, 194, 195*t*
X chromosome inactivation with, 276, 277*f*
Endonuclease
adenine methylation protecting DNA from cleavage by, 95
evolution of, in restriction-modification system, 76–77

in recombinant DNA technology, 85
in type I restriction-modification systems, 11, 13–14, 14*f*, 21, 75, 76*t*
of type II restriction-modification systems, 40, 53, 75
DNA binding of, 54
in inhibition of methylase assay, 41
on single-stranded DNA, 64–65
site recognition by, 55
star activity of, 58, 59
in type III restriction-modification systems, 11, 25–26, 25*f*, 75, 76*t*
Escherichia coli
5-azacytidine incorporation into DNA of, 168, 182
K-12, DNA methylation in, 81–102
Okazaki fragments of, methylation of, 117
type I restriction-modification systems of, 12–24
type II restriction-modification systems of, 41, 42, 49, 53–64
type III restriction-modification systems of, 25–29
Ethionine, 328
Ethylnitrosourea, 121
Eukaryotes
5-azacytidine incorporation into DNA of, 168
DNA methyltransferase inhibition by, 169–171
gene activation by, 165
mutagenicity and, 181
de novo methylation in, 133–134, 135
DNA methylation in, 5–6, 40, 79
via bacterial methylases, 65–66
for differentiation process, 111–112
DNA replication coupled to, 114, 116
formation of patterns of, 136–137, 138
gene activity and, 111, 222–223, 226, 230
maps of, 367–374*f*
for mismatch repair, 119, 123
for nascent strands, 113
pattern paradigms with, 358, 359*t*, 376*f*
site specificity in, 355
tissue specificity in, 355–356
gene expression in, 294
DNA topology with, 315–316
methylation pattern correlated to, 357–364
Okazaki fragments of, 116–117
Evolution
of *E. coli* K-12 cytosine methylation gene, 86–87
of restriction-modification systems, 76–77

F

Fetoprotein gene
methylation pattern for
cell population contamination and, 358
paradigm of, 359*t*
site specificity for, 355
Fluorocytidine, 5–169
Fluorouracil, 5–169
Formylguanylribosylurea, 165, 166

G

Genes
actin, *see* Actin gene
albumin, *see* Albumin gene
in antirestriction systems, 30, 32, 79
casein, 356, 359*t*, 366*f*
cloned, expression of, 310–311
collagen, *see* Collagen gene methylation
crystallin, *see* Crystallin gene methylation
in differentiation
DNA methylation of, 177–179, 178*t*, 190, 192
regulatory sequences of, 206–208
retroviruses and embryonal carcinoma cells for study of, 193
DNA methylation, for eukaryotic methylation, 5–6, 6*t*

Genes (*cont.*)
DNA methylation, for prokaryotice methylation, 4–5, 81–82, 82*t*
5-azacytidine and, 88–89
enzyme properties from, 90–91
identification of, 90
invisibility of, with other mutants, 99–101, 100*f*, 101*t*
methylase substrate sequence and, 86, 90
in mismatch repair, 91–93
mutagenic deamination and, 87
mutant phenotypes of, 86–87, 96
in protection of DNA, 95
in recombination and repair, 88
regulation of, 87–88, 93–94
replication and, 89, 94–95
in restriction-modification, 96
sensitivity of, to chemical and physical agents, 98–99
spontaneous mutability of, 96–98
in embryonic development, 157–158
expression of, chromatin structure in, 206, 293–334
core histones and nucleosomes in, 318–325
formation and maintenance of, 311–329
germ line, 333–334
hypersensitive sites, 312–315
in vivo correlations of, 301–305
propagation of, 330–334
protein composition and organization in, 318–326
protein-DNA interactions in, 318–326
sequence alterations and rearrangements, 306–311
topology of, 315–318
expression of, DNA methylation in, 111, 147–160, 221
chromatin structure and, 326–329
via DNA-protein interactions, 155–156, 221
patterns of, 353–364
selective regions for, 154–155
fetoprotein, 355, 358, 359*t*, 375*f*
globin, *see* Globin genes
growth hormones, *see* Growth hormone gene
immunoglobulin, *see* Immunoglobulin gene
inactivation of
DNA methylation in, 221–241
in globin gene clusters during development, 249–265
in X chromosome, 269–276
Z-DNA formation and, 290–291
at promoter regions, 228
regulatory signals and, 227–228
insulin, 224, 258, 356, 367*f*, 372*f*
metallothionein, 160, 305
oncogenes, 307
oocyte-specific *vs* embryo-specific, 210–211
ovalbumin, *see* Ovalbumin gene
RNA, 355, 359*t*, 360
steroid hormone activation of, 227
transfer of, into animal cells, 151–152
transferrin, 207
of type I restriction-modification systems, 12–16, 14*f*, 73, 75
of type II restriction-modification systems, 44, 46, 48–53, 49*f*, 74–75
of type III restriction-modification systems, 24–26, 25*f*, 73, 75
vitellogenin, *see* Vitellogenin gene
X-linked, *see* X chromosome
yeast, 306
Globin genes
5-azacytidine effect on, 176, 225
chicken, during development 255–257, 256*f*, 257*f*, 368*f*, 371*f*
cloned, expression of, 310
DNAse digestion of, 261–263, 299
enhancer sequences of, 250
human, during development, 251–255, 254*f*, 368*f*
methylation of
as causative effect *vs* gene activity reflection, 258–260
cell population contamination and, 358
chromatin structure and, 261–265
at cytosine bases, 259

developmental regulation and, 148, 152, 249–265
DNAse I-hypersensitivity sites and, 263
map of pattern of, 368*f*, 371*f*, 372*f*
paradigm for, 359*t*
patterns of, in gene clusters, 257–258
at promoter region, 240
site analysis for, 358
site specificity for, 355
transcription correlated to, 154, 232, 240, 261–265, 262*f*
of nonerythroid tissues, 253*f*, 258
nuclease sensitivity of, 298, 299, 302
nuclease-hypersensitive sites on, 300–301, 302
promoter sequences of, 249–250
rabbit, during development 250–251, 252*f*, 253*f*, 371*f*
regulatory sequences of, 206, 207
Growth hormone gene
cloned, expression of, 310
methylation pattern of
cell population contamination and, 358
map of, 374*f*
paradigm of, 359*t*, 360
site specificity for, 355
tissue expression of, 374*f*

H

Haemophilus methylases, 43*t*, 44, 46, 103
cloning for, 50–51
in type III restriction-modification systems, 24–29
Heparin, 21, 28–29
Hexamethylene bisacetamide, 177
Histones
antibodies to, 321
in chromatin compaction, 325
compositional changes in, with transcription, 293
depletion of, 325
DNA interactions of, 293, 295
DNA methylation and, 264
of nucleosome, 295
protein interactions of, 293

I

Immunoglobulin genes, 206, 207
DNAse digestion of, 299
enhancer element of, 309
methylation pattern of, 369*f*, 370*f*
paradigm for, 359*t*, 360
site analysis for, 358
sequence alterations in transcription of, 308–309
Insulin genes, 224, 258
methylation pattern for
map of, 367*f*, 372*f*
site specificity for, 355
nuclease-hypersensitive sites on, 301
Isoschizomers
in analysis of methylation patterns, 250, 354
restriction-modification systems and, 74

L

Lampbrush chromosomes, 297
Lesch-Nyhan chromosome, 272
Lymphomas, 307

M

Magnesium ions
in cytidine/uridine phosphorylation, 167
in DNA methylation inhibition, 132
in isolation of actively transcribed chromatin sequences, 322
Metallothionein gene, 160, 305
Methyltransferases, DNA, 3–4, 3*f*
for adenovirus type 2 DNA methylation, 234
5-azacytidine inhibition of, 168–173, 225
in eukaryotes, 169–171, 170*f*
in mechanism of, 170*f*, 171–173, 173*f*

Methyltransferases, DNA (*cont.*)
in prokaryotic systems, 88–89, 168
carcinogens effect on, 120–121
de novo vs maintenance, 3–4, 79, 132–133
in *E. Coli* K-12, 81–102
deficiency of, in mutants, 84–86
genes for, 81–82, 82*t*, 87–88, 90, 96–101
mutant phenotypes of, 86–87, 96–98, 99–101, 100*f*, 101*t*
in phage and plasmid dam and dcm methylation, 101–102
properties of, 90–91
strain and sequence specificity of, 82–83
substrate sequence, 86, 89–90
evolution of, in restriction-modification system, 76–77
function of, biological, 74
prokaryotic *vs* eukaryotic, 134, 135
sequence specificity of, 134*t*, 135
strain and species-specificity of, 82–83
substrate specificity of, 132, 132*f*, 133, 133*t*,
of type I restriction-modification systems, 14, 14*f*, 15, 21, 75, 76*t*
of type II restriction-modification systems, 11, 39–66
aberrant methylation and hemimethylated sites for, 59–64, 61–62*t*
assay of, 41
bacterial, expressed in eukaryotes, 65–66
catalysis by, 53, 57–58
of type II restriction-modification systems
methylation specificity of, 42–44, 43*t*
reaction mechanism of, 53–65
relaxation of specificity of, 58–59
for single-stranded DNA, 64–65
site recognition for, 53, 55–57, 56*f*
structure and properties of, 40, 44–48, 45*t*, 75, 76*t*
thermodynamics and kinetics of binding of, 54–55
of type III restriction-modification systems, 25–26, 25*f*, 75, 76*t*
Methyltransferases, RNA, 169
Methyl-N-nitrosourea, 120
Methylases, *see* Methyl transferases, DNA
Moloney murine leukemia virus gene expression
de novo methylation and, 202–206, 203*t*, 204*t*
demethylation and, 207
DNA methylation and, 199–201, 201*t*
in pre- *vs* postimplantation embryos, 196, 197*t*
tissue-specificity and, 194
Mycobacterium tuberculosis methylases, 103
Myosin gene, 159, 364

N

Neisseria gonorrhoeae methylases, 103–104
Netropsin, 64
Nicotinamide, 83, 117–118
Nitrogen mustard, 121
Nuclease
chromatin hypersensitive sites for, 300–301, 312–315
chromatin sensitivity to, 298–299, 302–305
on double- *vs* single-stranded DNA, 312-313
micrococcal, 295, 298–299, 322
in supercoiled DNA cleavage, 313, 314
Nucleosomes
B and Z-form conformations in assembly of, 319–320
chromatin structure in, 293, 295–296
constituents of, 295–296
DNA association of, 319–321
DNA methylation and, 264
multiple, in superbeads, 296
phasing of, 324
polymerase molecules and, 320–321
as regulatory element in chromatin function, 296

in ribosomal *vs* nonribosomal units, 321
supercoiling of DNA and, 315
T antigen and, 319

O

Okazaki fragments, 116–118
Oogenesis genes, 210–211
Ovalbumin gene
DNAse digestion of, 299
hormonal effects on, 305
methylation pattern of, 371*f*
map of, 371*f*
partial, tissue/site specificity for, 356
nuclease digestion of, 298, 299
nuclease-sensitive *vs* -hypersensitive sites in, 303

P

Phage
in antirestriction systems, 31*t*, 32, 79
dam and *dcm* methylation in, by *E. coli* enzymes, 101–102
in evolution of restriction-modification systems, 76–77
mismatch repair in, 92, 95
type III restriction-modification systems of, 24–29
Phosphoenolpyruvate carboxykinase gene, 368*f*
Plasmacytoma, 360
Plasmids
in gene manipulation and *in vitro* methylation, 226
methylation of *dam* and *dcm* sites in, 101–102
in transcription measurements, 152
type III restriction-modification systems of, 24–29
Z-form DNA in, 283, 284, 290
Polamines, 284
Prokaryotic DNA methylation, 4–5, 12, 24, 39
adenine base in, 81–82
chemical and physical agents affecting, 98–99, 98*t*
dam gene activity regulation and, 93–94
dam gene and, 90, 99–101, 101*f*, 101*t*
dam gene mutability and, 96–98
dam gene mutant phenotypes and, 96
methylase properties and, 90–91
methylase substrate sequence and, 89–90
for mismatch repair, 91–93
nicotinamide effects on, 117–118
protection of DNA and, 95
replication of DNA and, 94–95
restriction-modification and, 96
5-azacytidine inhibition of, 88–89, 168
cytosine base in, 81–82
dcm gene activity regulation and, 87–88
dcm gent mutant phenotype and, 86–87
dcm mutagenicity and, 87
methylase substrate sequence and, 86
recombination and repair with, 88
replication of DNA and, 89
demethylation and, 83
DNA replication coupled to, 114, 116
in *E. coli*, 81–102
endogenous restriction enzymes and, 122
formation of patterns of, 136, 136*f*
methylases in, 82–83, 86, 90–91
in mismatch repair, 119–121
in mutants of *E. coli* K-12, 84–86
methylase deficiency testing of, 85
recombinant DNA technology and, 85–86
for nascent strands, 113
of Okazaki fragments, 117–118
in phage and plasmid DNA by *E. coli* enzymes, 101–102
as postreplicative event, 83
regulatory protein binding correlated to changes in, 156

Prokaryotic DNA methylation (*cont.*)
  in species other than *E. coli,* 102–104
Prokaryotic gene expression, 294, 315
Proteins
  chromatin-associated, 325–326
  composition and organization of, in chromatin structure, 318–326
  DNA methylation and binding of, 7, 155–156, 221
  endonucleases, *see* Endonucleases
  histones, *see* Histones
  methyltransferases, *see* Methyltransferases, DNA
  phage T7 0.3, in antirestriction systems, 31*t*, 32
  phage lambda ral, in antirestriction systems, 31*t*, 32
  in transcription initiation, 294
Provirus, 194
  activation of
    in differentiated EC cells, 198
    in pre- *vs* postimplantation embryos, 196
  chromosomal position of, and genome expression, 194
  methylation of
    biological activity and, 198–202
    de novo, 203–205, 204*t*
  oncogenes and, 307
Pseudoisocytidine, 167, 172*f*, 177

R

Recombinant DNA
  in adenovirus DNA methylations, 229
  cytosine methylation and, 88
  gene manipulation and *in vitro* methylation with, 226
  historical perspectives on, 2
  in methylation mutants of *E. coli* K-12 testing, 85–86
  in mismatch genetic repair, 91–92
  restriction endonucleases in, type II, 40
  sequence alterations in transcription and, 308
  in transcriptional inhibition with methylation, 260
Restriction-modification systems
  antirestriction systems affecting, 30–32, 31*t*, 79–80
  evolution of, 76–77
  function of, biological, 73–74, 127
  reaction mechanisms of, 78–79
  recognition site *vs* cleavage site in, 78
  structure of genes and enzymes of, 74–76, 76*t*, 77–78
  systems opposing, 30–32, 31*t*
  type I, 11–12, 13*t*, 39–40
    AdoMet binding in, 19
    AdoMet hydrolase system and, T3, 30–31, 31*t*
    enzyme activation in, 19–20
    formation of initial complexes in, 20
    formation of recognition complexes in, 20–21
    function of, biological, 73
    prevalence of, 74
    reaction mechanisms of, 17–24, 18*f*
    recognition complexes and ATP, 21–22, 22*f*
    recognition sequences of, 16–17, 77
    structure of genes and enzymes of, 12–16, 14*f*
    substrates in, 78, 79
  type II, 11, 13*t*
    function of, biological, 74
    prevalence of, 74
    substrates in, 78, 79
  type III, 11–12, 13*t*, 39–40
    enzyme activation in, 28
    formation of recognition complexes in, 28–29
    function of, biological, 73
    reaction mechanisms of, 27–29
    recognition complexes and ATP in, 29
    recognition sequences of, 27, 77
    structure of genes and enzymes of, 24–26, 25*f*
    substrates in, 78, 79
Retroviruses
  DNAse digestion of genome of, 299

genome expression of, in mouse development, 194, 195*t*
de novo methylation and, 202
DNA methylation and, 199, 200*t*
level of differentiation influencing, 196–198, 198*t*
regulatory sequences and, 207–208
oncogenes and, 307
in study of embryonic development, 193
transcriptional competence of, 294
types of, in varying stages of murine development, 194, 195*t*
RNA
5-azacytidine incorporation into, 168, 169
genes for, methylation of, 355, 359*t*, 360
globin genes and splicing of, 250
RNA polymerase
DNA hypersensitive sites and, 312
DNA methylation effect on, 260
nucleosome assembly and, 320
in prokaryotic gene transcription, 294
RNA viruses, 176, 181
Rous sarcoma virus, 176

S

Sinefugin, 19, 21
Southern blot analysis of methylation patterns, 361, 365*f*
Spermine, 284
Streptococcus pneumoniae methylases, 103
Streptoverticillium, 165

T

Thymidine kinase gene, 235
Topoisomerases, 315, 316
Transcription
chromatin reorganization without, 301
DNA topology and conformation in, 293, 315–318
globin gene
developmental regulation of, 249–265
with promoter methylation, 240
of linear *vs* circular DNA, 316
measurement of, 152, 304
methylation effects on, 151, 155, 206
chromatin structure and, 326–329
gene inactivation from 224, 234–240
microscopy of, 297–298
nuclease-sensitive *vs* -hypersensitive sites in, 302
nucleosome assembly and, 321
of proviral genomes, 193–194
in embryonal cells, 198
methylation level *vs*, 199, 200*t*, 201*t*
of restriction-modification genes, 75
retroviral, 294
Transferrin gene, 207

U

Uridine-cytidine kinase, 167

V

Viruses
adenovirus, *see* Adenovirus
5-azacytidine effect on, 176, 181
DNA topology of, 316
gene methylation of
inverse correlations of, 223, 227
at promoter regions, 226, 235–236
Moloney murine leukemia, *see* Moloney murine leukemia virus
retroviruses, *see* Retroviruses
Vitellogenin genes
DNAse digestion of, 299
estrogen inducibility of, 224, 226, 304, 305
methylation of
gene expression and, 157, 225, 226, 232
map of pattern of, 373*f*
paradigm for, 359*t*, 361
protein appearance and, 363

Vitellogenin genes (*cont.*)
  nuclease-hypersensitive sites on, 301, 304

X

X chromosome
  active state of, 276
  inactivation of, 269
    cellular differentiation and, 276
    DNA methylation in, 271*t*, 272–273
    forms of, 274–275
    imprinting of, 275
    in males *vs* females, 273–274
    nonrandom, 274
    random, 269
    reactivation cycle and, 270
    sequence alterations and rearrangements in, 306
    stability of, in somatic cells, 271–272

Y

Yeast genes, 306

MIX
Papier aus verantwortungsvollen Quellen
Paper from responsible sources
FSC® C105338

If you have any concerns about our products,
you can contact us on
ProductSafety@springernature.com

In case Publisher is established outside the EU,
the EU authorized representative is:
Springer Nature Customer Service Center GmbH
Europaplatz 3, 69115 Heidelberg, Germany

Printed by Libri Plureos GmbH
in Hamburg, Germany